SURVEY OF MATHEMATICAL PROGRAMMING

SURVEY OF MATHEMATICAL PROGRAMMING

PROCEEDINGS OF THE 9TH INTERNATIONAL
MATHEMATICAL PROGRAMMING SYMPOSIUM

Budapest, August 23–27, 1976

VOL. 2

EDITED

by

A. PRÉKOPA

Professor of Mathematics, Technical University of Budapest,
Director of the Applied Mathematics Branch,
Computer and Automation Institute of the Hungarian Academy of Sciences

NORTH-HOLLAND PUBLISHING COMPANY

AMSTERDAM · OXFORD · NEW YORK

The distribution of this book is being handled by the following publishers:

for the U.S.A. and Canada
Elsevier/North-Holland, Inc.
52 Vanderbilt Avenue
New York, New York 10017

for the East European countries
Akadémiai Kiadó, The Publishing House of the Hungarian Academy of Sciences, Budapest

for all remaining areas
North-Holland Publishing Company
335 Jan van Galenstraat
P.O. Box 211, Amsterdam, The Netherlands
North-Holland Publishing Company ISBN 0-444-85033-3

A joint edition of
North-Holland Publishing Company, Amsterdam
and
Akadémiai Kiadó, Budapest
with the cooperation of
János Bolyai Mathematical Society, Budapest

Printed in Hungary

CONTENTS

VOLUME 2

PART 4

COMPLEMENTARITY AND FIXED POINTS

PART 5

STOCHASTIC AND DYNAMIC PROGRAMMING

PART 6

OPTIMIZATION IN NETWORKS

PART 7

GRAPHS AND COMBINATORICS

PART 8

DISCRETE PROGRAMMING

COMPLEMENTARITY AND FIXED POINTS

ON SOLVING LINEAR COMPLEMENTARITY PROBLEMS AS LINEAR PROGRAMS

R. W. COTTLE and YONG-SHI PANG

(Stanford, USA)

1. INTRODUCTION

It is a fairly well-known fact that if a linear complementarity problem has a solution, then it has a solution which is an extreme point of its "feasible set". This means that if an appropriate linear form were known, i.e., one whose minimum over the feasible set would necessarily occur at a complementary solution, then the linear complementarity problem could be solved as a linear program. Typically, one does *not* know an appropriate linear form in advance and can not rapidly find one. But there are exceptional cases, some of which have been noted in the literature. (See [7], [9], [18], [19], [26].) It is our contention that these linear complementarity problems solvable as linear programs are related to the theory of polyhedral sets with least elements. Examples of this relationship are made explicit by Cottle and Veinott [9] and by Tamir [26]. Some numerical experience based on this observation is reported in Cottle, Golub and Sacher [7].

More recently, Mangasarian [18], [19] has produced several additional examples of linear complementarity problems whose solutions can be obtained via linear programming (which incidentally is *not* intended here to imply the use of the simplex method or any of its derivatives). Mangasarian's results in this area are not explicitly based on least element arguments, but rather on a key lemma having to do with optimal dual variables. Our primary purpose in this paper is to demonstrate that Mangasarian's theory can be interpreted in terms of least elements of polyhedral sets. For the most part, our methods are matrix-theoretic. In the course of our investigation, we uncovered a few results of this type; they are included here because we believe them to be new and of independent interest.

The possibility suggested by the linear programming formulation of a linear complementarity problem raises the question of whether this approach can be recommended in practice. Hence our secondary purpose in this paper is to give at least a tentative answer by reporting the computational experience we have gathered in solving some linear programs—of the type that could arise from linear complementarity problems—by an iterative (relaxation) procedure rather than by the simplex method or any of its variants. Motivation for using an iterative method can be found in the size and structure of the matrices one might except to encounter in some potential applications of the linear complementarity problem.

The plan of the paper is the following. In Section 2, we cover a bit of background material. The section has two parts. The first part fixes our notation and gives some characterizations of matrices in terms of the linear complementarity problem. The second part is a synopsis of the main results Mangasarian obtained in [18] and [19]. In Section 3, we develop our least-element interpretation of the subject and present some incidental matrix-theoretic results. In the fourth and final section, we discuss our somewhat preliminary computational experience with solving linear complementarity problems as linear programs by relaxation methods.

2. BACKGROUND

2.1 Miscellaneous preliminaries

Throughout this paper, R^n_+ will denote the nonnegative orthant of the Euclidean n-space R^n and $R^{n \times m}$ will denote the class of real $n \times m$ matrices. We denote the i-th column (row) of a matrix $A \in R^{n \times m}$ by $A_i.$ ($A_{i.}$). A real matrix $A \in R^{n \times n}$ is said to be a *Z-matrix (P-matrix)* if it has non-positive off diagonal entries (positive principal minors). We shall call a matrix $A \in R^{n \times n}$ a *K-matrix* (or a *Minkowski matrix*) if it is both a *P*- and *Z*-matrix simultaneously. The classes of all real *Z*-, *P*- and *K*-matrices will be denoted by *Z*, *P* and *K* respectively. They are treated extensively by Fiedler and Pták [14].

For a vector $q \in R^n$ and a matrix $M \in R^{n \times n}$, the *linear complementarity problem*, denoted by (q, M), is that of finding $x \in R^n$ such that

$$(2.1) \qquad q + Mx \geqq 0, \quad x \geqq 0 \quad \text{and} \quad x^T(q + Mx) = 0.$$

By the *feasible set* for (q, M) we mean the polyhedral set

$$X(q, M) = \{x \in R^n : q + Mx \geqq 0, \; x \geqq 0\}.$$

We say that the problem (q, M) is *feasible* if $X(q, M)$ is nonempty. A subset S of R^n is said to be *bounded below* if there is a vector $x' \in R^n$ such that $x \geqq x'$ for all $x \in S$. The vector $\bar{x} \in S$ is the *least element* of S if $\bar{x} \leqq x$ for all $x \in S$. It is clear that the least element, if it exists, must be unique.

Minkowski matrices as well as *P*- and *Z*-matrices play very important roles in the linear complementarity problem. It is well known (see Samelson *et al.* [25]) that the problem (q, M) has a unique solution for every $q \in R^n$ if and only if $M \in P$. Tamir [26] characterized *Z*-matrices in the following way.

10

Theorem. *The matrix $M \in R^{n \times n}$ is a Z-matrix if and only if for each vector $q \in R^n$ for which the feasible set $X(q, M)$ is nonempty, there exists a least element $\bar{x}$ in $X(q, M)$ satisfying $x^T(q + Mx) = 0$.*

Cottle and Veinott [9] proved the following characterization of *K-matrices*.

Theorem. *The matrix $M \in R^{n \times n}$ is Minkowski if and only if for each $q \in R^n$, the feasible set $X(q, M)$ has a least element $\bar{x}$ which is the only vector in $X(q, M)$ satisfying $x^T(q + Mx) = 0$.*

Note that the characterizations of Z- and K-matrices are in terms of least elements of the feasible set $X(q, M)$. This feature is of fundamental importance in the present work.

Various methods for solving the linear complementarity problem (q, M) in the important special case where M is a Z-matrix have been considered intensively by a number of authors [3], [7], [8], [11], [23], [24]. While Mangasarian's proposal [18], [19] to solve linear complementarity problems (q, M) as linear programs is not entirely new, his results definitely appear to enlarge the class of problems to which this solution strategy is applicable. Specifically, he proved that for certain classes of matrices M, it is possible to find a vector p such that each solution of the linear program

$$(2.2) \qquad \text{minimize } p^T x \quad \text{subject to} \quad q + Mx \geqq 0, \quad x \geqq 0$$

solves the problem (q, M). We denote the linear program (2.2) by the triple (p, q, M). Its dual is equivalent to

$$(2.3) \qquad \text{minimize } q^T y \quad \text{subject to} \quad p - M^T y \geqq 0, \quad y \geqq 0,$$

which is just $(q, p, -M^T)$. We say that a linear complementarity problem (q, M) is *LP-solvable* if we can find a vector $p \in R^n$ such that each solution of the linear program (p, q, M) solves (q, M).

Recognizing that most LP-solvable linear complementarity problems arise from the discretization of (partial) differential equations (see [7], [8], [11], [23]) and that the properties of the matrices so obtained (e.g. Z-matrices) are not so conducive to efficient solution of the linear programs by the simplex method, Mangasarian proposed the use of relaxation methods ([1], [2], [13], [20]) for solving inequality systems. In particular, solving the linear program (2.2) is equivalent in a logical sense to solving the linear inequalities:

$$(2.4) \qquad \begin{pmatrix} M & 0 \\ 0 & -M^T \\ I & 0 \\ 0 & I \\ -p^T & -q^T \end{pmatrix} \begin{pmatrix} x \\ y \end{pmatrix} \geqq \begin{pmatrix} -q \\ -p \\ 0 \\ 0 \\ 0 \end{pmatrix}$$

which consist of primal and dual feasibilities and the reverse of the weak duality of the linear program (2.2). Presumably, the computational advantage offered by relaxation methods is their capacity for preserving matrix sparsity.

2.2 Mangasarian's results

Our purpose here is to summarize the principal results obtained by Mangasarian in the aforementioned papers. The fundamental theorem is the following:

Theorem 2.1. *Let the feasible set $X(q, M)$ be nonempty, and let M satisfy*

$$(2.5) \qquad\qquad MX = Y$$

$$(2.6) \qquad\qquad r^T X + s^T Y > 0 \quad \text{for some} \quad r, s \in R_+^n,$$

where $X, Y \in Z$. Then the linear complementarity problem (q, M) can be solved by solving the linear program (p, q, M) with $p = r + M^T s$.

The proof of the theorem depends heavily on the key lemma below.

Lemma 2.2. *If $\bar{x}$ solves the linear program (p, q, M) and there exists an optimal solution $\bar{y}$ of $(q, p, -M^T)$ such that $\bar{y} + p - M^T \bar{y} > 0$, then $\bar{x}$ solves the problem (q, M).*

The following corollary identifies some classes of matrices satisfying conditions (2.5) and (2.6) in Theorem 2.1.

Corollary 2.3. *Let the feasible set $X(q, M)$ be nonempty and let $e \in R^n$ be any positive vector. Then for each of the cases when*

(a)	$M = YX^{-1}$, $X \in K$, $Y \in Z$	$(p = r \geq 0, r^T X > 0)$
(b)	$M = YX^{-1}$, $X \in Z$, $Y \in K$	$(p = M^T s, s \geq 0, s^T Y > 0)$
(c)	$M \in Z$	$(p = e)$
(d)	$M^{-1} \in Z$	$(p = M^T e)$
(e)	$-M \in K$	$(p = -e \text{ or } p = M^T e)$
(f)	$-M^{-1} \in K$	$(p = -M^T e \text{ or } p = e)$,

the linear complementarity problem (q, M) has a solution which can be obtained by solving the linear program (p, q, M) with the indicated p.

The results above are drawn from the first of the two papers. In the second paper, Mangasarian extends the class of LP-solvable linear complementarity problems by establishing the following remarkable theorem.

Theorem 2.4. *Let the feasible set $X(q, M)$ be nonempty, and suppose there exist $X, Y \in R^{n \times n}$, $A \in R^{m \times m}$, $B, H \in R^{n \times m}$, $G \in R^{m \times n}$, $p \in R_+^n$ and $p_0 \in R_+^m$ such that*

$$(2.7) \qquad MX = Y + BG; \quad MH \geq BA; \quad X, \ Y, \ A \in Z; \quad G, \quad H \geq 0$$

$$(2.8) \qquad (p^T, p_0^T) \begin{bmatrix} X & -H \\ -G & A \end{bmatrix} > 0.$$

Then the linear complementarity problem (q, M) has a solution which can be obtained by solving the linear program (p, q, M).

By specializing Theorem 2.4, Mangasarian produced the following table.

Table 1

Matrix M of (1)	Conditions on M	Vector p of (2)	Conditions on p
1. $M = YX^{-1}$	$X \in K,\ Y \in Z$	p	$p \geqq 0,\ p^T X > 0$
2. $M = YX^{-1}$	$X \in Z,\ Y \in K$	$p = M^T s$	$s \geqq 0,\ s^T Y > 0$
3. M	$M \in Z$	p	$p > 0$
4. M	$M^{-1} \in Z$	$p = M^T e$	$e > 0$
5. $M = Y + ab^T$	$Y \in K,\ a \geqq 0,\ b > 0$	$p = b$	$b > 0$
6. $M = 2X - Y$	$X \in Z,\ Y \in K,\ X \geqq Y$ (componentwise)	$p = M^T p_0$	$p_0 > 0,\ p_0^T Y > 0$
7. $M \geqq 0$ (componentwise)	$m_{jj} > \sum\limits_{\substack{i=1 \\ i \neq j}}^{n} m_{ij}$ $\forall j = 1, \ldots, n$	$p = M^T e$	$e^T = (1, \ldots, 1) \in R^n$
8. $M \geqq 0$ (componentwise)	$m_{ii} > \sum\limits_{\substack{j=1 \\ j \neq i}}^{n} m_{ij}$ $\forall i = 1, \ldots, n$	$p = M^T p_0$	$p_0 > 0$ $p_0^T(-M + 2 \operatorname{diag} M) > 0$

3. CONNECTION WITH LEAST ELEMENTS

In this section we develop our least element interpretation of Mangasarian's theory. The cornerstone of our approach is a strengthening of Theorem 2.1. The new result (Theorem 3.9) makes it possible to invoke the theory of polyhedral sets with least elements. The desired relationship between the two theories is made explicit in Theorem 3.11. Except for the matrix-theoretic results mentioned earlier, the rest of the section is concerned with showing how Theorem 2.4 and the special cases enumerated in Table 1 can be related to Theorem 3.9 and thereby to the least element theory. We begin by reviewing a few more pertinent facts.

Definition 3.1. Let M be a square matrix. By a *principal rearrangement* of M, we mean a matrix $\overline{M} = P^T M P$ where P is a permutation matrix.

The following facts are well known (see e.g. [4]):

(i) The classes of Z-, P- and K-matrices are invariant under principal rearrangements.

(ii) The inverse of a P-matrix is a P-matrix.

(iii) The property of a matrix belonging to any one of the three classes Z, P and K is inherited by each of its principal submatrices.

In [14], Fiedler and Pták gave a list of thirteen equivalent conditions for a Z-matrix to be a K-matrix. Here we quote three which this paper needs later.

Proposition 3.2. *Let* $A \in Z$. *Then the conditions below are equivalent to each other:*

(iv) *there exists a vector* $x \geqq 0$ *such that* $Ax > 0$;

(v) *the inverse* A^{-1} *exists and* $A^{-1} \geqq 0$;

(vi) *the principal minors of* A *are positive.*

Definition 3.3. Let A be a nonsingular principal submatrix of a square matrix M. Let $\overline{M}$ be a principal rearrangement of M such that $\overline{M} = \begin{pmatrix} A & B \\ C & D \end{pmatrix}$. Then the *Schur complement of* A *in* M, denoted by (M/A), is the matrix $D - CA^{-1}B$.

Properties and applications of the Schur complements have been surveyed in Cottle [5]. A proof of the following proposition can be found in Crabtree [10].

Proposition 3.4. *Let* A *be a nonsingular principal submatrix of the Minkowski matrix.* M. *Then* (M/A) *is itself a Minkowski matrix.*

Definition 3.5. The matrix $A \in R^{n \times m}$ is said to be *Leontief* if it has exactly one positive element in each column and there is a vector $x \in R^m$ such that $x \geqq 0$ and $Ax > 0$.

Proposition 3.6. *Let* $A \in R^{n \times m}$ *be a Leontief matrix. Then there exists a submatrix* $B \in R^{n \times n}$ *such that* B *is itself Leontief. Furthermore,* B^{-1} *exists and is nonnegative.*

A proof of the preceding proposition is given in Dantzig [12]. It is based on an application of the simplex method of linear programming.

We are now ready to present our results. We first state an alternative proposition which is an immediate consequence of the well-known theorem of Kuhn–Fourier [15] on the solvability of a system of linear relations.

Proposition 3.7. *Let* $X, Y \in R^{n \times n}$. *Then the following two conditions are equivalent to each other:*

$$(3.1) \qquad r^T X + s^T Y > 0 \quad \textit{for some} \quad r, s \in R^n_+$$

and

$$(3.1)' \qquad \left. \begin{array}{r} Xu \leqq 0 \\ Yu \leqq 0 \\ u \geqq 0 \end{array} \right\} \Rightarrow u = 0.$$

14

The lemma below provides necessary and sufficient conditions for two Z-matrices X and Y to satisfy condition (3.1).

Lemma 3.8. *Let $X, Y \in R^{n \times n}$ be Z-matrices. Then (3.1) holds if and only if there exist a principal rearrangement with permutation matrix P and a partitioning of X and Y such that*

$$(3.2) \qquad P^T X P = \begin{pmatrix} X_{11} & X_{12} \\ X_{21} & X_{22} \end{pmatrix}, \quad P^T Y P = \begin{pmatrix} Y_{11} & Y_{12} \\ Y_{21} & Y_{22} \end{pmatrix}$$

$$(3.3) \qquad \begin{pmatrix} X_{11} & X_{12} \\ Y_{21} & Y_{22} \end{pmatrix} \in K.$$

Proof. *Sufficiency.* We shall show that condition (3.1)′ holds. Let u satisfy $Xu \leq 0$, $Yu \leq 0$ and $u \geq 0$. We then have

$$\begin{pmatrix} X_{11} & X_{12} \\ Y_{21} & Y_{22} \end{pmatrix} u \leq 0$$

which implies $u \leq 0$ (and thus $u = 0$) because

$$\begin{pmatrix} X_{11} & X_{12} \\ Y_{21} & Y_{22} \end{pmatrix}^{-1} \geq 0.$$

Necessity. Condition (3.1) can be rewritten as

$$(3.4) \qquad (X^T, Y^T) \begin{pmatrix} r \\ s \end{pmatrix} > 0 \quad \text{with} \quad \begin{pmatrix} r \\ s \end{pmatrix} \geq 0.$$

The matrix $A = (X^T, Y^T) \in R^{n \times 2n}$ has at most one positive element in each column. We may assume, without loss of generality, that it has exactly one in each column: otherwise we can always delete those columns of A consisting entirely of non-positive entries and delete at the same time the corresponding components of the vector $(r^T, s^T)^T$, then we are left with a smaller system satisfying (3.4) in which the matrix has exactly one positive element in each column and we can work with this smaller matrix. Now the matrix A is Leontief; hence by Proposition 3.6, there exists a submatrix $B \in R^{n \times n}$ such that B is itself Leontief. B^{-1} exists and is nonnegative. Note that the same column of X^T and Y^T cannot both simultaneously appear in B because B has exactly one positive element in each column. Hence by permuting the columns of B, if necessary, we may assume that the i-th column of B is either the i-th column of X^T or the i-th column of Y^T. Then B is a complementary submatrix of (X^T, Y^T). This suggests the permutation and the partitioning, thus proving (3.2) and (3.3).

Using this lemma, we give necessary and sufficient conditions for a matrix M to satisfy conditions (2.5) and (2.6) of Theorem 2.1.

Theorem 3.9. *Let M, X and Y be $n \times n$ matrices with X, Y in Z. Then*

$$(3.5) \qquad\qquad MX = Y$$

$$(3.6) \qquad\qquad r^T X + s^T Y > 0 \quad \text{for some} \quad r, \ s \geqq 0$$

if and only if there is a principal rearrangement and partitioning of M, X and Y such that

$$(3.7) \qquad \begin{pmatrix} M_{11} & M_{12} \\ M_{21} & M_{22} \end{pmatrix} \begin{pmatrix} X_{11} & X_{12} \\ X_{21} & X_{22} \end{pmatrix} = \begin{pmatrix} Y_{11} & Y_{12} \\ Y_{21} & Y_{22} \end{pmatrix}$$

$$(3.8) \qquad \begin{pmatrix} X_{11} & X_{12} \\ X_{21} & X_{22} \end{pmatrix}$$

is nonsingular.

$$(3.9) \qquad \begin{pmatrix} X_{11} & X_{12} \\ Y_{21} & Y_{22} \end{pmatrix} \in K.$$

Proof. We first note that if P is a permutation matrix, then (3.5) holds if and only if $(P^T M P)(P^T X P) = P^T Y P$. Therefore (3.5) holds if and only if it holds for every principal rearrangement of M, X and Y. Now, the sufficiency part of the theorem follows immediately from Lemma 3.8 and the observation above. For the necessity part, it remains to verify condition (3.8). Condition (3.9) implies that X_{11} is nonsingular. Solving for M_{21} in the equation

$$Y_{21} = M_{21} X_{11} + M_{22} X_{21}.$$

we obtain

$$M_{21} = (Y_{21} - M_{22} X_{21}) X_{11}^{-1}.$$

Hence

$$Y_{22} = M_{21} X_{12} + M_{22} X_{22} =$$

$$= (Y_{21} - M_{22} X_{21}) X_{11}^{-1} X_{12} + M_{22} X_{22},$$

or,

$$Y_{22} - Y_{21} X_{11}^{-1} X_{12} = M_{22} (X_{22} - X_{21} X_{11}^{-1} X_{12}).$$

Note that the matrix on the left side is just the Schur complement of Y_{22} in the Minkowski matrix

$$\begin{pmatrix} X_{11} & X_{12} \\ Y_{21} & Y_{22} \end{pmatrix}.$$

By Proposition 3.4, it is nonsingular. Therefore so is the matrix $X_{22} - X_{21} X_{11}^{-1} X_{12}$. Now (3.8) follows from Schur's determinental formula [5]

$$\det \begin{pmatrix} X_{11} & X_{12} \\ X_{21} & X_{22} \end{pmatrix} = \det X_{11} \det (X_{22} - X_{21} X_{11}^{-1} X_{12}).$$

This completes the proof of the theorem.

Notation. Let C denote the class of square matrices M satisfying conditions (3.5) and (3.6) for some $X, Y \in Z$.

It is clear that, using Theorem 3.9, we can readily construct matrices belonging to the class C starting with any Minkowski matrix. In the sequel, we shall focus our discussion on this class C. Our purpose is to establish a relationship between the class of linear complementarity problems (q, M) with $M \in C$ and the theory of polyhedral sets having least elements. In order to achieve this, we state and prove the following lemma.

Lemma 3.10. *Let $X, Y \in R^{n \times n}$ be Z-matrices and let $(s, q) \in R^n \times R^n$. Suppose that the polyhedral set*

$$V = \{v \in R^n : q + Yv \geq 0, \, s + Xv \geq 0\}$$

is nonempty and bounded below. Then there exists a least element $\bar{v} \in V$ satisfying $(q + Yv)^T(s + Xv) = 0$. Furthermore this least element can be obtained by solving the linear program

$$(3.10) \qquad \text{minimize } r^T v \quad \text{subject to} \quad q + Yv \geq 0, \quad s + Xv \geq 0$$

for any positive vector $r \in R^n$.

Proof. Consider the linear program (3.10) where $r \in R^n$ is positive. Since the constraint set V is bounded below and obviously closed, problem (3.10) has a solution, say $\bar{v}$. We want to prove that $\bar{v}$ is the least element of V. (This will imply that $\bar{v}$ solves the linear program (3.10) for any other choices of the positive vector r.) So let $v \in V$ and $v' = (v'_i)$ be the vector with $v'_i = \min(v_i, \bar{v}_i)$ for each i. Consider index k. We may assume, without loss of generality, that $v'_k = v_k$. Then we have

$$(q + Yv')_k = q_k + \sum_{\substack{l=1 \\ l \neq k}}^{n} Y_{kl}v'_l + Y_{kk}v'_k \geq$$

$$\geq q_k + \sum_{\substack{l=1 \\ l \neq k}}^{n} Y_{kl}v_l + Y_{kk}v_k \geq$$

$$\geq 0.$$

Similarly, we can deduce $(s + Xv')_k \geq 0$. These inequalities hold for $k = 1, \ldots, n$. Therefore $v' \in V$. By the definition of $\bar{v}$, it follows that

$$r^T \bar{v} \leq r^T v' \leq r^T \bar{v}.$$

Hence $\bar{v} = v' \leq v$. This shows that $\bar{v}$ is indeed the least element of V. It remains to verify that $\bar{v}$ satisfies the complementarity property. Clearly we have $(q + Y\bar{v})^T(s + X\bar{v}) \geq 0$ because $\bar{v} \in V$. Suppose $(q + Y\bar{v})_i > 0$ and $(s + X\bar{v})_i > 0$ for some index i. Let $\varepsilon > 0$ and consider the vector $v \equiv \bar{v} - \varepsilon e^i$ where e^i is the i-th unit vector. We have

$$q + Yv = q + Y\bar{v} - \varepsilon Ye^i.$$

Therefore

$$(q+Yv)_j \geqq 0 \quad \text{for every} \quad j \neq i,$$

and

$$(q+Yv)_i = q_i + (Y\bar{v})_i - \varepsilon Y_{ii}.$$

Similarly, we have

$$(s+Xv)_j \geqq 0 \quad \text{for every} \quad j \neq i,$$

and

$$(s+Xv)_i = (s+X\bar{v})_i - \varepsilon X_{ii}.$$

Clearly, $Y_{ii} \leqq 0$ $(X_{ii} \leqq 0)$ implies $(q+Yv)_i \geqq 0$ $((s+Xv)_i \geqq 0)$. If $Y_{ii} > 0$ and $X_i > 0$, then choose $\varepsilon > 0$ such that

$$0 < \varepsilon < \min \{(q+Y\bar{v})_i/Y_{ii}, (s+X\bar{v})_i/X_{ii}\}.$$

With this choice of ε, we see that $(q+Yv)_i \geqq 0$ and $(s+Xv)_i \geqq 0$. Hence it follows that $v \in V$. But $r^T v < r^T \bar{v}$, contradicting the fact that $\bar{v}$ solves (3.10).

Consider the problem (q, M) with $M \in C$. Then the inequalities

$$q + YX^{-1}x \geqq 0, \quad x \geqq 0$$

define the feasible set for (q, M). With $x = Xv$, the inequalities above can be expressed in the form

$$(3.11) \qquad \begin{pmatrix} X \\ Y \end{pmatrix} v \geqq \begin{pmatrix} 0 \\ -q \end{pmatrix}.$$

Let V be the set of all solutions of (3.11). Then it is clear that (q, M) is feasible if and only if $V \neq \emptyset$. Partitioning the vector

$$q = \begin{pmatrix} q_1 \\ q_2 \end{pmatrix}$$

according to Theorem 3.9, we see that

$$v \in V \quad \text{implies that} \quad \begin{pmatrix} X_{11} & X_{12} \\ Y_{21} & Y_{22} \end{pmatrix} v \geqq \begin{pmatrix} 0 \\ -q_2 \end{pmatrix}.$$

Since the matrix

$$\begin{pmatrix} X_{11} & X_{12} \\ Y_{21} & Y_{22} \end{pmatrix}$$

is Minkowski by condition (3.9) of Theorem 3.9, its inverse is nonnegative by Proposition 3.2. Therefore we have

$$v \in V \quad \text{implies} \quad v \geqq \begin{pmatrix} X_{11} & X_{12} \\ Y_{21} & Y_{22} \end{pmatrix}^{-1} \begin{pmatrix} 0 \\ -q_2 \end{pmatrix}$$

i.e., the set V is bounded below. Hence if $V \neq \emptyset$, it follows from Lemma 3.10 that V has a least element $\bar{v}$ satisfying $(q+Yv)^T(Xv) = 0$ and $\bar{v}$ can be obtained by solving the linear program

$$\text{minimize } r^T v \quad \text{subject to} \quad q+Yv \geqq 0, \quad Xv \geqq 0$$

for any positive vector $r \in R^n$. Letting $\bar{x} = X\bar{v}$, we see that $\bar{x}$ is a solution of (q, M) and it can be obtained by solving the linear program

$$(3.12) \qquad \text{minimize } (X^{-T}r)^T x \quad \text{subject to} \quad q + Mx \geqq 0, \quad x \geqq 0$$

which is just $(X^{-T}r, q, M)$. Summing this up, we have proved

Theorem 3.11. *Let $M \in C$ and suppose that the problem (q, M) is feasible. Then there exists a bijective, linear map $L : R^n \to R^n$ such that the feasible set $X(q, M)$ is mapped onto a polyhedral set V having a least element $\bar{v}$ whose preimage $\bar{x} = L^{-1}(\bar{v})$ solves the problem (q, M).*

It is this theorem which provides the desired relationship between the linear complementarity problems with matrices in class C and the theory of polyhedral sets having least elements. Together with Lemma 3.10, the theorem also provides an interpretation for the conditions imposed on the vector p chosen in the objective function of the corresponding linear programs.

Remark 1. When using the linear program (3.12) to obtain a solution to the linear complementarity problem (q, M), one must, first of all, solve the system of linear equations $p^T X = r^T$ where $r^T > 0$ is chosen arbitrarily, to get the vector p used in the objective function of the linear program. For large n, this problem of finding p is not an insignificant task.

Remark 2. For $M \in C$, Mangasarian shows in Theorem 2.4 that the problem (q, M) can be solved via the linear program (p, q, M) for any p belonging to the class

$$p_1 = \{p \in R^n : p^T = r^T + s^T M \quad \text{for some} \quad (r, s) \geqq 0 \quad \text{such that} \quad r^T X + s^T Y > 0\}.$$

Our least element argument above shows that such a vector p can be chosen arbitrarily in the class

$$p_2 = \{p \in R^n : p^T = r^T X^{-1} \quad \text{for some} \quad r > 0\}.$$

Here we demonstrate that $p_1 = p_2$. Note that

$$p \in p_1 \Leftrightarrow p^T = r^T + s^T M = (r^T X + s^T Y) X^{-1}$$

for some $r, s \geqq 0$ and $r^T X + s^T Y > 0$. Hence it suffices to show that for any $t \in R^n$, $t > 0$, there exist $r, s \geqq 0$ such that $(X^T, Y^T) \begin{pmatrix} r \\ s \end{pmatrix} = t$. We mentioned in the proof of Lemma 3.8 that we may assume without loss of generality that the matrix (X^T, Y^T) is Leontief. We also showed that there exists a complementary submatrix B of (X^T, Y^T) which has a nonnegative inverse. Clearly, $B^{-1}t > 0$. Now we define the vector (r^T, s^T) as follows:

$$r_i = \begin{cases} (B^{-1}t)_i & \text{if} \quad B_{.i} = (X^T)_{.i} \\ 0 & \text{otherwise} \end{cases}$$

and

$$s_i = \begin{cases} (B^{-1}t)_i & \text{if} \quad B_{.i} = (Y^T)_{.i} \\ 0 & \text{otherwise} \end{cases}$$

for each $i=1, \ldots, n$. Then it can readily be verified that (r^T, s^T) is the desired vector.

Remark 3. The proof of Theorem 3.11 shows that condition (3.6) implies that the polyhedral set V is bounded below. The converse is also true and is an immediate consequence of the duality theorem of linear programming.

Remark 4. For $M \in C$ and $q \in R^n$, the solution $\bar{x}$ of the problem (q, M) obtained in Theorem 3.11 need not be the least element of the feasible set $X(q, M)$ under the usual ordering of R^n, as the following example shows. However, it will be demonstrated in Pang [22] that $\bar{x}$ is always the least element under the partial ordering induced by the polyhedral cone

$$p = \{q \in R^n : X^{-1}q \geqq 0\}$$

where $M = YX^{-1}$ with X and Y satisfying conditions (3.5) and (3.6).

Example.

$$M = \begin{pmatrix} 1 & 1 \\ -2 & -3 \end{pmatrix} \quad X = \begin{pmatrix} 1 & -3 \\ -\frac{1}{2} & 1 \end{pmatrix} \quad Y = \begin{pmatrix} \frac{1}{2} & -2 \\ -\frac{1}{2} & 3 \end{pmatrix}$$

$$q^T = (-1, 6).$$

$$M = YX^{-1}; \quad Y \in K.$$

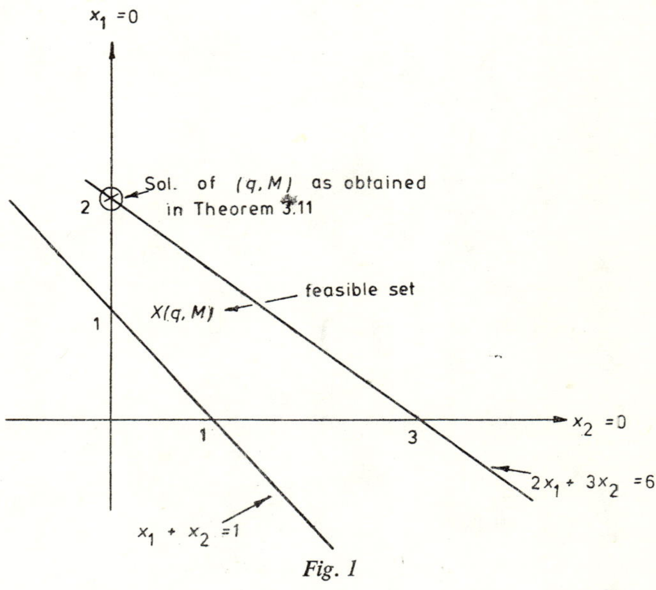

Fig. 1

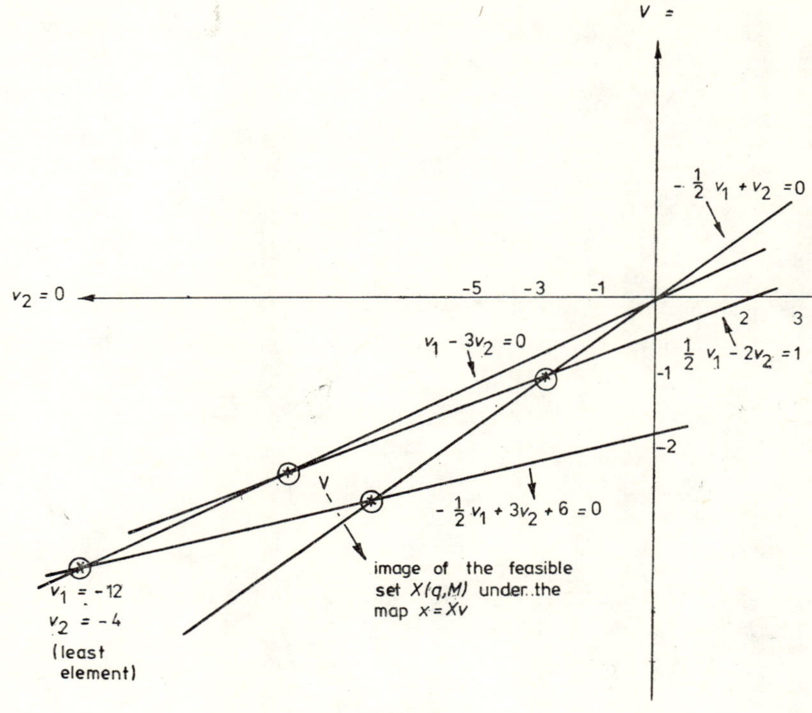

Fig. 2

Having established the desired relationship mentioned above, we proceed to investigate the classes of matrices introduced by Mangasarian. It is clear that all the classes of matrices in Corollary 2.3 are subclasses of C.

In [19], Mangasarian introduced the "slack linear complementarity problem" of finding $x, y \in R^n_+$ such that

(3.13)
$$\begin{pmatrix} u \\ v \end{pmatrix} = \begin{pmatrix} q \\ 0 \end{pmatrix} + \begin{pmatrix} M & B \\ 0 & A \end{pmatrix} \begin{pmatrix} x \\ y \end{pmatrix} \geq 0$$

and
$$u^T x = v^T y = 0$$

in order to extend the class of LP-solvable linear complementarity problems. It is clear that if A satisfies the condition that $x^T A x \neq 0$ for all $0 \neq x \geq 0$, and if (3.13) has a solution $(\bar{x}, \bar{y})$, it necessarily follows that $\bar{y} = 0$. So $\bar{x}$ solves (q, M). Mangasarian then shows that if M satisfies conditions (2.7) and (2.8), the matrix $\begin{pmatrix} M & B \\ 0 & I \end{pmatrix}$ satisfies conditions (2.5) and (2.6). He then invokes Theorem 2.1 and the above observation to obtain Theorem 2.4.

Here we intend to derive Theorem 2.4 using a different approach. We want to employ the above established relationship. Our eventual aim is to show that a matrix satisfying conditions (2.7) and (2.8) belongs to the class C.

It has long been an open problem in the theory of linear complementarity problems to characterize the class $\varkappa$ of matrices for which the feasibility of the linear complementarity problems implies their solvability. Although this class $\varkappa$ is still a mystery, various subclasses have been explored and studied intensively. (See e.g. Lemke [17].) Theorem 3.11 shows that the class C is a subclass of $\varkappa$. The following example shows that it is a *proper* subclass of $\varkappa$.

Example. Let $M=\begin{pmatrix}1 & 1\\ 1 & 1\end{pmatrix}$. Then it can easily be shown that the problem (q, M) always has a solution for every $q\in R^2$. We claim that $M\notin C$. Suppose not, then there exist $X, Y, \in R^{2\times 2}$, both Z-matrices satisfying conditions (3.5) and (3.6). We have

$$\begin{pmatrix}1 & 1\\ 1 & 1\end{pmatrix}\begin{pmatrix}x_{11} & x_{12}\\ x_{21} & x_{22}\end{pmatrix}=\begin{pmatrix}y_{11} & y_{12}\\ y_{21} & y_{22}\end{pmatrix}$$

which implies

$$x_{11}+x_{21}=y_{11}$$

and

$$x_{11}+x_{21}=y_{21}.$$

Thus $y_{11}=y_{21}\leqq 0$. Similarly, $y_{22}=y_{12}\leqq 0$. Lemma 3.8 would then imply $X\in K$. In particular, $\det X=x_{11}x_{22}-x_{12}x_{21}>0$. However,

$$0<x_{11}=y_{21}-x_{21}\leqq -x_{21}$$

and

$$0<x_{22}=y_{22}-x_{12}\leqq -x_{12}.$$

Therefore, $x_{11}x_{22}\leqq x_{12}x_{21}$ which is a contradiction. This proves that $M\notin C$.

As the last four subclasses of matrices in Table 1 are all defined so explicitly in terms of well-known classes of matrices, it is natural to ask whether any of these will perhaps add to our knowledge about the class $\varkappa$. In the sequel, we shall study these special classes separately and show they all belong to P which is of course already well-known in the theory of linear complementarity problems. (See [4], [6], [25].)

Proposition 3.12. *Let* $Y\in K\cap R^{n\times n}$ *and* $a, b\in R^n_+$. *Then the matrix* $Y+ab^T\in P$.

Proof. The hypothesis is inherited by principal submatrices, hence it suffices to show $\det (Y+ab^T)>0$. Clearly,

$$\det (Y+ab^T)=(\det Y)\ \left(\det \left(I+(Y^{-1}a)b^T\right)\right).$$

Using the formula

$$\det (xy^T-\lambda I)=(-1)^n\lambda^{n-1}(\lambda-x^Ty)$$

which holds for all $x, y\in R^n$ and $\lambda\in R$, we obviously have (substituting $\lambda=-1$, $x=Y^{-1}a$ and $y=b$)

$$\det (Y+ab^T)=(\det Y)(1+(Y^{-1}a)^Tb)\geqq \det Y>0$$

because $Y^{-1}\geqq 0$ and $a, b\geqq 0$. Therefore $Y+ab^T\in P$.

Proposition 3.13. *If $M = Y + ab^T$ where $Y \in K$ and $a, b \geq 0$, then there exists a matrix $X \in K$ such that $M = YX^{-1}$.*

Proof. Since $Y \in K$ and $a \geq 0$, it follows that $\bar{a} \equiv Y^{-1}a \geq 0$. We may write $M = Y(I + \bar{a}b^T)$. Let $X \equiv (I + \bar{a}b^T)^{-1}$. Then $M = YX^{-1}$. It remains to show that $X \in K$. An easy calculation shows that

$$X = I - \frac{1}{1 + b^T\bar{a}} \bar{a}b^T \in Z.$$

It follows from Proposition 3.12 that $X^{-1} = I + \bar{a}b^T \in P$; hence $X \in P$ by fact (ii) mentioned earlier. Therefore $X \in K$. This completes the proof.

Corollary 3.14. *If $M = Y + ab^T$ where $Y \in K$ and $a, b \geq 0$, then $M \in C$.*

Thus we have shown that the class of matrices

$$\{M \in R^{n \times n} : M = Y + ab^T, \ Y \in K \quad \text{and} \quad a, b \in R^n_+\}$$

is a subclass of C. Furthermore, the matrices X and Y in the factorization $M = YX^{-1}$ are Minkowski.

Remark. In [19], Mangasarian treated the above class of matrices with the assumption that the vector b is strictly positive. Here, we have relaxed this assumption slightly and merely required b to be nonnegative.

We next proceed to another class. The proof of the following lemma can be found in Fiedler and Pták [14].

Lemma 3.15. *Let $A \in Z$, $B \in K$ and $A \geq B$, then*

(vii)	$A \in K$
(viii)	$A^{-1}B \in K, \quad BA^{-1} \in K$
(ix)	$A^{-1}B \leq I, \quad BA^{-1} \leq I.$

Lemma 3.16. *Let $A \in Z$, $B \in K$, $I \geq B$ and $A \geq B$. Then the matrix $I + A - B \in P$.*

Proof. The hypothesis are inherited by principal submatrices, hence it suffices to show $\det(I + A - B) > 0$. Let $C = A - B \geq 0$. Then $B = A - C$ belongs to K by assumption. We have

$$(I + C)(A - C) = A - C + CB = A - C(I - B) \leq A.$$

Hence $(I + C)(A - C) \in Z$. Since $I + C \geq I$ and $A - C \in K$, it follows readily from condition (iv) in Proposition 3.2 that $(I + C)(A - C) \in K$. Hence

$$\det(I + A - B) = \det(I + C)$$

$$= \det(I + C)(A - C)/\det(A - C) > 0.$$

This completes the proof.

Lemma 3.17. *Let* $Z_1, Z_2, Z_3 \in Z$. $Z_1 \geq Z_2 \geq Z_3$ *and* $Z_3 \in K$. *Then the matrix* $Z_1 + Z_2 - Z_3 \in P$.

Proof. Again, it suffices to show $\det (Z_1 + Z_2 - Z_3) > 0$. We write

$$Z_1 + Z_2 - Z_3 = Z_1(I + Z_1^{-1}Z_2 - Z_1^{-1}Z_3).$$

Lemma 3.15 implies the assumptions of Lemma 3.16 are satisfied with $A = Z_1^{-1}Z_2$ and $B = Z_1^{-1}Z_3$. Therefore $\det (I + Z_1^{-1}Z_2 - Z_1^{-1}Z_3) > 0$. Lemma 3.15 also implies that $\det Z_1 > 0$. Hence $\det (Z_1 + Z_2 - Z_3) > 0$.

Corollary 3.18. *Let* $A \in Z$, $B \in K$ *and* $A \geq B$. *Then the matrix* $2A - B \in P$.

Remark. As a matter of fact, following the proof of Lemma 3.17, we may deduce that if A, B satisfy the assumptions in Corollary 3.18, then the matrix $\lambda A - B \in P$ for all $\lambda \geq 2$.

Corollary 3.19. *If* $A \in Z$, $B \in K$ *and* $A \geq B$, *then* $\det (2A - B) \geq \det B$.

Proof. According to Theorem 1 of Ostrowski [21], if $Y \in Z$ has positive diagonals and M is a matrix such that $|m_{ii}| \geq y_{ii}$ and $|m_{ij}| \leq -y_{ij}$ for all $i \neq j$, then $|\det M| \geq \geq \det Y$. So choosing $Y = B$, $M = 2A - B$, we have $|\det M| = \det M$ and $|m_{ii}| = m_{ii} = = 2a_{ii} - b_{ii} \geq b_{ii}$ and $0 \geq 2a_{ij} \geq 2b_{ij}$ for all $i \neq j$. Hence $-b_{ij} \geq 2a_{ij} - b_{ij} \geq b_{ij}$, i.e., $|2a_{ij} - b_{ij}| \leq -b_j$. Ostrowski's theorem applies and the proof is complete.

Proposition 3.20. *Let* $M = 2A - B$ *where* $A \in Z$, $B \in K$ *and* $A \geq B$. *Then there exist* $Y \in Z$ *and* $X \in K$ *such that* $M = YX^{-1}$.

Proof. Lemma 3.15 implies A^{-1} exists. We may write

$$M = 2A - B = (2B - BA^{-1}B)(B^{-1}A) = (2B - BA^{-1}B)(A^{-1}B)^{-1}.$$

Let $X = A^{-1}B$ and $Y = 2B - BA^{-1}B$. Then Lemma 3.15 implies that $X \in K$ and $X \leq I$. We have

$$Y = 2B - BA^{-1}B = A - (A - B)(I - A^{-1}B) \leq A.$$

Hence $Y \in Z$ and the proof is complete.

Corollary 3.21. *Let* $M = 2A - B$ *with* $A \in Z$, $B \in K$ *and* $A \geq B$. *Then* $M \in C$.

Corollary 3.22. *Let M satisfy either of the following conditions:*

$$(3.14) \qquad m_{ii} > \sum_{\substack{j=1 \\ j \neq i}}^{n} |m_{ij}|, \quad i = 1, \ldots, n$$

$$(3.15) \qquad m_{jj} > \sum_{\substack{i=1 \\ i \neq j}}^{n} |m_{ij}|, \quad j = 1, \ldots, n$$

then $M \in C$.

Proof. It suffices to show that M satisfies the assumptions in Corollary 3.21. Define the matrix $A = (a_{ij})$ as follows:

$$a_{ij} = \begin{cases} m_{ii} & \text{if } j = i \\ m_{ij} & \text{if } j \neq i \text{ and } m_{ij} \leq 0 \\ 0 & \text{if } j \neq i \text{ and } m_{ij} > 0. \end{cases}$$

Let $B = 2A - M$; clearly, $M = 2A - B$ and $A \geq B$. We obviously have $A \in Z$, thus $B \in Z$. It remains to show that $B \in K$. If $B = (b_{ij})$ then by its definition,

$$b_{ij} = \begin{cases} m_{ii} & \text{if } j = i \\ m_{ij} & \text{if } j \neq i, \, m_{ij} \leq 0 \\ -m_{ij} & \text{if } j \neq i, \, m_{ij} > 0, \end{cases}$$

so that letting $e^T = (1, \ldots, 1) \in R^n$, we have for $i = 1, \ldots, n$,

$$(Be)_i = \sum_{j=1}^{n} b_{ij} = m_{ii} + \sum_{\substack{j=1 \\ j \neq i, \, m_{ij} \leq 0}}^{n} m_{ij} + \sum_{\substack{j=1 \\ j \neq i, \, m_{ij} > 0}}^{n} (-m_{ij}).$$

Thus, $Be > 0$ if condition (3.14) is satisfied. Similarly, we may deduce that $B^T e > 0$ if condition (3.15) is satisfied. Therefore, in either case, it follows that $B \in K$. This completes the proof.

Remark 1. The proof above and Lemma 3.20 show that a matrix satisfying conditions (3.14) and (3.15) belongs to the class P.

Remark 2. In [19], Mangasarian treated the classes of matrices M satisfying conditions (3.14) and (3.15) with the additional assumption that M is nonnegative. The corollary above enlarges these classes by omitting the nonnegativity assumption and shows that the corresponding linear complementarity problems are still related to polyhedral sets with least elements.

Up to this point, we have shown that all matrices in Mangasarian's Table 1 belong to the class C. Furthermore, we have established that the last four subclasses belong to P, thus to the class $\varkappa$ mentioned earlier. Therefore they do not extend our knowledge about the class $\varkappa$.

Remark. If M belongs to any of these four subclasses, the problem (q, M) has a solution for every $q \in R$ because M must necessarily be a P-matrix. Thus the feasibility assumption can be removed.

As all matrices in Table 1 are obtained by specializing the conditions (2.7) and (2.8) in Theorem 2.4, it is natural to ask whether a matrix satisfying these two conditions alone belongs to C. We answer this question by establishing the following theorem which summarizes some of the previous results.

Theorem 3.23. *Let $M \in R^{n \times n}$. Suppose there exist $X, Y \in R^{n \times n}$, $A \in R^{m \times m}$, $B, H \in R^{n \times m}$, $G \in R^{m \times n}$, $p \in R_{+}^{n}$ and $p_0 \in R_{+}^{m}$ satisfying*

$$(3.16) \qquad MX = Y + BG; \quad MH \geqq BA; \quad X, Y, A \in Z; \quad \text{and} \quad G, H \geqq 0;$$

$$(3.17) \qquad (p^T, p_0^T) \begin{bmatrix} X & -H \\ -G & A \end{bmatrix} > 0.$$

Then

(3.18) M has the representation

$$M = \overline{Y}\,\overline{X}^{-1}$$

where

$$\overline{Y} \equiv Y - \overline{H}A^{-1}G, \quad \overline{X} \equiv X - HA^{-1}G, \quad \text{and} \quad \overline{H} \equiv MH - BA;$$

(3.19) furthermore, $\overline{Y} \in Z$ and $\overline{X} \in K$.

Proof. The matrix $\begin{bmatrix} X & -H \\ -G & A \end{bmatrix}$ belongs to class Z, thus condition (3.17) implies that it belongs to K. In particular A^{-1} exists. We have

$$\overline{Y} \equiv Y - \overline{H}A^{-1}G = Y - (MH - BA)A^{-1}G =$$
$$= Y + BG - M(HA^{-1}G) =$$
$$= M(X - HA^{-1}G) = M\overline{X}.$$

The matrix $\overline{X} \equiv X - HA^{-1}G$ is the Schur complement of A in the Minkowski matrix $\begin{bmatrix} X & -H \\ -G & A \end{bmatrix}$, hence $\overline{X}$ is itself Minkowski by Proposition 3.4 and in particular, nonsingular. Thus $M = \overline{Y}\,\overline{X}^{-1}$, establishing (3.18). It remains to verify that $\overline{Y} \in Z$. We have $\overline{Y} \equiv Y - \overline{H}A^{-1}G$ where $\overline{H} = MH - BA \geqq 0$ and $G \geqq 0$ by assumption; moreover, $A^{-1} \geqq 0$ because $A \in K$. Therefore $\overline{Y} \in Z$. This completes the proof of the theorem.

Corollary 3.24. *Suppose $M \in R^{n \times n}$ satisfies conditions (3.16) and (3.17). Then $M \in C$.*

26

Remark. In the factorization $M = \overline{Y}\overline{X}^{-1}$ above, the matrix $\overline{Y} \equiv Y - \overline{H}A^{-1}G$ contains the matrix $\overline{H}$ which is defined in terms of M, i.e. the factorization involves M itself implicitly. This somewhat awkward situation can be remedied by solving for M using condition (3.16) to obtain $M = (Y + BG)X^{-1}$. Substituting into the definition of $\overline{H}$, we see that M is no longer involved in the factors.

4. COMPUTATIONAL EXPERIENCE

Methods for solving linear complementarity problems for various classes of matrices have been proposed and investigated intensively. Among these are the principal pivoting method [4], [5] and Lemke's almost complementarity pivoting algorithm [6], [16]. These methods (and some others) work rather satisfactorily for matrices of reasonable size. But in many applications of the linear complementarity problems to partial differential equations, the matrices are often large, sparse and specially structured. See [21] for example. Methods like those mentioned above seem to be inefficient when applied to these problems. For one thing, most of the nice properties (especially the sparsity which is a very important factor for efficiency) that the matrices originally possess will be destroyed when the problems are being processed. Recognizing this disadvantage, one would like to use iterative (relaxation) procedures which, presumably, have the computational advantage of preserving matrix sparsity.

In [18], Mangasarian proposed formulating linear complementarity problems as linear programs and solving them by applying relaxation methods to the linear inequality system (2.4). In this section, we discuss our somewhat preliminary computational experience using this solution strategy and attempt to answer the question of whether this approach can be recommended in practice.

For the sake of clarity, we first review part of the theory of relaxation methods for solving linear inequality systems [1], [2], [13], [20].

We want to find a vector $z \in R^n$ satisfying the system of linear inequalities: $Az \leq b$ where $A \in R^{m \times n}$ and $b \in R^m$. Let $A \equiv A_i$. be the i-th row of the matrix A. The *relaxation method* (due to Eremin [13]) constructs a sequence $\{z^k\}$ in the following manner:

(i) Choose $z^0 \in R^n$ arbitrarily. Let $k = 0$.

(ii) If $Az^k \leq b$, the procedure terminates.

If not, then some linear inequality is violated. Let i be the smallest index of the most violated constraints:

$$A_i z^k - b_i = \max_{1 \leq i \leq n} \{A_j z^k - b_j\}.$$

(iii) Define

$$(4.1) \qquad z^{k+1} = z^k - \lambda^k \left(\frac{A_i z^k - b_i}{\|A_i\|^2} \right) A_i^T.$$

Theorem (Eremin [13]).

The sequence $\{z^k\}$ defined by the relaxation method under the assumptions

(i) $\lambda^k \in (0, 2]$, $k = 1, 2, \ldots$

(ii) $\inf \lambda^k > 0$.

Converges to one of the solutions of the linear inequality system $Az \leqq b$ if the latter is consistent.

We now apply the method to solve the linear program (p, q, M), or equivalently, to find vectors $x, y \in R^n$ such that the following system of linear inequalities is satisfied.

$$(4.2) \qquad \begin{pmatrix} -M & 0 \\ 0 & M^T \\ -I & 0 \\ 0 & -I \\ p^T & q^T \end{pmatrix} \begin{pmatrix} x \\ y \end{pmatrix} \leqq \begin{pmatrix} q \\ p \\ 0 \\ 0 \\ 0 \end{pmatrix}.$$

Remark. The vectors x and y are primal and dual variables, respectively.

We now present an algorithm based on the relaxation method for solving (4.2). Let $\varepsilon > 0$ be some preassigned positive tolerance. The algorithm starts by choosing

$$z^0 = \begin{pmatrix} x^0 \\ y^0 \end{pmatrix}$$

arbitrarily. Initially $l = 0$.

Step 1. Let

$$z^l = \begin{pmatrix} x^l \\ y^l \end{pmatrix} \text{ and define}$$

$$(4.3) \qquad d(z^l) = \max \left\{ \max_{1 \leq i \leq n} \left(-\sum_{j=1}^{n} m_{ij} x_j^l - q_i \right), \; \max_{1 \leq j \leq n} \left(\sum_{k=1}^{n} m_{kj} y_k^l - p_j \right), \right.$$

$$\left. \max_{1 \leq i \leq n} (-x_i^l), \; \max_{1 \leq j \leq n} (-y_j^l), \; \sum_{i=1}^{n} p_i x_i^l + \sum_{j=1}^{n} q_j y_j^l \right\}.$$

Step 2. If $d(z^l) \leqq \varepsilon$, stop.

Otherwise, construct $z^{l+1} = \begin{pmatrix} x^{l+1} \\ y^{l+1} \end{pmatrix}$ according to (4.1) and return to step 1 with $l+1$ replacing l.

28

The main work involved in each iteration cycle of the algorithm is the computation of the quantity $d(z^l)$ and the updating of the new iterates. When actually programmed, $d(z^l)$ is computed as in (4.3). While in constructing the new approximate solution z^{l+1}, an index is set to check where the maximizing term comes from, in order to avoid computing the same components repeatedly. For instance, if the maximizing term is in $\left\{-\sum_{j=i}^{n} m_{ij}x_j^l - q_i\right\}_{i=1}^{n}$, then the vector y^l need not be updated; on the other hand, if the maximizing term comes from $\{-x_i^l\}_{i=1}^{n}$, then only one component of x^l will be changed while the other components of z^l will remain unchanged in the next iteration. These features of the system are important in reducing the computational effort of the algorithm.

There is a variation of the algorithm that one might want to consider when actually coding it. In (4.2) where the new iterate z^{l+1} is computed, it is necessary to divide by the 2-norm of the gradient of the most violated constraint. If the number of iterations is large compared to $4n+1$ which is the number of constraints in (4.1), it would be more economical to reduce these divisions if possible. A way to achieve this is to work with the "normalized system" which is obtained from (4.1) by normalizing each constraint separately. The division steps in (4.2) are then no longer needed. An obvious disadvantage of this "normalized system" is the need for more storage space for the whole matrix and also for the constant vectors. But if the problem itself were of moderate size so that no storage problem would occur, then one might try to use this "normalized system" instead of the original system (4.1).

We performed several experiments to solve the problem (q, M) where $q \in R^n$ was randomly chosen and

$$
M = \begin{pmatrix}
2 & -1 & & & & & & \\
-1 & 2 & -1 & & & & & \\
& -1 & 2 & -1 & & & & \\
& & \cdot & \cdot & \cdot & & & \\
& & & \cdot & \cdot & \cdot & & \\
& & & & \cdot & \cdot & \cdot & \\
& & & & & -1 & 2 & -1 \\
& & & & & & -1 & 2
\end{pmatrix}.
$$

We chose $p^T = (1, \ldots, 1) \in R^n$. The relaxation parameters $\{\lambda^k\}$ were chosen to be the same in each iteration. The tolerance ε was generously chosen to be 10^{-4}. The algorithm was then applied to the inequality system (4.1). All the computation was done on the IBM 370/168 using FORTRAN H with Opt=2. The results are summarized in the following tables.

Inputs: $n=5$. $q^T=(2, -1, -4, 6, -5)$;
    starting iterate $x^0=y^0=0$; original system.

Solution of (q, M): $x^T=(0, 2, 3, 0, 2.5)$.

Table 2

Value of λ	No. of Iterations	Execution time (sec)
.7	1737	.58
.8	1380	.50
.9	1108	.45
1.1	749	.38
1.2	649	.37
1.3	505	.33
1.4	432	.33
1.5	358	.31
1.6	176	.27
1.7	949	.42
1.8	668	.37

Inputs: $n=5$. $q^T=(2, -1, -4, 6, -5)$
    starting iterate $x^0=y^0=0$; normalized system.

Solution of (q, M): $x^T=(0, 2, 3, 0, 2.5)$.

Table 3

Value of λ	No. of Iterations	Execution time (sec)
1.0	669	.38
1.5	150	.28
1.6	887	.41
1.8	965	.43

Inputs: $n=5$. $q^T=(2, -1, -3, 4, -5)$;
starting iterate: $x^0=(3, 3, 3, 3, 3)^T$,
$y^0=(2, \ldots, 2)^T$; normalized system.

Solution of (q, M): $x^T=(0, 2, 3, 1, 3)$.

Table 4

Value of λ	No. of Iterations	Execution time (sec)
.8	17,364	3.12
.9	14,263	2.63
1.0	11,664	2.19
1.1	9,663	1.88
1.2	7,926	1.57
1.3	6,571	1.36
1 4	5,626	1.19
1.5	4,381	1.21
1.6	3.352	.81
1.7	2,353	.65
1.8	1,490	.50
1.9	664	.37

Inputs: $n=7$. $q^T=(2, -1, -4, 6, -5, 3, -2)$;

starting iterate $x^0=y^0=0$; original system.

Solution of problem (q, M): $x^T=(0, 2, 3, 0, 2.75, .5, 1.25)$.

Table 5

Value of λ	No. of Iterations	Execution time (sec)
1.95	2,042	.74
1.8	9,820	2.63
1.7	14,831	3.87
1.6	19,935*	4.60

* 20,000 is the maximumr numbe of iterations that we allow.

Inputs: $n=7$. $q^T=(2, -1, -4, 6, -5, 3, -2)$;
starting iterate: $x^0=(2, \ldots, 2)^T$, $y^0=(1, \ldots, 1)$ original system; solution of (q, M) given in Table 5.

Table 6

Value of λ	No. of Iterations	Execution time (sec)
1.95	1,362	.57
1.85	5,910	1.68
1.75	10,597	2.67
1.65	14,313	3.65
1.55	19,219	4.95

It was conjectured that the disappointing computational results reported above could be accounted for by the implicit presence of the equality constraint $p^Tx+q^Ty=0$. Further experiments revealed that this alone cannot be the reason. In cases like those studied here (i.e., where $M\in K$ and $q\neq 0$) other equality constraints can be identified *ab initio*. To eliminate variables would entail at least partial use of a direct method and would generate density where it did not previously exist.

The problems solved in these experiments are far from the size and difficulty one would except to encounter in practice. Nevertheless, they suggest that the relaxation method described above lacks the efficiency required for solving more realistic problems. It is conceivable that by some clever modification the method could be made more attractive.

Acknowledgements

The authors wish to thank Mr. Hung-Po Chao for suggesting the method of proof used in Lemma 3.16. We are also indebted to Professor Hans Schneider for indicating the applicability of Ostrowski's theorem in Corollary 3.19. In a private communication, Professor Schneider suggested an interpretation of Corollary 3.21 in terms of a well-known class of matrices, namely the class $\mathcal{K}$ of real square matrices M such that $\hat{M}\in K$ where $\hat{m}_{ii}=m_{ii}$ and $\hat{m}_{ij}=-|m_{ij}|$. It is known that $M\in K$ if and only if $M=2A-B$ for suitable matrices A and B such that $A\in Z$, $B\in\mathcal{K}$ and $A\cong B$. Thus our Corollary 3.21 can be rephrased as the assertion $\mathcal{K}\subset C$.

In addition, the first author wishes to thank Professor Olvi Mangasarian for several stimulating discussions some of which were held while visiting the Applied Mathematics Division of the Argonne National Laboratory.

REFERENCES

[1] Agmon, S.: The relaxation method for linear inequalities, *Canadian J. of Math.* 6 (1954), 382–392.
[2] Bregman, C. M.: The method of successive projection for finding a common point of convex sets, *Soviet Math. Doklady* (transl.) 6 (1965), 688–692.

32

[3] Chandrasekaran, R.: A special case of the complementarity pivot problem, *Opsearch 7* (1970), 263–268.

[4] Cottle, R. W.: Principal pivoting method for quadratic programming, *Mathematics of the Decision Sciences,* Part I (G. B. Dantzig and A. F. Veinott, Jr., eds.) Amer. Math. Society, Providence, R. I. 1968.

[5] Cottle, R. W.: Manifestations of the Schur complement, *Linear Algebra and Its Applications, 8* (1974), 189–211.

[6] Cottle, R. W. and Dantzig, G. B.: Complementarity pivot theory of mathematical programming, *Linear Algebra and Its Applications 1* (1968), 103–125.

[7] Cottle, R. W., Golub, G. H. and Sacher, R. S.: On the solution of large, structured, linear complementarity problems: III, Tech. Report 74–7. Department of Operations Research, Stanford University, 1974.

[8] Cottle, R. W. and Sacher, R. S.: On the solution of large, structured linear complementarity problems: I, Tech. Report 73–4, Department of Operations Research, Stanford University, 1973.

[9] Cottle, R. W. and Veinott, A. F. Jr.: Polyhedral sets having a least element, *Math. Programming 3* (1969), 238–249.

[10] Crabtree, D. E.: *Applications of M-matrices to nonnegative matrices, Duke Math. J. 33* (1966), 197–208.

[11] Cryer, C. W.: The solution of a quadratic programming problem using systematic overrelaxation, *SIAM J. Control 9* (1971), 385–392.

[12] Dantzig, G. B.: Optimal solution of a dynamic Leontief model with substitution, *Econometrica 23* (1955), 295–302.

[13] Eremin, I. I.: On systems of inequalities with convex functions in the left sides, *Soviet Math. Doklady* (transl.) *6* (1965), 219–222.

[14] Fiedler, M. and Pták, V.: On matrices with nonpositive off-diagonal elements and positive principal minors, *Czech. J. Math. 12* (1962), 382–400.

[15] Kuhn, H. W.: Solvability and consistency for linear equations and inequalities, *Amer. Math. Monthly 63* (1956), 217–232.

[16] Lemke, C. E.: Bimatrix equilibrium points and mathematical programming, *Management Science, 11,* No. 7 (1965), 681–689.

[17] Lemke, C. E.: Recent results on complementarity problems, in *Symposium on Nonlinear Programming,* ed. by J. B. Rosen, O. L. Mangasarian and K. Ritter, Acad. Press 1970.

[18] Mangasarian, O. L.: Linear complementarity problems solvable by a single linear program, Tech. Report No. 237, Computer Sciences Dept., Univ. of Wisconsin, Madison, Jan. 1975.

[19] Mangasarian, O. L.: Solution of linear complementarity problems by linear programming, Tech. Report No. 257, Computer Sciences Dept., Univ. of Wisconsin, Madison, June 1975.

[20] Motzkin, Th. and Schoenberg, I. J.: The relaxation method for linear inequalities, *Canadian J. Math. 6* (1954), 393–404.

[21] Ostrowski, A.: Über die Determinanten mit überwiegender Hauptdiagonale, *Comment Math. Helv. 10* (1937/1938), 69–96.

[22] Pang, J. S.: Least element complementarity theory, Ph. D. dissertation, Department of Operations Research, Stanford University, in preparation.

[23] Sacher, R. S.: On the solution of large, structured complementarity problems, Ph. D. dissertation, Dept. of Operations Research, Stanford University, 1974.

[24] Saigal, R.: A note on a special linear complementarity problem, *Opsearch 7* (1970), 175–183.

[25] Samelson, H., Thrall, R. M. and Wesler, O.: A partitioning theorem for Euclidean n-space, *Proceedings Amer. Math. Soc. 9* (1958), 805–807.

[26] Tamir, A.: The complementarity problem of mathematical programming, Ph. D. dissertation, Department of Operations Research, Case Western Reserve University, June, 1973.

LINEAR COMPLEMENTARITY PROBLEM AND A TREE SEARCH ALGORITHM FOR ITS SOLUTION

G. R. JAHANSHAHLOU and G. MITRA

(Tehran, Iran)   (Uxbridge, U.K.)

1. INTRODUCTION

Consider the linear complementarity problem,

$$(1) \qquad w = q + Mz$$

$$(2) \qquad w, \quad z \geqq 0$$

$$(3) \qquad zw = 0$$

w and z are vectors of n variables, q is a given n element vector, M is a given $n \times n$ matrix.

The above problem involves $2n$ variables, restricted to be non-negative, where (w_i, z_i), $i = 1, \ldots, n$, is a complementary pair; and w_i and z_i are complement of one another.

The special cases of linear complementarity problem are linear and quadratic programming problems, the problem of finding equilibrium points in bimatrix games and some engineering problems; for these and other applications see [1].

The two prominent methods of solution for the problem (1), (2), (3) are the principal pivoting method and Lemke's method. The method proposed by Lemke [3] can be considered to be a generalization of Dantzig's self-dual parametric method (see [2]), and its generalization for convex quadratic programming. This motivated S. R. McCammon [6] to develop his parametric pivoting method. Lemke has proven that his method finds a solution to the problem or else the solution comes to an unbounded ray and there is no solution to the given problem if M belongs to a class of matrices called copositive plus.

The principal pivoting method was developed by Cottle and Dantzig [1]. This method is applicable to the matrices, which have positive principal minors (in particular to positive definite matrices). The modified form of principal pivoting method can be applied to positive semi-definite matrices.

The method of solution of the problem (1), (2), (3) is generally dependent on the matrix M. In section 2, therefore, after introducing the relevant notation, some properties of pivotal transformation and different types of M matrices are considered. The authors have reviewed some of the important developments in this field in a report [5].

Lemke's method and the principal pivoting method may not produce the solution to the problem (1), (2), (3) even if such a solution exists; section 3 illustrates such a situation. The method proposed by the authors is then put forward in this section. Section 4 contains some remarks on the computational experiences of the authors.

2. SOME PRELIMINARY NOTATION AND MATHEMATICAL BACKGROUND

2.1 Notation

Let $R^{n \times n}$ denote the set of $n \times n$ matrices with real coefficients, let $M \in R^{n \times n}$, $M_{i\cdot}$ and $M_{\cdot j}$ denote the i-th row and the i-th column of M and m_{ij} denote the element of M in row i and column j. Further, let e denote the sum vector $(1, \ldots, 1)^T \in R^{n \times 1}$ and e_i denote the unit vector whose i-th component is unity and the other components are zero. The bar above a variable (say $\bar{z}_j$ or $\bar{w}_j$) denotes the explicit value of the variable.

2.2 Tableau Representation and Pivotal Transformation

In (1), the components of z are nonbasic variables, while the elements of w comprise the basic variables. A solution of problem (1) is any pair $(\bar{w}, \bar{z})$ satisfying (1).

If for some $\bar{z} \geqq 0$, $\bar{w} = q + M\bar{z} \geqq 0$, then the pair $(\bar{w}, \bar{z})$ provides a feasible solution to the problem (1), i.e. a solution which satisfies (1) and (2). A solution of (1) satisfying (3) is a complementary solution.

If every solution $(\bar{w}, \bar{z})$ of the problem (1) contains not more than n zero components among the $2n$ variables (w, z), then the problem (1) is nondegenerate. In the present discussion only such nondegenerate problem are considered.

Assume that the element

$$m_{ij} \neq 0,$$

then using the element $m_{ij} \neq 0$ a "pivotal transformation" may be carried out on the form

$$(4) \qquad\qquad w = q + Mz.$$

This transformation consists of

(a) solving the ith equation of (4) for the variable z_j, this requires dividing by the pivot element m_{ij},

(b) replacing z_j by the resulting expression in each of the remaining $(n-1)$ equations.

Upon completion of a pivotal transformation, z_j becomes basic, while w_i becomes nonbasic. (w_i, z_j) is the pivot pair, and by specifying that this pair must be exchanged, the pivot is completely determined. The result of a sequence of pivotal transformations after t steps may be expressed as

$$(5) \qquad w^t = q^t + M^t z^t,$$

where w^t denotes the set of basic variables, while z^t denotes the set of nonbasic variables.

For completeness of notation the results of carrying out one pivotal transformation is summarized below. Given the tableau-0

	1	$-z_1^t$	$-z_2^t \ldots -z_n^t$
w_1^t	q_1^t	$-m_{11}^t$	$-m_{12}^t \ldots -m_{1n}^t$
w_2^t	q_2^t	$-m_{21}^t$	$-m_{22}^t \ldots -m_{2n}^t$
			$\ldots$
w_n^t	q_n^t	$-m_{n1}^t$	$-m_{n2}^t \ldots -m_{nn}^t$

Tableau–0

The next tableau is constructed by the following relationship:

1) $(-m_{ij}^{t+1}) = (1)/(-m_{ij}^t)$

2) $(-m_{ik}^{t+1}) = (-m_{ik}^t)/(-m_{ij}^t) \quad 1 \leq k \leq n \quad k \neq j$

3) $(-m_{lj}^{t+1}) = -(-m_{lj}^t)/(-m_{ij}^t) \quad 1 \leq l \leq n \quad l \neq i$

4) $(-m_{lk}^{t+1}) = (-m_{lk}^t) - (-m_{lj}^t)(-m_{ik}^t)/(-m_{ij}^t)$.

5) Replace the variables such that w_i^{t+1} is the variable in the i-th row (z_j^t is renamed) and z_j^{t+1} is the variable in the j-th column. (w_i^t is renamed).

3. BRANCHING PROCEDURE FOR SOLVING THE LINEAR COMPLEMENTARITY PROBLEM

It has been pointed out earlier that principal pivoting algorithm can be applied to solve the Fundamental Problem only if M is positive definite, or more generally when M is a P-matrix [3]. Further a modified form of this method can be used to obtain a solution to the Fundamental Problem if the system

$$(6) \qquad w = q + Mz,$$

$$z \geq 0, \quad w \geq 0,$$

has a solution and M is positive semi-definite.

If Lemke's method is applied to solve the Fundamental Problem, and the procedure terminates in an unbounded ray and M does not belong to the class of $\mathcal{L}$ matrices, in this case the method does not provide any information concerning the solvability of the problem. For an illustration consider the problem, stated below:

Example 1

$$(7) \quad \begin{cases} w_1 = 10 - 2z_1 + 3z_2 - z_3 \\ w_2 = -1 + z_1 - 2z_2 + z_3 \\ w_3 = 3 - z_1 + 2z_2 + 3z_3 \\ w_1, w_2, w_3, z_1, z_2, z_3 \geqq 0 \\ w_i z_i = 0 \quad (i = 1, 2, 3). \end{cases}$$

Note that in this case the matrix

$$M = \begin{pmatrix} -2 & 3 & -1 \\ 1 & -2 & 1 \\ -1 & 2 & 3 \end{pmatrix},$$

is not a P-matrix, therefore principal pivoting method cannot be applied to solve this problem.

Now Lemke's method is applied as follows:

$$(8) \quad \begin{cases} w_1 = 10 + z_0 - 2z_1 + 3z_2 - z_3 \\ w_2 = -1 + z_0 + z_1 - 2z_2 + z_3 \\ w_3 = 3 + z_0 - z_1 + 2z_2 - 3z. \end{cases}$$

In tableau representation this can be set out in Tableau 3–1, in this tableau z_0 is set to 1 following the ratio test.

	1	$-z_0$	$-z_1$	$-z_2$	$-z_3$
w_1	11	-1	2	-3	1
w_2	0	-1	-1	2	-1
w_3	4	-1	1	-2	-3

Tableau 3–1

	1	$-w_2$	$-z_1$	$-z_2$	$-z_3$
w_1	11	-1	3	-5	2
z_0	1	-1	1	-2	1
w_3	4	-1	2	-4	-2

Tableau 3–2

38

In Tableau 3–2 z_2 is the complement of variable w_2, and cannot be made a basic variable taking non-negative value. Therefore the procedure terminates in an unbounded ray. This means neither Lemke's method nor principal pivoting method can be used to establish the solvability of the problem.

It is shown later on that this problem has the following feasible solutions.

$$(9) \qquad z' = \begin{pmatrix} 17 \\ 8 \\ 0 \end{pmatrix}, \quad w' = \begin{pmatrix} 0 \\ 0 \\ 2 \end{pmatrix} \quad \text{and} \quad z'' = \begin{pmatrix} \dfrac{33}{7} \\ 0 \\ \dfrac{4}{7} \end{pmatrix}, \quad w'' = \begin{pmatrix} 0 \\ \dfrac{30}{7} \\ 0 \end{pmatrix}.$$

In Lemke's method the components of the artificial vector do not necessarily have to be unity. Therefore by suitable choice of these components different paths of complementary solution may be followed, this idea is illustrated later on in this section by means of two examples. One may then naturally ask if by following such possible paths a complementary feasible solution may be obtained to the problem if it exists. By means of the following examples it is shown that this assumption is invalid in the general case.

Consider the problem in the following example

Example 2

$$(10) \qquad \begin{cases} w_1 = 2 + 2z_1 - z_2 - 3z_3 + 4z_4 \\ w_2 = -4 + 10z_1 + z_2 - z_3 + z_4 \\ w_3 = 3 - z_1 - 2z_2 + z_3 - 2z_4 \\ w_4 = -6 + 20z_1 + 3z_2 - z_3 - 3z_4 \\ w_i \geqq 0, \quad z_i \geqq 0 \quad (i = 1, \ldots, 4) \\ w_i z_i = 0 \quad i = 1, \ldots, 4. \end{cases}$$

By introducing z_0 as an artificial variable where $e' = (1, 1, 1, 1)$ the problem becomes:

$$(11) \qquad \begin{cases} w_1 = 2 + z_0 + 2z_1 - z_2 - 3z_3 + 4z_4 \\ w_2 = -4 + z_0 + 10z_1 + z_2 - z_3 + z_4 \\ w_3 = 3 + z_0 - z_1 - 2z_2 + z_3 - 2z_4 \\ w_4 = -6 + z_0 + 20z_1 + 3z_2 - z_3 - 3z_4 \\ w_i \geqq 0, \quad z_i \geqq 0, \quad \text{and} \quad z_i w_i = 0 \quad i = 1, 2, 3, 4. \end{cases}$$

	1	$-z_0$	$-z_1$	$-z_2$	$-z_3$	$-z_4$
w_1	8	-1	-2	1	3	-4
w_2	2	-1	-10	-1	1	-1
w_3	9	-1	1	2	-1	2
w_4	0	-1	-20	-3	1	3

Tableau 3–3

In Tableau 3–3 z_0 is set to 6 to make $q_3 + e'_3 z_0 = 0$.

	1	$-w_4$	$-z_1$	$-z_2$	$-z_3$	$-z_4$
w_1	8	-1				-7
w_2	2	-1				-4
w_3	9	-1	not	up	dated	-1
z_0	6	-1				-3

Tableau 3–4

In Tableau 3–4, z_4 cannot be made basic variable, therefore the procedure terminates in an unbounded ray.

Now choose $e' = (1, \frac{1}{2}, 1, 2)$, the corresponding representation tableau is shown in Tableau 3–5:

	1	$-z_0$	$-z_1$	$-z_2$	$-z_3$	$-z_4$
w_1	10	-1	-2	1	3	-4
w_2	0	$-\frac{1}{2}$	-10	-1	1	-1
w_3	11	-1	1	2	-1	2
w_4	10	-2	-20	-3	1	3

Tableau 3–5

The value of z_0 in Tableau 3–5 is 8

	1	$-w_2$	$-z_1$	$-z_2$	$-z_3$	$-z_4$
w_1	10	-2	18	3	1	-2
z_0	8	-2	20	2	-2	2
w_3	11	-2	21	4	-3	4
w_4	10	-4	20	1	-3	7

Tableau 3–6

z_2 is the complement of the variable w_2, which is made basic in this step. The procedure is then followed until Tableau 3–9.

	1	$-w_2$	$-z_1$	$-w_3$	$-z_3$	$-z_4$
w_1	$\dfrac{7}{4}$	$-\dfrac{1}{2}$	$\dfrac{9}{4}$	$-\dfrac{3}{4}$	$\dfrac{13}{4}$	-5
z_0	$\dfrac{10}{4}$	-1	$\dfrac{38}{4}$	$-\dfrac{2}{4}$	$-\dfrac{1}{2}$	0
z_2	$\dfrac{11}{4}$	$-\dfrac{1}{2}$	$\dfrac{21}{4}$	$\dfrac{1}{4}$	$-\dfrac{3}{4}$	1
z_4	$\dfrac{29}{4}$	$-\dfrac{7}{2}$	$\dfrac{59}{4}$	$-\dfrac{1}{4}$	$-\dfrac{9}{4}$	6

Tableau 3–7

	1	$-w_2$	$-z_1$	$-w_3$	$-w_1$	$-z_4$
z_3	$\dfrac{7}{13}$	not	$\dfrac{9}{13}$	not	$\dfrac{13}{4}$	not
z_0	$\dfrac{36}{13}$	up	$\dfrac{128}{13}$	up	$-\dfrac{1}{2}$	up
z_2	$\dfrac{411}{13}$	dated	$\dfrac{75}{13}$	dated	$-\dfrac{3}{4}$	dated
w_4	$\dfrac{110}{13}$		$\dfrac{212}{13}$		$-\dfrac{9}{4}$	

Tableau 3–8

	1	$-w_2$	$-z_0$	$-w_3$	$-w_1$	$-z_4$
z_3	$\dfrac{11}{32}$					
z_1	$\dfrac{9}{32}$	not	up	dated		
z_2	$\dfrac{49}{32}$					
w_4	$\dfrac{124}{32}$					

Tableau 3–9

So the solution is

$$(12) \qquad w = \begin{pmatrix} 0 \\ 0 \\ 0 \\ \dfrac{124}{32} \end{pmatrix}, \qquad z = \begin{pmatrix} \dfrac{9}{32} \\ \dfrac{49}{32} \\ \dfrac{11}{32} \\ 0 \end{pmatrix}.$$

In another example (see below) it is shown that all the possible paths lead to unbounded rays.

Example 3

$$(13) \qquad \begin{cases} w_1 = 2 + 2z_1 - z_2 - 3z_3 + 4z_4 \\ w_2 = -4 - z_1 + 2z_2 - z_3 + z_4 \\ w_3 = 3 + 2z_1 - 2z_2 + z_3 - 2z_4 \\ w_4 = -6 + 4z_1 + 3z_2 - z_3 - 3z_4 \\ w_i \geqq 0, \ z_i \geqq 0, \ w_i z_i = 0 \ \text{for all} \ (i = 1, \ldots, 4). \end{cases}$$

First z_0 is introduced with corresponding column $e' = (1, 1, 1, 1)$, so the problem can be written as:

$$(14) \qquad \begin{cases} w_1 = 2 + z_0 + 2z_1 - z_2 - 3z_3 + 4z_4 \\ w_2 = -4 + z_0 - z_1 + 2z_2 - z_3 + z_4 \\ w_3 = 3 + z_0 - 2z_1 - 2z_2 + z_3 - 2z_4 \\ w_4 = -6 + z_0 + 4z_1 + 3z_2 - z_3 - 3z_4 \\ w_i \geqq 0, \ z_i \geqq 0, \ w_i z_i = 0 \quad (i = 1, \ldots, 4). \end{cases}$$

	1	$-z_0$	$-z_1$	$-z_2$	$-z_3$	$-z_4$
w_1	8	-1	-2	1	3	-4
w_2	2	-1	1	-1	1	-1
w_3	9	-1	-2	2	-1	2
w_4	0	-1	-4	-3	1	3

Tableau 3–10

In tableau (3–10) z_0 is set to 6.

	1	$-w_4$	$-z_1$	$-z_2$	$-z_3$	$-z_4$
w_1	8	-1	0	4	2	-7
w_2	2	-1	3	2	0	-4
w_3	9	-1	0	5	-2	-1
z_0	6	-1	4	3	-1	-3

Tableau 3–11

z_4 is the complement of the variable w_4, and cannot be made basic variable, therefore the procedure terminates in unbounded ray.

Now, if the column associated with the artificial variable is introduced as $e' = (1, \frac{1}{2}, 1, 2)$, and Lemke's method is applied, the following tableaux are obtained

	1	$-z_0$	$-z_1$	$-z_2$	$-z_3$	$-z_4$
w_1	10	-1	-2	1	3	-4
w_2	0	$-\frac{1}{2}$	1	-1	1	-1
w_3	11	-1	-2	2	-1	2
w_4	10	-2	-4	-3	1	3

Tableau 3–12

In Tableau 3–12 the value of z_0 is 8.

	1	$-w_2$	$-z_1$	$-z_2$	$-z_3$	$-z_4$
w_1	10	-2	-4	3	1	-2
z_0	8	-2	-2	2	-2	2
w_3	11	-2	-4	4	-3	4
w_4	10	-4	-8	1	-3	7

Tableau 3–13

	1	$-w_2$	$-z_1$	$-w_3$	$-z_3$	$-z_4$
w_1			-1	$-\frac{3}{4}$	$\frac{13}{4}$	not
z_0	not	up	0	$-\frac{2}{4}$	$-\frac{2}{4}$	up
z_2	dated	dated	-1	$\frac{1}{4}$	$-\frac{3}{4}$	dated
w_4			-7	$-\frac{1}{4}$	$-\frac{9}{4}$	

Tableau 3–14

	1	$-w_2$	$-z_1$	$-w_3$	$-w_1$	$-z_4$
z_3			$-\frac{4}{13}$			$\frac{4}{13}$
z_0	not	up	$-\frac{2}{13}$	not	up	$\frac{2}{13}$
z_2	dated	dated	$-\frac{16}{13}$	dated	dated	$\frac{3}{13}$
w_4			$-\frac{100}{13}$			$\frac{9}{13}$

Tableau 3–15

44

z_1, the complement of the variable w_1 cannot be made a basic variable. Again the procedure terminates in unbounded ray. It can be seen that this problem has a complementary feasible solution and it is

$$(15) \qquad w = \begin{pmatrix} 0 \\ 0 \\ 0 \\ \dfrac{40}{7} \end{pmatrix}, \qquad z = \begin{pmatrix} 1 \\ \dfrac{19}{7} \\ \dfrac{3}{7} \\ 0 \end{pmatrix}.$$

Since these two well-known methods and their variants may fail to provide a solution to the linear complementarity problem in the general case an alternative algorithm is suggested. This algorithm is based on an algorithm proposed by the authors for finding all the vertices of a convex polyhedron (see [4] G. R. Jahanshahlou and G. Mitra). This algorithm generates only a small subset of all the vertices of the problem defined by (1), (2) and further this subset contains all the solutions of the linear complementarity problem. The generality of the procedure is attractive in as much as it makes no assumption about the problem matrix M.

Before stating the algorithm the following terms are defined.

"Kilter number", K, is the number of complementary pairs of variables which are in the basis.

A variable is said to be "starred" if in all the subsequent tableaux it is forced to remain non-basic, similarly a variable and its associated row is said to be "flagged" if in all the subsequent tableaux the variable is forced to stay in the basis.

The steps of the algorithm may be stated as follows:

Step 1. Apply the phase I of the simplex method to the system

$$(16) \qquad w = q + Mz, \quad z \geqq 0, \quad w \geqq 0,$$

to find a basic feasible solution. If there is no basic feasible solution to (16), then there is no solution to the Fundamental Problem, and go to step 5, otherwise number the tableau associated with the basic feasible solution Tableau-0, set $N=0$, $L=0$, $K_L = K$ (K_L is the kilter number in the current tableau).

Step 2. Pick tableau N from the stack of the tableaux, go to auxiliary sequence. If a pivotal transformation is carried out set $L = L+1$, number the new tableau as tableau L, and add it to the stack of the tableaux and go to step 3. If the auxiliary sequence ends in the terminal step, i.e., step f, go to step 4.

Step 3. Pick tableau L, out of the stack, go to auxiliary sequence. If a pivotal transformation has taken place put $L = L+1$, number the new tableau as tableau L, and add it to the stack of the tableaux, go to step 3. If the auxiliary sequence ends in the terminal step, i.e. step f, go to step 2.

Step 4. Set $N=N+1$, if $N>L$ go to step 5. If $N\leq L$ and the tableau N is marked, go to step 4 and if the tableau N is not marked go to step 2.

Step 5. Tree search is completed.
Auxiliary sequence.
In this sequence if possible a pivotal transformation is carried out on the given tableau.
The first three steps are for column choice.
Initial step. If the kilter number of the tableau is zero go to step a, otherwise go to step b.

Step a. Out of the "nonstarred" nonbasic variables choose a column q with variable z_r or w_r which admits a row i ($i \notin F$ where F is the set of row indices which are flagged) with a positive coefficient, i.e. $-m_{iq}>0$. Go to step c, otherwise no pivotal transformation can be carried out and go to terminal step f.

Step b. If in the given tableau there exists a pair of nonbasic complementary variables, which are starred no column should be chosen and go to terminal step f. If this is not the case define two sets of column indices

$Q_1=\{l|\, l$ with one unstarred variable z_r or w_r and z_r, w_r both nonbasic$\}$
$Q_2=\{l|\, l$ with unstarred variables z_r, w_r and both nonbasic$\}$

(i) choose $q \in Q_1$ such that the associated variable z_r or w_r can be made basic, i.e. it admits a row i, $i \notin F$ and $-m_{iq}>0$, and go to step c else,
(ii) choose $q \in Q_2$, such that the associated variable z_r or w_r can be made basic as in (i) above. If such a q does not exist go to step a.

Step c. (Row choice). Out of the rows not "flagged" find a row index p such that

$$\beta_p/(-m_{pq})=\min_{i \notin F} \{\beta_i/(-m_{iq}) \mid -m_{iq}>0\}.$$

Step d. Pivotal Transformation and flagging and starring. There may be four possible cases in each of which pivotal transformation is carried out on the element $-m_{pq}>0$.

Case (i). Let z_r be the basic variable in row p, $p \notin F$ and w_r be the nonbasic variable in column q. [See Tableau a–1, Tableau a–2 and Tableau a–3.] In this case in the original tableau z_r, and row p are "flagged" and w_r is starred w_r^*, Tableau a–2, and in the new tableau L, z_r is starred z_r^* and w_r is flagged $\bar{w}_r$ and the row p is also flagged, Tableau a–3.

46

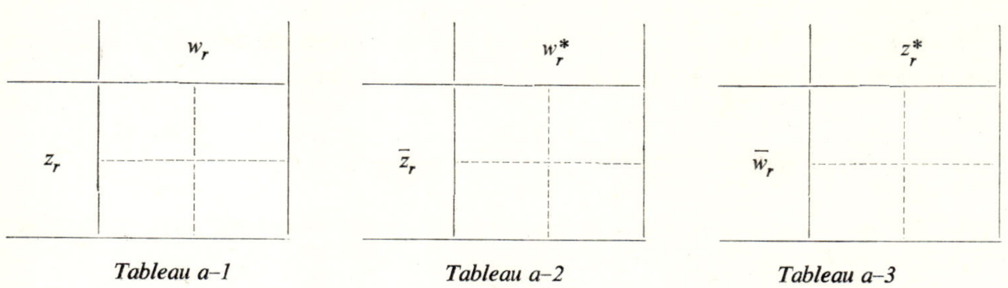

| | Tableau a–1 | | Tableau a–2 | | Tableau a–3 |

Case (ii). In this case let w_r be the nonbasic variable which is in column q (z_r is also nonbasic) and z_s is the variable in row p [See Tableau a–4], then in the original tableau w_r is starred w_r^*, Tableau a–5, and in the new tableau L, $\bar{w}_r$ and row p are "flagged"; and z_r is starred if it has not been already "starred" (Tableau a–6).

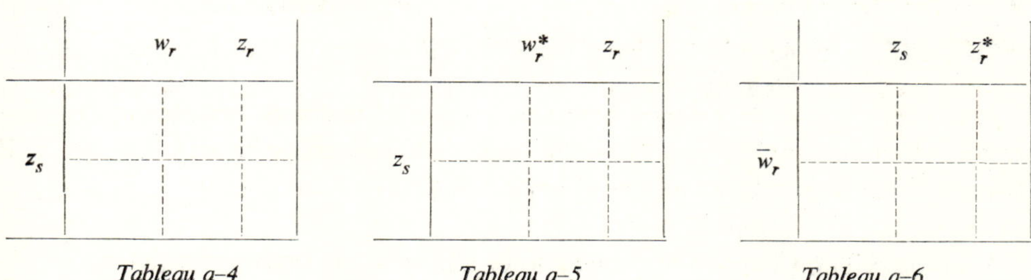

| | Tableau a–4 | | Tableau a–5 | | Tableau a–6 |

Case (iii). In this case let w_r in column q be the nonbasic variable, whereas z_r is basic. Let z_s be the variable in row p (Tableau a–7). Then in the

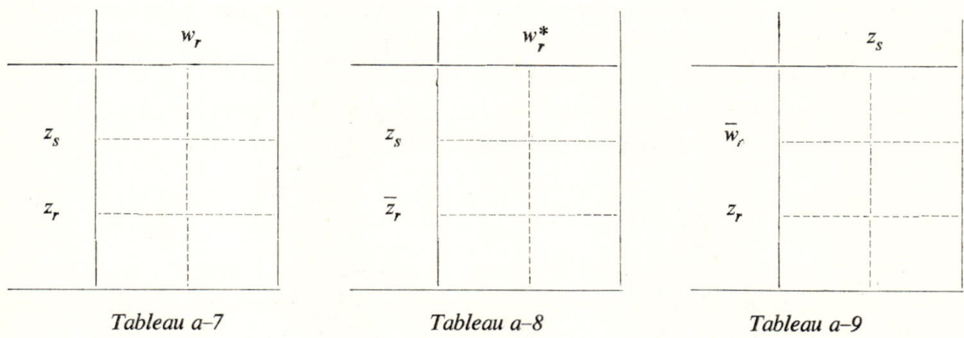

| | Tableau a–7 | | Tableau a–8 | | Tableau a–9 |

original Tableau w_r is starred and z_r and the associated row are "flagged" (Tableau a–8), and in the new tableau, L, $\bar{w}_r$ and row p, are flagged (Tableau a–9).

Case (iv). In this case let w_r in column q be the nonbasic variable, whereas z_s in row p and w_s are both basic, further w_s and the associated row are "flagged" [see Tableau a–10].

47

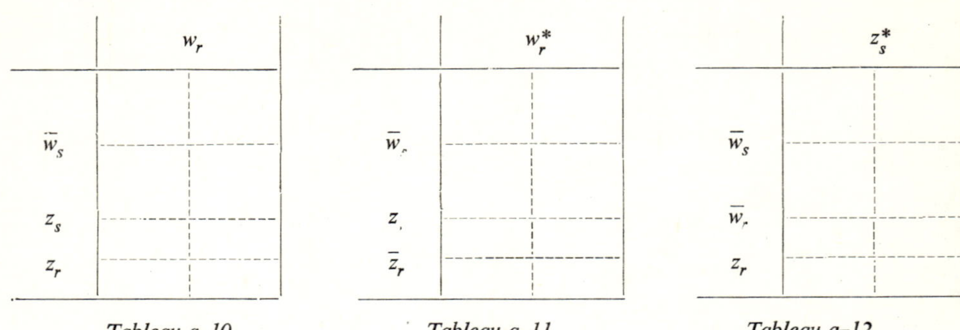

Tableau a–10   Tableau a–11   Tableau a–12

Then in the original Tableau z_r and the associated row are flagged and w_r is starred w_r^* (Tableau a–11). In the new tableau L, z_s is starred z_s^* and w_r and row p are flagged, (Tableau a–12).

Step e. If in a tableau its kilter number is zero and the values of the basic variables are non-negative, this tableau represents a feasible complementary solution. Return to the calling step.

Step f. (Terminal.) No pivotal transformation is carried out, mark the tableau and return to the calling step.

A set description is now introduced to explain the applicability of the algorithm and the theorem which follows. Let

S_P be the set of all the possible bases of (1) and (2),
S_T be the set of all the bases generated by the algorithm in [4],
S_F be the set of all feasible bases (i.e. vertices) of (1) and (2),
S_C be the set of all complementary bases of (1) and (2),
S_{CF} be the set of all complementary feasible bases of (1) and (2).

These sets and their relations are diagrammatically illustrated in Fig. 1.

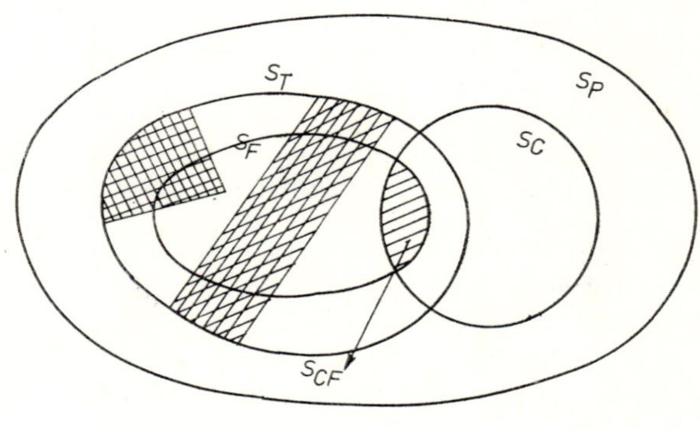

Fig. 1

The double shaded areas represent those subsets of S_T which are not generated by the present algorithm. Later on in the theorem it is shown that these must be subsets of the set $(S_T - S_C)$.

Theorem. *The above-mentioned algorithm generates all the complementary feasible bases of the set defined by* (1) *and* (2) *provided such vertices exist i.e. the set* $(S_F \cap S_C)$ *is nonempty.*

Proof. To prove this theorem, it is first noted that the algorithm in [4] generates the set S_T, which contains the set S_F i.e. all the feasible bases of (1) and (2). The modification introduced in the present algorithm leaves out certain subsets of S_T. It is now shown that these subsets are not contained in S_C. The possible cases are considered in turn:

Case a. In the Tableau a–13 w_q is a potential variable to become a basic variable and to generate some bases of (1) and (2). The kilter number of this and all the sub-

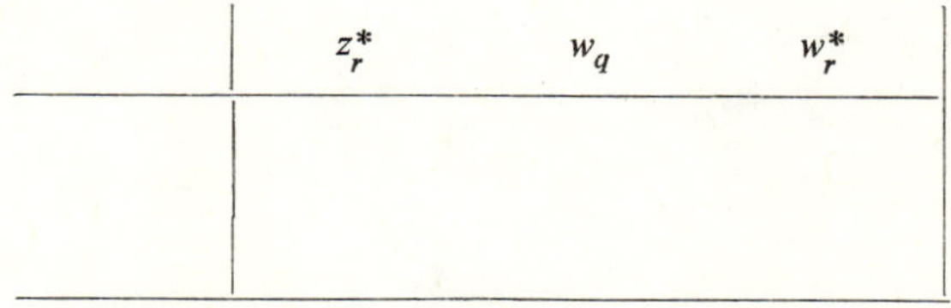

	z_r^*	w_q	w_r^*

Tableau a–13

sequent tableaux which follow are at least one, because z_r and w_r are forced to remain nonbasic. Therefore no complementary vertices are lost, if this tableau is marked, and the associated branch is terminated in the tree search.

Case b. Consider the possible cases mentioned in step d of the auxiliary sequence. Starring the variables and flagging the rows whereby the subproblems which follow exclude some possible enumerations. In the following it is shown that by these actions no complementary vertices are lost. These are considered in turn:

In Case (i), z_r and row p are flagged; this is not done in the algorithm in [4]. If this variable and the row p are not flagged, it might become a nonbasic variable in a subsequent step, and as w_r is forced to remain nonbasic (w_r is a starred variable), therefore in all the subsequent tableaux which might be obtained from this tableau, the kilter number must be at least one. Similarly if z_r is not starred in the new tableau L, and it could be pivoted into the basis, as w_r is forced to be the basic variable, so in all the subsequent tableaux obtained from this tableau, their kilter number must be one or more. Therefore no complementary vertices are lost in this case [see Tableau a–2 and Tableau a–3].

By similar argument it can be shown that no complementary vertices are lost as a result of additional starring of nonbasic variables of flagging of basic variables and their associated rows of Case (iii) and Case (iv). Since all other possible bases which may be generated by the algorithm in [4] are considered the bases excluded by this algorithm belong to the set $(S_T - S_C)$. Therefore the set of bases generated by this algorithm contains the subset $S_C \cap S_F$ if this is nonempty.

Example 4. Here Example 1 is solved by the proposed algorithm. It has been shown that Lemke's method as well as the principal pivoting method failed to produce a CFS solution to the problem.

The problem is restated here

$$(17) \quad \begin{cases} 2z_1 - 3z_2 + z_3 + w_1 = 10 \\ z_1 - 2z_2 + z_3 - w_2 = 1 \\ z_1 - 2z_2 - 3z_3 + w_3 = 3 \\ w_i \geqq 0, \ z_i \geqq 0 \quad (i = 1, 2, 3). \end{cases}$$

By introducing an artificial variable corresponding to the second equation of the system (17), and applying the Phase I of the simplex method the Tableau 3–16 containing the basic feasible solution is obtained. In this tableau kilter number is 1. The rest of the steps of the algorithm as related to this problem are illustrated below

	1	$-z_2$	$-z_3$	$-w_2$
w_1	8	①	-1	2
z_1	1	-2	1	-1
w_3	2	0	-4	1

(pivot element is circled)

Tableau 3–16

	1	$-z_2^*$	$-z_3$	$-w_2$
w_1	8	1	-1	2
z_1	1	-2	1	-1
w_3	2	0	-4	1

Tableau 3–16a

	1	$-w_1$	$-z_3$	$-w_2^*$
$\overline{z}_2$	8	1	-1	2
z_1	17	2	-1	3
w_3	2	0	-4	1

Tableau 3–17

The new position of the original Tableau 3–16 is shown in Tableau 3–16a. In this tableau z_2 is starred. In Tableau 3–17 which has been obtained by pivotal transformation z_2 and row 1 are flagged and w_2 is starred, which contains a feasible complementary solution.

50

	1	$-w_1^*$	$-z_3$	$-w_2^*$
$\bar{z}_2$	8	1	-1	2
$\bar{z}_1$	17	2	-1	3
w_3	2	0	-4	1

Tableau 3–17a

	1	$-z_1^*$	$-z_3$	$-w_2^*$
$\bar{z}_2$	$-\dfrac{1}{2}$	$-\dfrac{1}{2}$	$\dfrac{1}{2}$	$\dfrac{1}{2}$
$\bar{w}_1$	$\dfrac{17}{2}$	$\dfrac{1}{2}$	$-\dfrac{1}{2}$	$\dfrac{3}{2}$
w_3	2	0	-4	1

Tableau 3–18

Tableau 3–17a is the new position of Tableau 3–17 in which z_1 and row 2 are flagged and w_1 is starred. By carrying out a pivotal transformation on Tableau 3–17, Tableau 3–18 is obtained, in which w_1 and row 2 are flagged and z_1 is starred.

Tableau 3–18 is marked, because no column can be chosen. Now tableau 3–16a is picked up, w_2 the complement of z_2, which is a starred variable is chosen to become a basic variable and a pivot step is carried out as shown below.

	1	$-z_2^*$	$-z_3$	$-w_2^*$
w_1	8	1	-1	2
z_1	1	-2	1	-1
w_3	2	0	-4	1

Tableau 3–16b

	1	$-z_2^*$	$-z_3$	$-w_3$
w_1	4	1	7	-2
z_1	3	-2	-3	1
$\bar{w}_2$	2	0	-4	1

Tableau 3–19

Tableau 3–16b is the new position of the Tableau 3–16a, in which w_2 is starred. Tableau 3–19 is obtained by carrying out a pivotal transformation from Tableau 3–16a. In Tableau 3–19 w_2 and row 3 are flagged. It should be noted that all three Tableaux 3–16, 3–16a and 3–16b are associated with $N=0$, i.e. these three tableaux have been considered as one tableau, but for the purpose of illustration they have been considered separately.

Now pick Tableau 3–19 and carry out a pivotal transformation on the pivot element 7. Having done this operation the following tableaux are obtained.

	1	$-z_2^*$	$-z_3^*$	$-w_3$
w_1	4	1	7	-2
z_1	3	-2	-3	1
$\bar{w}_2$	2	0	-4	1

Tableau 3–19a

	1	$-z_2^*$	$-w_1$	$-w_3$
$\bar{z}_3$	$\dfrac{4}{7}$	$\dfrac{1}{7}$	$\dfrac{1}{7}$	$-\dfrac{2}{7}$
z_1	$\dfrac{33}{7}$	$-\dfrac{11}{7}$	$\dfrac{3}{7}$	$\dfrac{1}{7}$
$\bar{w}_2$	$\dfrac{30}{7}$	$\dfrac{4}{7}$	$\dfrac{4}{7}$	$-\dfrac{1}{7}$

Tableau 3–20

In Tableau 3–19a which is the new position of the Tableau 3–19 z_3 is starred and in Tableau 3–20 which is obtained from Tableau 3–19 z_3 and row 1 are flagged, this tableau also contains a complementary feasible solution.

From Tableau 3–20 the following are obtained:

	1	$-z_2^*$	$-w_1^*$	$-w_3$
$\bar{z}_3$	$\dfrac{4}{7}$	$\dfrac{1}{7}$	$\dfrac{1}{7}$	$-\dfrac{2}{7}$
$\bar{z}_1$	$\dfrac{33}{7}$	$-\dfrac{11}{7}$	$\dfrac{3}{7}$	$\dfrac{1}{7}$
$\bar{w}_2$	$\dfrac{30}{7}$	$\dfrac{4}{7}$	$\dfrac{4}{7}$	$-\dfrac{1}{7}$

Tableau 3–20a

	1	$-z_2^*$	$-z_1^*$	$-w_3$
$\bar{z}$	-1	$\dfrac{2}{3}$	$-\dfrac{1}{3}$	$-\dfrac{1}{3}$
$\bar{w}_1$	11	$-\dfrac{11}{3}$	$\dfrac{7}{3}$	$\dfrac{1}{3}$
$\bar{w}_2$	-2	$\dfrac{8}{3}$	$-\dfrac{4}{3}$	$-\dfrac{1}{3}$

Tableau 3–21

Tableau 3–20a is the new position of Tableau 3–20. In this tableau z_1 and row 2 are flagged and w_1 is starred. In Tableau 3–21 w_1 and row 2 are flagged and z_1 is starred.

No pivotal transformation can be carried out on Tableau 3–21, therefore it is marked. Both w_2 and z_2 are starred in Tableau 3–16b, so this tableau is also marked. No column can be chosen from Tableaux 3–17a and 3–18, therefore they are also marked. The Tableau 3–19a is picked up, and from this the Tableaux 3–19b and 3–22 are obtained.

	1	$-z_2^*$	$-z_3^*$	$-w_3^*$
w_1	4	1	7	-2
z_1	3	-2	-3	1
$\bar{w}_2$	2	0	-4	1

Tableau 3–19b

	1	$-z_2^*$	$-z_3^*$	$-z_1$
w_1	10	-3	1	2
$\bar{w}_3$	3	-2	-3	1
$\bar{w}_2$	-1	2	-1	-1

Tableau 3–22

In the new position of Tableau 3–19a, i.e. in Tableau 3–19b w_3 is starred, and in Tableau 3–22 which is obtained from Tableau 3–19a w_3 and row 2 are flagged. The Tableaux 3–20a, 3–21 and 3–19b are marked, since no pivotal transformation can be carried out. The Tableau 3–22 is chosen. In this tableau z_1 is chosen to pivot against w_1. Carrying out a pivot on the pivot element z leads to the following two tableaux.

	1	$-z_2^*$	$-z_3^*$	$-z_1^*$
$\bar{w}_1$	10	-3	1	2
$\bar{w}_3$	3	-2	-3	1
$\bar{w}_2$	-1	2	-1	-1

Tableau 3–22a

	1	$-z_2^*$	$-z_3^*$	$-w_1^*$
$\bar{z}_1$	5	$-\dfrac{3}{2}$	$\dfrac{1}{2}$	1
$\bar{w}_3$	-2	$-\dfrac{1}{2}$	$-\dfrac{7}{2}$	$-\dfrac{1}{2}$
$\bar{w}_2$	4	$\dfrac{1}{2}$	$-\dfrac{1}{2}$	$\dfrac{1}{2}$

Tableau 3–23

As no pivotal transformation can be carried out on the Tableaux 3–22a and 3–23, they are marked, so the search is complete. Table 1 shows a summary of the steps of the algorithm as related to this problem.

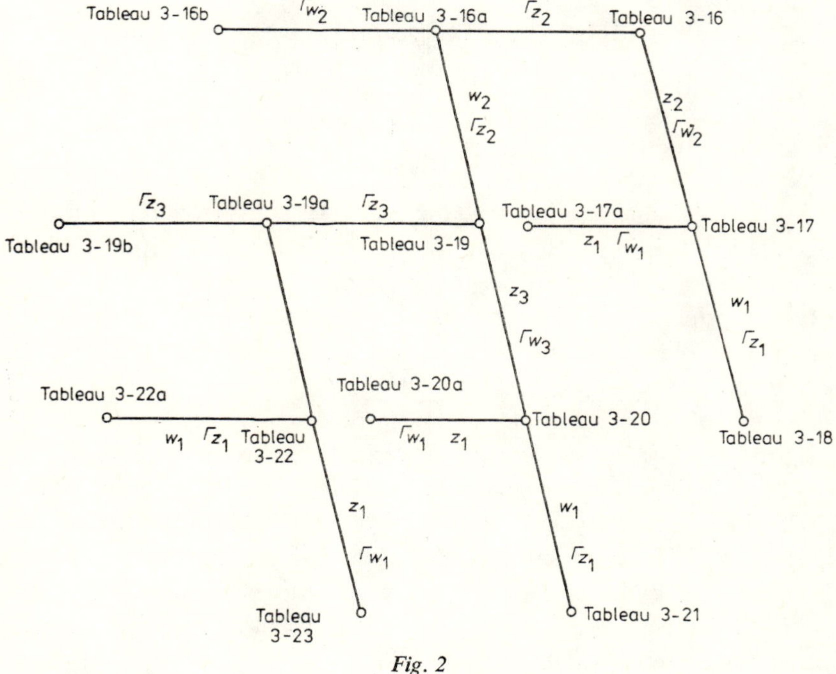

Fig. 2

Table 1

Iteration Number	The feasibility F = feasible N = not feasible	Complementarity C = complementary NC = not complementary	Tableaux generated	L	N
0	F	NC	3–16	0	0
*1	F	C	3–16a, 3–17	1	0
2	N	C	3–17a, 3–18	2	0
3	F	NC	3–16b, 3–19	3	1
*4	F	C	3–19a, 3–20	4	1
5	N	C	3–20a, 3–21	5	1
6	N	C	3–19b, 3–22	6	2
7	N	C	3–22a, 3–23	7	2

After iteration 7 N increases and L remains fixed until $N = 7$ when the search is complete

The tree developed by this method is shown in Fig. 2, and the sequence of tableaux which are generated are set out in Tableau 3–16 up to Tableau 3–23.

In Fig. 2 Γz_i, Γw_i indicates that the corresponding variable is not in the basis, and z_i or w_i indicate that the corresponding variable is in the basis.

5. DISCUSSION AND SOME REMARKS ON THE COMPUTATIONAL EXPERIENCE

Application of the phase I of the simplex method to the problem, leads to the conslusion that either there is a basic feasible solution to the problem or otherwise. If there is a basic feasible solution, then the branching starts from the node associated with this solution. It is interesting that in each iteration the size of the problem in the branch is reduced at least by one row and one column or two columns and one row. In the case of principal pivoting or the case of the step e in the auxiliary sequence one row and one column are flagged and starred, i.e. the size of the problem is reduced by one row and one column in each branch.

While studying Lemke's method another problem suggested itself, namely, what other vectors associated with the artificial variable z_0 may be introduced instead of a vector e' with all non-negative components as considered in this paper. The motivation for finding such a vector is that one may be able to follow n different paths starting from the initial basic solution. The author's ideas are described in [5].

Both Lemke's method, and the algorithm proposed by the authors have been programmed in FORTRAN IV. These programs have been used to solve ten problems. The results are set out in Table 2.

54

Table 2

Problem No.	Order of M	$N(S_P) \leq$	$N(S_T)$	$N(S_F)$	Lemke's method F = failed S = succeed	$N(S_A)$	$N(S_{CF})$
1	3	20	14	7	F	8	2
2	4	70	17	4	F	7	2
3	4	70	36	12	F	13	4
4	4	70	31	12	F	13	4
5	5	252	106	25	F	26	2
6	5	252	103	26	F	24	3
7	5	252	129	31	F	25	3
8	6	924	264	21	F	50	1
9	6	924	248	21	F	49	1
10	6	924	312	34	F	43	3

S_A the set of all bases generated by the present algorithm; $N(S)$ = cardinality of the set S.

REFERENCES

[1] Dantzig, G. B. and Cottle, R. W.: *On the Complementary Pivot Theory*, in [3], pp. 115–136.

[2] Dantzig, G. B.: *Linear Programming and Extensions*, Princeton University Press, 1963.

[3] Dantzig, G. B. and Veinott, A. F. (eds.): *Mathematics of the Decision Sciences*. Published by the American Mathematical Society, 1968.

[4] Jahanshahlou, G. R. and Mitra, G.: Two Algorithms for Finding all the Vertices of a Convex Polyhedron, Colloquia Mathematica Societatis János Bolyai 12. *Progress in Operations Research*, Eger (Hungary), 1974.

[5] Jahanshahlou, G. R. and Mitra, G.: A Review of the Linear Complementarity Problem and a Tree Search Algorithm for its Solution, Technical Report STR/12, Department of Statistics & O.R., Brunel University, February 1976.

[6] MacCammon, S. R.: On the Complementary Pivoting, Ph. D. Thesis, Rennsselaer Polytechnic Institute, Tory, N.S. (1970).

CUTTING-PLANES FOR COMPLEMENTARITY CONSTRAINTS

R. G. JEROSLOW

(Atlanta. USA)

(*Note*. This is an abstract for [9].)

We describe two simple rules of cutting-plane generation for the complementarity constraints

$$Ax \geqq b$$
$$x \geqq 0.$$

(CMP)
$$\sum_{h=1}^{t} \prod_{K \in J_h} \left(\sum_{k \in K} x_k \right) = 0$$

and we show that these rules generate all (and only) the valid cuttingplanes for (CMP), if there is some b' for which $\{x \geqq 0 \mid Ax \geqq b'\}$ is non-empty and bounded. We also describe the simple modifications needed for the unbounded case.

In (CMP), $x = (x_1, \ldots, x_r)$, and J_h is a non-empty set of non-empty subsets K of $\{1, \ldots, r\}$. The problem (CMP) includes the linear complementarity problem and bivalent integer programming, along with many other constraint sets which impose logical restrictions on linear inequalities. For example, an obvious specific instance of (CMP) is

(GLC)
$$Ay + Bz \geqq b$$
$$y, z \geqq 0$$
$$y \cdot z = 0$$

in which $r = 2s$ for an integer $s \geqq 1$, $y = (x_1, \ldots, x_s)$ and $z = (x_{s+1}, \ldots, x_{2s})$; also, $J_h = \{\{j\}, \{j+s\}\}$ for $h = 1, \ldots, s$ and $t = s$.

The problem (GLC) is itself a generalization of the linear complementarity problem, since we do not require that

$$A = \begin{bmatrix} M \\ -M \end{bmatrix} \quad (M \text{ a square matrix})$$

$$B = \begin{bmatrix} -I_s \\ I_s \end{bmatrix}$$

$$b = \begin{bmatrix} q \\ -q \end{bmatrix}$$

as would be required for linear complementarity constraints [6], [7], [11], [12], nor do we require that M have a specific matrix property (e.g., co-positive-plus).

For another straightforward application of (CMP), the constraints of a bivalent integer program

$$\text{(IP)} \qquad \begin{aligned} Ax &\geqq d \\ x_j = 0 \text{ or } 1, \quad &j = 1, \ldots, s \end{aligned}$$

can trivially be rewritten as

$$\text{(IP)}' \qquad \begin{aligned} Ax &\geqq d \\ x_j + z_j = 1, \quad &j = 1, \ldots, s \\ x_j, \ z_j \geqq 0, \quad &j = 1, \ldots, s \\ \sum_{j=1}^{r} x_j z_j &= 0 \end{aligned}$$

which is of the form (GLC).

To state our result, we shall consider "deductions" in tree form. The tree shall be spread out at the "top" and narrow to one node at the "bottom". The propositions of our logical calculus shall be linear inequalities. The propositions occurring at the top of deductions are called "assumptions".

Where a given node has several others connected to it by an edge and just above it, we require that the inequality assigned to this node "follow from" the inequalities assigned to the nodes just above, in the sense that it is the *conclusion* of one of the *rules of deduction* of the logical system and the inequalities above are the *premisses* of this rule.

For instance, one of our rules of deduction shall be

$$\text{(LC)} \qquad \frac{a_1 v_1 + \ldots + a_n v_n \geqq a_0, \ b_1 v_1 + \ldots + b_n v_n \geqq b_0}{(\lambda a_1 + \Theta b_1) v_1 + \ldots + (\lambda a_n + \Theta b_n) v_n \geqq c_0}$$

and for the application of (LC) we require that λ, $\Theta \geqq 0$ and that $c_0 \leqq \lambda a_0 + \Theta b_0$. The premisses of (LC) are $a_1 v_1 + \ldots + a_n v_n \geqq a_0$ and also $b_1 v_1 + \ldots + b_n v_n \geqq b_0$; the conclusion of (LC) is $(\lambda a_1 + \Theta b_1) v_1 + \ldots + (\lambda a_n + \Theta b_n) v_n \geqq c_0$. The *parameters* of (LC) are Θ, λ and c_0. The rule (LC) is understood (as is any rule) as "saying", that if its premisses have already been "deduced", one is entitled to "deduce" its conclusion. For the generic variables $v_1, \ldots, v_n$ one may employ any of the variables $x_1, x_2, \ldots, x_r, y_1, y_2, \ldots, z_1, \ldots$

When (LC) occurs in a tree, as part of it, that part looks like

where the top two nodes correspond to the premisses, and the bottom node corresponds to the conclusion. An instance of (LC), with the corresponding inequalities underlined, and $\lambda = \Theta = 1$, is

$$2x_1 - x_2 \geqq 7 \qquad\qquad 5x_1 + 3x_2 \geqq -2$$
$$7x_1 + 2x_2 \geqq 5$$

which is often abbreviated

$$2x_1 - x_2 \geqq 7 \qquad\qquad\qquad 5x_1 + 3x_2 \geqq -2 .$$
$$7x_1 + 2x_2 \geqq 5$$

For details on these concepts of derivation trees, formulae, etc., the reader may wish to consult [13], [14], or [15]. Chvátal [4] and Blair [2] have used these concepts earlier in an optimization context.

In addition to (LC), we shall have one more type of rule of deduction, and it is given in Fig. 1. This rule is added for $h=1, \ldots, t$, making $(t+1)$ rules in all. In stating this rule (CMC)$_h$, we let $h'=t(h)$ denote the number of elements $|J_h|$ in J_h, and without loss of generality we assume $h'\geqq 2$. Also, we have used the notation $J_h= =\{K_{h(1)}\ldots, K_{h(h')}\}$, so that $K_{h(j)}$ is the j-th subset (of $\{1, \ldots, r\}$) that is a member of J_h.

$$\frac{a_{11}x_1 + \ldots + a_{1r}x_r \geqq a_0, \ldots, a_{j1}x_1 + \ldots + a_{jr}x_r \geqq a_0, \ldots, a_{h'1}x_1 + \ldots + a_{h'r}x_r \geqq a_0}{a_1 x_1 + \ldots + a_r x_r \geqq a_0}$$

$$(h' = t(h))$$

where

$$a_{jk} = a_k \quad \text{if} \quad k \notin K_{h(j)}$$

Fig. 1. The Rule (CMC)$_h$ for J_h

The rule (CMC)$_h$ is obviously valid in that model m consisting of all $x \in R^r$ that satisfy the constraints (CMP), if $m \neq \emptyset$. Indeed, suppose that it is the condition $x_k = 0$, $k \in K_{h(j)}$ which is satisfied by a point $(x_1, \ldots, x_r) \in m$. Whenever $a_{j1}x_1 + \ldots + a_{jr}x_r \geqq a_0$ is also satisfied, then so is $a_1 x_1 + \ldots + a_r x_r \geqq a_0$, provided that a_k is a_{jk} if $k \notin K_{h(j)}$, while a_k can be arbitrary for $k \in K_{h(j)}$ since $x_k = 0$. The stipulation on scalar quantities in (CMC)$_h$ insures this proviso. Therefore, the use of the rules (LC) and (CMC)$_h$ will yield some of the valid cutting-planes for the constraints (CMP). (By a "cutting-plane" we simply mean a linear inequality valid for all solutions to (CMP)).

As an example of an application of the rule (CMC)$_h$, we have, with $r=4$ and $J_h = \{\{1, 2\}, \{1, 3, 4\}, \{1, 4\}\}$:

$$\frac{x_1 + 2x_2 + 3x_3 + 4x_4 \geqq 5, \; 6x_1 + 7x_2 + 8x_3 + 9x_4 \geqq 5, \; 10x_1 + 7x_2 + 3x_3 + 11x_4 \geqq 5}{12x_1 + 7x_2 + 3x_3 + 4x_4 \geqq 5}$$

In this example, we have made coefficients distinct whenever possible. E.g., x_1 has the new coefficient "12" in the conclusion, since the index "1" appears in $K_{h(1)}= =\{1, 2\}$, $K_{h(2)}=\{1, 3, 4\}$, and $K_{h(3)}=\{1, 4\}$, indicating that there are no restrictions on a_1 in (CMC)$_h$. Similarly, since $3 \notin K_{h(1)}$ and $3 \notin K_{h(3)}$, while $3 \in K_{h(2)}$, the coefficient of x_3 in the conclusion must match that in the leftmost and rightmost premisses, and these two coefficients must themselves agree (they are all "3").

The result we have to announce is, in essence, the following: if $\{x \geqq 0 \mid Ax \geqq b\}$ is bounded and non-empty, then the use of the rules (LC) and $(CMC)_h$ $(h=1, \ldots, t)$ provide all (not just some!) of the valid cutting-planes for (CMP). In fact, we can even obtain information on the form of proof of any given cutting-plane, as described below.

Main Result: The assumptions $Ax \geqq b$, $x \geqq 0$ together with rules of deduction (LC) and $(CMC)_h$, $h=1, \ldots, t$ yield all the cutting-planes for the model $\mathcal{M}$ of (CMP) consisting of all real vectors $(x_1, \ldots, x_r)$ satisfying (CMP), provided that $\{x \geqq 0 \mid Ax \geqq b\}$ is bounded and $\mathcal{M} \neq \emptyset$.

In fact with this proviso, there are finitely many derivations $\Sigma_1, \ldots, \Sigma_s$ with assumptions among $Ax \geqq b$, $x \geqq 0$, such that any linear inequality statement P that is valid in $\mathcal{M}$ has a deduction of the form

$$(\mathrm{N}_{\leqq}) \qquad \qquad \frac{\Sigma_1 \ldots \Sigma_s}{P}$$

in which the last rule of deduction is (LC).

Furthermore, in going from any topmost node of Σ_q, $q=1, \ldots, s$ downward by arcs of the tree to its bottommost node, one encounters, in the following specified order, these rules of deduction: one application of (LC), followed by one application each of $(CMC)_1, \ldots, (CMC)_t$, between each of which is one or more applications of (LC).

Also, if Σ is a subderivation Σ_q, $q=1, \ldots, s$ obtained by selecting a node from the tree Σ_q and detaching the subtree of nodes and arcs above and including this node, and if the last inference in Σ is an application of the rule $(CMC)_h$, then the endformula of Σ provides a facet or singular inequality of K_h. Moreover, any facet or singular inequality of K_h arises as the endformula of exactly one such subderivation Σ of the derivation (NF) with last inference $(CMC)_h$.

If $\mathcal{M} = \emptyset$, but $\{x \geqq 0 \mid Ax \geqq b'\}$ is bounded and non-empty for some b', then from the axioms and rules of deduction cited, any inequality can be proven.

For a proof of this result and the others below, the reader may write for a copy of [9].

A very similar result holds if $\{x \geqq 0 \mid Ax \geqq b\}$ is unbounded. One simply appends an "infinitely large" quantity M and makes the problem bounded by means of the constraint

$$(BD) \qquad \qquad x_1 + \ldots + x_r \leqq M.$$

Details are in [9]. In essence, one adds in the "phantom" of infinity as an intermediate device, and then uses only those derived cutting-planes in which it does not occur in the bottommost node. For similar applications of infinite quantities, see [3] and [8].

If we apply the Main Result to the case (GLC), the rule $(CMC)_j$ simplifies to $(ii)_j$ below, and we obtain the following, which was announced in [10].

60

Corollary. *If*

$$\{(x, y)\mid Ax + Bz \geqq d', \, x \geqq 0, \, z \geqq 0\}$$

is bounded and non-empty for some d', then any valid cutting-plane for the complementarity constraints

(GLC) $$Ax + Bz \geqq d, \quad x \geqq 0, \quad z \geqq 0, \quad x \cdot z = 0,$$

is obtained by starting from the linear defining inequalities

$$Ax + Bz \geqq d, \quad x \geqq 0, \quad z \geqq 0$$

and applying, finitely often, the following two rules (the second for $j = 1, \ldots, r$):

(i) Take non-negative combinations of given inequalities, and possibly weaken the right-hand side.

(ii) Having already obtained two inequalities

$$\alpha_1 x_1 + \ldots + u x_j + \ldots + \alpha_r x_r + \beta_1 z_1 + \ldots + t z_j + \ldots + \beta_r z_r \geqq \alpha_0,$$
$$\alpha_1 x_1 + \ldots + u' x_j + \ldots + \alpha_r x_r + \beta_1 z_1 + \ldots + t' z_j + \ldots + \beta_r z_r \geqq \alpha_0,$$

one may deduce

$$\alpha_1 x_1 + \ldots + u x_j + \ldots + \alpha_r x_r + \beta_1 z_1 + \ldots + t' z_j + \ldots + \beta_r z_r \geqq \alpha_0.$$

Conversely, any inequality thus obtained is valid for the constraints (GLC). ∎

We remark that the boundedness hypothesis of the Corollary is also removed by means of "regularizations" like (BD).

Our motivation for rules such as $(\text{CMC})_h$ is that they generalize a rule of Blair (see the rule $(\text{BR})_j$ below) for constraint sets of the form (IP). In fact, we can derive Blair's result as a consequence of our previous Corollary.

Corollary. [2, Chapter 3]

Any cutting-plane for the constraints (IP) *is obtained by starting from the linear inequalities*

$$\begin{aligned}
Ax &\geqq d, \\
x_j &\geqq 0, \quad j = 1, \ldots, r \\
-x_j &\geqq -1, \quad j = 1, \ldots, r
\end{aligned}$$

and applying, finitely often, the following two rules (the second for $j = 1, \ldots, r$):

(i) Take non-negative combinations of given inequalities, and possibly weaken the right-hand side.

(BR)$_j$ *Having already obtained two inequalities*

$$\Theta_1 x_1 + \ldots + u x_j + \ldots + \Theta_r x_r \geqq P,$$
$$\Theta_1 x_1 + \ldots + w x_j + \ldots + \Theta_r x_r \geqq T,$$

one may deduce

$$\Theta_1 x_1 + \ldots + (w + P - T) x_j + \ldots + \Theta_r x_r \geqq P.$$

Conversely, any inequality thus obtained is valid for (IP). ∎

Our main result also generalizes a result of Balas [1], and the interrelations are discussed in [9].

REFERENCES

[1] Balas, E.: Disjunctive Programming: Facets of the Convex Hull of Feasible Points, Man. Sci. Res. Rep. No. 348, Carnegie-Mellon University, July 1974.

[2] Blair, C. E. Topics in Integer Programming, Ph. D. Dissertation, Carnegie-Mellon University, April 1975, 27 pp. (Chapter 3 will shortly appear in the SIAM *Journal on Applied Mathematics*).

[3] Charnes, A. and Cooper, W. W.: *Management Models and Industrial Applications of Linear Programming,* vols 1 and 2, Wiley, New York, 1961.

[4] Chvátal, V.: Edmonds Polytopes and a Hierarchy of Combinatorial Problems, *Discrete Mathematics* 4 (1973), 305–337.

[5] Cottle, R. W.: Complementarity and Variational Problems, Tech. rep. SOL 74–6, May 1974, Systems Optimization Laboratory, Stanford University.

[6] Cottle. R. W. and Dantzig, G. B.: Complementary Pivot Theory of Mathematical Programming, *Linear Algebra and Its Applications* 1 (1968), 103–125.

[7] Eaves, B. C.: The Linear Complementarity Problem, *Management Science* 17 (1971), 612–634.

[8] Jeroslow, R.: Asymptotic Linear Programming, *Operations Research* 21 (1973), 1128–1141.

[9] Jeroslow, R.: Cutting-Planes for Complementarity Constraints, MSRR no. 394, Carnegie-Mellon University, June 1976.

[10] Jeroslow, R.: Cutting-Planes for Complementarity Constraints, *Notices of the AMS* 23 (1976), A-364.

[11] Lemke, C. E.: Bimatrix Equilibrium Points and Mathematical Programming, *Management Science* 11 (1965), 681–689.

[12] Lemke, C. E. and Howson, J. T.: Equilibrium Points of Bimatrix Games, *Journal of Soc. for Ind. and Appl. Math* 12 (1964), 413–423.

[13] Mendelson, E.: *Introduction to Mathematical Logic,* D. van Nostrand, New York, 1964. 271+pp.

[14] Prawitz, D.: *Natural Deduction: A Proof-Theoretical Study,* Stockholm Studies in Philosophy 3, Almqvist and Wiksell, Stockholm, 1965. 105+pp.

[15] Shoenfield, J. R.: *Mathematical Logic,* Addison-Wesley, Reading, Mass., 1967. 336+pp.

AN EFFICIENT IMPLEMENTATION
OF THE LEMKE ALGORITHM AND ITS EXTENSION
TO DEAL WITH UPPER AND LOWER BOUNDS

R. W. H. SARGENT

(London, U.K.)

1. INTRODUCTION

The linear complementarity problem is concerned with the system

$$w = Mz + q, \tag{a}$$
$$(1) \qquad w \geqq 0, \quad z \geqq 0, \tag{b}$$
$$w^T z = 0, \tag{c}$$

where it is required to find vectors $w \in R^n$, $z \in R^n$ satisfying (1) for a given vector $q \in R^n$ and a given matrix $M \in R^{n \times n}$.

The problem arises in a variety of contexts, as described for example in [1], [2] and [8], and its properties have been widely studied. The earliest algorithm proposed for its solution was the "principal pivoting method" of Cottle and Dantzig [3], followed by an algorithm due to Lemke [8], [9], which uses "almost complementary" pivots. Lemke's algorithm is logically simpler than that of Cottle and Dantzig, and it also processes the linear complementarity problem for a wider class of problems. However, in the context of quadratic programming, principal pivots have a simple geometric interpretation as changes in the active set of constraints, and they also preserve the special structure of M which allows savings in computation and storage. This has given rise to so-called "complementary variants" of Lemke's algorithm, notably by McCammon [10] and van de Panne [12], which preserve principal pivots. However, all the principal pivoting methods have major and minor cycles, with non-complementary pivots during the minor cycles, resulting in a principal block-pivot only at the end of each major cycle. Such complication is necessary to deal with a wide class of matrices, M, but largely destroys the advantages of principal pivots.

In this paper we first describe a version of Lemke's algorithm which uses the inverse or orthogonal factorization of a submatrix of M, rather than transforming the whole of M at each iteration. We then give a rearrangement of this algorithm, applicable only to a restricted class of matrices, which uses an elementary principal pivot at each iteration. We then develop a hybrid algorithm from these two, using orthogonal factorizations, which yields an efficient numerically stable algorithm, applicable to the same class of matrices as the basic Lemke algorithm.

Finally, we discuss the quadratic programming problem, leading to a generalized complementarity problem with upper and lower bounds on the variables, and the extension of the above algorithm to deal with this problem.

2. THE BASIC LEMKE ALGORITHM

In view of the non-negativity conditions on the variables, the "complementarity condition" (equation (1c)) implies that for each $i = 1, 2, \ldots, n$, we have $w_i z_i = 0$ and hence either $w_i = 0$ or $z_i = 0$ (or both). The corresponding pairs (w_i, z_i) are therefore known as complementary pairs.

It is immediately obvious from (1) that if $q \geqq 0$ a solution to the problem is $w = q$, $z = 0$. More generally, if a solution exists it must be possible to partition w and z into two sets of complementary pairs (w_1, z_1), (w_2, z_2) with $w_1 = 0$ and $z_2 = 0$, such that (1a) and (1b) are satisfied. Then the non-zero w_2 and z_1 may be obtained by solving (1a) for them.

The difficulty is, of course, to find the appropriate partition, and Lemke [8] gave three schemes for doing this, corresponding to different structures of the matrix M. We are here concerned with his Scheme I, for which problem (1) is extended by the addition of an extra variable and column:

$$
\begin{aligned}
w &= M_0 z_0 + M z + q, \quad M_0 > 0, & \text{(a)} \\
w &\geqq 0, \quad z_0 \geqq 0, \quad z \geqq 0, & \text{(b)} \\
w^T z &= 0. & \text{(c)}
\end{aligned}
$$
(2)

Clearly, any solution of (2) with $z_0 = 0$ is a solution of the original problem (1).

Setting $z = 0$ in (2) it is also clear that $w > 0$ for all sufficiently large positive values of z_0. Lemke's algorithm starts by finding the smallest value of z_0 such that $w \geqq 0$, for which of course at least one element is exactly zero. We select row r as the row of smallest index containing such a zero element, $w_r = 0$. We may then solve this equation for z_0 and substitute its value into the remaining equations to give a transformed system of the form

$$
w' = M_0' w_r + M' z' + q',
$$
(3)

where $z_0 \in w'$. This is, of course, the familiar pivot operation of linear programming, and we adopt the standard convention of referring to the left-hand side (non-zero) variables w' as basic variables, and the right-hand side (zero) variables as non-basic.

With $w_r = 0$ we may now increase its complementary variable z_r from zero without violating the non-negativity and complementarity conditions. If any of the elements M_{ir}' (in the column of M' corresponding to z_r) are negative, then the corresponding

64

elements w_i will decrease and eventually one or more of them is reduced to zero, blocking further increase of z_r without violating the non-negativity conditions. In this operation z_r is called the "driving variable", while the element of w' (of smallest index) which first reaches zero is called the "blocking variable". As before we can interchange the blocking and driving variables by a pivot operation, and since the blocking variable is now zero (non-basic) the iteration can be repeated using its complementary variable as the new driving variable.

If at some stage z_0 becomes the blocking variable, the resulting pivot operation yields the solution to the original problem (1) since we then have $z_0=0$ and the algorithm has throughout maintained the non-negativity and complementarity conditions.

Alternatively, we may reach a stage for which all the column elements M'_{ir} are non-negative, so that the driving variable can increase to infinity without reducing any of the basic variables (including z_0) to zero. Lemke showed that this indicates that problem (1) has no solution if M belongs to the class of copositive-plus matrices, and Eaves [4] extended this result to the more general class of $\mathcal{L}$-matrices.

The algorithm may therefore be compactly stated as follows:

Algorithm I.

Start: Set $c=0$

1. Find $\{r \,|\, M'_{ic}z'_c + q'_i \geqq M'_{rc}z'_c + q'_r = 0\}$
   If r does not exist, STOP (no solution).

2. Pivot w'_r and z'_c.

3. If $w'_r \equiv z_0$, STOP (solution found).
4. Set $c=r$ and return to step 1.

Here primes denote the current transformed values of M, q, while w', z' are the current basic and non-basic variables respectively (regardless of their original identity).

It is worth noting that for the first iteration step 1 requires the condition

$$(4a) \qquad z_0 = -\frac{q_r}{M_{r0}} \geqq -\frac{q_i}{M_{i0}} \quad \text{for all } q_i < 0,$$

while in subsequent iterations the condition becomes

$$(4b) \qquad z'_c = -\frac{q'_r}{M'_{rc}} \leqq -\frac{q'_i}{M'_{ic}} \quad \text{for all } M'_{ic} < 0.$$

It follows that at each iteration only two vectors (q' and $M'_{.c}$) of the transformed system are used, so we can look for economies in computation by generating these vectors from the original data, rather than by transforming the whole system at each iteration.

The following table indicates possible results of the first three iterations of the algorithm, where r in step 1 takes the successive values r_1, r_2, r_3 (the pivoted variables are underlined):

	Basic Variables				**Non-basic Variables**				
Start	w_{r_1}	w_{r_2}	w_{r_3}	$\cdots$	z_0	z_{r_1}	z_{r_2}	z_{r_3}	$\cdots$
1.	$\underline{z_0}$	w_{r_2}	w_{r_3}	$\cdots$	$\underline{w_{r_1}}$	z_{r_1}	z_{r_2}	z_{r_3}	$\cdots$
2.	z_0	$\underline{z_{r_1}}$	w_{r_3}	$\cdots$	w_{r_1}	$\underline{w_{r_2}}$	z_{r_2}	z_{r_3}	$\cdots$
3.	z_0	z_{r_1}	$\underline{z_{r_2}}$	$\cdots$	w_{r_1}	w_{r_2}	$\underline{w_{r_3}}$	z_{r_3}	$\cdots$

It can be seen that after every iteration the members (w_i, z_i) of each complementary pair are on opposite sides of the equation, with the exception of one non-basic pair, say (w_r, z_r), which involves the blocking variable just pivoted. Thus a typical iteration corresponds to a partitioning of the original system of the form

$$
(5a) \qquad
\begin{bmatrix} w_1 \\ \hline w_r \\ \hline w_2 \end{bmatrix}
=
\left[\begin{array}{cc|cc}
M_{10} & M_{11} & M_{1r} & M_{12} \\
\hline
M_{r0} & M_{r1} & M_{rr} & M_{r2} \\
\hline
M_{20} & M_{21} & M_{2r} & M_{22}
\end{array}\right]
\begin{bmatrix} z_0 \\ z_1 \\ \hline z_r \\ z_2 \end{bmatrix}
+
\begin{bmatrix} q_1 \\ \hline q_r \\ \hline q_2 \end{bmatrix} .
$$

We shall write this in the shorthand form:

$$
(5b) \qquad
\begin{bmatrix} \bar{w}_1 \\ \bar{w}_2 \end{bmatrix}
=
\begin{bmatrix} \overline{M}_{11} & \overline{M}_{12} \\ \overline{M}_{21} & \overline{M}_{22} \end{bmatrix}
\begin{bmatrix} \bar{z}_1 \\ \bar{z}_2 \end{bmatrix}
+
\begin{bmatrix} \bar{q}_1 \\ \bar{q}_2 \end{bmatrix} ,
$$

with the corresponding current transformed system:

$$
(5c) \qquad
\begin{bmatrix} \bar{z}_1 \\ \bar{w}_2 \end{bmatrix}
=
\begin{bmatrix} \overline{M}'_{11} & \overline{M}'_{12} \\ \overline{M}'_{21} & \overline{M}'_{22} \end{bmatrix}
\begin{bmatrix} \bar{w}_1 \\ \bar{z}_2 \end{bmatrix}
+
\begin{bmatrix} \bar{q}'_1 \\ \bar{q}'_2 \end{bmatrix} .
$$

Equations (5b) and (5c) are related by

$$
\overline{M}_{11}\overline{M}'_{11}=I, \quad \overline{M}_{11}\overline{M}'_{12}=-\overline{M}_{12}, \quad \overline{M}_{11}\bar{q}'_1=-\bar{q}_1,
$$

$$
(6)
$$

$$
\overline{M}'_{21}=\overline{M}_{21}\overline{M}'_{11}, \quad \overline{M}'_{22}=\overline{M}_{22}+\overline{M}_{21}\overline{M}'_{12}, \quad \bar{q}'_2=\bar{q}_2+\overline{M}_{21}\bar{q}'_1.
$$

The vector q' and the relevant column $M'_{.c}$ required for test (4b) may thus be obtained from (6) without generating the full transformed system. The work may be further reduced either by storing and updating the inverse matrix $\overline{M}'_{11}$, using the recursion formulae given by Sargent and Murtagh [11] (see Appendix I), or preferably by storing and updating the orthogonal factorization of $\overline{M}_{11}$, as described later in Section 4.

3. THE PRINCIPAL PIVOTING VERSION

Suppose that, after determining row r in the first iteration of the basic Lemke algorithm, we pivot w_r with its complementary variable z_r rather than with z_0. The pivot element is then a diagonal element of M and the operation is called a "principal pivot". To carry out step 1 in the following iteration, we should then have to pivot z_r with z_0 in order to compute q', $M'_{.c}$, but it turns out that the test can be made without carrying out this operation. With this modification we have after a typical iteration the transformed system

$$(7a) \qquad \begin{bmatrix} z_1 \\ w_2 \end{bmatrix} = \begin{bmatrix} M'_{10} \\ M'_{20} \end{bmatrix} z_0 + \begin{bmatrix} M'_{11} & M'_{12} \\ M'_{21} & M'_{22} \end{bmatrix} \begin{bmatrix} w_1 \\ z_2 \end{bmatrix} + \begin{bmatrix} q'_1 \\ q'_2 \end{bmatrix},$$

corresponding to the original system

$$(7b) \qquad \begin{bmatrix} w_1 \\ w_2 \end{bmatrix} = \begin{bmatrix} M_{10} \\ M_{20} \end{bmatrix} z_0 + \begin{bmatrix} M_{11} & M_{12} \\ M_{21} & M_{22} \end{bmatrix} \begin{bmatrix} z_1 \\ z_2 \end{bmatrix} + \begin{bmatrix} q_1 \\ q_2 \end{bmatrix}.$$

Again these systems are related by the equations

$$M_{11}M'_{11}=I, \quad M_{11}M'_{12}=-M_{12}, \quad M_{11}M'_{10}=-M_{10}, \quad M_{11}q'_1=-q_1,$$

$$(8) \quad M'_{21}=M_{21}M'_{11}, \quad M'_{22}=M_{22}+M_{21}M'_{12}, \quad M'_{20}=M_{20}+M_{21}M'_{10}, \quad q'_2=q_2+M_{21}q'_1.$$

To obtain the appropriate system for test (4b) we must pivot z_0 with z'_c, the last driving variable, yielding the vectors

$$(9) \qquad \begin{aligned} M''_{c0} &= (M'_{c0})^{-1}, & q''_i &= -q'_c M''_{c0}, \\ M''_{i0} &= M'_{i0} M''_{c0}, & q''_i &= q'_i + M'_{i0} q''_i, \quad i \neq c. \end{aligned}$$

The result of this pivot operation is to put z_0 in row c on the left-hand side, and z'_c multiplies column $M''_{.0}$, so the vectors in (9) are those required for test (4b) and we may write the algorithm in the form:

Algorithm II.

Start: Set $c=0$

1. Find $\{r \mid M''_{i0}z'_i + q''_i \geqq M''_{r0}z'_i + q''_r = 0\}$

   (a) If r does not exist, STOP (no solution).

   (b) If $r=c$, STOP (solution found).

2. Pivot z'_r with w'_r.

3. Set $c=r$ and return to step 1.

As noted above, the computation of $M''_{.0}$ and q'' can be avoided, for we have from (9):

$$(10) \qquad -\frac{q''_i}{M''_{i0}} = q'_c - \frac{q'_i}{M''_{i0}}, \quad M''_{i0} = \frac{M'_{i0}}{M'_{c0}}, \quad i \neq c; \quad -\frac{q''_c}{M''_{c0}} = q'_c$$

and hence

$$(11) \qquad \bar{r} = \arg\min_{i \neq c} \left\{ -\frac{q_i''}{M_{i0}''} \,\middle|\, M_{i0}'' < 0 \right\} = \arg\min_{i \neq c} \left\{ \frac{q_i}{|M_{i0}'|} \,\middle|\, M_{i0}' \operatorname{sgn} M_{c0}' < 0 \right\}.$$

We then have one of the following situations:

i) If $M_{i0}' \geqq 0$, all i, then $M_{i0}'' \geqq 0$ for all i and r does not exist.

ii) If $M_{c0}' < 0$ and $q_r' \geqq 0$, then $(-q_{r0}''/M_{r0}'') \geqq (-q_c''/M_{c0}'')$ and we may take $r = c$.

iii) Otherwise $(-q_{r0}''/M_{r0}'') \leqq (-q_i''/M_{i0}'')$ for all $M_{i0}'' < 0$ and hence $r = \bar{r}$.

Using these tests to find r in step 1 of Algorithm II, all that is required is to compute $M_{\cdot 0}'$ and q' using (8).

Of course this requires that M_{11} should be non-singular, and equations (9), (10) and (11) assume that M_{c0}' is non-zero. However, since the Lemke algorithm itself ensures the existence of $M_{\cdot 0}''$ and q'', a sufficient condition is that all the principal minors M_{11} involved in successive iterations be non-singular. This is more restrictive than Eaves' class of $\mathcal{L}$-matrices, but it does include the matrices arising from regular quadratic programming problems, and Algorithm II has the advantage of preserving the special structure of the principal minors of M. As in the case of Algorithm I, the implementation can update either the inverse or the orthogonal factorization of M_{11}.

4. THE HYBRID ALGORITHM

The updating of $\bar{M}_{11}$ for Algorithm I involves one of four possible situations:

i) Pivot w_r with z_c — the addition of row r and column c to $\bar{M}_{11}$.

ii) Pivot z_r with w_c — the deletion of row c and column r from $\bar{M}_{11}$.

iii) Pivot w_r with w_c — the replacement of row c by row r in $\bar{M}_{11}$.

iv) Pivot z_r with z_c — the replacement of column r by column c in $\bar{M}_{11}$.

The updating of M_{11} for Algorithm II is simpler since it involves only principal pivots, which require either the addition or the deletion of a row and column in M_{11}. If we use the orthogonal (QR) factorization, rather than its inverse, then the factorization always exists (and its formation is numerically stable) whether or not M_{11} or its update is singular. Thus the restriction that M must have non-singular principal minors can be relaxed. However, if M_{11} is singular there are complications in computing the vectors for finding the blocking variable, so in practice we use a hybrid of Algorithm I and II.

Since $\bar{M}_{11}$ of Algorithm I is always non-singular, we can use the reduction

$$(12) \qquad\qquad V\bar{M}_{11} = R,$$

where R is unit upper-triangular and V is a square matrix satisfying

$$(13) \qquad V^T D V = I$$

for some positive diagonal matrix D. Thus $D^{\frac{1}{2}}V$ is an orthogonal matrix and $(V^T D^{\frac{1}{2}})(D^{\frac{1}{2}}R)$ is the orthogonal factorization of $\overline{M}_{11}$.

Now if at a given iteration w_r is the blocking variable, then z_r is the next driving variable and from equations (5) and (6) the vectors $\overline{M}'_{\cdot r}$, $\bar{q}'$ will be given by

$$\overline{M}_{11}\overline{M}'_{1r} = -\overline{M}_{1r} \qquad \overline{M}_{11}\bar{q}'_1 = -\bar{q}_1,$$

$$(14a)$$

$$\overline{M}'_{2r} = \overline{M}_{2r} + \overline{M}_{21}\overline{M}'_{1r}, \qquad \bar{q}'_2 = \bar{q}_2 + \overline{M}_{21}\bar{q}'_1.$$

Using (12) and (13) this yields

$$R\overline{M}'_{1r} = -V\overline{M}_{1r}, \qquad R\bar{q}'_1 = -V\bar{q}_1 \qquad \text{(i)}$$

$$(15a) \qquad \overline{M}'_{2r} = \overline{M}_{2r} + \overline{M}_{21}M'_{1r}, \quad \bar{q}'_2 = \bar{q}_2 + \overline{M}_{21}\bar{q}'_1 \qquad \text{(ii)}$$

$$R^T D(V\overline{M}_{1r}) = \overline{M}_{11}^T\overline{M}_{1r}, \qquad R^T D(V\bar{q}_1) = \overline{M}_{11}^T\bar{q}_1. \qquad \text{(iii)}$$

It is therefore convenient to adjoin $\overline{M}_{1r}$ and $\bar{q}_1$ to $\overline{M}_{11}$ to give

$$(16) \qquad V[\overline{M}_{11}, \overline{M}_{1r}, \bar{q}_1] = [R, R_{\cdot m+1}, R_{\cdot m+2}].$$

Then $R_{\cdot m+1}, R_{\cdot m+2}$ can be generated by solving equations (15a, iii) or updating the relation (16), and $\overline{M}'_{1r}, \bar{q}'$ are then given by (15a, i and ii). Comparing (16) with (5a), we see that updating (16) corresponds to adding the r-th row and column to $[M_{10}, M_{11}]$, as in Algorithm II, but the triangular factor R corresponds to $\overline{M}_{11}$, as is Algorithm I.

Conversely, if z_r is the blocking variable, then w_r is the next driving variable and the analogous relations are

$$R\overline{M}'_{1r} = -Ve_r, \qquad R\bar{q}_1 = -V\bar{q}_1 \qquad \text{(i)}$$

$$(15b) \qquad \overline{M}'_{2r} = \overline{M}_{21}\overline{M}'_{1r} \qquad \bar{q}'_2 = \bar{q}_2 + \overline{M}_{21}\bar{q}'_1 \quad \text{(ii)}$$

$$R^T D(Ve_r) = \overline{M}_{11}^T e_r, \qquad R^T D(V\bar{q}_1) = \overline{M}_{11}^T\bar{q}_1. \qquad \text{(iii)}$$

Thus we can again use R from (16) to compute Ve_r from (15b, iii) and then $\overline{M}_{\cdot r}, \bar{q}'$ from (15b, i and ii). Finally (16) is updated by dropping the r-th row and column. Of course, if z_0 is the blocking variable, it suffices to compute $\bar{q}'_1 = z_1$ and $\bar{q}'_2 = w_2$, yielding the solution.

The revised algorithm can therefore be written:

Algorithm III.

Start: Set $c = 0$

1. Find $\{r | \overline{M}'_{ic}z'_c + \bar{q}'_i \geq \overline{M}'_{rc}z'_c + \bar{q}'_r = 0\}$
   If r does not exist, STOP (no solution).

2. Pivot w_r and z_r, computing $\overline{M}'_{1r}$ and $\bar{q}'$.

If $w'_r \equiv z_0$, STOP (solution found).

3. Set $c = r$ and return to step 1.

It remains to consider how (16) is updated, which makes extensive use of ideas developed by Gill et al. [6, 7].

4.1 Adding a Row and Column

For this section it is convenient to simplify the notation and write (16) in the form

$$(17) \qquad V[M, M._{m+1}, M._{m+2}] = [R, R._{m+1}, R._{m+2}].$$

We wish to find the factors for the relation $\overline{V}M = \overline{R}$ corresponding to the addition of a row and column, and hence consider

$$(18)$$

$$\overline{M} = \begin{bmatrix} M & M._{m+1} & M._{m+2} & M._{m+3} \\ M_{m+1.} & M_{m+1,m+1} & M_{m+1,m+2} & M_{m+1,m+3} \end{bmatrix} =$$

$$= \begin{bmatrix} V^T & \cdot \\ \cdot & 1 \end{bmatrix} \begin{bmatrix} D & \cdot \\ p^T & p_{m+1} \end{bmatrix} \begin{bmatrix} R & R._{m+1} & R._{m+2} & R._{m+3} \\ \cdot & 1 & R_{m+1,m+2} & R_{m+1,m+3} \end{bmatrix}.$$

Comparing partitions in (18) we obtain

$$(19)$$

$$M = V^T D R \qquad M._j = V^T D R._j \qquad j = 1, 2, \ldots, m+3$$

$$R^T p = M^T_{m+1.} \qquad M_{m+1,j} = p^T r._j + p_{m+1} R_{m+1,j}, \qquad j = 1, 2, \ldots, m+3$$

$$p_{m+1} = M_{m+1,m+1} - p^T R._{m+1}.$$

Now, using relations (I. 3) from Appendix I in (18) we have

$$(20) \qquad \begin{bmatrix} V^T & \cdot \\ \cdot & 1 \end{bmatrix} \hat{V}^T \hat{D} \begin{bmatrix} U & p_{m+1}\beta \\ \cdot & 1 \end{bmatrix} \begin{bmatrix} R & R._{m+1} & R._{m+2} & R._{m+3} \\ \cdot & 1 & R_{m+1,m+2} & R_{m+1,m+3} \end{bmatrix}.$$

Thus we obtain the required factorization by defining

$$\overline{M} = \overline{V}^T \overline{D} \overline{R}, \qquad \overline{V}^T \overline{D} \overline{V} = I$$

$$(21)$$

$$\overline{V} = \begin{bmatrix} V^T & \cdot \\ \cdot & 1 \end{bmatrix} \hat{V} \begin{bmatrix} I & \cdot \\ \cdot p_{m+1}^{-1} & \cdot \end{bmatrix}, \qquad \overline{D} = \hat{D} \begin{bmatrix} I & \cdot \\ \cdot & p_{m+1} \end{bmatrix},$$

$$\overline{R} = \begin{bmatrix} U & p_{m+1}\beta \\ \cdot & 1 \end{bmatrix} \begin{bmatrix} R & R._{m+1} & R._{m+2} & R._{m+3} \\ \cdot & \cdot & R_{m+1,m+2} & R_{m+1,m+3} \end{bmatrix}.$$

The required quantities in (21) are obtained from the recursions (I. 4) and (I. 7), but to use these in Step 2 of Algorithm III we need $R._{m+3} = VM._{m+3}$, corresponding to the new column to be added. This is obtained from

$$(22) \qquad R^T D R._{m+3} = M^T M._{m+3},$$

(cf. equation 15a, iii) first solving for $DR._{m+3}$ by successive substitution, then dividing

70

by the elements D_i. We further note from (19) that $R^Tp = M_{m+1}^T$, and hence it follows from (I.7) that $p_i = v_{i-1,i}$. As a result, we can carry out all the required recursions simultaneously, giving:

Algorithm (ADD).

1. Set $t_0 = 1$;

$$u_{0j} = \sum_{i=1}^{m} M_{ij} M_{i,m+3}, \quad j = 1, 2, \ldots, m;$$

$$v_{0j} = M_{m+1,j}, \quad j = 1, 2, \ldots, (m+3).$$

2. For $i = 1, 2, \ldots, m$: $\quad p_i = v_{i-1,i}, \quad R_{i,m+3} = u_{i-1,i}/D_i.$

$$t_i = t_{i-1} + p_i^2/D_i$$

$$\overline{D}_i = D_i t_i/t_{i-1}$$

$$\beta_i = p_i/D_i t_i$$

$$u_{ij} = u_{i-1,j} - u_{i-1,i} R_{ij}, \quad j = (i+1), \ldots, m$$

$$v_{ij} = v_{i-1,j} - v_{i-1,i} R_{ij}, \quad j = (i+1), \ldots, (m+3)$$

$$\overline{R}_{ij} = R_{ij} + \beta_i v_{ij}, \quad j = (i+1), \ldots, (m+3).$$

3. $p_{m+1} = v_{m,m+1}, \quad \overline{D}_{m+1} = p_{m+1}^2/t_m, \quad \overline{R}_{m+1,j} = v_{mj}/p_{m+1}, \quad j = m+2, m+3.$

It then remains to compute $\overline{M}'_{\cdot r}$ and $\bar{q}'$ from (15a, i and ii) using the updated system

4.2 Dropping a Row and Column

Here we should start by computing $M'_{\cdot r}$ and $\bar{q}'$ from the $(m+1) \times (m+3)$ system $\overline{M} = \overline{V}^T \overline{D} \overline{R}$, then obtain the factorization corresponding to dropping the r-th row and column from among the first $(m+1)$ rows and columns of $\overline{M}$. However, it saves some computation to defer the calculation of $\overline{M}'_{\cdot r}$ and $\bar{q}'$, and the first step is to permute the r-th row and column to the $(m+1)$-th position and obtain the corresponding factorization:

$$(23) \qquad \tilde{M} = \begin{bmatrix} M & M_{\cdot r} & M_{\cdot m+2} & M_{\cdot m+3} \\ M_{r \cdot} & M_{srr} & M_{r,m+2} & M_{r,m+3} \end{bmatrix} = \tilde{V}^T \tilde{D}[\tilde{R}, \tilde{R}_{\cdot m+2} \tilde{R}_{\cdot m+3}]$$

with $\tilde{V}^T \tilde{D} \tilde{V} = I$, and $\tilde{R}$ unit upper-triangular.

Let P be the post-multiplying permutation matrix which extracts column r and places it in column $(m+1)$. Then the permuted system before reduction is of the form

$$(24) \qquad \tilde{M} - P^T V^T P \begin{bmatrix} \overline{D}' & . \\ . & \overline{D}_r \end{bmatrix} \begin{bmatrix} \overline{R}' & \overline{R}'_{\cdot r} & \overline{R}'_{\cdot m+2} & \overline{R}'_{\cdot m+3} \\ \overline{R}'_r & \overline{R}'_{rr} & \overline{R}'_{r,m+2} & \overline{R}'_{r,m+3} \end{bmatrix},$$

where the prime denotes deletion of the r-th row and/or column from the corresponding vector/matrix. $\bar{R}'$ is already unit upper-triangular, so for the reduction it suffices to consider the system:

$$(25) \qquad \begin{bmatrix} \bar{D}' & \cdot \\ \cdot & \bar{D}_r \end{bmatrix} \begin{bmatrix} \bar{R}' & \bar{R}'_{\cdot r} & \bar{R}'_{\cdot m+2} & \bar{R}'_{\cdot m+3} \\ \bar{R}'_{r\cdot} & \bar{R}_{rr} & \bar{R}'_{r,m+2} & \bar{R}'_{r,m+3} \end{bmatrix} =$$
$$= \begin{bmatrix} \bar{D}' & \cdot \\ p^T & p_{m+1} \end{bmatrix} \begin{bmatrix} \bar{R}' & \bar{R}'_{\cdot r} & \bar{R}'_{\cdot m+2} & \bar{R}'_{\cdot m+3} \\ \cdot & 1 & \bar{R}''_{m+1,m+2} & \bar{R}''_{m+1,m+3} \end{bmatrix}.$$

As with the system in (18), we can use recursions (I.4) and (I.7) to reduce (25), generating p at the same time. However, since $\bar{R}'_{rj}=0, j=1, 2, \ldots, (r-1)$, the recursion can start with row r, leaving the first $(r-1)$ rows of $\bar{D}'$ and the $\bar{R}'$-matrix unchanged:

Algorithm (DROP)

1. Set $t_{r-1}=1;\ v_{r-1,j}=\bar{D}_r\bar{R}'_{rj}, j=r, \ldots, (m+3)$

2. For $i=r, \ldots, m$: $\quad p_i=v_{i-1,i},$

$$t_i=t_{i-1}+p_i^2/\bar{D}_i$$
$$\tilde{D}_i=\bar{D}_i t_i/t_{i-1}$$
$$\beta_i=p_i/\bar{D}_i t_i$$
$$v_{ij}=v_{i-1,j}-p_i\bar{R}_{ij}, \quad j=(i+1), \ldots, (m+3),$$
$$\tilde{R}_{ij}=R_{ij}+\beta_i v_{ij}, \qquad j=(i+1), \ldots, (m+3).$$

3. Set $p_{m+1}=v_{m,m+1},\ \tilde{D}_{m+1}=p_{m+1}^2/t_m,\ \tilde{R}_{m+1,j}=v_{mj}/p_{m+1},\ j=m+2, m+3$.

This yields the system (23). We now calculate $\bar{M}'_{\cdot s}$ and $\bar{q}'$ from this system using (15b). We note that in doing this we shall compute a vector $\tilde{p}$ satisfying

$$(26) \qquad \tilde{R}^T\tilde{p}=\tilde{M}^T_{m+1\cdot},$$

and hence also

$$\tilde{V}^T\tilde{p}=e_{m+1}.$$

From (23) and (26) we obtain

$$(27) \qquad \begin{bmatrix} M & M_{\cdot s} & M_{\cdot m+2} & M_{\cdot m+3} \\ 0 & 0 & 0 & 0 \end{bmatrix} = (I-e_{m+1}e_{m+1}^T)\tilde{M}$$
$$= (\tilde{V}^T\tilde{D}\tilde{V}-\tilde{V}^T\tilde{p}\,\tilde{p}^T\tilde{V})\tilde{M}$$
$$= \tilde{V}^T(\tilde{D}-\tilde{p}\,\tilde{p}^T)[\tilde{R}, \tilde{R}_{\cdot m+2}, \tilde{R}_{\cdot m+3}].$$

Equations (23) and (26) also give $\tilde{p}^T\tilde{D}^{-1}\tilde{p}=1$, so we can apply (I.5), which we write in the form

$$(28) \qquad \tilde{D}-\tilde{p}\tilde{p}^T=\hat{V}^T\hat{D}\begin{bmatrix} U & \tilde{p}_{m+1}\beta \\ \cdot & \cdot \end{bmatrix}, \quad \hat{V}^T\hat{D}\hat{V}=\tilde{D},$$

with $\hat{D}_{m+1}=1$. We note from (I. 5) that $\hat{V}^T e_{m+1}=\tilde{p}$, and hence from (26) and (28) we have

$$\tilde{V}^T\hat{V}^T e_{m+1}=\tilde{V}^T\tilde{p}=e_{m+1} \quad \text{and} \quad \tilde{V}^T\hat{V}^T\hat{D}\hat{V}\tilde{V}=I,$$

so that $\tilde{V}^T\hat{V}^T$ must be of the form:

$$(29) \qquad \tilde{V}^T\hat{V}^T=\begin{bmatrix} V^T & \cdot \\ \cdot & 1 \end{bmatrix}.$$

Thus from (27), (28) and (29) we obtain the desired factorization:

$$(30) \qquad [M, M_{\cdot r}, M_{\cdot m+2}, M_{\cdot m+3}]=V^TDR=V^TD[U,\tilde{p}_{m+1}\beta][\tilde{R}, \tilde{R}_{\cdot m+2}, \tilde{R}_{\cdot m+3}]$$

with $D_i=\hat{D}_i$, $i=1, 2, \ldots, m$, and the required quantities are obtained by combining the recursions (I. 6) and (I. 7) as follows:

Algorithm (DROP) — continued

4. Compute $\tilde{p}$, $\overline{M}'_{\cdot}$, and $\bar{q}'$ using (26) and (15b).
5. Set $t_m=\tilde{p}^2_{m+1}/\tilde{D}_{m+1}$,
6. For $i=m, m-1, \ldots, 1$:
$$\begin{aligned}
t_{i-1}&=t_i+p_i^2/D_i \\
D_i&=\tilde{D}_i t_i/t_{i-1} \\
\beta_i&=-\tilde{p}_i/\tilde{D}_i t_i \\
v_{i,i+1}&=\tilde{p}_{i+1} \\
v_{ij}&=v_{i+1, j}+\tilde{p}_{i+1}R_{i+1, j}, \quad j=i+2..(m+3) \\
R_{ij}&=\tilde{R}_{ij}+\beta_i v_{ij}, \qquad\quad j=i+1..(m+3).
\end{aligned}$$

In practice, we do not need to compute $R_{\cdot m+1}$ in this algorithm, since this is the column to be dropped.

5. QUADRATIC PROGRAMMING

The standard form of the quadratic programming problem is

$$(31) \qquad \min_{x} \{F(x)|f(x)\geqq 0, \quad x\geqq 0\},$$

where $x\in R^n$, $F(x): R^n\to R$, $f(x): R^n\to R^m$,

$$(32) \qquad \begin{aligned} F(x)&=F_0+g^Tx+\tfrac{1}{2}x^THx, \\ F(x)&=f_0+G^Tx, \end{aligned}$$

and the $n\times n$ matrix H may be taken as symmetric. If there exists an interior feasible point, any solution satisfies the Kuhn-Tucker conditions, which require the existence of vectors $u\in R^n$, $\lambda\in R^m$ such that

$$(33) \qquad \begin{aligned} g+Hx&=G\lambda+Iu \\ f\geqq 0, \quad x\geqq 0, &\quad \lambda\geqq 0, \quad u\geqq 0 \\ \lambda^Tf=0, &\quad u^Tx=0. \end{aligned}$$

It is well known [1] that these relations may be written in the form of the linear complementarity problem (1) by writing

$$(34) \qquad w = \begin{bmatrix} \mu \\ f \end{bmatrix}, \quad z = \begin{bmatrix} x \\ \lambda \end{bmatrix}, \quad q = \begin{bmatrix} g \\ f_0 \end{bmatrix}, \quad M = \begin{bmatrix} H & -G \\ G^T & 0 \end{bmatrix}.$$

However, variants of this standard form often arise in practice; we may wish to relax non-negativity constraints on some elements of x, put upper and lower bounds on some elements of x or f, or make the constraints on some elements of f equality constraints. All these variants are included as special cases of the general problem

$$(35) \qquad \min_x \{F(x) | a \leq f(x) \leq b, \quad c \leq x \leq d\}.$$

Here a constraint can be relaxed by allowing the corresponding bound to tend to plus or minus infinity as appropriate, while an equality constraint is imposed if the upper and lower bounds on a variable are made equal.

The Kuhn–Tucker conditions for problem (35) are

$$
\begin{aligned}
g + Hx &= G\lambda + \mu, \\
a \leq f(x) &\leq b, \quad c \leq x \leq d, \\
\lambda_i(a_i - f_i(x)) \geq 0, \quad \lambda_i(b_i - f_i(x)) &\geq 0, \quad i = 1, 2, \ldots, m, \\
\mu_i(c_i - x_i) \geq 0, \quad \mu_i(d_i - x_i) &\geq 0, \quad i = 1, 2, \ldots, n.
\end{aligned}
\tag{36}
$$

These conditions may be written as a generalized linear complementarity problem, but in using the notation of (34) it is convenient to retain the partitioning into primal and dual variables:

$$
\begin{aligned}
\begin{array}{c} w_1 \\ w_2 \end{array} &= \begin{bmatrix} M_{11} & M_{12} \\ M_{21} & M_{22} \end{bmatrix} \begin{bmatrix} z_1 \\ z_2 \end{bmatrix} + \begin{bmatrix} q_1 \\ q_2 \end{bmatrix}, \\
a_1 = b_1 &= 0, \quad c_1 \leq z_1 \leq d_i, \\
a_2 \leq w_2 &\leq b_2, \quad c_2 = d_2 = 0, \\
(w_i - a_i)(d_i - z_i) \geq 0, \quad (b_i - w_i)(z_i - c_i) &\geq 0, \quad \text{all} \quad i.
\end{aligned}
\tag{37}
$$

6. SOLUTION OF THE GENERALIZED LINEAR COMPLEMENTARITY PROBLEM

The generalized problem (37) can be converted into a standard linear complementarity problem (1) by using the substitutions:

$$
\begin{aligned}
z_1' = z_1 - c_1 &\geq 0, \quad z_1'' = d_1 - z_1 \geq 0, \\
w_1' \geq 0, \quad (w_1')^T z_1' = 0, \quad w_1'' &\geq 0, \quad (w_1'')^T z_1'' = 0, \\
w_2' = w_2 - a_2 &\geq 0, \quad w_2'' = b_2 - w_2 \geq 0, \\
z_2' \geq 0, \quad (z_2')^T w_2' = 0, \quad z_2'' &\geq 0, \quad (z_2'')^T w_2'' = 0, \\
w_1 = w_1' - w_1'', \quad z_2 &= z_2' - z_2''.
\end{aligned}
\tag{38}
$$

74

The corresponding system for the application of the Lemke algorithm is then:

$$(39) \quad
\begin{bmatrix} w_1' \\ w_2' \\ w_2'' \\ z_1'' \end{bmatrix}
=
\begin{bmatrix} M_{10}' \\ M_{20}' \\ M_{20}'' \\ M_{10}'' \end{bmatrix} z_0
+
\begin{bmatrix}
M_{11} & M_{12} & -M_{12} & I \\
M_{21} & M_{22} & -M_{22} & 0 \\
-M_{21} & -M_{22} & M_{22} & 0 \\
-I & 0 & 0 & 0
\end{bmatrix}
\begin{bmatrix} z_1' \\ z_2' \\ z_2'' \\ w_1'' \end{bmatrix}
+
\begin{bmatrix}
q_1 + M_{11}c_1 \\
(q_2 + M_{21}c_1) - a_2 \\
b_2 - (q_2 + M_{21}c_1) \\
d_1 - c_1
\end{bmatrix}.$$

However, it is not necessary to deal directly with this enlarged problem, for we may set $M_{10}'=0$, $M_{20}''=M_{20}$ and hence use the system

$$
\begin{aligned}
(40) \qquad & w = M_0 z_0 + M \cdot \delta z + (q + M\bar{z}), \\
& w = \bar{w} + \delta w, \quad z = \bar{z} + \delta z, \quad \delta w^T \delta z = 0. \\
& \sigma_i \delta w_i \geq 0, \quad \sigma_i \delta z_i \geq 0, \quad i = 1, 2, 3, \ldots.
\end{aligned}
$$

Here $(\bar{w}_i, \bar{z}_i)$ represent complementary bounds, and $\sigma_i = +1$ if the primal variable is at its lower bound or $\sigma_i = -1$ if it is at its upper bound.

It is not necessary to start the algorithm as in (39), with each primal z_i at its lower bound, and we can in fact make an arbitrary choice of initial σ_i values for these variables. So that the appropriate row in (40) agrees with (39) we must then set $M_{i0} = \sigma_i \bar{M}_{i0}$, with $\bar{M}_{i0} > 0$, and this choice of M_{i0} remains fixed thereafter.

Each row corresponding to a primal w_i in (40) is used for both w_2' and w_2'' and again we have $M_{i0}\sigma_i \bar{M}_{i0}$ with $\bar{M}_{i0} > 0$, but here σ_i must correspond to the interpretation as w_2' or w_2'' and both possibilities must be tested at each iteration. We note, however, that if a primal w_i becomes non-basic at a given bound, this fixes σ_i and hence the sign of M_{i0}, so there is no ambiguity in updating the inverse or factorization of $\bar{M}_{11}$ pfor such a pivot.

Thus to start the algorithm we choose an airbtrary set of σ_i for the primal z_i variables and evaluate the vector $\bar{q} = q + M\bar{z}$. (It is unnecessary to specify initial σ_i for dual z_i variables since in any case $\bar{z}_i = 0$). The first blocking variable is then found by checking both upper and lower bounds (where they exist) for each w_i, yielding

$$(41a) \qquad z_0 = \frac{\bar{w}_r - \bar{q}_r}{M_{r0}} \geq \frac{\bar{w}_i - \bar{q}_i}{M_{i0}}, \quad \sigma_i = \pm 1, \text{ all } i.$$

Obviously, if this yields $z_0 \leq 0$ the solution is given immediately by setting $z_0 = 0$.

For a general iteration the driving variable z_c' is changed from its bound in the feasible direction $(\sigma_c' \delta z_c' > 0,)$ and the blocking variable is found by checking the relevant bound for each basic variable:

$$(41b) \qquad \sigma_c' \delta z_c' = \frac{\bar{w}_r' - \bar{q}_r'}{M_{rc}' \sigma_c'} \leq \frac{\bar{w}_i' - \bar{q}_i'}{M_{ic}' \sigma_c'}, \quad \text{all} \quad \sigma_i' M_{ic}' \sigma_c' < 0.$$

Note that if the driving variable is subject to upper and lower bounds, its opposite bound must be tested for the blocking variable. If z_c' is an element of w_2 this is included in (41b) with $\sigma_i' = -\sigma_c'$ (since we only have $w_2 = w_2' + a_2 = b_2 - w_2''$ at the solution with $z_0 = 0$), but if z_c' is an element of z_1 we have $\sigma_c' \delta z_c' = d_c - c_c$ (since $M_{10}'' = 0$ and $z_1 = z_1' + c_1 = d_1 - z_1''$ at each iteration).

With these modifications in determining the blocking variable, any of the versions of the Lemke algorithm described earlier may be used to solve the generalized linear complementarity problem. In effect we are using the Lemke algorithm on the expanded system (39), and if the corresponding expanded matrix is in Eaves' class $\mathcal{L}$, termination in failure to find a blocking variable implies that the original problem has no solution. This may be expressed as a condition on the matrix M by defining a new class of matrices $\mathcal{L}' = \mathcal{L}_1' \cap \mathcal{L}_2'$ with the following definitions of $\mathcal{L}_1'$ and $\mathcal{L}_2'$:

Definition. $M \in \mathcal{L}_1'$ if and only if for each $z \neq 0$ there is an element $z_i \neq 0$ such that $z_i(Mz)_i \geqq 0$.

Definition. $M \in \mathcal{L}_2'$ if, for each $z \neq 0$ such that $z_i(Mz)_i = 0$ for all i, there exist diagonal matrices $\Lambda \geqq 0$, $\Gamma \geqq 0$ such that $\Gamma z \neq 0$ and $\Lambda Mz = -M^T \Gamma z$.

In Appendix II an extension of a theorem due to Garcia [10] is given, which shows that if $M \in \mathcal{L}'$ then termination in failure to find a blocking variable implies that elements in the row $M_{c\cdot}'$ corresponding to z_0 satisfy the condition $M_{ci}' \sigma_i' \geqq 0$. Clearly this implies that no feasible values of the $z_i'(\sigma_i' \delta z_i' \geqq 0)$ can reduce z_0 to zero and hence the original problem has no solution. As suggested by van de Panne [7], in case of failure the elements of row $M_{c\cdot}'$ can be tested directly for this condition, thus extending somewhat the class of matrices for which the generalized problem can be successfully processed.

If one of the bounds of a primal variable becomes infinite, the corresponding σ_i remains one-signed throughout, and the candidate vectors z in the definitions of $\mathcal{L}_1'$ and $\mathcal{L}_2'$ can then be restricted to those with z_i of the appropriate sign, thus extending the class $\mathcal{L}'$. If no variables are subject to upper and lower bounds, the variables can always be defined so that they have only a lower bound and then classes $\mathcal{L}_1'$ and $\mathcal{L}_2'$ become Eaves' classes $\mathcal{L}_1$ and $\mathcal{L}_2$ respectively.

The algorithm deals with equality constraints without special provisions. However, once an equality-constrained primal variable becomes non-basic the row and column corresponding to it and its dual variable may be ignored thereafter, with some saving of computation, since the dual variable is unconstrained and hence can never become a blocking variable.

7. DISCUSSION

The table below gives the number of multiplications for the pivot step (including calculation of $\bar{M}_{\cdot r}'$ and $\bar{q}'$) for the various versions of the algorithm. The number of additions is not given since this number is always of the same order but less than the number of multiplications. The averages are based on equal frequencies of occurrence of all types of update.

Type of Pivot	Algorithm I Inverse	Algorithm II Inverse	Algorithm II $V^T DR$	Algorithm III $V^T DR$
$w_r - z_c$ Add row and column	$2n(m+1)$ $+3m^2+3m+1$	$2n(m+1)$ $+3m^2+3m+1$	$2n(m+1)$ $+\frac{1}{2}(3m^2+15m+4)$	$2n(m+1)$ $+\frac{1}{2}(3m^2+15m+4)$
$z_r - w_c$ Drop row and column	$2nm+m$	$2nm+m^2+m$	$2nm$ $+1/6(5m^2+75m+53)$	$2n(m+1)$ $+1/6(5m^2+69m+47)$
$w_r - w_c$ Replace row	$2nm+2m^2+m$	—	—	—
$z_r - z_c$ Replace column	$2nm+m^2+m$	—	—	—
Average	$n(2m+\frac{1}{2})$ $+\frac{1}{4}(6m^2+6m+1)$	$n(2m+1)$ $+\frac{1}{2}(4m^2+4m+1)$	$n(2m+1)$ $+1/12(14m^2+120m+65)$	$2n(m+1)$ $+1/12(14m^2+114m+59)$

The results given for the factorization version of Algorithm II assume that columns M_{10} and q_1 are adjoined to M_{11} and updated with it, as in Algorithm III. Algorithm III can be regarded as the factorization version of Algorithm I.

Comparing the quadratic terms it can be seen from the table that the factorization version is in each case more efficient than the version based on updating the inverse, while the computational requirements of the factorization versions of Algorithms II and III are similar. Updating the whole matrix M at each iteration involves (n^2+n) multiplications for all types of pivot, so the factorization algorithms have comparable computing requirements for an average value of $m \simeq 0.4n$ and in most cases one would expect a lower value than this. For example, for a linear programme m at the solution will not exceed the number of non-trivial constraints and is often much smaller.

In terms of storage, the basic Lemke algorithm updating M requires $n^2 + O[n]$ locations, since the transformed matrix can overwrite M. The versions based on inverses require an additional $m^2 + O[m]$ locations for the inverse, while the factorization versions require $\frac{1}{2}m^2 + O[m]$ for R and D (V is not required). However these comparisons are based on full matrices, and if M is sparse or its elements need only be generated as required there will be significant savings in storage.

The further savings in computation when M is sparse require more study, for they depend on the relative fill-in of classical tableau pivots compared with the updating of R in orthogonal factorizations, but for some special structures of non-zero elements in M it is clear that the gains can be significant.

A further important advantage of using factorizations rather than classical pivots or updated inverses is their numerical stability, which becomes more important as the size of the problem increases.

Algorithm III and the factorization version of Algorithm II have essentially the same storage and computational requirements. Since Algorithm III processes a wider class of problems it is to be preferred, unless there are particular advantages to be gained from preserving the structure of M_{11}.

The use of Algorithm III to solve the corresponding generalized linear complementarity problem is an efficient method of solving a linear or quadratic programming problem, offering gains in both storage and computation time for large sparse problems and with the additional important advantage that it does not require an initial feasible point.

REFERENCES

[1] Cottle, R. W. and Dantzig, G. B.: Complementary Pivot Theory of Mathematical Programming, *Linear Algebra and its Applications* 1 (1968), 103–125.

[2] Cottle, R. W. and Sacher, R. S.: On the Solution of Large Structured Linear Complementarity Problems — I, Technical Report 73-4 (Department of Operations Research, Stanford University, 1973).

[3] Dantzig, G. B. and Cottle, R. W.: Positive Semi-Definite Programming, ORC 63–18 (RR), (Operations Research Center, University of California, Berkeley, 1963).

[4] Eaves, B. C.: The Linear Complementarity Problem, *Management Science* 17 (1971), 612–634.

[5] Garcia, C. B.: A Note on a Complementary Variant of Lemke's Method, *Mathematical Programming* 10 (1976), 134–136.

[6] Gill, P. E., Golub, G. H., Murray, W. and Saunders, M. A.: Methods for Modifying Matrix Factorizations, NPL Report NAC29 (National Physical Laboratory, 1972).

[7] Gill, P. E., Murray, W. and Saunders, M. A.: Methods for Computing and Modifying the LDV Factors of a Matrix, *Mathematics of Computation* 29 (132) (1975), 1051–1077.

[8] Lemke, C. E.: On Complementary Pivot Theory, in G. B. Dantzig and A. F. Veinott, Eds., *Mathematics of the Decision Sciences, Part I* (American Mathematical Society, 1968) 95–114.

[9] Lemke, C. E. and Howson, J. T. Jr.: Equilibrium Points of Bimatrix Games, *SIAM J. Appl. Math.* 12 (1964), 413–423.

[10] McCammon, S. R.: On Complementary Pivoting, Ph. D. Thesis, Rensselaer Polytechnic Institute, Troy, N. S. (1970).

[11] Sargent, R. W. H. and Murtagh, B. A.: Projection Methods for Nonlinear Programming, *Mathematical Programming* 4 (1973), 245–268.

[12] van de Panne, C.: A Complementary Variant of Lemke's Method for the Linear Complementarity Problem, *Mathematical Programming* 7 (1974), 283–310.

APPENDIX I

Factorization

I. Gill, Murray and Saunders [7] prove the following result:

$$(I.1) \qquad \begin{bmatrix} Im & \cdot \\ \bar{p}^T & 1 \end{bmatrix} = \bar{V}^T \bar{D} \begin{bmatrix} \bar{U} & \beta \\ \cdot & 1 \end{bmatrix},$$

where $\bar{V}^T \bar{D} \bar{V} = I$, $\bar{U}$ is unit upper-triangular with $\bar{U}_{ij} = \beta_i \bar{p}_j$, $i < j$, and $\bar{D}$ is a positive diagonal matrix. The quantities $\bar{D}_i$, β_i are given by the recursion:

$$\text{(I.2)} \qquad
\begin{aligned}
&1.\ \bar{t}_0 = 1 \\
&2.\ \text{For } i = 1, 2, \ldots, m: \qquad
\begin{aligned}
\bar{t}_i &= \bar{t}_{i-1} + \bar{p}_i^2 \\
\bar{D}_i &= \bar{t}_i / \bar{t}_{i-1} \\
\beta_i &= \bar{p}_i / \bar{t}_i
\end{aligned} \\
&3.\ \bar{D}_{m+1} = 1/\bar{t}_m.
\end{aligned}$$

This is readily generalized for any positive diagonal as follows:

$$
\begin{bmatrix} D & \cdot \\ p^T & p_{m+1} \end{bmatrix}
=
\begin{bmatrix} D^{\frac{1}{2}} & \cdot \\ \cdot & p_{m+1}^{\frac{1}{2}} \end{bmatrix}
\begin{bmatrix} I_m & \cdot \\ \bar{p}^T & 1 \end{bmatrix}
\begin{bmatrix} D^{\frac{1}{2}} & \cdot \\ \cdot & p_{m+1}^{\frac{1}{2}} \end{bmatrix}
=
$$

$$
=
\begin{bmatrix} D^{\frac{1}{2}} & \cdot \\ \cdot & p_{m+1}^{\frac{1}{2}} \end{bmatrix}
\bar{V}^T
\begin{bmatrix} D^{\frac{1}{2}} & \cdot \\ \cdot & p_{m+1}^{-\frac{1}{2}} \end{bmatrix}
\bar{D}
\begin{bmatrix} D & \cdot \\ \cdot & p_{m+1} \end{bmatrix}
\begin{bmatrix} D^{-\frac{1}{2}}\bar{U}D^{\frac{1}{2}} & D^{-\frac{1}{2}}\bar{\beta}p_{m+1}^{\frac{1}{2}} \\ \cdot & 1 \end{bmatrix}.
$$

This is of the form

$$\text{(I.3)} \qquad
\begin{bmatrix} D & \cdot \\ p^T & p_{m+1} \end{bmatrix}
= \hat{V}^T \hat{D}
\begin{bmatrix} U & p_{m+1}\beta \\ \cdot & 1 \end{bmatrix}, \qquad
\hat{V}^T \hat{D} \hat{V} =
\begin{bmatrix} D & \cdot \\ \cdot & p_{m+1} \end{bmatrix}
$$

where again U is unit upper-triangular with $U_{ij} = \beta_i p_j$, $i < j$, and $\hat{D}$ is a positive diagonal matrix. We have

$$
p_i = \bar{p}_i (D_i p_{m+1})^{\frac{1}{2}}, \qquad \beta_i = \bar{\beta}_i (D_i p_{m+1})^{-\frac{1}{2}}, \qquad t_i = \bar{t}_i p_{m+1}
$$

and the recursion becomes

$$\text{(I.4)} \qquad
\begin{aligned}
&1.\ t_0 = 1 \\
&2.\ \text{For } i = 1, 2, \ldots, m: \qquad
\begin{aligned}
t_i &= t_{i-1} + p_i^2 / D_i \\
\hat{D}_i &= D_i t_i / t_{i-1} \\
\beta_i &= p_i / D_i t_i
\end{aligned} \\
&3.\ \hat{D}_{m+1} = p_{m+1} / t_m.
\end{aligned}$$

II. Gill, Murray and Saunders also give the following result:

$$\text{(I.5)} \qquad
\begin{aligned}
D &= pp^T =
\begin{bmatrix} U & p_{m+1}\beta \\ \bar{p}^T & p_{m+1} \end{bmatrix}^T
\begin{bmatrix} \hat{D} & \cdot \\ \cdot & 1 \end{bmatrix}
\begin{bmatrix} U & p_{m+1}\beta \\ \cdot & \cdot \end{bmatrix} \\[2mm]
\text{where} \quad &
\begin{bmatrix} U & p_{m+1}\beta \\ \bar{p}^T & p_{m+1} \end{bmatrix}^T
\begin{bmatrix} \hat{D} & \cdot \\ \cdot & 1 \end{bmatrix}
\begin{bmatrix} U & p_{m+1}\beta \\ \bar{p}^T & p_{m+1} \end{bmatrix} = D,
\end{aligned}$$

$p^T D^{-1} p = 1$, $p^T = [\bar{p}^T, p_{m+1}]$, U is a unit upper-triangular matrix with $U_{ij} = \beta_i p_j$, $i < j$, and D, $\hat{D}$ are positive diagonal matrices. The quantities $\hat{D}_i$, β_i are given by the recursion:

$$\text{(I.6)} \qquad
\begin{aligned}
&1.\ t_m = p_{m+1}^2 / D_{m+1} \\
&2.\ \text{For } i = m, m-1, \ldots, 2, 1: \qquad
\begin{aligned}
t_{i-1} &= t_i + p_i^2 / D_i \\
\hat{D}_i &= D_i t_i / t_{i-1} \\
\beta_i &= -p_i / D_i t_i
\end{aligned}
\end{aligned}$$

III. Finally we derive a recursion formula for computing $\bar{R}=UR$ when U is of the special form appearing in (I.3) and (I.5), and R is a general $m\times n$ matrix $(m\leqq n)$:

$$\bar{R}_{ij}=\sum_{k=1}^{m} U_{ik}R_{kj}=R_{ij}+\beta_i\sum_{k=i+1}^{m} p_k R_{kj}, \quad i=1,2,\ldots,(m-1); \quad j=1,2,\ldots,n$$

Define: $\quad v_{ij}=\sum_{k=i+1}^{m} p_k R_{kj}, \quad i=0,1,2,\ldots,(m-1); \quad j=1,2,\ldots,n.$

This yields the recursion:

$$
\begin{aligned}
&\text{1. Set } v_{0j}=(R^T p)_j, \quad && j=1,2,\ldots,n\\
\text{(I.7)}\quad &\text{2. For } i=1,2,\ldots,(m-1):\ v_{ij}=v_{i-1,j}-p_i R_{ij}, \quad && j=1,2,\ldots,n\\
& \qquad\qquad\qquad\qquad\qquad\quad R_{ij}=R_{ij}+\beta_i v_{ij}, \quad && j=1,2,\ldots,n\\
&\text{3.} \qquad\qquad\qquad\qquad\qquad\quad \bar{R}_{mj}=R_{mj}, \quad && j=1,2,\ldots,n.
\end{aligned}
$$

We note that if R is unit upper-triangular, then $\bar{R}$ is also unit upper-triangular and the range of j in steps 2 and 3 can be restricted to $j=i+1,\ldots,n$.

Gill, Murray and Saunders noted that when U is generated from (I.4) we have $1-\beta_i p_i=D_i/\hat{D}_i$ and hence $0\leqq 1-\beta_i p_i\leqq 1$. When this quantity is small the recursions (I.7) can be unstable, so they recommend that if $\hat{D}_i/D_i>4$ then $\bar{R}_{ij}$ should be computed from the rearranged formula:

$$\text{(I.8)}\qquad\qquad\qquad \bar{R}_{ij}=(D_i/\hat{D}_i)R_{ij}+\beta_i v_{i-1,j}.$$

This is not used systematically, because of the additional multiplications involved.

When U is generated from (I.6) we have $1-\beta_i p_i=D_i/\hat{D}_i\geqq 1$, so no instability occurs in the use of (I.7).

Updating Inverses

Comparing partitions in

$$\text{(I.9)}\qquad \begin{bmatrix} M & M_{.c}\\ M_{r.} & M_{rc}\end{bmatrix}\begin{bmatrix}\bar{R} & R_{.c}\\ R_{r.} & R_{rc}\end{bmatrix}=\begin{bmatrix}\bar{R} & R_{.c}\\ R_{r.} & R_{rc}\end{bmatrix}\begin{bmatrix}M & M_{.c}\\ M_{r.} & M_{rc}\end{bmatrix}=I_{m+1},$$

with $MR=I_m$, we obtain:

1. Add Row and Column:

$$
\begin{aligned}
\text{(I.10)}\qquad\qquad R_{rc}&=(M_{rc}-M_{r.}RM_{.c})^{-1}\\
R_{r.}&=-R_{rc}(M_r R)\\
R_{.c}&=-R_{rc}(RM_{.c})\\
\bar{R}&=R-(RM_{.c})R_r.
\end{aligned}
$$

2. Drop Row and Column:

$$\text{(I.11)}\qquad\qquad\qquad \bar{R}=R-R_{.c}R_{r.}/R_{rc}.$$

From the Shermann–Morrison formula we obtain:

80

3. Change Row:

$$(I.12) \qquad \bar{R} = R - \frac{R_{r.}(\varDelta M_{r.}R)}{1 + (\varDelta M_{r.}R)R_{.r}}$$

4. Change Column:

$$(I.13) \qquad \bar{R} = R - \frac{(R\varDelta M_{.c})R_{.c}}{1 + R_{.c}(R\varDelta M_{.c})}.$$

APPENDIX II

Here we consider the termination properties of the extended Lemke algorithm applied to the generalized linear complementarity problem (37), and for this purpose write (40) in the form:

$$(II.1) \qquad \delta w = M_0 \delta z_0 + M \delta z + \bar{q}, \quad M_{i0} = \sigma_i \bar{M}_{i0}, \quad \bar{M}_{i0} > 0,$$
$$\sigma_i \delta w_i \geqq 0, \quad \sigma_i \delta z_i \geqq 0, \quad \delta w_i \delta z_i = 0, \quad \text{all} \quad i,$$

where $\bar{q} = M_0 \bar{z}_0 + M \bar{z} + q - \bar{w}$, with bounds $(\bar{w}_i, \bar{z}_i)$ determined by the σ_i and of course $\bar{z}_0 = 0$. We note that initially σ_i is specified for each primal z_i, and (II.1) must be valid for $\sigma_i = \pm 1$ for each dual z_i.

After iteration we have a transformed system:

$$(II.2) \qquad \delta w' = M_0' \delta z_0' + M' \delta z' + \bar{q}'$$
$$\sigma_i' \delta w_i' \geqq 0, \quad \sigma_i' \delta z_i' \geqq 0, \quad \delta w_i' \delta z_i' = 0, \quad \text{all} \quad i,$$

where $w_k' \equiv z_0$, $(z_0', z_k') = (w_k, z_k)$ is the non-basic complementary pair with z_0' the driving variable, and $(w_i', z_i') = (w_i, z_i)$, $i \neq k$, are the remaining complementary pairs. From the properties of the algorithm (z_0', z', w') is a solution of both (II.1) and (II.2), and we have $\bar{q}' = M_0' \bar{z}_0' + M' \bar{z}' + q' - \bar{w}'$ with $\sigma_i \bar{q}_i' \geqq 0$, and $\bar{q}_k' > 0$ unless (II.2) represents the solution of the original problem.

We note that σ_k' is fixed for the non-basic pair, and the blocking variable is uniquely defined as the basic variable of smallest index which first reaches its bound as $\sigma_k' \delta z_0'$ increases. Thus, just as for the basic Lemke algorithm, the extended algorithm may be represented as a graph whose nodes correspond to specified bounds for the non-basic variables, and each node is of degree two with the initial ray infinite $(z_0 \to \infty)$. It follows that the algorithm cannot cycle and must terminate either at a solution or in a second infinite ray, distinct from the initial ray.

Unsuccessful termination implies that $\delta w'$ in (II.2) remains feasible as $\sigma_k' \delta z_0' \to + \infty$ and hence that $\sigma_i' M_{i0}' \sigma_k' \geqq 0$. If w_i' is a primal variable subject to finite upper and lower bounds this is true for $\sigma_i' = \pm 1$ and hence $M_{i0}' = 0$.

Theorem. *If $M \in \mathscr{L}'$ and the extended Lemke algorithm applied to system* (II.1) *terminates unsuccessfully with system* (II.2), *then there exists a scalar $\alpha > 0$ such that*

$$\text{(II.3)} \qquad M'_{k0} = 0, \quad 0 \leqq M'_{kk}\sigma'_k \leqq \alpha, \quad 0 \leqq M'_{ki}\sigma_i \leqq \alpha\sigma'_i M'_{i0}\sigma'_k, \quad i \neq k.$$

Proof. Suppose that (z'_0, z', w') is a solution of (II.1) with $\delta z = 0$. Then the terminal ray corresponds to $\delta z_0 \to \infty$ and hence δw_i has the same sign as M_{i0} for all i. Thus all primal $\bar{z}_i$ are the initially set bounds $(\sigma'_i = \sigma_i)$ and of course all dual $\bar{z}_i$ are zero. But this implies that the terminal ray is the initial ray, which is impossible, so we must have $\delta z \neq 0$.

Now $\delta z' = 0$, $\delta z'_0 > 0$, $\delta w' = M'_0 \delta z'_0$ is a solution of the system

$$\delta w = M_0 \delta z_0 + M \delta z$$

$$\text{(II.4)}$$

$$\sigma'_i \delta w_i \geqq 0, \quad \sigma'_i \delta z_i \geqq 0, \quad \delta w_i \delta z_i = 0, \quad \text{all} \quad i,$$

and hence

$$\text{(II.4a)} \qquad \delta z_i \delta w_i = \delta z_i M_{i0} \delta z_0 + \delta z_i (M\delta z)_i = 0.$$

But $\delta z \neq 0$ and $M \in \mathscr{L}'_1$, so there exists a $\delta z_i \neq 0$ such that $\delta z_i (M\delta z)_i \geqq 0$. If this δz_i is a primal variable then it must have an infinite bound and hence $\sigma'_i = \sigma_i$ (otherwise $\delta z_i = M'_{i0}\delta z'_0 = 0$, a contradiction); if it is a dual variable then (II.4) must hold for $\sigma'_i = \pm 1$ and hence for $\sigma'_i = \sigma_i$. Thus in either case $\delta z_i M_{i0} > 0$, and since $\delta z_0 \geqq 0$ it follows from (II.4a) that $\delta z_0 = 0$, and hence $M'_{k0} = 0$. Thus from (II.4) we have $\delta w = M\delta z$ with $\delta z_i \delta w_i = 0$, all i, and since $\delta z \neq 0$ and $M \in \mathscr{L}'_2$ there exist diagonal matrices $\Lambda \geqq 0$, $\Gamma \geqq 0$ such that

$$\text{(II.5)} \qquad \delta\hat{w} = \Lambda\delta w, \quad \delta\hat{z} = \Gamma\delta z \neq 0, \quad \delta\hat{w} = -M^T\delta\hat{z}.$$

We note that Λ and Γ can be scaled so that all diagonal elements λ_i of Λ satisfy $0 \leqq \lambda_i \leqq 1$.

Defining $\delta\hat{w}_0 = -M_0^T\delta\hat{z}$, we obtain from (II.1), (II.2) and (II.5):

$$[\delta\hat{w}_0, \delta\hat{w}^T] = -\delta\hat{z}^T[M_0, M]$$

$$\text{(11.6)} \qquad [\delta\hat{z}'_k, (\delta\hat{w}')^T] = -[\delta\hat{z}'_1, \ldots, \delta\hat{z}'_{k-1}, \delta\hat{w}_0, \delta\hat{z}'_{k+1}, \ldots, \delta\hat{z}']^T[M'_0, M'],$$

where $(\delta\hat{w}'_i, \delta\hat{z}'_i) = (\delta\hat{w}_i, \delta\hat{z}_i)$, all i, with permutations between basic and non-basic variables corresponding with those associated with (II.2). But $\delta z' = 0$, so $\delta\hat{z}' = \Gamma'\delta z' = 0$ and from (II.6) we have $\delta\hat{w}'_i = -\delta\hat{w}_0 . M'_{ki}$, all i. Using (II.5) this yields

$$\text{(II.7)} \qquad \delta\hat{w}'_k = -\delta\hat{w}_0 M'_{kk} = \lambda'_k \delta z'_0; \quad \delta\hat{w}'_i = -\delta\hat{w}_0 M'_{ki} = \lambda'_i M'_{i0}\delta z'_0, \quad i \neq k.$$

Now $\delta\hat{w}_0 = -M_0^T\delta\hat{z}$, $\delta\hat{z} \neq 0$, and for each $\delta\hat{z}_i \neq 0$ we have $M_{i0}\delta\hat{z}_i = \gamma_i M_{i0}\delta z_i > 0$, so $\delta\hat{w}_0 < 0$. We showed earlier that $M'_{k0} = 0$, and if we define $\alpha = -\sigma'_k \delta z'_0/\delta\hat{w}_0 > 0$, (II.7) gives the remaining relations of (II.3).

Q.E.D.

82

FIXED POINTS, FAIR SHARINGS AND MATHEMATICAL PROGRAMMING

HOANG TUY

(Hanoi, Vietnam)

Fixed point theorems play a prominent role in pure and applied mathematics.

Indeed, solving an equation $Tx=0$, where T is a given mapping from some linear space to itself, amounts to finding a fixed point of the mapping $I+T$, with I being the identity mapping. As is well known, this is the basis for the application of fixed point methods to differential equations, integral equations, and, more generally, operator equations.

On the other hand, in applied mathematics, we have often to deal with systems whose input x and output y are interdependent, in such a way that $y \in f(x)$, $x \in g(y)$ with f, g being some set-valued mappings. For such systems, in an equilibrium state one must have $x \in g(f(x))$, i.e. must be a fixed point of the mapping $F=g \circ f$. Thus, fixed point theorems naturally emerge as necessary tools in various problems of mathematical economics, theory of games, control systems theory, etc.

The purpose of the present paper is to discuss some questions of fixed point methods related to economics and mathematical programming. First, in §1, we shall introduce a general notion of fixed point which contains Brouwer fixed point notion as a special case. Then, in §2, we shall provide a method for finding such a fixed point. This method is a direct generalization of Scarf's combinatorial algorithm ([1], [6]) and is intended to overcome a major difficulty of the latter, which has been noticed by many authors (see e.g. [2], [4]). Finally, in §3, we shall consider the applications of our method to variational inequalities and convex programming. Many of the basic ideas in this paper are essentially a continuation of the ideas developed previously in the work of Scarf, Kuhn, Eaves, etc. (see e.g. [3] for a more complete bibliography).

§1. FIXED POINTS AS FAIR SHARINGS

The classical Brouwer fixed point theorem and its generalization by Kakutani are now of frequent use in nonlinear optimization theory (necessary conditions, stability questions), theory of games (saddle point theorems) and mathematical economics (theory of general equilibrium). In the last years, via the theories of variational inequalities and complementarity problems, fixed point methods have also been introduced in mathematical programming.

We shall show in this section that Brouwer and Kakutani's fixed point theorems can be viewed as special cases of a more general proposition which can be interpreted as a theorem on "fair sharing". Hopefully this will shed some new light on the logical structure that underlies many real equilibrium situations and problems.

1. Formulation of the main theorem

Let $S=[a^1, \ldots, a^n]$ be a closed $(n-1)$ simplex in R^{n-1}, with vertices $a^1, \ldots, a^n$. For every $x \in S$ we shall denote by x_i the barycentric i-th coordinate of x in this simplex $\left(\text{so } x_i \geq 0, \sum_{i=1}^{n} x_i = 1\right)$. Let $F_i = \{x \in S : x_i = 0\}$ be the i-th face of S (the face opposite to vertex a^i) and let $\mathring{F}_i = \{x \in S : x_i = 0, x_j > 0 \text{ for all } j \neq i\}$.

Consider now a symmetric n-ary predicate L over S, i.e. a mapping from the set of all unordered systems of n elements of S to the set {true, false}. A unordered system $U = (u^1, \ldots, u^n)$ of n distinct elements of S is said to be a L-system if $L(U) = L(u^1, \ldots, u^n) =$ true. A subset E of S is said to be a L-set if it contains a L-system. A point $x^* \in S$ is called a *fixed point* of the predicate L if every neighbourhood of x^* in S is a L-set.

We shall be concerned with the conditions under which a given predicate L has a fixed point. Before stating these conditions, let us introduce a convenient terminology and notation. For any two sets of same cardinality, if $U' = (U \backslash \{u\}) \cup \{u'\}$, we shall say that U' obtains from U by the *pivoting* u/u' and shall write $U' = U(u/u')$, or $U \xrightarrow{u/u'} U'$ or simply $U - U'$, if there is no need to precise the pivoting. Here it is not excluded that $u' = u$ in which case $U' = U$.

The conditions we had in view can be formulated as follows:

(i) For every L-system U and for every $u' \in S \backslash U$ there exists a unique $u \in U$ such that $U' = U(u/u')$ is also a L-system.

(ii) The vertex set of S: $(a^1, \ldots, a^n)$ is a L-system, but no proper face of S is a L-set.

(ii') Every system $U = (u^1, \ldots, u^n)$ with $u^i \in \mathring{F}_i$ is a L-system, but no system $U = (u^1, \ldots, u^n)$ with $|U \cap \mathring{F}_i| > 1$ (for some i) is a L-system.

The predicate L is said to be *proper* if it satisfies:

(i) and (ii), or (i) and (ii').

Theorem 1. *Every proper predicate L has a fixed point, which can be computed with any prescribed accuracy. More precisely, given any $\varepsilon > 0$ one can find by a finite procedure a L-set U with $\operatorname{diam} U < \varepsilon$.*

Proof. The first statement follows from the second one. Indeed, if the second statement is true, one can find for every positive integer k a L-system $U_k = (u^{k,1}, \ldots, u^{k,n})$ with $\operatorname{diam} U_k < \varepsilon_k$, where $\varepsilon_k \downarrow 0$. Using the compactness of S, one can assume, by taking subsequences if necessary, that $u^{k,1} \to x^* \in S$. Since $\operatorname{diam} U_k \to 0$, we then have $u^{k,i} \to x^*$ for all $i = 1, \ldots, n$. This means that x^* is a fixed point of L.

Thus, it is enough to prove the second statement of the Theorem. We shall do it by pointing out in §2 an algorithm for computing the desired L-system.

Note 1. In many applications, the predicate L of concern is not proper, but one can easily find a proper predicate L' implying L (i.e. such that every L'-system is a L-system). In these cases, Theorem 1 obviously remains valid.

Note 2. If condition (i) holds, the condition (ii′) is equivalent to the following one:

(ii*) Every system $U = (u^1, \ldots, u^n)$ with $u^i \in \mathring{F}_i$ is a L-system; moreover, if U is a L-system, and if $u \in \mathring{F}_i \cap U$ and $u' \in \mathring{F}_i \backslash U$, then $U' = U(u/u')$ is also a L-system.

Indeed, assuming (i) and (ii′) to hold, let U be a L-system and $u \in \mathring{F}_i \cap U$, $u' \in \mathring{F}_i \backslash U$. Then there is an element $v \in U$ such that $U(v/u')$ is a L-system. One must have $v = u$ otherwise $U(v/u')$ would contain two elements $u, u' \in \mathring{F}_i$ contrary to (ii′). Therefore (ii*) holds. Conversely, assuming (i) and (ii*) to hold, let $U = (u^1, \ldots, u^n)$ be a L-system such that $U \cap \mathring{F}_i$ contains two distinct elements u, v. Then for any $u' \in \mathring{F}_i \backslash U$ both $U(u/u')$ and $U(v/u')$ are L-system, contrary to (i).

2. Corollaries: The Brouwer's fixed point theorem

We shall see later (§3) that Theorem 2 can be used to derive easily Kakutani's fixed point theorem. Here let us derive from Theorem 1 two propositions which are known to be equivalent forms of Brouwer's fixed point theorem.

Corollary 1. *Let $S = [a^1, \ldots, a^n]$ be a closed $(n-1)$-simplex whose i-th face is denoted by F_i. If $L_1, \ldots, L_n$ are closed sets such that: 1) $S \subset \bigcup\limits_{i=1}^{n} L_i$; 2) $a^i \in L_i$, $F_i \cap L_i = \emptyset \, (\forall i)$, then there exists in S a point $x^* \in \bigcap\limits_{i=1}^{n} L_i$.*

Proof. Using the hypotheses: $S \subset \bigcup\limits_{i=1}^{n} L_i$, $a^i \in L_i$, we can define a mapping $l: S \to \{1, \ldots, n\}$ such that $l(a^i) = i$ and $l(x) = i$ only if $x \in L_i$. Let L denote the symmetric n-ary predicate over S, such that a set U of n distinct elements of S is a L-system if and only if $l(U)=\{1, \ldots, n\}$. Then it can be easily verified that L fulfils conditions (i) and (ii). Hence, by Theorem 1, there exists in S a point x^*, every neighbourhood of which contains n points $u^i \in L_i$ $(i=1, \ldots, n)$. Since each L_i is closed this implies $x^* \in L_i$ $(i=1, \ldots, n)$, as required.

Corollary 2. *Let $S=[a^1, \ldots, a^n]$ be a closed $(n-1)$ simplex, whose i-th face is denoted by F_i. If $L_1, \ldots, L_n$ are closed sets such that: 1) $S \subset \bigcup_{i=1}^{n} L_i$; 2) $F_i \subset L_i$ $(\forall i)$, then there exists in S a point $x^* \in \bigcap_{i=1}^{n} L_i$.*

Proof. Using the hypothesis $S \subset \bigcup_{i=1}^{n} L_i$ we can define a mapping $l: S \to \{1, \ldots, n\}$ such that $l(x)=i$ implies $x \in L_i$. Let L denote the symmetric n-ary predicate over S, such that a set U of n distinct elements of S is a L-system if and only if $l(U)=\{1, \ldots, n\}$. Then it is readily seen that L fulfils conditions (i) and (ii'). The proof can be completed just in the same way as in the previous case.

Brouwer's fixed point theorem follows easily from any of the preceding corollaries. Indeed, assume for example that Corollary 2 holds and consider a continuous mapping $f: S \to S$. Setting $L_i=\{x \in S : x_i \leqq f_i(x)\}$ ($f_i(x)$ denotes barycentric i-th coordinate of $f(x)$ in S), one can verify that $L_1, \ldots, L_n$ are closed sets satisfying conditions 1) and 2) of Corollary 2. Hence, there is in S a point $x^* \in \bigcap_{i=1}^{n} L_i$, i.e. such that $x_i^* \leqq f_i(x^*)$ for every $i=1, \ldots, n$. Since $\sum_{i=1}^{n} x_i^* = \sum_{i=1}^{n} f_i(x^*)=1$, this implies $x^*=f(x^*)$.

3. *"Economic" interpretation*

Assume that some utility (which may be positive, like a profit, or negative, like a loss) is to be shared among a group of n persons $1, 2, \ldots, n$, and we are looking for a "fair sharing", i.e. a sharing acceptable for everybody (in some sense to be made precise).

Let us represent every sharing by a point x of the simplex $S=[a^1, \ldots, a^n]$, such that the barycentric i-coordinate x_i of x is equal to the share of person i in this sharing.

Imagine that the same utility is to be shared not just once, but n times, successively. Then each sequence $U=(u^1, \ldots, u^n)$ of such sharings may be fair or not, and we can describe the family of all fair systems U by giving a symmetric n-ary predicate L over S, such that a (unordered) set $U=(u^1, \ldots, u^n)$ of n sharings $u^1, \ldots, u^n$ is fair if and only if U is a L-system. Having thus defined the notion of fairness for systems of n sharings, we can accept as fair any sharing x^*, in every neighbourhood of which, as small as we like, there are n sharings $u^1, \ldots, u^n$ forming a fair system. In other words, a fair sharing is a fixed point of the predicate L, in the sense defined above.

86

Theorem 1 can thus be interpreted as pointing out the conditions under which a fair sharing exists.

Condition (i) expresses a rather common feature of many real situations. Assume, for example, that each person i, has chosen a set $L_i \subset S$, which represents the set of all sharings good for him, and that a set U of n sharings is regarded as fair if and only if for each $i=1, \ldots, n$ the set U contains just one $u^i \in L_i$. In that case condition (i) reduces to

$$(i^*) \qquad \bigcup_{i=1}^{n} L_i = S, \quad L_i \cap L_j = \varphi \text{ for } i \neq j$$

which means that: every sharing is good for exactly one person in the group.

Indeed, if (i^*) holds then for every fair system $U=(u^1, \ldots, u^n)$ and every $v \in S \backslash U$, we have $u^i \in L_i$ $(i=1, \ldots,) n$ and $v \in L_{i_0}$ for just one i_0 hence u^{i_0} is the only element of U for which $U(u^{i_0}/v)$ is again a fair system. Therefore (i^*) implies (i). Conversely, if (i) holds, then it is easily seen that each $v \in S$ belongs to exactly one L_i, i.e. (i^*) holds.

So, in this particular case, condition (i) amounts to requiring simply that each possible sharing is good for just one person. Perhaps the requirement $L_i \cap L_j = \emptyset (i \neq j)$ is a too stringent one, however, if $\bigcup_{i=1}^{n} L_i = S$, one can always find subsets $L_i' \subset L_i$ such that $\bigcup_{i=1}^{n} L_i' = S$ and $L_i' \cap L_j' = \emptyset$ $(i \neq j)$, and by this way it is possible to define a proper predicate L' implying L. As was pointed out in Note 1, this will suffice to ensure the existence of a fair sharing.

Conditions (ii) and (ii') concern the cases when the utility to be shared is a positive or a negative one, respectively.

If the utility is positive, then $a^i = (0, \ldots, \underset{i}{1}, \ldots, 0)$ represents for person i the best sharing, while every $x \in F_i$ represents for him the worst one, so it is natural that the system $(a^1, \ldots, a^n)$ is fair, and that any system U contained in a face F_i cannot be fair: this is just what is stated in Condition (ii). On the other hand, if the utility is negative, then a^i represents for person i the worst sharing, while every $x \in F_i$ represents for him the best one, so it is reasonable to assume, as was expressed in Condition (ii'), that any system $(u^1, \ldots, u^n)$ with $u^i \in \mathring{F}_i$ is fair, but any system $(u^1, \ldots, u^n)$ having more than one element in some face F_i (i.e. such that some person i gets the best sharing at least twice) is not fair.

Thus, the conditions under which there is a fair sharing according to Theorem 2, appear to be quite natural.

It is worth while noticing also that, in the context of the interpretation given above, Corollaries 1 and 2 could be restated as follows. Assume that each person i has chosen a closed set $L_i \subset S$ representing the collections of all sharings acceptable for him. Then, for the case of a positive utility, Corollary 1 says that a fair sharing always exists, provided every possible sharing is acceptable for at least someone is the group (Condition 1) and every sharing in which a person i has whole part is acceptable for him, while every sharing in which he has no part is not acceptable (Condition 2). For the case of a negative utility, Corollary 2 says that a fair sharing always exists, provided every possible sharing is acceptable for at least someone in the group (Condition 1) and every sharing in which a person i has no part is acceptable for him (Condition 2).

As we see, a common sense underlies Brouwer fixed point theorem. Perhaps it is this common sense that makes fixed point methods so useful in the study of various equilibrium models.

§2. AN ALGORITHM FOR FINDING THE "FAIR SHARING"

To complete the proof of Theorem 1, we proceed now to describe a finite algorithm for finding, for a given proper predicate L and a given number $\varepsilon > 0$, a L-set U with diam $U < \varepsilon$.

We may restrict ourselves to the case where the predicate L satisfies Conditions (i) and (ii'), because the case where L satisfies (i) and (ii) can be reduced to the previous one.

The main idea of the method is to take a finite grid Q, fine enough, of the set S and to define two families of subsets of cardinality $n+1$ of the set $\bar{Q} = Q \cup \{\bar{0}, \bar{1}, \ldots, \bar{n}\}$, where $\bar{0}, \bar{1}, \ldots, \bar{n}$ are arbitrary elements not belonging to Q. These subsets will be called "primitive sets" and "complete sets", respectively. To each subset V of $\bar{Q}$ is associated a simplex $\Delta(V) \subset S$, which will be a L-set if V is complete, and will have a diameter less then ε if, moreover, V is primitive. The problem reduces to finding a set V which is both primitive and complete (so that $\Delta(V)$ will be the desired L-set). This can be done by a so-called "pivotal" procedure, consisting of a finite number of operations similar to the pivot steps in the simplex method for solving linear programs.

1. Primitive sets

Consider $n+1$ arbitrary orderings $\underset{i}{\leqq}$ $(i=0, 1, \ldots, n)$ on the set Q, and define, additionally, for every $i=0, 1, \ldots, n$ and for every $x \in Q$:

$$\text{(1)} \qquad \bar{i} \underset{i}{<} x \underset{i}{<} \overline{(i+1)} \underset{i}{<} \ldots \underset{i}{<} \bar{n} \underset{i}{<} \bar{0} \underset{i}{<} \bar{1} \underset{i}{<} \ldots \underset{i}{<} \overline{(i-1)}.$$

In this way we obtain $n+1$ orderings (denoted also by $\underset{i}{\leqq}$) on the set $\bar{Q} = Q \cup \{\bar{0}, \bar{1}, \ldots, \bar{n}\}$.

88

A set V of $n+1$ distinct elements of $\bar{Q}$ is said to be *primitive* if $(\forall x \in Q)(\exists i)(\forall v \in V)$ $x \underset{i}{\leqq} v$, or equivalently, if there is no $x \in Q$ such that $(\forall i) x \underset{i}{>} c^i(V)$, where $c^i(V) =$ $= i - \min V$, the minimal element of V in the i-th ordering.

For example, the set $V_0 = (q, \bar{1}, \ldots, \bar{n})$, with $q = 0 - \max Q$, the maximal element of Q in the 0-ordering, is obviously primitive.

For our purpose the most important property of primitive sets is the following.

Lemma 1. *Let V be a primitive set and let v be an element of V. If $(V \setminus \{v\}) \cap Q \neq \emptyset$, there is exactly one $u \in \bar{Q}$ such that $V(v/u)$ is again primitive; otherwise, there is no such u.*

The proof of this Lemma is similar to that of the replacement theorem in Scarf's theory ([1], or [6]).

In the sequel we shall assume the n orderings $i = 1, \ldots, n$ to be such that for all $x, x' \in Q$:

$$(2) \qquad\qquad x \underset{i}{\leqq} x' \Rightarrow x_i \leqq x'_i.$$

For example, one might define, as did Scarf in [5]:

$$x \underset{i}{<} x' \Leftrightarrow (x_i, \ldots, x_n, x_1, \ldots, x_{i-1}) < (x'_i, \ldots, x'_n, x'_1, \ldots, x'_{i-1})$$

in the lexicographical sense.

It follows from the definition that if V is a primitive set, then $V = (c^0(V), \ldots, c^n(V))$. Indeed, if $c^k(V) = c^j(V)$ for some $k \neq j$, there would exist an element $v \in V$ such that $v \underset{i}{>} c^i(V)$ for every i; this would imply, in view of (1), $v \in Q$ and would contradict the definition of a primitive set.

Let us associate to each subset V of $\bar{Q}$ a simplex $\Delta(V)$, which is the smallest simplex, positively homothetic to S, containing all elements of $V \cap Q$, and meeting every $\mathring{F}$, such that $i > 0$, $\bar{i} \in V$. If V is primitive, then $\Delta(V)$ is just the simplex, positively homothetic to S, with the i-th face passing through $c^i(V)$ if $c^i(V) \in Q$ or contained in the i-th face of S if $c^i(V) = \bar{i}$ (note that if $c^i(V) \notin Q$ i.e. if $c^i(V) \in \{\bar{0}, \bar{1}, \ldots, \bar{n}\}$, then, in view of (1), necessarily $c^i(V) = \bar{i}$).

Lemma 2. *If V is a primitive set containing $\bar{0}$ and if Q is a η-net of S, then diam $\Delta(V) < N \cdot \eta$, where N is some positive constant (depending only upon S).*

(By a η-net of S we mean a set Q such that every ball of radius η around any point of S contains at least one point of Q; since S is compact, such a η-net exists for any given $\eta > 0$.)

Proof. If V is a primitive set containing $\bar{0}$, then it is easily seen that the simplex $\Delta(V)$ contains no point of Q in its interior. Therefore, the greatest ball contained in

$\Delta(V)$ contains no point of Q and hence, must have a radius less than η. Let N denote the ratio of the diameter of S to the radius of the gratest ball contained in S. Then, obviously, diam $\Delta(V) < N \cdot \eta$.

2. Complete sets

A set V of $n+1$ distinct elements of $\bar{Q}$ is said to be *complete* if V contains the element $\bar{0}$ and if every set obtained from $V \backslash \{\bar{0}\}$ by replacing each $v \in V$ of the form $v = \bar{i}$ $(i = 1, \ldots, n)$ by an arbitrary element of $\mathring{F}_i$ is a L-system. Clearly, if V is complete, then $\Delta(V)$ is a L-set.

For example. the set $V_0^* = (\bar{0}, \bar{1}, \ldots, \bar{n})$ is obviously complete [see condition (ii*)]. This complete set and the primitive set $V_0 = (q, \bar{1}, \ldots, \bar{n})$ with $q = 0 - \max Q$ will play a special role in the algorithm to be described below.

Lemma 3. *For every complete set V and for every element $v' \in \bar{Q} \backslash V$ there is exactly one $v \in V$ such that $V(v/v')$ is again complete.*

Proof. Denote by $\gamma : \bar{Q} \to S$ and arbitrary injective mapping such that $\gamma(x) = x$ for every $x \in Q$ and $\gamma(\bar{i}) \in \mathring{F}_i$ for every $i = 1, \ldots, n$. It is easy to see that a set V containing $\bar{0}$ $(V \subset \bar{Q}, |V| = n+1)$ is complete if and only if $U = \gamma(V \backslash \{\bar{0}\})$ is a L-system. Indeed, suppose U to be a L-system and consider an arbitrary set $\tilde{U}$ obtained from $V \backslash \{\bar{0}\}$ by replacing each $v \in V$ of form $v = \bar{i}$ by an element $\tilde{v} \in \mathring{F}_i$. For each $v \in V$ of form $v = \bar{i}$ we have $\gamma(v) \in \mathring{F}_i$ and so, by (ii*) the pivoting $\gamma(v)/\tilde{v}$ performed on U yields again a L-system. Since $\tilde{U}$ obtains from U by a finite sequence of pivotings of the form just described, we see that $\tilde{U}$ is a L-system. Hence, if U is a L-system, then V is complete. The converse being obvious, our assertion is established.

Let now V be any complete, and let $v' \in \bar{Q} \backslash V$. Then $v' \neq \bar{0}$ (because $\bar{0} \in V$), $u' = \gamma(v') \in$ $\in S \backslash U$, where $U = \gamma(V \backslash \{\bar{0}\})$. Since U is a L-system, there is, by virtue of property (i), just one $u \in U$ such that $U(u/u')$ is a L-system. Hence, $v = \gamma^{-1}(u)$ is the only element of V such that $V(v/v')$ is complete. This proves the Lemma.

3. The algorithm

We shall say that (V, V^*) is a *p.c. couple* if V is primitive, while V^* is complete $V^* = V(u/\bar{0})$ for some $u \in V$. If for two p.c. couples $(V, V^*) \neq (V', V'^*)$ we have and

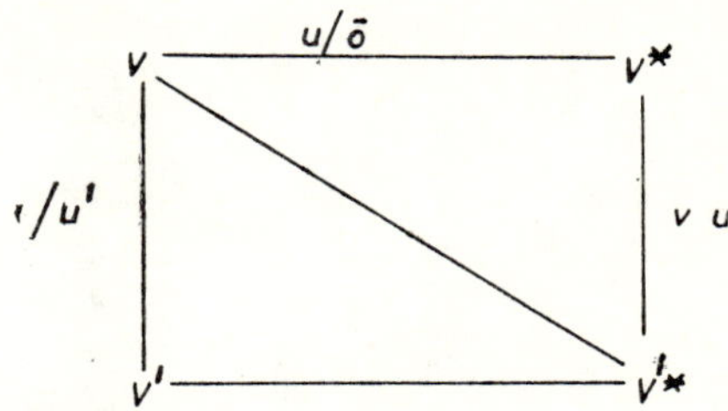

we shall say that (V', V'^*) *follows* (V, V^*) *by* u.

Lemma 4. *For every p.c. couple (V, V^*) such that V is not complete and $V \neq V^0$ or $V^* = V_0^*$ there is exactly one p.c. couple following it.*

Proof. Assume that (V, V^*) is a p.c. couple and V is not complete. We have $V^* = = V(u/\bar{0})$. Then $u \neq \bar{0}$, so $u \notin V^*$ and hence, by Lemma 3, there is a uniquely determined $v \in V^*$ such that $V'^* = V^*(v/u)$ is complete. We cannot have $v = \bar{0}$, since then $V = = V'^*(\bar{0}/v) = V'^*$, i.e. V would be complete. Thus $v \neq \bar{0}$ and hence $v \in V$. Since $V'^* = = V(v/\bar{0})$ and $v \neq \bar{0}$, if $V \setminus \{v\} \subset \{\bar{0}, \bar{1}, \ldots, \bar{n}\}$, then necessarily $V \setminus \{v\} = \{\bar{1}, \ldots, \bar{n}\}$, and so $V'^* = V_0^*$ (hence $V^* \neq V_0^*$) and, by Lemma 1, $V = V_0$. Therefore, if $V \neq V_0$ or $V^* = V_0^*$, then $(V \setminus \{v\}) \cap Q \neq \emptyset$ and by Lemma 1 there is a uniquely determined u' such that $V' = V(v/u')$ is primitive. Obviously, $V'^* = V'(u'/\bar{0})$ and $V' \neq V$, $V'^* \neq V^*$, so the Lemma is proved.

Theorem 2. *There exists a uniquely determined sequence of distinct p.c. couples*

$$(*) \qquad (V_0, V_0^*), (V_1, V_1^*), \ldots, (V_m, V_m^*)$$

such that (V_{i+1}, V_{i+1}^*) follows (V_i, V_i^*) $(i = 0, 1, \ldots, m-1)$ and V_m is complete, while no V_i $(i = 0, 1, \ldots, m-1)$ is complete.

Proof. Since (V_0, V_0^*) is a p.c. couple and V_0 is not complete (because $\bar{0} \notin V_0$), it follows from Lemma 4 that there exists a uniquely determined p.c. couple (V_1, V_1^*) following (V_0, V_0^*). We now observe that in any sequence of successive p.c. couples of form $(*)$ all V_i are distinct. Indeed, assuming the contrary, let k denote the smallest integer such that $V_k = V_h$ for some $h < k$. Since $V_k^* = V_h(u^k/\bar{0})$, $V_h^* = V_h(u^h/\bar{0})$ we have $V_k^* = V_h^*(u^k/u^h)$. ($u^k \neq u^h$ as can easily be seen.) But $V_{h+1}^* = V_h^*(v^{h+1}/u^h)$ for some v^{h+1}, hence, by Lemma 3, $u^k = v^{h+1}$, $V_k^* = V_{h+1}^*$. The latter equality implies $h+1 < k-1$, since $V_i^* \neq V_{i+1}^*$ for every i. On the other hand, $V_{h+1} = V_h(v^{h+1}/u^{h+1})$, whereas $V_{k-1} = V_k(u^k/v^k)$, and since $V_k = V_h$, $u^k = v^{h+1}$, it follows from Lemma 1 that $V_{k-1} = V_{h+1}$. This contradicts the definition of k, because $h+1 < k-1$. Therefore, all V_i are distinct. In particular, $V_i \neq V_0$ for every $i > 0$, and so, by Lemma 4, if the sequence $(*)$ has been constructed up to (V_i, V_i^*) and if V_i is not complete, then the sequence can be continued in a unique way to (V_{i+1}, V_{i+1}^*). Since there are only finitely many primitive sets (Q being finite), the sequence must terminate at some (V_m, V_m^*) with V_m complete.

Theorem 2 provides the following algorithm for solving our problem.

Algorithm. Take a grid $Q \subset \operatorname{int} S$, which is a η-net of S, with $\eta < \dfrac{\varepsilon}{N}$ (N-being the constant mentioned in Lemma 2). Start from the p.c. couple (V_0, V_0^*), where $V_0 = (q, \bar{1}, \ldots, \bar{n})$ with $q = 0 - \max Q$, $V_0^* = (\bar{0}, \bar{1}, \ldots, \bar{n})$. At step k, one has a p.c. couple (V_k, V_k^*), such that $V_k^* = V_k(u^k/\bar{0})$. If V_k is not complete. determine the following p.c. couple (V_{k+1}, V_{k+1}^*) by the rule: $V_{k+1}^* = V_k^*(v^k/u^k)$, where v^k is determined by u^k according to Lemma 3; then $V_{k+1} = V_k(v^k/u^{k+1})$, where u^{k+1} is determined by v^k according to Lemma 1. Otherwise, V_k is complete and the procedure terminates: since V_k is complete, the set $U = \Delta(V_k)$ is a L-set; since V_k is primitive diam $(U) \leq$ diam $(V_k) < N \cdot \eta < \varepsilon$ (Lemma 2).

Note 3. The effectiveness of the algorithm highly depends upon the *choice of the 0-ordering*, in particular, upon the choice of the maximal element of Q in this ordering. In practice, prior information may suggest to seek the L-set in the proximity of some point $x^0 \in \operatorname{int} S$. In this case x^0 should be included in the grid Q and the 0-ordering should be defined so that $x^0 = q = 0 - \max Q$, and $x' \underset{0}{\leq} x$ only if $|x' - x^0| \geq |x - x^0|$. So the starting primitive set should be $V_0 = (x^0, \bar{1}, \ldots, \bar{n})$.

Note 4. If the 0-ordering on Q is taken to be the same as one of the i-orderings ($i = 1, \ldots, n$), i.e. such that $x' \underset{0}{\leq} x$ implies $x_i' \leq x_i$, then the above algorithm, as applied to the Brouwer fixed point problem, will coincide with the algorithm of Scarf ([1] or [6]). In this respect the above algorithm can be considered as an extension of the algorithm of Scarf. There is, however, an essential difference between the two approaches: whereas in Scarf's one the procedure always begins near to a vertex of the simplex S (which constitutes, as is well known, a serious difficulty of this method), our approach allows the procedure to start from *any* point one likes in the region where the fixed point is expected to be. This improvement is possible owing to the generalization of the replacement therem in Scarf's theory, provided by Lemma 1. On the basis of this crucial lemma a 0-ordering has been introduced on the set Q, allowing a large freedom in the choice of the starting primitive set V_0.

§3. APPLICATIONS TO VARIATIONAL INEQUALITIES AND CONVEX PROGRAMMING

We shall consider in this section some applications of the previous results.

1. A particular class of proper predicates

Let us first prove a lemma pointing out an important class of proper predicates.

As before, $S=[a^1, \ldots, a^n]$ denotes a closed $(n-1)$-simplex with vertices $a^1, \ldots, a^n$; $\mathring{F}_i$ denotes the relative interior of the face of S opposite to a^i.

Lemma 5. *Let b be an interior point of S, $B_i S \to R^{n-1}$ a mapping such that $B(x)=a^i$ for every $x \in \mathring{F}_i$. Let L be a symmetric n-ary predicate over S, such that a set U of n distinct elements of S is a L-system if and only if the vectors $\{B(u), u \in U\}$ span a $(n-1)$-simplex containing b. Then L is implied by a proper predicate over S (and hence, by Theorem 1, has a fixed point).*

Proof. To each $x \in R^{n-1}$ associate $x' = \begin{pmatrix} x \\ 1 \end{pmatrix} \in R^n$. Then we have a vector $b' \in R^n$ and a mapping $B' : S \to R^n$ such that $\{B(u), u \in U\}$ span a $(n-1)$-simplex containing b if and only if the system

$$(3) \qquad \sum_{u \in U} t(u)B'(u)=b', \quad t(u) \geqq 0 \quad (u \in U)$$

has a unique solution. Define a predicate L' over S such that a set U of n distinct elements of S is a L'-system if and only if for all $\varepsilon > 0$ small enough the system

$$(4) \qquad \sum_{u \in U} t(u)B'(u)=b'+ \sum_{i=1}^{n} \varepsilon^i(a^i)', \quad t(u) > 0 \quad (u \in U)$$

has a unique solution. Then L' obviously implies L and it is easy to see that L' is proper.

Indeed, if $u^i \in \mathring{F}_i$ $(i=1, \ldots, n)$, then $U=(u^1, \ldots, u^n)$ is a L'-set since $B(u^i)=a^i$ and b is by hypothesis an interior point of S. Moreover, if $|U \cap \mathring{F}_i| > 1$, then $B'(U)$ has at least two equal elements and so $B'(U)$ cannot be a set of n independent vectors in R^n, i.e. U cannot be a L'-system. Thus L' satisfies condition (ii'). On the other hand, if U is a L'-system and $v \notin U$ then the system

$$(5) \qquad \begin{aligned} \sum_{u \in U} t(u)B'(u)+t(v)B'(v)=b'+ \sum_{i=1}^{n} \varepsilon^i(a^i)' \\ t(u) \geqq 0 \quad (u \in U), \quad t(v) \geqq 0 \end{aligned}$$

admits $\{B'(u), u \in U\}$ as a nondegenerate feasible basis (i.e. (5) has a unique solution, such that $t(v)=0$, $t(u) > 0$ for all $u \in U$). Since (5) implies

$$\sum_{u \in U} t(u)+t(v)=1+ \sum_{i=1}^{n} \varepsilon^i,$$

the set of solutions to (5) is bounded; hence, by a known result of linear programming theory, there is a unique $w \in U$ such that $\{B'(u), u \in U(w/v)\}$ will be a nondegenerate feasible basis for (5), i.e. such that $U(w/v)$ is a L'-system (note the fact, established in linear programming theory, that for all $\varepsilon > 0$ small enough the polytope (5) is nondegenerate). Thus L' satisfies also condition (i), and the lemma is proved.

Consider the following problem which arises in many fields of applied mathematics.

Given a set C in a finite dimensional euclidean space X and a set-valued mapping $f: C \to 2^X$ find a point $x^* \in C$ for which there exists y^* satisfying

$$(6) \qquad y^* \in f(x^*), \quad (\forall x \in C) \quad \langle x - x^*, y^* \rangle \geqq 0$$

where, as usually, $\langle \ldots \rangle$ denotes the inner product.

Such a point x^* is called a solution of the *variational inequality* (6).

For each $x \in X$ let $T(x)$ denote the set of all $y \in X$ such that $(\forall x' \in C) \ \langle x' - x, y \rangle \leqq 0$. Then solving the variational inequality (6) amounts to finding a point $x^* \in C$ satisfying the inclusion

$$(7) \qquad 0 \in f(x^*) + T(x^*).$$

We shall assume that: 1) C is a compact convex set; 2) for every $x \in C$ the set $f(x)$ is nonempty, compact and convex; 3) f is an upper semicontinuous set-valued mapping. Under these assumptions we shall show that a solution x^* to (6) always exists and can be found by using the method presented in the previous section.

Let $X = R^k$. Clearly, without loss of generality, we may assume, additionnally, that C has a nonempty interior and is contained in the interior of the simplex $S = [a^0, a^1, \ldots, a^k]$ with $a^0 = 0 \in R^k$, $a^i =$ the i-th unit vector in R^k. Let $\mathring{F}_i = \{x \in S : x_i = 0, x_j > 0$ for $j \neq i\}$, where $x_i (i = 1, 2, \ldots, k)$ is the i-th coordinate of x and $x_0 = 1 - \sum_{j=1}^{k} x_j$. Let $e = (1/\sqrt{k}, 1/\sqrt{k}, \ldots, 1/\sqrt{k}) \in R^k$.

Consider now a mapping $B : S \to X$ satisfying:

1) $B(x) = -a^i$ for $x \in \mathring{F}_i$ $(i = 1, \ldots, k)$, $B(x) = e$ for $x \in \mathring{F}_0$;
2) $B(x) \in f(x)$ for $x \in C$; $B(x) \in T(x)$ and $|B(x)| = 1$ for $x \in S \backslash C$.

Since $-a^i \in T(x)$ for $x \in \mathring{F}_i$ $(i = 1, \ldots, k)$ and $e \in T(x)$ for $x \in \mathring{F}_0$, conditions 1) and 2) are consistent. Furthermore, the simplex generated by $\{-a^1, -a^2, \ldots, -a^k, e\}$ contains 0 in its interior, and so one can associate with B a $(k+1)$-ary predicate L as described in Lemma 5.

Theorem 3. *Every fixed point of the predicate L just defined is a solution of the variational inequality* (6).

Proof. Let x^* be any fixed point of L (x^* exists, by Lemma 5). Then there is a sequence of L-systems $U_\nu = (u^{0,\nu}, u^{1\nu} t^\nu, \ldots, u^{k,\nu}) \subset S$ such that $u^{i\nu} \to x^*$ as $\nu \to \infty$ $(i = 0, 1, \ldots, k)$. Denote by N_ν the set of all i for which $u^{i\nu} \in C$.

By definition of L-systems, there exist, for each v, numbers t_{iv} satisfying

$$t_{iv} \geqq 0, \quad \sum_{i=0}^{k} t_{iv} = 1, \quad \sum_{i=0}^{k} t_{iv} B(u^{iv}) = 0.$$

Since C is compact and f is upper semicontinuous, $f(C)$ is compact and hence, the set $B(S)$ is bounded. We may then assume, by taking subsequences if necessary,

$$N_v = N \quad \text{for all} \quad v;$$
$$t_{iv} \to t_{i*}, \quad B(u^{iv}) \to v^{i*} \quad \text{as} \quad v \to \infty.$$

Obviously,

$$(8) \qquad t_{i*} \geqq 0, \quad \sum_{i=0}^{k} t_{i*} = 1, \quad \sum_{i=0}^{k} t_{i*} v^{i*} = 0.$$

We contend that $\Theta = \sum_{i \in N} t_{i*} > 0$. Indeed, if $\Theta = 0$, then we have from (8): $\sum_{i \notin N} t_{i*} v^{i*} = 0$ and hence,

$$(9) \qquad (\forall x \in C) \quad \sum_{i \notin N} t_{i*} \langle x - x^*, v^{i*} \rangle = 0.$$

On the other hand, the set-valued mapping $x \mapsto T(x)$ being closed (as can be easily proved), it follows that $v^{i*} \in T(x^*)$ for all $i \notin N$, i.e.

$$(\forall x \in C) \quad \langle x - x^*, v^{i*} \rangle \leqq 0 \quad (i \notin N).$$

Hence, for every $i \notin N$,

$$(\forall x \in C) \quad \langle x - x^*, t_{i*} v^{i*} \rangle = 0.$$

But, the set C having a nonempty interior can be contained in none of the manifolds $\langle x - x^*, t_{i*} v^{i*} \rangle = 0$, unless $t_{i*} v^{i*} = 0$ for all $i \notin N$. Since the numbers t_{i*} $(i \notin N)$ sum up to one, there must be at least one $i_0 \notin N$ such that $t_{i_0} > 0$. Then $v^{i_0 *} = 0$, which conflicts with the fact $v^{i_0 *} = \lim_{v \to \infty} B(u^{i_0 v}) \, |B(u^{i_0 v})| = 1$.

Thus we have proved that $\Theta > 0$. Let now

$$y^* = \sum_{i \in N} (t_{i*}/\delta) v^{i*}, \quad z^* = \sum_{i \notin N} (t_{i*}/\Theta) v^{i*},$$

so that, from (8), $0 = y^* + z^*$. Since, for $i \in N$, $u^{iv} \in C$, $u^{iv} \to x^*$, $B(u^{iv}) \in f(u^{iv})$, $B(u^{iv}) \to v^{i*}$, it follows from the closedness of the set C and the mapping f that $x^* \in C$, $v^{i*} \in f(x^*)$ for $i \in N$ and hence, $y^* \in f(x^*)$, by the convexity of $f(x^*)$. Furthermore, for $i \notin N$, the relations $u^{iv} \to x^*$, $B(u^{iv}) \to v^{i*}$, $B(u^{iv}) \in T(u^{iv})$ imply, in view of the closedness of T, that $v^{i*} \in T(x^*)$. Hence, $z^* \in T(x^*)$, because $T(x^*)$ is a convex cone. Therefore,

$$0 \in f(x^*) + T(x^*),$$

as was to be proved.

Note 5. The implementation of the procedure described above requires an effective method for constructing at each point $u \in S \backslash C$ a vector $B(u) \in T(u)$, i.e. a vector satisfying

$$(\forall x \in C) \quad \langle x - u, B(u) \rangle \leqq 0.$$

But this is an easy task, at least in usual cases. Indeed, assume C to be given by a system of inequalities of the form

$$(10) \qquad\qquad g_i(x) \leqq 0, \quad i = 1, 2, \ldots, m,$$

where each g_i is a continuous convex function. Then $u \notin C$ implies $g_i(u) > 0$ for some i. Let $\partial g_i(u)$ be the subdifferential of g_i at point u (as it is known, $\partial g_i(u)$ is nonempty) and let $B(u)$ be an arbitrary element of the set $\partial g_i(u)$. Since $x \in C$ only if $g_i(x) \leqq 0$, we have, for all $x \in C$:

$$\langle x - u, B(u) \rangle \leqq g_i(x) - g_i(u) < 0.$$

Note 6. The previous theorem can be used to derive easily *Kakutani's fixed point theorem*. Indeed, if $F : C \to 2^C$ is a set-valued mapping satisfying all conditions in Kakutani's theorem, then the mapping $f(x) = x - F(x)$ will satisfy all conditions in Theorem 3. Applying the latter theorem yields a point $x^* \in C$ such that, for some $z^* \in F(x^*)$, $(\forall x \in C) \ \langle x - x^*, x^* - z^* \rangle \geqq 0$ hence, in particular, $\langle z^* - x^*, x^* - z^* \rangle \geqq 0$, which implies $x^* = z^* \in F(x^*)$.

3. Fixed points and convex programming

To conclude the paper, let us consider the *convex programming* problem

$$(11) \qquad\qquad \min F(x) : x \in C,$$

where C is a compact convex set in R^k, $F(x)$ is a continuous convex function defined in some open set $C' \supset C$.

Under these hypotheses the subdifferential $\partial F(x)$ of F at every $x \in C$ is a nonempty convex set and, as can be easily shown, the set-valued mapping $x \mapsto \partial F(x)$ is upper semicontinuous. Therefore, the previous algorithm can be applied to find a point $x^* \in C$ solving the variational inequality

$$y^* \in F(x^*), \quad (\forall x \in C) \quad \langle x - x^*, y^* \rangle \geqq 0.$$

Since $F(x) - F(x^*) \geqq \langle x - x^*, y^* \rangle \geqq 0$ for every $x \in C$, it follows that x^* is an optimal solution of (11).

It should be pointed out that, in the case where C is given by a system of inequalities of the form (10), this algorithm does not require the functions F and g_i to be differentiable.

96

REFERENCES

[1] Arrow, K. J. and Hahn, H. H.: *General Competitive Analysis*. Oliver and Boyd, Edinburgh, 1971.

[2] Eaves, B. C.: Homotopies for Computation of Fixed Points, *Mathematical Programming*, 3 (1972), 1–22.

[3] Eaves, B. C.: A Short Course in Solving Equations with PL Homotopies, *SIAM–AMS Proceedings*, 9 (1976), 73–143.

[4] Kuhn, H. W. and Mac Kinnon, J. G.: Sandwich Methods for Finding Fixed Points. *Journal of Optimization Theory and Applications*, 17 (1975), 189–204.

[5] Scarf, H. E.: The Approximation of Fixed Points of a Continuous Mapping. *SIAM Journal on Appl. Math.* 15 (1967), 1323–1343.

[6] Scarf, H. E. with the collaboration of Hansen, T. *The Computation of Economic Equilibrium.* Yale University Press, New Haven, 1973.

PART 5

STOCHASTIC AND DYNAMIC PROGRAMMING

EXPERIENCE IN STOCHASTIC PROGRAMMING MODELS

J. DUPAČOVÁ

(Prague, Czechoslovakia)

When solving a real stochastic program we are aware of the fact that a suitable model is to be chosen and an efficient algorithm is to be developed. As a rule, complete knowledge of the distribution of random coefficients is supposed.

The well-known dependence of the analytical form and the optimal solution of the deterministic equivalent on the assumed type of distribution together with the fact that the assumption of complete knowledge of the joint distribution is not fulfilled in many practical situations has led to various game-theoretical approaches, in which the assumption of known distribution has been weakened.

Let us consider a two-stage stochastic program

$$(1) \qquad \text{maximize } E\{g(x; A, b, c)\} \quad \text{on a set} \quad X \subset E_n.$$

Here, $g(x; A, b, c) = c^T x - \varphi(x; A, b)$, where $A(m, n)$, $b(m, 1)$, $c(n, 1)$ are random matrices and the violation of constraints $Ax = b$ (in the original linear program) is penalized by a penalty function $\varphi(x; A, b)$.

Denote by $R \subset E_{mn+m+n}$ the set of all possible realizations of the random coefficients in (1) and consider the two-person zero-sum game in the normal form

$$(2) \qquad G_1 = (X, R, g).$$

Obviously, pure optimal strategies of the first player in G_1 are of the greatest interest.

Let $\mathcal{R}$ denote the set of all mixed strategies ϱ of the second player (the Nature). Consider the following three possibilities:

(i) The mixed strategy $\varrho_0 \in \mathcal{R}$ used by the second player is known. Then we have a common two-stage stochastic program

$$\text{maximize } \int [c^T x - \varphi(x; A, b)] \, d\varrho_0 \quad \text{on the set } X.$$

The same applies to the case when the first player does not know the mixed strategy ϱ_0 but he is able to estimate it on the basis of past data or additional observations (e.g. in cases of hydrological or technological data). If additional measurements are needed, their cost should be considered and, eventually, incorporated with the model.

(ii) Suppose that our knowledge of the mixed strategy used by the second player is not complete but limited to the fact that ϱ belongs to a set $\mathcal{R}_0 \subset \mathcal{R}$. Denote by $\mathcal{F}$ the set of distribution functions corresponding to $\mathcal{R}_0$ and consider the following two-person zero-sum game in the normal form ([9])

$$(3) \qquad\qquad G_2 = (X, \mathcal{F}, H)$$

where for $x \in X$, $F \in \mathcal{F}$, the pay-off function

$$H(x, F) = E_F\{c^T x - \varphi(x; A, b)\}.$$

The resulting deterministic program

$$(4) \qquad \text{maximize } \inf_{F \in \mathcal{F}} E_F\{c^T x - \varphi(x; A, b)\} \quad \text{on the set } X$$

has as a rule a rather simple structure in comparison with the deterministic equivalents corresponding to given continuous distribution functions. The reason is that "the worst" distribution function F^* is often a discrete one. In the sequel, the solution of (4) will be called *the minimax solution* of the stochastic program (1).

Even in the considered case, the minimax solution depends on the choice of the set $\mathcal{F}$ that is based on our *a priori experience*. Selected results concerning the analytical form of (4) for special choices of $\mathcal{F}$ and φ are listed in Table 1 (for detailed discussion see [3], [4]).

Anyway, minimax solutions (together with maximax ones, eventually) provide easily computable boundaries for the optimal value of the objective function:

$$\max_{x \in X} \min_{F \in \mathcal{F}} E_F\{c^T x - \varphi(x; A, b)\} \leq \max_{x \in X} E_F\{c^T x - \varphi(x; A, b)\} \leq$$

$$(5)$$

$$\leq \max_{x \in X} \max_{F \in \mathcal{F}} E_F\{c^T x - \varphi(x; A, b)\} \quad \text{for every } F \in \mathcal{F}.$$

If the penalty function $\varphi(x; A, b)$ is convex in A, b and the set $\mathcal{F}$ contains distribution functions that have, i.e., fixed mean values EA, Eb, Ec then in addition to (5)

$$(6) \qquad \max_{x \in X} \max_{F \in \mathcal{F}} E_F\{c^T x - \varphi(x; A, b)\} \leq \max_{x \in X} \{(Ec)^T x - \varphi(x; EA, Eb)\}$$

according to Jensen's inequality.

100

Table 1

$\mathcal{F}$	$\varphi(x, A, b)$	$\displaystyle \inf_{F \in \mathcal{F}} E_F\{c^T x - \varphi(x, A, b)\}$
$\alpha'_{ij} \leqq a_{ij} \leqq \alpha''_{ij}$ $\beta'_i \leqq b_i \leqq \beta''_i$ $\gamma'_j \leqq c_j \leqq \gamma''_j$ $1 \leqq i \leqq m,\ 1 \leqq j \leqq n$	$\psi\left(\left(\sum_{j=1}^{n} a_{ij} x_j - b_i\right)^+,\ 1 \leqq i \leqq m\right)$ ψ nondecreasing	$\sum_{j=1}^{n} \gamma_j x_j + \psi\left(\left(\sum_{j=1}^{n} \alpha''_{ij} x_j - \beta'_i\right)^+,\ 1 \leqq i \leqq m\right)$
$\beta'_i \leqq b_i \leqq \beta''_i$ $Eb_i = \beta_i,\ \beta'_i < \beta_i < \beta''_i$ $Ec_j = \gamma_j$ $1 \leqq i \leqq m,\ 1 \leqq j \leqq n$	$\sum_{i=1}^{m} \varphi_i\left(\sum_{j=1}^{n} a_{ij} x_j - b_i\right)$ φ_i convex, $1 \leqq i \leqq m$	$\sum_{j=1}^{n} \gamma_j x_j - \sum_{i=1}^{m}\left[\lambda_i \varphi_i\left(\sum_{j=1}^{n} a_{ij} x_j - \beta'_i\right) + (1-\lambda_i)\varphi_i\left(\sum_{j=1}^{n} a_{ij} x_j - \beta''_i\right)\right],$ where $\lambda_i = \dfrac{\beta''_i - \beta_i}{\beta''_i - \beta'_i},\quad 1 \leqq i \leqq m$
$Eb_i = \beta_i$ $\mathrm{var}\, b_i = \sigma_i^2 > 0,\ 1 \leqq i \leqq m$ $Ec = \gamma_j,\ 1 \leqq j \leqq n$	$\sum_{i=1}^{n} q_i\left(\sum_{j=1}^{n} a_{ij} x_j - b_i\right)^+$ $q_i \geqq 0,\ 1 \leqq i \leqq m$	$\sum_{j=1}^{n} \gamma_j x_j - \sum_{i=1}^{m} \frac{q_i}{2}\left\{\sqrt{\left(\sum_{j=1}^{n} a_{ij} x_j - \beta_i\right)^2 + \sigma_i^2} + \left(\sum_{j=1}^{n} a_{ij} x_j - \beta_i\right)\right\}$
$\beta'_i \leqq b_i \leqq \beta''_i$ $Eb_i = \frac{1}{2}(\beta'_i + \beta''_i)$ F_i unimodal, continuous, independent, $1 \leqq i \leqq m$ $Ec_j = \gamma_j;\ 1 \leqq j \leqq n$	$\sum_{i=1}^{m} \varphi_i\left(\sum_{j=1}^{n} a_{ij} x_j - b_i\right)$ φ_i convex, $1 \leqq i \leqq m$	$\sum_{j=1}^{n} \gamma_j x_j - \sum_{i=1}^{m} \int_{\beta'_i}^{\beta''_i} \varphi_i\left(\sum_{j=1}^{n} a_{ij} x_j - b_i\right) \frac{db_i}{\beta''_i - \beta'_i}$

Table 2

Random variable	Range	Mean value
b_1	$\langle 60,200 \rangle$	130
b_2	$\langle 120,480 \rangle$	300
b_3	$\langle 160,600 \rangle$	380
b_4	$\langle 120,600 \rangle$	360

Example. The numerical data contained in Table 2 concern a two-stage stochastic linear program with complete recourse. The four random coefficients b_1, b_2, b_3, b_4 represent the random demand on spare parts for oil pumps. The considered production process is subject to deterministic constraints and the total cost of production process is to be minimized (so that, in case of known distribution, the corresponding wo-stage stochastic program would be $\min_{x \in X} E\{c^Tx + \varphi(x; A, b)\}$). The lower bounds for the random demand are given through fixed contracts. Using past records it was possible to estimate the mean values of random demand; its upper bounds were fixed ad hoc (see [7]). Let $\mathcal{F}_1$ be the set of all distribution functions with mean values and ranges given in Table 2. Then we have

$$\min_{x \in X} \max_{F \in \mathcal{F}_1} E_F\{c^Tx + \varphi(x; A, b)\} = 159{,}991.81$$

$$\min_{x \in X} \min_{F \in \mathcal{F}_1} E_F\{c^Tx + \varphi(x; A, b)\} = 140{,}029.75$$

so that for all $F \in \mathcal{F}_1$ we have the optimal value of (4) within the interval $\langle 140{,}029.7, 159{,}991.81 \rangle$.

If the set of distribution functions with ranges and mean values given in Table 2 contains unimodal continuous distribution functions only, denote it $\mathcal{F}_2$, we have

$$\min_{x \in X} \max_{F \in \mathcal{F}_2} E_F\{c^Tx + \varphi(x; A, b)\} = 150{,}536.74$$

and

$$\min_{x \in X} \inf_{F \in \mathcal{F}_2} E_F\{c^Tx + \varphi(x; A, b)\} = 140{,}029.75$$

so that for all $F \in \mathcal{F}_2$ we have the optimal value of (4) within the interval $\langle 140{,}029.75, 150{,}536.74 \rangle$.

Another description of the set of mixed strategies $\mathcal{R}_0$ stems out of an a priori information concerning the ordering of probabilities of distinct realizations. The class of distributions is described as

$$\mathcal{R}_0 = \left\{ p \in E_K : Bp \geq d, p \geq 0, \sum_{k=1}^{K} p_k = 1 \right\}$$

(see e.g. [2]). The question of the source and completeness of the a priori information appears again.

102

(iii) Let us return to the game $G_1=(X, R, g)$ and suppose that the second player (the Nature) uses a fixed mixed strategy ϱ_0 and that this strategy is unknown to the first player. We shall deal with the situation when the possibility to estimate the unknown mixed strategy ϱ_0 in advance does not exist or is too cumbersome or expensive. On the other hand, we shall assume that the same game G_1 is repeated sequentially and that the mixed strategy ϱ_0 resp. the distribution function of random coefficients in the corresponding stochastic program remains fixed in the whole series. The problem is, *how to use experience based on past observations for to construct an optimal sequence of strategies of the first player*. As we shall see, this sequence can be formed by strategies optimal with respect to empirical distributions constructed on the basis of observations in previous plays, i.e., on the basis of past experience.

Let us describe the model precisely: Assume that

(a) A suitable probability space $[\Omega, \delta, \mu]$ is given.
(b) The space X of strategies of the first player is nonempty and compact.
(c) The pay-off function $g(x; r)$ is bounded and uniformly continuous on $X \times R$.
(d) The second player uses an unknown fixed mixed strategy $\varrho_0 \in \mathcal{R}$.
(e) The random pure strategies $r_n(\omega)$ of the second player in the n-th play (corresponding to the mixed strategy ϱ_0) are independent and observable, $n=1, 2, \ldots$.

Under the above assumptions, the game $G_1=(X, R, g)$ is bounded and separable from the right. Denote

$$H_1(x, \varrho)=\int\limits_R g(x; r) \, d\varrho.$$

Define a sequence of random mixed strategies $\{\varrho_n\}$ of the second player in the following way (see [8, Theorem 6.1]):

$$\varrho_n(E, \omega)=\tilde{\varrho}(E), \quad n=1, \ldots, k$$

$$\varrho_n(E, \omega)=\frac{1}{n-k} \sum_{i=1}^{n-k} \chi_E(r_i(\omega)), \quad E \subset R; \omega \in \Omega,$$
$$n=k+1, \ldots,$$

where k is a fixed integer and $\tilde{\varrho} \in \mathcal{R}$ is a deliberately chosen mixed strategy (e.g., the minimax decision rule can be used) and construct a sequence of random pure strategies $\{x_n(\omega)\}$ of the first player:

$$\max_{x \in X} H_1(x, \varrho_n)=H_1(x_n, \varrho_n).$$

Denote further by x_0 the maximum point of $H_1(x, \varrho_0)$ on the set X. The following assertions hold true:

1. There exists $\Omega_0 \subset \Omega$ with $\mu(\Omega_0)=1$ such that $H_1(x, \varrho_n) \to H_1(x, \varrho_0)$ uniformly in X for all $\omega \in \Omega_0$.

2. The average pay-offs in N games

$$\frac{1}{N} \sum_{k=1}^{N} g\big(x_k(\omega), r_k(\omega)\big) \to \max_{x \in X} H_1(x, \varrho) \quad \text{a.s.}$$

3. $\displaystyle \max_{x \in X} H_1(x, \varrho_n) \to \max_{x \in X} H_1(x, \varrho_0) \quad \text{a.s.}$

4. Let $g(x; r)$ be strictly concave in x; then $x_n(\omega) \to x_0 \quad$ a.s.

These results follow from theory of experience in games [8, Theorems 3.4, 5.5, 6.2]. The assertions 3 and 4 were proved in [5] under more general assumption of ergodicity of $\{r_k(\omega)\}$. The concavity assumption can be obviously weakened. Further possible generalization is to suppose that $\{r_k(\omega)\}$ is a stationary Markov sequence with a finite number of observable stages and an unknown matrix of transition probabilities.

The assumption of observability of $r_k(\omega)$ is of great importance for the above model. In many practical cases, we are given inaccurate (or even deliberately destroyed) data so that the observability assumption is not often met by practice. The simplest situation of this nature occurs when the increasing number of observations is presented in the form of class representation:

Let n be the number of observations and let $\{R_{in}\}$ be an ε_n-net in R. Let r_{in} denote the representative of the class R_{in} and let m_{in} denote the frequency of the class R_{in}, $\sum_i m_{in} = n$. Construct a random pure strategy of the first player $x_n(\omega)$ in the following way:

$$\max_{x \in X} \sum_i m_{in} g(x; r_{in}) = \sum_i m_{in} g(x_n; r_{in}).$$

To prove the convergence of the sequence of the average pay-offs in N plays

$$\frac{1}{N} \sum_{k=1}^{N} g\big(x_k(\omega); r_k(\omega)\big) \to \max_{x \in X} H(x, \varrho_0)$$

with probability 1, it is necessary to suppose that the sequence $\{\varepsilon_n\} \to 0$ ([8, Theorem 6.5]), what is an unrealistic assumption. Further related results were presented in [6].

The model (a)–(e) described above may be of use, e.g., in problems of demand-supply type or in the problem considered in [1] where the random coefficients represent numbers of disposable working hours in a farm.

Furthermore, the above results can serve as a starting point for forming the theoretical basis for using empirical distributions instead of theoretical ones. This idea seems to be quite natural when compared with situations when the assumed known continuous distribution (often found as a result of fitting to the data) has to be for purposes of algorithm replaced by a discrete or piecewise uniform one.

104

REFERENCES

[1] Bočvarova, C. E.: Ob odnoj stochastičeskoj modeli selskochozjajstvennovo proizvodstva, *Ekonomika i matem. metody,* XI (1975), 716–722.

[2] Bühler, W.: Characterization of the Extreme Points of a Class of Special Polyhedra. A Comment to Kofler's Paper "Entscheidungen bei teilweise bekannter Verteilung der Zustände", *Z. für Oper. Res.,* 19 (1975), 131–137.

[3] Dupačová, J.: On minimax decision rule in stochastic linear programming, in: A. Prékopa ed., Studies on Mathematical Programming. Akadémiai Kiadó, Budapest 1979.

[4] Dupačová, J.: Minimaxová úloha stochastického lineárního programování a momentový problém. *Ekonomicko–matematický Obzor,* 13 (1977), 279–307.

[5] Kaňková, V.: Optimum Solution of a Stochastic Optimization Problem with Unknown Parameters, in: Transactions of the 7-th Prague Conference and of the European Meeting of Statisticians 1974. Academia Praha 1978. Vol B, pp. 239–244.

[6] Stahl, J.: On the Two-stage Stochastic Linear Programming Problem, Conferenc on Mathematical Programming, Mátrafüred 1973. Unpublished.

[7] Tloušť, R.: Simulační studie ve stochastickém lineárním programování, Diploma work, Charles University (Prague, 1976).

[8] Winkelbauer, K.: Experience in Games of Strategy and in Statistical Decision, in: J. Kožešník ed., *Transactions of the 1-st Prague Conference* 1956 (Czechoslovak Academy of Sciences Prague 1957) pp. 297–354.

[9] Žáčková, J.: On minimax solutions of stochastic linear programming problems, *Časopis pro pěstování matematiky,* 91 (1966), 423–430.

THE STOCHASTIC QUASIGRADIENT METHODS AND THEIR APPLICATION TO THE STOCHASTIC PROGRAMMING PROBLEMS WITH NON-SMOOTH FUNCTIONS

YU. M. ERMOL'EV

(Kiev, USSR)

Since 1960 in the Institute of Cybernetics of Kiev much work has been done to solve complex extremum problems with non-smooth functions and with a great number of variables. By the present time many methods have been developed and we have much experience with the solution of those problems.

In this report we will deal only with the most difficult problems — the problems of stochastic programming. The basis of this report is the review by the author [1] and the monograph [2].

1. THE STOCHASTIC PROGRAMMING PROBLEM. THE MAIN DIFFICULTIES

A rather general problem of stochastic programming can be formulated as follows: minimize the function

$$(1) \qquad F^0(x)$$

under the conditions

$$(2) \qquad F^i(x) \leqq 0, \quad i = \overline{1, m},$$

$$(3) \qquad x \in X \subseteq R^n,$$

where

$$(4) \qquad F^l(x) = Ef^l(x, \omega), \quad l = \overline{0, m}.$$

The problem (1)–(4) is more difficult than the common nonlinear programming problem. The first main difficulty of this problem is that the objective function $F^0(x)$ and the constraint functions $F^i(x)$, $i = \overline{1, m}$ look like as in (4), i.e. the calculation of the values $F^l(x)$, $l = \overline{0, m}$ at any point x requires the calculation of an integral with respect to the measure $P(d\omega)$. The examples considered below show that this circumstance, as a rule, does not allow us to calculate the precise values of the functions $F^l(x)$. Instead of $F^l(x)$ we can usually calculate only the random values $f^l(x, \omega)$, so the principal object of the development of numerical methods of stochastic programming is to create on the basis grounds of this information a procedure that solves problem (1)–(4). Another essential difficulty of the problem often being found in practice is that the function $F^l(x)$, as a rule, has no continuous derivatives.

Let us analyse in detail the mentioned specificity of problem (1)–(4) via concrete examples.

2. EXAMPLES

(a) The two-stage problem. It has been studied by Dantzig, Madansky, Dempster, Wets, Rockafeller, etc.). The objective function of this problem is the following:

$$(5) \qquad F^0(x)=Ef^0(x, \omega)=(c, x)+E \min_{y \geq 0} (d(\omega), y).$$
$$A(\omega)x+ D(\omega)y=b(\omega).$$

Here $F^0(x)$ is a convex function, often non-smooth, for instance in case the distribution $P(d\omega)$ is not smooth. It is easy to calculate $f^0(x, \omega)$. To calculate $F^0(x)$ it is necessary to find the distribution of

$$(d(\omega), y(x, \omega))= \min_{y \geq 0} (d(\omega), y)$$
$$A(\omega)x+ D(\omega)y=b(\omega)$$

as a function of x and then to take the corresponding integral. This is only rarely possible.

(b) The chance-constraint problems have constraints

$$(6) \qquad P\{g^i(x, \omega) \leq 0\} \geq p_i, \quad i=\overline{1, m}$$

come to the constraints of the type (2), if we assume that

$$f^i(x, \omega)=\begin{cases} P_i-1, & g^i(x, \omega) \leq 0, \\ P_i, & g^i(x, \omega) > 0 \end{cases}$$

Here the functions $F^i(x)$ will be often discontinuous as is shown in Fig. 1,

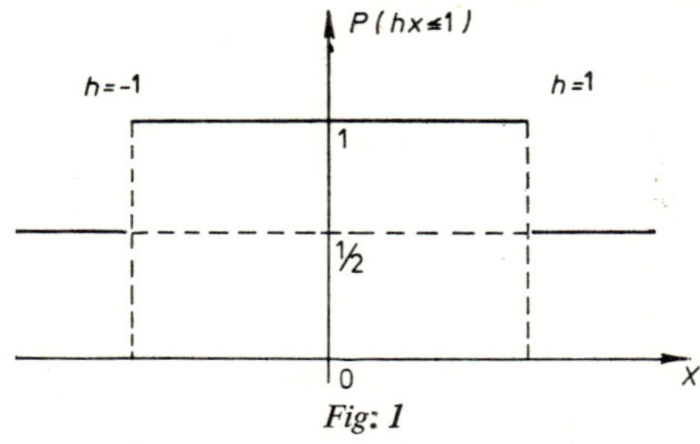

Fig: 1

where x is scalar, h a random value, equal to ± 1 with probability $\frac{1}{2}$.

108

(c) In the simplest stochastic min–max problems the objective function looks as follows

$$(7) \qquad F^0(x)=Ef^0(x,\omega)=E\max_{1\le i\le m}\left(\sum_{j=1}^{m}a_{ij}(\omega)x_j-b_i(\omega)\right).$$

Here it is also easy to calculate $f^0(x,\omega)$, but it is very difficult to deal with $F^0(x)$. The function $F^0(x)$ is a convex function often non-smooth, since under the symbol of mathematical expectation E there is the operation max.

(d) The same difficulties are inherent in the stochastic problems of the theory of optimum control for which at $x=(x(0),\ldots,x(N-1))$ we have

$$(8) \qquad F^0(x)=E\varphi\big(Z(0),\ldots,Z(N),x(0),\ldots,x(N-1)\big),$$

$$(9) \qquad Z(k+1)=g(Z(k),x(k),\omega,k),Z(0)=Z^0,\quad k=0,1,\ldots,N-1.$$

In particular $F^0(x)$ may appear as

$$(10) \qquad F^0(x)=E\max_{k}\|Z(k)-Z^*(k)\|.$$

The stochastic quasigradient (SQG)-methods as shown in [2] allow to solve successfully the above-mentioned problems with the rather arbitrary practically interesting measures.

3. THE GENERAL IDEA OF THE METHODS

Let us consider the problem

$$(11) \qquad F^0(x)=\min$$

$$(12) \qquad F^i(x)\le 0,\quad i=\overline{1,m},$$

$$(13) \qquad x\in X.$$

In the meanwhile we assume that $F^l(x)$, $l=\overline{0,m}$ are convex functions, i.e. we have the inequality:

$$(14) \qquad F^l(Z)-F^l(x)\ge(\hat F^l_x(x,)\,Z-x)$$

where $\hat F^l_x(x)$ is subgradient. The set X is convex. In the SQG-methods the sequence of approximate solutions $x^0,x^1,\ldots$ is formed by means of the random vectors $\xi^l(s)$ and random values $\Theta_l(s)$ in accordance with the relations

$$(15) \qquad E\big(\xi^l(s)/x^0,\ldots,x^s\big)=\hat F^l_x(x^s)+b^l(s),$$

$$(16) \qquad E\big(\Theta_l(s)/x^0,\ldots,x^s\big)=F^l(x^s)+d_l(s)$$

where $b^l(s)$ are some vector, $d_l(s)$ is scalar, depending on $(x^0,\ldots,x^s)$. That is, the

$_r$andom values $\Theta_l(s)$ substitute the functions and the vectors $\xi^l(s)$ substitute the $_s$ubgradients

$$\Theta_l(s)\sim F^l(x^s),\quad \xi^l(s)\sim \hat{F}^l_x(x^s).$$

For further understanding it is important to size up that the random values $\Theta_i(s)$ and vectors $\xi^l(s)$ are easy calculated in the problems of the stochastic programming. For example:

$$F^l(x)=Ef^l(x,\omega),$$

for

$$\Theta_l(s)=f^l(x^s,\omega^s),$$

where ω^s is the result of the random trial of ω with measure $P(.)$ at the s-iteration. For the objective function (5) of the two-stage problem the vector $\xi^0(s)$ is as follows:

$$\xi^0(s)=C-A^T(\omega^s)U^s,$$

where U^s are the dual variables, corresponding to the second-stage optimal plan $y(x^s,\omega^s)$. It can be shown that under this condition

$$E\big(\xi^0(s)/x^s\big)=\hat{F}^0_x(x^s).$$

For the objective function (7) of the stochastic minimax problem the vector $\xi^0(s)= = \big(\xi^0_1(s),\ldots,\xi^0_n(s)\big)$ is calculated by the formula

$$\xi^0_j(s)=a_{i_s,j}(\omega^s),\quad j=\overline{1,n}$$

where i_r is defined by the relation:

$$\sum_{j=1}^{n} a_{i_s,j}(\omega^s)x^s_j-b_{i_s}(\omega^s)=\max_{1\le i\le m}\left(\sum_{j=1}^{n} a_{ij}(\omega^s)x^s_j-b_i(\omega^s)\right).$$

Let us consider the SQG-methods in which instead of

$$F^l(x),\ \hat{F}^l_x(x^s)\quad\text{we use}\quad \Theta_l(s),\ \xi^l(s).$$

4. OPTIMIZATION WITHOUT CONSTRAINTS

Let us minimize the function $F^0(x)$ for $x\in X$.

(a) The stochastic quasigradient projection method. This method is defined by the relations:

(17)
$$x^{s+1}=\pi_X\big(x^s-\varrho_s\xi^0(s)\big),\quad s=0,1,\ldots,$$

(18)
$$E\big(\xi^0(s)/x^0,\ldots,x^s\big)=\hat{F}^0_x(x^s)+b^0(s),$$

where $\pi_X(.)$ is the projecting operation on X, ϱ_s is the step multiplier.

110

If $\xi^0(s)=\hat{F}^0_x(x^s)$, we obtain the well-known method of generalized gradients, proposed by M.Z. Shor in 1962. For the first time the general conditions of its convergence have been obtained by Yu.M. Ermol'ev [3] and independently by B.T. Polyak [4].

If

$$\xi^0(s)=f^0_x(x^s, \omega^s)$$

or

$$\xi^0(s)=\sum_{j=1}^{n} \frac{f^0(x^s+\Delta_s e^j, \omega^s)-f^0(x^s, \omega^s)}{\Delta_s} e^j,$$

then the method (17)–(18) coincides with the stochastic approximation method.

If the values $F^0(x)$ are known, i.e. the common non-linear programming problem is considered, then $\xi^0(s)$ can be determined as follows:

$$(19) \qquad \xi^0(s)=\frac{3}{2} \frac{F^0(x^s+\Delta_s h^s)-F^0(x^s)}{\Delta_s} h^s,$$

where h^s is the random vector realization $h=(h_1, \ldots, h_n)$ with independent and uniformly distributed components within the interval $[-1, 1]$. If $F^0(x)$ has restricted second derivatives in the set X, then

$$E\big(\xi^0(s)/x^s\big)=F^0(x^s)+b^0(s)$$

and

$$(20) \qquad \|b^0(s)\|\leqq \mathrm{Const}\,\Delta_s.$$

The method (17)–(18) has been proposed and developed in the works [5]–[8]. The characteristic requirements, under which the sequence $\{x^s\}$ with the probability 1 converges to the solution, are:

$$(21) \qquad E\big(\|\xi^0(s)\|^2/x^0, \ldots, x^s\big)\leqq C_B, \quad \|x^k\|\leqq B, \quad k=0, 1, \ldots, s,$$

where B, C_B are certain constants;

$$(22) \qquad \varrho_s\geqq 0, \quad \sum_{s=0}^{\infty} \varrho= \infty, \quad \sum_{s=0}^{\infty} E\big(\varrho_s^2+\varrho_s\|b^0(s)\|\big)< \infty.$$

Particularly, if ϱ_s are determinate values independent of $(x^0, \ldots, x^s)$, then, under (21) and (22), for the method using the random directions (19), we obtain, that

$$\sum_{r=0}^{\infty} \varrho_s\Delta_s< \infty.$$

The methods, which we shall consider below, converge under conditions approximately analogous to (21)–(22), so later on we shall not dwell on that.

(b) The linearization stochastic method. This method has been studied in the works [12], [2] and is defined by the relations:

$$(23) \qquad x^{s+1}=x^s+\varrho_s(\bar{x}^s-x^s),$$

$$(24) \qquad \begin{cases} (v^0(s),\bar{x}^s)=\min_{x\in X}\,(v^0(s),x), \\ v^0(s+1)=v^0(s)+\delta_s(\xi^0(s)-v^0(s)), \end{cases}$$

where $s=0,1,\ldots.$ The function $F^0(x)$ is continuously differentiable,

$$\delta_s\geq 0;\quad \sum_{r=0}^{\infty}\delta_s=\infty,\quad \sum_{r=0}^{\infty}E(\delta_s^2+\delta_s^2\|b^0(s)\|)<\infty.$$

5. THE CONSTRAINED OPTIMIZATION

Let us now have the general problem of the minimization of the function $F^0(x)$ under the conditions

$$F^i(x)\leq 0,\quad i=\overline{1,m},\quad x\in X.$$

(a) The penalty functions stochastic method. The constraints (as (2)) we can take into account by means of the penalty functions and instead of the general problem we can consider, for example, the minimization of the function

$$F(x)=F^0(x)+C\sum_{i=1}^{m}\min\,(0,F^i(x))$$

in the set X.

Since it is practically impossible to calculate $F^i(x)$ in the problems of the stochastic programming, i.e. it is impossible to find $\min\,(0,F^i(x))$, the work [2] studied the method defined by the relations

$$(25) \qquad \begin{cases} x^{s+1}=\pi_x[x^s-\varrho_r(\xi^0(s))+C\sum_{i=1}^{m}\min\,(0,Z_i(s))\xi^i(s)], \\ Z_i(s+1)=Z_i(s)+\delta_s(\Theta_i(s)-Z_i(s)),\quad i=\overline{1,m}, \end{cases}$$

where $s=0,1,\ldots.$ Now

$$\delta_s\geq 0,\quad \sum_{s=0}^{\infty}\delta_s=\infty,\quad \sum_{s=0}^{\infty}E(\delta_s^2+\delta_s\|d(s)\|)<\infty.$$

(b) The Arrow–Hurvitz stochastic method. This method is characterized by the relations

$$(26) \qquad \begin{cases} x^{s+1}=\pi_x\left[x^s-\varrho_r\left(\xi^0(s)+\sum_{i=1}^{m}U_i(s)\xi^i(s)\right)\right], \\ U_i^{s+1}=\max\,\{0,U_i(s)+\gamma_s\Theta_i(s)\}. \end{cases}$$

112

For the strictly convex function $F^0(x)$ the convergence of the method has been studied in the works [2], [6], and for the convex function $F^0(x)$ in the work [16].

(c) Besides the enumerated methods there are a lot of others that cannot be mentioned in such a short report. In particular, in [14] the method, characterized by the relations:

$$x^{s+1} = \pi_X(x^s - \varrho_s \beta(s)), \quad s = 0, 1, \ldots,$$

$$\beta(s) = \begin{cases} \xi^0(s), & Z_{i_s}(s) = \max_{1 \le i \le m} Z_i(s) \le 0, \\ \xi^{i_s}(s), & Z_{i_s} > 0 \end{cases}$$

is studied where the values $Z_i(s)$ are defined by the relations (25).

6. THE NON-CONVEX FUNCTIONS

In the Institute of Cybernetics of Kiev results concerning the solution problem (1)–(4) with non-convex and non-smooth functions were obtained. E. A. Nurminsky [9]–[10] was the first to begin the study of the convergence of the stochastic quasigradient methods for the functions satisfying the following condition:

$$F^l(Z) - F^l(x) \ge (F^l_x(x), Z - x) + 0(\|z - x\|).$$

Moreover, he studied the most important features of such functions and proposed the new proof technique of convergence, based on ad absurdum arguments, a technique adopted to the study of convergence of non-relaxation numerical methods to minimize non-smooth non-convex function. Later on this technique was widely used to prove the convergence of various algorithms.

The interesting results of the minimization of the almost-everywhere differentiable functions have been obtained by Shor, Bajenov (see [18]).

Especially, we should like to note the investigation of minimization of almost-everywhere-differentiable functions and discontinuous functions by Gupal [14]–[15]. In brief, to minimize such a function, for example to minimize the almost-everywhere-differentiable function $F^0(x)$, in the works by Gupal there has been studied the convergence of the methods

$$(27) \qquad x^{s+1} = x^s - \varrho_r \sum_{j=1}^{n} \frac{F^0(\bar{x}^s + \Delta_s e^j) - F(\bar{x}^s)}{\Delta_s} e^j,$$

where e^j is the jth unit vector and the point $\bar{x}^s$ is randomly choosen in the neighbourhood of the point x^s radius τ_s; and $\tau_s \to 0$, $s \to \infty$ (see also 2).

7. THE LIMIT EXTREMUM PROBLEMS

The procedures, as (27), are based on the general ideas of solving limit extremum problems, which began to be developed by Ermol'ev and Nurmansky [13]. In brief, these ideas are the following: let us minimize the function $F^0(x)$ having complex nature, for example it may have discontinuous derivatives. We consider the sequence of "good" functions $F^0(s, x)$, for instance smooth, converging to $F^0(x)$ as $s \to \infty$, i.e.

$$F^0(s, x) \to F^0(x)$$

and we also consider procedures

$$x^{s+1} = x^s - \varrho_s \operatorname{grad} F^0(s, x^s), \quad s = 0, 1, \ldots.$$

Under rather general conditions we can prove that here

$$F^0(x^s) \to \min F^0(x).$$

In the present paper we did not dwell on a lot of other investigations, in particular, the determinate methods, stated in the review paper by Shor [18].

In the Institute of Cybernetics of Kiev we have a tradition of solving practical problems. For the more effective control of the step-multipliers ϱ_s the calculations often are carried out, using the display at the interactive regime.

REFERENCES

[1] Ermol'ev, Yu. M.: The models and methods of the stochastic programming, *J. Kibernetika*, 4 (1975), (in Russian).

[2] Ermol'ev, Yu. M.: The stochastic programming methods. *Nauka*, Moscow (1976).

[3] Ermol'ev, Yu. M.: The methods of the solving of the nonlinear extremum problems. *J. Kibernetika*, 4, (1966).

[4] Polyak, B. T.: On a general method of the solving of extremum problems, *Doklady of Academy of Sciences of the USSR*, 174, I (1967).

[5] Ermol'ev, Yu. M., Nekrylova Z. V.: On the certain methods of the stochastic optimization, *J. Kibernetika*, 6 (1966).

[6] Ermol'ev, Yu. M., Nekrylova, Z. V.: The stochastic quasigradient method and its application, In *Theory of Optimal Decisions*, I, published by the Institute of Cybernetics; Kiev, 1967.

[7] Ermol'ev, Yu. M., Shor, N. Z.: The random search method for the two-stage stochastic programming problem and its generalization, *J. Kibernetika*, I, (1968).

[8] Ermol'ev, Yu. M.: On the generalized stochastic gradients method and the stochastic quasi-Fejér sequences, *J. Kibernetika*, 2, (1969).

[9] Nurminsky, E. A.: On one-class functions properties. In *Theory of optimal decisions*, published by the Institute of Cybernetics, Kiev, (1972).

[10] Nurminsky, E. A.: The convergence conditions of nonlinear programming algorithms. *J. Kibernetika*, 3 (1973).

[11] Nurminsky, E. A.: The quasi-gradient method for the solving of the nonlinear programming problem, *J. Kibernetika,* I (1973).

[12] Gupal, A. M., Bajenov, L. G.: The stochastic linearization method, *J. Kibernetika,* 3 (1972).

[13] Ermol'ev, Yu. M., Nurminsky, E. A.: The limit extremum problems, *J. Kibernetika,* 4 (1973).

[14] Gupal, A. M.: The search algorithm of the non-differentiable function extremums with constraints, Report 76–8, The Institute of Cybernetics, Kiev, 1976.

[15] Gupal, A. M., Norkin, B. U.: The discontinuous functions minimization algorithm, *J. Kibernetika,* 6, 1976.

[16] Nurminsky, E. A., Verchenko, P. I.: The saddle-points finding algorithm, *J. Kibernetika,* 6, 1976.

[17] Nurminsky, E. A., Jelihovsky, A. A.: ε-quasigradient methods, *J. Kibernetika,* 5, 1976.

[18] Shor, N. Z.: The generalized gradient methods of the non-smooth function minimization and their application for the mathematical programming problems, *J. Ekonomika i mathematicheskie metody,* 2, 1976.

ON APPROXIMATIVE SOLUTIONS OF STOCHASTIC PROGRAMMING PROBLEMS BY MEANS OF STOCHASTIC DOMINANCE AND STOCHASTIC PENALTY METHODS

K. MARTI

(Zürich, Switzerland)

1. INTRODUCTION

In this paper we present two new methods for the iterative solution of stochastic optimization problems of the type

$$\min E_\omega\big(c_1(\omega)'x + c_2'y(\omega)\big)$$

(1.1)
$$\text{s. t. } A(\omega)x + My(\omega) = b(\omega) \text{ a. s.}$$
$$x \in U, \quad y(\omega) \geqq 0 \text{ a. s.}$$

and

(1.2)
$$\min E_\omega p\big(v(\omega) - T(\omega)x\big)$$
$$\text{s. t. } x \in U$$

where ω is an element of a probability space $(\Omega, \mathcal{A}, P)$, E_ω denotes the expectation operator, U is a convex subset of $\mathbf{R}^n$, $p : \mathbf{R}^{m+1} \to \mathbf{R}$ is a sublinear function (i.e. $p(tz) = tp(z)$, $t \geqq 0$, $z \in \mathbf{R}^{m+1}$ and $p(z_1 + z_2) \leqq p(z_1) + p(z_2)$), M is a given (m, r)-matrix, c_2 a given r-vector, such that $M\mathbf{R}'_+ = \mathbf{R}^m$ ($\mathbf{R}'_+$ is the positive orthant of $\mathbf{R}'$) and $c_2 \geqq 0$, $(A(\cdot), b(\cdot), c_1(\cdot))$ is a random variable on Ω with values in $\mathbf{R}^{(m \cdot n + m + n)}$ and $(T(\cdot), v(\cdot))$ is a random variable with values in $\mathbf{R}^{(m+1) \cdot n + (m+1)}$. We assume that $E_\omega\|(A(\omega), b(\omega), c_1(\omega))\|^{\bar\alpha} < +\infty$ for a number $\bar\alpha \geqq 1$. Clearly, (1.1) is a stochastic linear program with complete fixed recourse [3]: x resp. $y(\omega)$ denotes the decision of the first resp. second stage (after the random element ω has been revealed). It can be shown [4] that (1.1) can be transformed into a problem of the type (1.2) by setting $p(t, z) = t + q(z)$, $t \in \mathbf{R}$, $z \in \mathbf{R}^m$, where

$$q(z) = \inf \{c_2'y : My = z, y \geqq 0\}, \quad z \in \mathbf{R}^m$$

and $T(\omega) = (c(\omega), A(\omega)')'$, $v(\omega) = (0, b(\omega)')'$. Let us still define the numbers

$$\|p\| = \sup \{|p(z)| : \|z\| = 1\}, \quad \mathbf{p} = \inf \{p(z) : \|z\| = 1\}$$

related to p. It is a considerable facilitation in proving convergence theorems for iterative procedures for the solution of programming problems if the set U of admissible (first stage) decisions x may be restricted from the beginning to bounded sets. Of course, this is allowed e.g. if $R(x) \to + \infty$ for $\|x\| \to + \infty$, where R is the convex function on $\mathbf{R}^n$ defined by

$$R(x) = E_\omega p(v(\omega) - T(\omega)x), \quad x \in \mathbf{R}^n.$$

By the considerations in [5], [7], we know that in the case $\mathbf{p} \geqq 0$

$$R(x) \geqq \mathbf{p} \cdot (N(x) - E_\omega \|v(\omega)\|) \geqq \mathbf{p} \cdot (\mathbf{N}\|x\| - E_\omega \|v(\omega)\|), \quad x \in \mathbf{R}^n,$$

where $N(x) = E_\omega \|T(\omega)x\|$, $\mathbf{N} = \inf\{N(x) : \|x\| = 1\}$. Therefore $\mathbf{p} > 0$, $\mathbf{N} > 0$ is a sufficient condition for $R(x) \to + \infty$, $\|x\| \to + \infty$. While $\mathbf{p} > 0$ is a condition concerning the cost-function of the underlying decision problem, the second condition $\mathbf{N} > 0$ depends on the stochastics of the problem. In the following theorem we state some conditions for $\mathbf{N} > 0$.

Theorem 1.1. a) *If* $T(\omega)x \neq 0$ *a. s. for each* $x \neq 0$, *then* $\mathbf{N} > 0$; b) *If* $(m+1) \geqq n$ *and* $T_{\text{reg}}(\omega)$ *is a regular* (n, n)-*submatrix of* $T(\omega)$, *then* $\mathbf{N} \geqq (\text{ess.}_\omega \sup \|T_{\text{reg}}(\omega)^{-1}\|)^{-1}$ *and a similar condition holds of course for* $m + 1 < n$; c) *If* $\|T(\omega)\| \leqq c$ *a.s. for a fixed positive constant* $c > 0$, *then* $\mathbf{N} \geqq (1/c)\vartheta_{\min}(E_\omega T(\omega)'T(\omega)) \geqq E_\omega \vartheta_{\min}(T(\omega)'T(\omega))$, *where* $\vartheta_{\min}(Q)$ *denotes the least eigenvalue of a square matrix* Q.

Our iterative schemes mentioned in the title are based on the following:

(A) Using some properties of the preference relations $P_1 \leqq {}_C (<{}_C) P_2$ between two probability measures P_1, P_2 on the Borel σ-field $\mathscr{B}^{m+1}$ of $\mathbf{R}^{m+1}$ and defined according to

$$P_1 \leqq {}_C (<{}_C) P_2 \text{ if and only if } \int \text{ud}\, P_1 \leqq (<) \int \text{ud}\, P_2 \text{ for all } u \in C,$$

where C is a set of convex functions $u : \mathbf{R}^{m+1} \to \mathbf{R}$ on $\mathbf{R}^{m+1}$.

(B) Let $1 \leqq \alpha < \bar{\alpha}$ be a fixed number and let $\|\cdot\|_E$ denote the Euclidean norm. It is known e.g. by [11] that the optimal second stage decision $y^0 = y^0(\omega)$ is a piecewise linear function of $b(\omega) - A(\omega)x$. Hence by our assumptions concerning the integrability of $A(\cdot)$ and $b(\cdot)$ we may assume that $E_\omega \|y(\omega)\|^\alpha < + \infty$; This will be assumed (tacitly) in this paper. The second method presented here consists, then in the

Embedding of problem (1.1) into the family of $L_\alpha(\Omega, \mathcal{A}, P)$-space-problems

$$(1.1)_s \qquad \min E_\omega \big(c_1(\omega)'x + c_2'y(\omega) + (1/s)\|A(\omega)x + My(\omega) - b(\omega)\|_E^\alpha\big)$$

$$\text{s. t. } x \in U, \quad y(\omega) \geqq 0 \text{ a. s.,}$$

where $s > 0$ is a positive (small) number; $(1.1)_s$ is called the *s*-problem related to (1.1).

Note. (a) Clearly, we may incorporate also the constraint $y(\omega) \geqq 0$ a.s. into $(1.1)_s$ by adding the term $(1/s)E\|y(\omega)^+\|_E^\alpha$, where $y^+ = (y_1^+, \ldots, y_r^+)'$, $y_i^+ = \max(0, y_i)$; (b) This approach works of course also for nonlinear problems of the (recourse-) type (1.1), but for the sake of simplicity we confine us to linear problems; (c) Procedure

(A) is an outgrowth of the author's considerations in portefolio theory, method (B) is a generalization of the well-known penalty function method for the iterative solution of programming problems (see[2]) to the L_α-space problem (1.1). A somewhat related procedure for solving problems of optimal control has been studied by Balakrishnan [1].

(d) An increasing importance in the discussion of stochastic optimization problems with recourse has the explicit treatment of the NON-ANTICIPATIVITY-constraint (concerning the first stage decision x) contained implicitly in the problem formulation (1.1): "First we have to select $x \in U$, then ω is revealed and y, but not x, may be a function of ω". Of course, to this constraint belongs also some price-system ϱ. The study of ϱ and related theoretical questions has been started by Rockafellar and Wets in a series of papers [8], [9], [12]. It turned out now that our approach (B) may be used to provide (for the first time by our knowledge) a constructive approach to this price-system ϱ. This can be established by working with the following modification of problem (1.1)

$$\min E_\omega\big(c_1(\omega)'x(\omega)+c_2'y(\omega)\big)$$

$(1.1)^{na}$
$$\text{s. t. } A(\omega)x(\omega)+My(\omega)=b(\omega) \text{ a. s.}$$
$$x(\omega)\in U, \quad y(\omega)\geqq 0 \text{ a. s.}$$
$$x(\omega)=x_0 \text{ a. s.}$$

and of problem $(1.1)_s$

$$\min E_\omega\big(c_1(\omega)'x(\omega)+c_2'y(\omega)+(1/s)\|A(\omega)x(\omega)+My(\omega)-b(\omega)\|_E^\alpha+(1/s)\|x(\omega)-x_0\|_E^\pi\big)$$
$(1.1)_s^{na}$

$$\text{s.t. } x(\omega)\in U, \quad y(\omega)\geqq 0 \text{ a.s.,}$$

where x_0 is an n-vector, depending on $x(\cdot)$, and $x=x(\omega)$ is—for a suitable number $\pi \geqq 1$—an element of the space $L_\pi^n(\Omega, \mathcal{A}, P)$ of measurable functions $x: \Omega \to \mathbf{R}^n$ such that $E_\omega\|x(\omega)\|^\pi < +\infty$.

While the appearance of the constant x_0 (with respect to ω!) in $(1.1)^{na}$ is simply an other formulation for the non-dependence of the first stage decision x on ω, we must be careful in the choice of x_0 within the s-problem $(1.1)_s^{na}$, since x_0 has to be related to $x(\cdot)$ in some way, i.e. $x_0=\Delta(x(\cdot))$, and the resulting problem $(1.1)_s^{na}$ should remain mathematically simple.

Lemma 1.1. *A suitable choice, guaranteeing convexity, for x_0 in the above problems $(1.1)^{na}$, $(1.1)_s^{na}$ is $x_0=\Delta(x(\cdot)):=\int x(\omega)\mu(d\omega)$, where μ is a (n, n)-matrix-valued measure on the measurable space $(\Omega, \mathcal{A})$ such that $\int \mu(d\omega)=I$ and I is the (n, n)-identity matrix.*

Proof. If $x(\omega)=d$ a.s., where d is a fixed n-vector, i.e. x is an a.s. constant first-stage decision function, then $x_0=\Delta(x(\cdot))=\int x(\omega)\mu(d\omega)=(\int \mu(d\omega))d=Id=d$. Hence $x_0=x(\omega)$ a.s. and therefore the penalty term $(1/s)E_\omega\|x(\omega)-x_0\|_E^\pi$ vanishs. Secondly, since Δ is linear in $x(\cdot)$ we find that the mapping $x(\cdot)\to E_\omega\|x(\omega)-\Delta(x(\omega))\|_E^\pi$ is convex.

2. CONSTRUCTION OF DIRECTIONS OF DECREASE FOR (1.1)
BY MEANS OF STOCHASTIC DOMINANCE

In this case we have that $R(x)=c'x+Q(x)$, where

$$c=E_\omega c_1(\omega) \quad \text{and} \quad Q(x)=E_\omega q\big(b(\omega)-A(\omega)x\big).$$

The relation between our optimization problem (1.1) and the preference relations "$\leqq_c$", "$<_c$" defined in Section 1, (A) is given in the following lemma.

Lemma 2.1. *Let P_{b-Ax} resp. P_{Ax} denote the distribution of $b-Ax$ resp. Ax and let C be a set of convex functions on $\mathbf{R}^m$. a) If $q\in C$ and $P_{b-Ay}\leqq_c(<_c)P_{b-Ax}$, then $Q(y)\leqq$ $\leqq(<)Q(x)$; b) Let A, b be independent random variables, $Q_0. Q_0(z)=E_\omega q(b(\omega)-z))\in C$ and $P_{Ay}\leqq_c(<_c)P_{Ax}$, then $Q(y)\leqq(<)Q(x)$.*

Note. If in addition to the above relations between P_{b-Ay}, P_{Ay} and P_{b-Ax}, P_{Ax} we have that $c'y\leqq(<)c'x$, then $R(y)\leqq(<)R(x)$ (according to the strictness of the inequalities). If $R(y)<R(x)$, then $h=y-x$ is a direction of decrease (obtained by means of stochastic dominance) of R at x, since R is obviously a convex function on $\mathbf{R}^n$.

Properties of "$\leqq_c$", "$<_c$" which allow an algorithmic application are given in the following, see also [10].

Theorem 2.1. *Let $L: \mathbf{R}^m \to \mathbf{R}^{\bar{m}}$ be an affin function and $L(P)$, the image of a probability measure P on $\mathcal{B}^m$ with respect to L. Furthermore let C resp. Γ denote a set of convex functions on $\mathbf{R}^m$ resp. $\mathbf{R}^{\bar{m}}$ such that $\{u\circ L : u\in C\}\subset\Gamma$. Then we have that $L(P_1)\leqq$ $\leqq_\Gamma(<_\Gamma)L(P_2)$, provided that $P_1\leqq_c(<_c)P_2$.*

Corollary 2.1. *Let $c\in\mathbf{R}^m$ be a fixed vector and $L(z)=z+c$, $z\in\mathbf{R}^m$. Then $P_1\leqq_{C_m}(<_{C_m})P_2$ if and only if $L(P_1)\leqq_{C_m}(<_{C_m})L(P_2)$, where C_m denotes the set of all convex functions on $\mathbf{R}^m$.*

Theorem 2.2. *Let U, V, W be subsets of C_{m_1}, C_{m_2}, $C_{m_1+m_2}$, such that for all $w\in W$ either $\int w(x,\cdot)\lambda_1(dx)\in V$, $\int w(\cdot,y)\mu_2(dy)\in U$ or $\int w(.,y)\mu_1(dy)\in U$, $\int w(x,.)\lambda_2(dx)\in V$ for probability measures λ_1, λ_2 on $\mathcal{B}^{m_1}$ and μ_1, μ_2 on $\mathcal{B}^{m_2}$. a) If $\lambda_1\leqq_U\lambda_2$, $\mu_1\leqq_V\mu_2$, then $\lambda_1\otimes\mu_1\leqq_W\lambda_2\otimes\mu_2$ ($\otimes$ denotes the product of two measures); b) If $\lambda_1\leqq_U\lambda_2$, $\mu_1<_V\mu_2$ or $\lambda_1<_U\lambda_2$, $\mu_1\leqq_U\mu_2$, then $\lambda_1\otimes\mu_1<_W\lambda_2\otimes\mu_2$.*

Theorem 2.3. *Let $L_a(z)=az$, $z\in\mathbf{R}^m$, where a is a given number. If $0\leqq a<b$, then $L_a(P)<_c L_b(P)$, provided that $\int u(az)P(dz)<+\infty$ for all $u\in C$ and P is not concentrated at $z=0$.*

Theorem 2.4. *Let $L(z)=Bz$, where B is an (m,m)-matrix. If $\lambda\leqq_{C_m}L(\lambda)$ for all probability measures λ such that $\int z\lambda(dz)=0$, then B must have the form $B=(1+s)I$, where $s\geqq 0$ and I is the identity matrix.*

120

With help of these theorems we get an algorithmic applicable version of the relation $P_{b-Ay}<_c P_{b-Ax}$ in Lemma 2.1 under the following assumptions concerning the random elements of $(-A, b)=(-\bar{A}, \bar{b})+\Xi$, where $\bar{A}, \bar{b}$ are the expectations of A, b: For each $i=1, 2, \ldots, m$ let s_{ij}, $j\in J_i$ ($=$index set) be a partition of $\{1, 2, \ldots, n+1\}$ such that for the $|s_{ij}|$-column vector $(\xi_{ik})_{k\in s_{ij}}$ ($|s_{ij}|$ denotes the cardinality of s_{ij}) we have that

$$(\xi_{ik})_{k\in s_{ij}}=\Gamma_{ij}(\xi^0_{ik})_{k\in s_{ij}}, \quad i=1, 2, \ldots, m, \quad j\in J_i,$$

where Γ_{ij} denotes a $(|s_{ij}|, |s_{ij}|)$-matrix for all $i=1, 2, \ldots, m, j\in J_i$ and $\{\xi^0_{ik}: 1\le i\le m, 1\le k\le n+1\}$ are independent random variables with zero mean, such that $P_{\xi^0_{ik}}$ ($=$distribution of ξ^0_{ik})$=\lambda_{ij}$ for all $k\in s_{ij}$, where the Fourier transform $\hat{\lambda}_{ij}$ of λ_{ij} is given by

$$\hat{\lambda}_{ij}(t)=\exp(-|t|^{\alpha_{ij}}), \quad t\in\mathbf{R}, \quad 1<\alpha_{ij}\le 2,$$

i.e. ξ^0_{ik}, $k\in s_{ij}$ has a stable, symmetric distribution of the type $(0, \alpha_{ij})$; We observe that $\int t\,\lambda_{ij}(dt)\in\mathbf{R}$. By using the convenient notation $x_{n+1}\equiv 1$ and denoting then $(x'\ 1)'$ again by x (which should cause no confusion) we find that $-A(\cdot)x+b(\cdot)=(-\bar{A}, \bar{b})x+ \Xi x$, the coordinates $\Xi_i x$ of Ξx are stochastically independent and they have a distribution $P_{\Xi_i x}$ which is given by the convolution $P_{\Xi_i x}= \underset{j\in J_i}{*} L_{\|\Gamma'_{ij}x\|_{\alpha_{ij}}}(\lambda_{ij})$ $(L_a(z)=az,$ $z\in\mathbf{R}^m$ for fixed $a\in\mathbf{R})$, where $\Gamma'_{ij}x=\Gamma'_{ij}(x_k)_{k\in s_{ij}}$, $\|(w_k)_{k\in s_{ij}}\|_{\alpha_{ij}}=\left(\sum_{k\in s_{ij}}|w_k|^{\alpha_{ij}}\right)^{1/\alpha_{ij}}$. Based on the Theorems 2.1, 2.2, 2.3 and with the above assumptions we obtained in [6], [7] the next Theorem 2.5 by adopting the following notation: Let (i_0, j_0) be a fixed pair $1\le i_0\le m$, $j_0\in J_{i_0}$ and $K^{i_0 j_0}_1$, $K^{i_0 j_0}_2$ a partition of $\{(ij): 1\le i\le m, j\in J_i\}\setminus\{(i_0 j_0)\}$, $x, y\in\mathbf{R}^n$ and let e_{ij} denote for each $j\in J_i$ the i-th unit vector of $\mathbf{R}^m$. If finally

$$q^{x,y}_{K^{i_0 j_0}_1 K^{i_0 j_0}_2}(w_{i_0 j_0})=\int q\Big(-\bar{A}x+\bar{b}+\sum_{(ij)\in K^{i_0 j_0}_1}\|\Gamma'_{ij}x\|_{\alpha_{ij}}w_{ij}e_{ij}+w_{i_0 j_0}e_{i_0 j_0}+$$

$$+\sum_{(ij)\in K^{i_0 j_0}_2}\|\Gamma'_{ij}y\|_{\alpha_{ij}}w_{ij}e_{ij}\Big)\prod_{(ij)\ne(i_0 j_0)}\lambda_{ij}(dw_{ij}),$$

then we have this theorem.

Theorem 2.5. *Let $x, y\in\mathbf{R}^n$. Then the following implications hold.*

(a) *The relations $c'y\le c'x$, $\bar{A}y=\bar{A}x$, $\|\Gamma'_{ij}y\|_{\alpha_{ij}}\le\|\Gamma'_{ij}x\|_{\alpha_{ij}}$ for all (ij) imply $R(y)<R(x)$;*

(b) *If the assumptions in* (a) *hold, where in addition $\|\Gamma'_{i_0 j_0}y\|_{\alpha_{i_0 j_0}}<\|\Gamma'_{i_0 j_0}x\|_{\alpha_{i_0 j_0}}$ for at least one $(i_0 j_0)$, then $R(y)<R(x)$, provided that $q^{x,y}_{K^{i_0 j_0}_1 K^{i_0 j_0}_2}$ is strictly convex for a partition $K^{i_0 j_0}_1$, $K^{i_0 j_0}_2$;*

(c) *If the assumptions in* (a) *hold and $c'y<c'x$, then $R(y)<R(x)$.*

We observe that this theorem contains sufficient conditions for the validity of the following decisive implication which can be handeled rather easily compared with the assumptions in Lemma 2.1.

$$(2) \qquad c'y\le c'x, \quad \bar{A}y=\bar{A}x, \quad \|\Gamma'_{ij}y\|_{\alpha_{ij}}\le\|\Gamma_{ij}x\|_{\alpha_{ij}} \quad \text{for all } (ij) \text{ imply} \quad R(y)\le R(x)$$

and we have that $R(y) < R(x)$ if "$<$" holds at least once in the left hand side of the implication.

Corollary 2.5. *Let (up to the assumptions in section 1) $b(\cdot)$ be an arbitrary random m-vector independent on $A(\cdot)$ and let $A(\cdot) = \bar{A} + \Xi$ be distributed according to the above definitions (note that $x_{n+1} = 1$ is then no more present). If Q_0, defined by $Q_0(z) = E_\omega q(b(\omega) - z)$, $z \in \mathbf{R}^m$, is strictly convex, then (2) holds.*

Proofs of the above and following results: See [6], [7].

The implication (2) is now the key to the algorithm $\mathcal{A}$ developed in the following. For this purpose we consider first the convex program

$$(3)_{c_{ij}, z^1, z^2} \qquad \min \varrho c'x + \sigma \sum_{(ij)} \|\Gamma'_{ij}x\|_{\alpha_{ij}}$$

$$\text{s.t.} \quad \|\Gamma'_{ij}x\|_{\alpha_t} \leq c_{ij} \quad \text{for all} \quad (ij)$$

$$c'x \leq z^1, \quad \bar{A}x = z^2, \quad x \in U,$$

where ϱ, σ are fixed positive numbers, $c_{ij} \geq 0$, $z^1 \in \mathbf{R}$, $z^2 \in \mathbf{R}^m$.

Theorem 2.6. *(Necessary condition for a solution x^0 of (1.1).) Let the implication (2) be valid and x^0 a solution of* (1.1). *Then x^0 solves also the program* $(3)_{c_{ij}, z^1, z^2}$ *with $c_{ij} = \|\Gamma'_{ij}x^0\|_{\alpha_{ij}}$, $z^1 = c'x^0$, $z^2 = \bar{A}x^0$.*

Note. As it is shown in [7], this theorem may be used for a further characterization of x^0 in the case $A(\cdot) = (\bar{A}_{\mathrm{I}}, A_{\mathrm{II}}(\cdot))$.

Now we can begin with the description of our algorithm.

Definition 2.1.1. For any $x \in U$ define $\Phi(x)$ by $\Phi(x) = \{y \in U : y$ solves the program $(3)_{c_{ij}, z^1, z^2}$ with $c_{ij} = \|\Gamma'_{ij}x\|_{\alpha_{ij}}$, $z^1 = c'x$, $z^2 = \bar{A}x\}$.

Lemma 2.2. *If $y \in \Phi(x)$, $x \notin \Phi(x)$, then $R(y) < R(x)$, provided that (2) holds.*

Since $x \in \Phi(x)$ is possible (see Theorem 2.6 and $x = x^0$) we have to supplement this descent-direction-giving map $x \to -x + \Phi(x)$.

Definition 2.1.2. For any $x \in U$ let $\Psi(x)$ be defined by $\Psi(x) = \{y \in U : y$ is a solution of $\min R'_+(x, y - x)$ s.t. $y \in U\}$, where $R'_+(x, h)$ denotes the directional derivative of R at x and in direction h.

Lemma 2.3. a) *x^0 is a solution of (1.1) if and only if $x^0 \in \Psi(x^0)$;* b) *If $y \in \Psi(x)$, $x \notin \Psi(x)$, then $h = y - x$ is a direction of decrease for R at x.*

Definition 2.2. For $x \in U$ define $D(x)$ (see also the modifications in Theorem 2.7) by $D(x) = -x + \Phi(x)$ if $x \notin \Phi(x)$ *(No derivatives have to be computed!)* and $D(x) = -x + \Psi(x)$ if $x \in \Phi(x)$.

122

Corollary 2.3. *If $x \in U$ is not a solution of (1.1), then $D(x)$ contains directions of decrease for R at x, provided that (2) holds.*

Definition 2.3. $m(x, h) = \{x + \tau h : R(x + \tau h) = \min \{R(x + th) \text{ s. t. } 0 \leq t \leq 1\}$, $x, h \in \mathbf{R}^n$.
Now we can define the algorithm:

Definition 2.4. (Algorithm $\mathcal{A}$.)
$\mathcal{A}$ (1) Determine $x^1 \in U$, put $k = 1$:

$\mathcal{A}(2)$ Select $h^k = y^k - x^k \in D(x^k)$: If $y^k \in \Psi(x^k)$ and $R'_+(x^k, h^k) = 0$, then x^k is a solution of (1.1), otherwise go to

$\mathcal{A}$ (3) Compute $x^{k+1} \in m(x^k, h^k)$ and go to ($\mathcal{A}(2)$) putting $k \to k + 1$.

Theorem 2.7. *(Convergence of $\mathcal{A}$). Let U be a closed convex subset of $\mathbf{R}^n$ and let the implication (2) be valid. If (x^k) is a bounded sequence generated by $(\mathcal{A})$ such that $h^{k_\nu} \in \Psi(x^{k_\nu})$, $\nu = 1, 2, \ldots$, then any accumulation point x^0 of the subsequence (x^{k_ν}) is a solution of (1.1).*

Note. a) According to the condition $h^{k_\nu} \in \Psi(x^{k_\nu})$, $\nu = 1, 2, \ldots$ we may in general not use at any step (if it would be possible at all) the relatively simple map $x \to -x + \Phi(x)$. This is clear, since Φ depends on U and the stochastics of problem (1.1) but *not* on the costfunction q; b) Also by $h^{k_\nu} \in \Psi(x^{k_\nu})$ it is clear that only at comparatively "few" steps $k_1, k_2, \ldots$ we must compute derivatives what causes large numerical difficulties!

3. STOCHASTIC PENALTY METHODS

For the sake of simplicity we assume besides the linearity of problem (1.1) also that $\alpha = \pi = 2$. For any $x \in L_2^n$, $y \in L_2^r$ and $\omega \in \Omega$ we define the functions $f_s = f_s(\omega, x, y)$ and $F_s = F_s(x, y)$, $s > 0$ by

$$f_s(\omega, x, y) = c_1(\omega)' x(\omega) + c_2' y(\omega) + (1/s) \| A(\omega) x(\omega) + My(\omega) - b(\omega) \|_E^2 +$$

$$+ (1/s) \| x(\omega) - \Delta(x(\cdot)) \|_E^2$$

and

$$F_s(x, y) = E_\omega f_s(\omega, x, y).$$

Obviously, the last term of f_s and F_s vanishs if we restrict x from the beginning to the subset $\mathbf{R}^n$ of a. s. constant decision functions. Since $f_s(\omega, \cdot, \cdot)$, $F_s(\cdot, \cdot)$ are convex functions and F_s is real by assumption we find that

$$F'_{s+}(x, y, g, h) = E_\omega f'_{s+}(\omega, x, y, g, h) \quad \text{for all} \quad x, g \in L_2^n \quad \text{and} \quad y, h \in L_2^r,$$

where $F'_{s+}(x, y, g, h)$ resp. $f'_{s+}(\omega, x, y, g, h)$ denotes the directional derivative of F_s resp. $f_s(\omega, \cdot, \cdot)$ at (x, y) with respect to (g, h). Furthermore we have the equation

$$f'_{s+}(\omega, x, y, g, h) = c_1(\omega)'g(\omega) + c_2'h(\omega) + (2/s)(A(\omega)x(\omega) + My(\omega) - b(\omega))' \times$$
$$\times (A(\omega)g(\omega) + Mh(\omega)) + (2/s)(x(\omega) - \Delta(x(\cdot)))'(g(\omega) - \Delta(g(\cdot))) =$$
$$= (c_1(\omega) + (2/s)A(\omega)'(A(\omega)x(\omega) + My(\omega)) + (2/s)(x(\omega) -$$
$$(4) \qquad - \Delta(x(\cdot))))'g(\omega) + (c_2 + M'(A(\omega)x(\omega) + My(\omega) - b(\omega)))'h(\omega) -$$
$$- (2/s)(x(\omega) - \Delta(x(\cdot)))'\Delta(g(\cdot)).$$

Since $(1.1)_s^{na}$ is a convex optimization problem, a solution $(x_s(\cdot), y_s(\cdot))$ of $(1.1)_s^{na}$ is characterized by the relation

$$(5) \qquad F'_{s+}(x_s, y_s, x - x_s, y - y_s) \geqq 0 \quad \text{for all } x, y \text{ such that } x(\omega) \in U, \ y(\omega) \geqq 0 \text{ a. s.}$$

Combining the relations (4) and (5) we obtain the following relations (I), (II), assuming for the moment that they admit a solution $(x(.), y(.)) \in L_2^n \times L_2^r$:

$$(I) \qquad c_2 + (2/s)M'(A(\omega)x_s(\omega) + My_s(\omega) - b(\omega)) \geqq 0, \quad y_s(\omega) \geqq 0 \text{ a. s.}$$
$$(c_2 + (2/s)M'(A(\omega)x_s(\omega) + My_s(\omega) - b(\omega)))'y_s(\omega) = 0 \text{ a. s.},$$

what is equivalent to the requirement that $y_s = y_s(\omega)$ solves the program

$$(I') \quad \min c_2'y + (1/s)\|A(\omega)x_s(\omega) + My - b(\omega)\|_E^2$$
$$\text{s.t.} \quad y \geqq 0;$$

$$(II) \quad E_\omega(c_1(\omega) + (2/s)A(\omega)'(A(\omega)x_s(\omega) + My_s(\omega) - b(\omega)) + (2/s)(x_s(\omega) - \Delta(x_s(\cdot))))'g(\omega)$$
$$\geqq (2/s)(E_\omega x_s(\omega) - \Delta(x_s(\cdot)))'\Delta(g(\cdot))$$

for all $g = x - x_s$ such that $x(\omega) \in U$ a. s.

We can state the following simple facts.

Lemma 3.1. *If* $\Delta(x(\cdot)) = \int x(\omega)P(d\omega) = E_\omega x(\omega)$, *then the left hand side of* (II) *is equal to* 0.

We consider then measures μ (see also Lemma 1.1) such that $\mu \ll P$, i.e. there is a measurable (n, n)-matrix-valued function $K = K(\omega)$ on Ω, such that

$$(6) \qquad \Delta(x(\cdot)) = \int x(\omega)\mu(d\omega) = \int K(\omega)x(\omega)P(d\omega) = E_\omega K(\omega)x(\omega).$$

Lemma 3.1.1. *If* $\Delta(x(\cdot))$ *is defined by* (6), *then* (II) *is equivalent to*

$$(7) \quad E_\omega(c_1(\omega) + (2/s)A(\omega)'(A(\omega)x_s(\omega) + My_s(\omega) - b(\omega)) + (2/s)(x_s(\omega) - E_\omega K(\omega)x_s(\omega)) +$$
$$- ((2/s)K(\omega)'(E_\omega x_s(\omega) - E_\omega K(\omega)x_s(\omega))))'g(\omega) \geqq 0$$

for all $g = x - x_s$ *such that* $x(\omega) \in U$ a. s.

Note. (a) The last term in (7) is not present if $K(\omega) = I$ a. s.; (b) In the special, but important case $U = \mathbf{R}_+^n$ condition (7) is equivalent to

$$(8) \qquad c_1(\omega) + (2/s)A(\omega)'(A(\omega)x_s(\omega) + My_s(\omega) - b(\omega)) + (2/s)(x_s(\omega) - E_\omega K(\omega)x_s(\omega)) +$$
$$- (2/s)K(\omega)'(E_\omega x_s(\omega) - E_\omega K(\omega)x_s(\omega)) = z_s(\omega) \geqq 0 \text{ a. s.}$$
$$z_s(\omega)'x_s(\omega) = 0, \quad x_s(\omega) \geqq 0 \text{ a. s.}$$

124

Let now in the sequel for the sake of simplicity be $K(\omega)=I$ a.s., $U=\mathbf{R}^n_+$, $c_1(\omega)=$ $=c_1>0$ (pointwise) a. s. and use for $x=x(\omega)$ the description

$$x(\omega)=\bar{x}+X(\omega),$$

where $\bar{x}=E_\omega x(\omega)$ and $E_\omega X(\omega)=0$; Of course, the constraint $x(\omega)\in U$ a. s. is equivalent to $\bar{x}\in U$ and $\bar{x}+X(\omega)\in U$ a. s. In this case (8) has the form

$$(9)\quad c_1+(2/s)A(\omega)'\big(A(\omega)\bar{x}_s+A(\omega)X_s(\omega)+My_s(\omega)-b(\omega)\big)+(2/s)X_s(\omega)\equiv z_s(\omega)\geqq 0 \text{ a. s.}$$
$$z_s(\omega)'\big(\bar{x}_s+X_s(\omega)\big)=0,\quad \bar{x}_s\geqq 0,\quad \bar{x}_s+X_s(\omega)\geqq 0 \text{ a. s.}$$

By solving (I) or (I') for y we find that

$$(10)\qquad\qquad y=Y\big(A(\omega)x_s(\omega)-b(\omega)\big),$$

where Y is piecewise linear, putting this into (9) we get $\bar{x}_s$ and $X_s(\omega)$; Using then again (10) we obtain also $y_s(\omega)$; For the details: See [7]. In this paper it is shown that the (existing) variables $(\bar{x}_s,X_s(\cdot),y_s(\cdot))$ lie in a bounded subset of $\mathbf{R}^n\times L^n_2\times L^r_2$, hence the family $\big((\bar{x}_s,X_s,y_s)\big)_{s>0}$ admits weak accumulation points $\bar{x}^0,X^0(\cdot),y^0(\cdot)$, satisfying the constraints of (1.1).

Definition 3.1. The Lagrangian related to problem $(1.1)^{na}$ is defined by

$$L(x(\cdot),y(\cdot),u(\cdot),\varrho(\cdot))=E_\omega\big(c_1'x(\omega)+c_2'y(\omega)+u(\omega)'\big(A(\omega)x(\omega)+My(\omega)-b(\omega)\big)+$$
$$+\varrho(\omega)'\big(x(\omega)-E_\omega x(\omega)\big)\big),$$

where $x(\cdot),\varrho(\cdot)\in L^n_2$ and $y\in L^r_2$, $u\in L^m_2$.

Note that also here the last term of the formula vanishs if we consider only a. s. constant decision functions $x(\cdot)=\bar{x}$.

By [7] we have this next result (since $F'_{s+}(x_s,y_s,\cdot,\cdot)=L'_{s+}(x_s,y_s,\cdot,\cdot)$).

Lemma 3.2. *For any $s>0$ let $u_s(.)$ and $\varrho_s(.)$ be defined by*

$$u_s(\omega)=(2/s)\big(A(\omega)x_s(\omega)+My_s(\omega)-b(\omega)\big),$$
$$\varrho_s(\omega)=(2/s)\big(x_s(\omega)-E_\omega x_s(\omega)\big)$$

and put $L_s(x(.),y(.))=L(x(.),y(.),u_s(.),\varrho_s(.))$. Then (x_s,y_s) solves also the optimization problem

$$\min L_s\big(x(.),y(.)\big) \ s. \ t. \ x(\omega)\geqq 0,\quad y(\omega)\geqq 0 \text{ a. s.}$$

We observe that $E_\omega\varrho_s(\omega)=0$, $s>0$, hence any weak accumulation point ϱ^0 of the family $(\varrho_s)_{s>0}$ satisfies also the equation $E_\omega\varrho^0(\omega)=0$.

From this Lemma 3.2 we obtain the following properties of the approximative Lagrange multipliers $u_s(.)$, $\varrho_s(.)$.

Lemma 3.3. *For any $s>0$ the first dual variable $u_s(.)$ satisfies the relation c_2+ $+M'u_s(\omega)\geqq 0$ a. s.*

Corollary 3.1. *Since $c_2 \geqq 0$ and $M\mathbf{R}_+^r = \mathbf{R}^m$ (according to our assumptions in section 1), we have that $\|u(_s\omega)\|_E = (2/s)\|A(\omega)x_s(\omega) + My_s(\omega) - b(\omega)\|_E \leqq \eta$ a. s., where η is a fixed positive number.*

Since $\|A(\omega)x_s(\omega) + My_s(\omega) - b(\omega)\|_E \leqq (2/s)\|A(\omega)x_s(\omega) + My_s(\omega) - b(\omega)\|_E$ for $s < 2$, we have also that $\|A(\omega)x_s(\omega) + My_s(\omega) - b(\omega)\|_E \leqq \eta$ for small $s > 0$.

Lemma 3.3.1. *For any $s > 0$ the approximative price-system (related to the nonanticipativity constraint) ϱ_s satisfies the relations*

$$c_1 + A(\omega)'u_s(\omega) + \varrho_s(\omega) \geqq 0 \text{ a. s.,}$$
$$(c_1 + A(\omega)'u_s(\omega) + \varrho_s(\omega))'x_s(\omega) = 0, \quad x_s(\omega) \geqq 0 \text{ a. s.}$$

Corollary 3.1.1. *It is $\|\varrho_s(\omega)\|_E \leqq \|c_1\|_E + \eta\|A(\omega)\|_E$ a. s. for $s > 0$.*

By this results we see that also the dual variables $u_s(\cdot)$, $\varrho_s(\cdot)$ are contained in a (fixed) bounded subset of L_2^m, L_2^n, hence the families $(u_s(\cdot))_{s>0}$, $(\varrho_s(\cdot))_{s>0}$ have weak accumulation points $u^0(\cdot)$, $\varrho^0(\cdot)$, where $E_\omega\varrho^0(\omega) = 0$.

By means of the above results we can infer (see [7]) our final result.

Theorem 3.1. *Let $(x^0(\cdot), y^0(\cdot), u^0(\cdot), \varrho^0(\cdot))$ be a weak accumulation point of the family $(x_s, y_s, u_s, \varrho_s)$, $s > 0$ (The existence is guaranteed by the above considerations). Then it is $x^0(\omega) \geqq 0$, $y^0(\omega) \geqq 0$ a. s., $E_\omega\varrho^0(\omega) = 0$ and $(x^0, y^0, u^0, \varrho^0)$ is a saddle point of the Lagrangian $L = L(x, y, u, \varrho)$ with respect to the constraints $x(\omega) \geqq 0$, $y(\omega) \geqq 0$ a. s. and $E_\omega\varrho(\omega) = 0$.*

Corollary 3.2. *(x^0, y^0) is then a solution of (1.1) and u^0, ϱ^0 are optimal Lagrange multipliers related to the constraints of $(1.1)^{na}$.*

Note. The full potential of this constructive approach is manifested if the approximative Lagrange multipliers $u_s(\cdot)$, $\varrho_s(\cdot)$ are calculated explicitly, since from these calculations the accumulation points u^0, ϱ^0 may guessed often rather easily. We observe once more that also non linear problems of the type (1.1) may be treated in this way!

REFERENCES

[1] Balakrishnan, A. V.: On a new computing technique in optimal control and its application to minimal-time flight profile optimization. *JOTA* 4, (1969), 1–21.

[2] Bertsekas, D. P.: On penalty and multiplier methods for constrained minimization. *SIAM J. Control and Optimization* 14, (1976), 216–235.

[3] Kall, P.: *Stochastic Linear Programming*. Berlin–Heidelberg–New York: Springer Verlag 1976, Series in Econometrics and Operations Research Vol. 21.

[4] Marti, K.: Entscheidungsprobleme mit linearen Aktionen- und Ergebnisraum. *Z. Wahrscheinlichkeitstheorie verw. Geb.* 23 (1972), 133–147.

[5] Marti, K.: Approximationen der Entscheidungsprobleme mit linearer Ergebnisfunktion und positiv homogener, subadditiver Verlustfunktion. *Z. Wahrscheinlichkeitstheorie verw. Geb.* 31, (1975), 203–233.

[6] Marti, K.: Stochastische Dominanz und Stochastische Lineare Programme. To appear in: *Methods of Operations Research* 1977, A. Hain, Meisenheim.

[7] Marti, K.: Approximations for Stochastic Optimization Problems and related Stability Questions. Zürich 1976: Habilitation thesis (in preparation).

[8] Rockafellar, R. T., Wets, R.: Stochastic convex programming: Kuhn–Tucker conditions. *J. Mathematical Economics* 2, (1975), 349–370.

[9] Rockafellar, R. T., Wets, R.: Nonanticipativity and L^1-martingales in stochastic optimization problems. In: R. Wets (ed.): *Stochastic Systems: Modeling, Identification and Optimization,* North-Holland Publ. Co., 1976.

[10] Stoyan, D.: Monotonieeigenschaften stochastischer Modelle. *ZAMM* 52, (1972), 23–30.

[11] Walkup, D. W., Wets, R.: A Note on Decision Rules for Stochastic Programs, *J. Computer and System Sciences* 2, (1968), 305–311.

[12] Wets, R.: Stochastic multipliers, induced feasibility and nonanticipativity in stochastic programming. In: M. Dempster (ed.): *Proceedings of the International Symposium in Stochastic Programming,* Academic Press, Oxford 1974.

A NONLINEAR PROGRAMMING METHOD FOR THE SOLUTION OF A STOCHASTIC PROGRAMMING MODEL OF A. PRÉKOPA

J. MAYER

(Budapest, Hungary)

1. INTRODUCTION

The nonlinear programming algorithm presented in this paper is designed for the solution of a stochastic programming problem of A. Prékopa [6], [8], [10], but it can also be applied for nonlinear programming problems with only one nonlinear constraint. It is a combination of Zoutendijk's P1 method [14], and the reduced gradient method of Wolfe [13], Abadie and Carpentier [1]. The algorithm works using feasible directions, but for handling the linear constraints, the very effective techniques of the reduced gradient method are applied.

In the next section the stochastic programming problem of A. Prékopa is discussed from a nonlinear programming point of view. Section 3 contains some numerical experiences with GRG and the motivation of our method. In the subsequent section the algorithm is described in details, and its convergence investigated. The last section contains some numerical experiences with the method.

2. THE STOCHASTIC PROGRAMMING MODEL OF A. PRÉKOPA

We consider the question of solving the following stochastic programming mode of A. Prékopa:

$$
\begin{aligned}
\max f(\mathbf{x}) \\
\mathcal{G}(\mathbf{x}) &\geq p \\
A\mathbf{x} &= \mathbf{b} \\
\mathbf{x} &\geq \mathbf{0},
\end{aligned}
\tag{1}
$$

where $\mathbf{x} \in R^n$, $\mathbf{b} \in R^m$, A is $m \times n$, and $m < n$, $r(A) = m$, matrix A is of full rank. The objective function is assumed to be a concave, continuously differentiable function and the nonlinear constraint has the following form: $\mathcal{G}(\mathbf{x}) = P\{S\mathbf{x} \geq \boldsymbol{\beta}\}$, where S is an $r \times n$ matrix, $\boldsymbol{\beta} = (\beta_1, \ldots, \beta_r)$ is a random vector — variable, i.e. $\mathcal{G}(\mathbf{x})$ is the probability of the event, that the system of linear inequalities $S\mathbf{x} \geqq \boldsymbol{\beta}$ with random right-hand sides is fulfilled.

From the nonlinear programming point of view (1) is a nonlinear programming problem of a special type, it contains only one nonlinear constraint. The nonlinear programming approaches to the solution of the problem are all based on the following fundamental result of A. Prékopa: [6], [7], [8].

Theorem 1. *Let $\beta_1, \ldots, \beta_r$ be random variables with a joint probability density function of the following form: $h(\mathbf{z}) = e^{-Q(\mathbf{z})}$, where $\mathbf{z} \in R^r$, $Q(\mathbf{z})$ is a convex function in the entire space R^r. Then the function $\mathcal{G}(\mathbf{x})$ in (1) is a logarithmic concave function on the entire space R^n.*

In this paper it is assumed throughout that $\beta_1, \ldots, \beta_r$ are in general not independent random variables with a normal joint probability density function. It follows at once, that $\mathcal{G}(\mathbf{x})$ has the following attractive properties:

Lemma 1. *Under the above assumptions:*

(a) *$\mathcal{G}(\mathbf{x})$ is a logarithmic concave function.*

(b) *$\nabla \mathcal{G}(\mathbf{x})$ is Lipschitz-continuous, i.e. there exists a contant $L > 0$, such that for every $\mathbf{x} \in R^n$, $\mathbf{y} \in R^n$*

$$\| \nabla \mathcal{G}(\mathbf{x}) - \nabla \mathcal{G}(\mathbf{y}) \| \leqq L \cdot \| \mathbf{x} - \mathbf{y} \|$$

holds.

(c) *If there exists a feasible $\mathbf{y}$ such that $\mathcal{G}(\mathbf{y}) > p$, then for problem (1) the Kuhn–Tucker conditions are necessary and sufficient conditions of optimality.*

Proof. Statement (a) is a direct consequence of Theorem 1, and (b) follows easily from the nice properties of the normal probability distribution function. For (c) Prékopa proved [9], that the existence of a vector $\mathbf{y} \in R^n$, for which $\mathcal{G}(\mathbf{y}) > p$ holds implies that the Kuhn–Tucker constraint qualification holds on the whole set of feasible solutions of problem (1). From this fact it is easy to derive (c) using the classical theory of optimality conditions.

3. MOTIVATION OF THE ALGORITHM

In this section we summarize our computational experiences with GRG reported in [5], and give some motivation of the algorithm presented in the next section.

There are two main difficulties which one encounters trying to implement some nonlinear programming technique to solve problem (1).

(a) The evaluation of $\mathcal{G}(\mathbf{x})$ requires the calculation of the r-dimensional normal probability distribution function, i.e. numerical integration in r-dimensional space. From the numerical study made by Deák [2] it turns out, that the best way to compute

130

$\mathcal{G}(\mathbf{x})$ is using simulation techniques. We also used his subroutine for the calculation of the stochastic constraint in our numerical experimentation.

(b) The nonlinear programming problem (1) contains a constraint, which can be calculated only approximately, and its calculation is relatively time-consuming. The purpose of this paper is to discuss this second difficulty.

Our first attempt to solve (1) was using a general purpose GRG code, a failure. Namely at every iteration GRG first generates a point, which is in general not feasible, then it tries to return to the feasible surface by initiating a procedure based on Newton's method. However, in the necessary matrix inversion computing errors get out of control, and afterwards it is impossible to return to the feasible surface.

Next we developed a specialized version of GRG, which is based on the product form inverse-base technique for the linear constraints, and the critical step, i.e. returning to the feasible surface, reduces to the problem of computing the intersection of a line, completely determined by the linear constraints, and the surface $\mathcal{G}(\mathbf{x})=p$. This is a much more stable approach, but the algorithm requires usually too much evaluations of the stochastic constraint in locating the surface. As the calculation of $\mathcal{G}(\mathbf{x})$ is time-consuming, the code uses too much computing time.

Guided by this experiences we decided to modify the reduced gradient method. In the new method only feasible points are generated, and the efficient way of handling linear constraints, inherent in the reduced gradient method, is preserved. To explain the idea, let us assume that at some point $\mathbf{x}$, if we take the partition of the variables $\mathbf{x}=(\mathbf{y}, \mathbf{z})$, $\mathbf{y}\in R^m$, and the same partition for A as $A=(B, C)$, where B is $m\times m$, then B is nonsingular and $\mathbf{y}>0$.

We consider two different ways for determination a direction of move at $\mathbf{x}$. Let us denote the direction by $\mathbf{w}$, and take the same partition as for $\mathbf{x}$: $\mathbf{w}=(\mathbf{u}, \mathbf{v})$, with $\mathbf{u}\in R^m$.

The corresponding subproblem in Zoutendijk's P1 method:

$$\max \zeta$$
$$\nabla_y^T f(\mathbf{x})\mathbf{u} + \nabla_z^T f(\mathbf{x})\mathbf{v} \geqq \zeta$$
$$\nabla_y^T \mathcal{G}(\mathbf{x})\mathbf{u} + \nabla_z^T \mathcal{G}(\mathbf{x})\mathbf{v} \geqq \vartheta\zeta, \quad \text{if} \quad \mathcal{G}(\mathbf{x})=p$$
$$B\mathbf{u} + C\mathbf{v} = 0$$
$$v_i \geqq 0, \quad \text{if} \quad z_i=0, \quad i=1, 2, \ldots, n-m.$$
$$\|(\mathbf{u}, \mathbf{v})\| \leqq 1,$$

where $\vartheta>0$ is a fixed number during the procedure. The subproblem in GRG:

$$\max \zeta$$
$$\nabla_y^T f(\mathbf{x})\mathbf{u} + \nabla_z^T f(\mathbf{x})\mathbf{v} \geqq \zeta$$
$$\nabla_y^T \mathcal{G}(\mathbf{x})\mathbf{u} + \nabla_z^T \mathcal{G}(\mathbf{x})\mathbf{v} \geqq 0, \quad \text{if} \quad \mathcal{G}(\mathbf{x})=p$$
$$B\mathbf{u} + C\mathbf{v} = 0$$
$$v_i \geqq 0, \quad \text{if} \quad z_i=0, \quad i=1, \ldots, n-m.$$
$$\|\mathbf{v}\| \leqq 1.$$

In the new method the following subproblem is used determining the direction of move:

$$\max \zeta$$
$$\nabla_y^T f(\mathbf{x})\mathbf{u} + \nabla_z^T f(\mathbf{x})\mathbf{v} \geqq \zeta$$
$$\nabla_y^T \mathcal{G}(\mathbf{x})\mathbf{u} + \nabla_z^T \mathcal{G}(\mathbf{x})\mathbf{v} \geqq \vartheta\zeta, \quad \text{if} \quad \mathcal{G}(\mathbf{x})=p$$
$$\mathbf{B}\mathbf{u} + \mathbf{C}\mathbf{v} = \mathbf{0}$$
$$v_i \geqq 0, \quad \text{if} \quad z_i=0, \quad i=1,\ldots,n-m$$
$$\|\mathbf{v}\| \leqq 1.$$

This form enables a similar reduction like in the reduced gradient method, and in addition the algorithm will generate feasible points. So the linear part is handled by storing and updating B^{-1}, performing a change of the base if necessary, while feasibility is maintained. The algorithm constructed this way would be subject to jamming so an anti zig-zagging prevention must be incorporated.

4. ALGORITHM FOR THE SOLUTION OF THE STOCHASTIC PROGRAMMING PROBLEM

In this section we describe a nonlinear programming method for the solution of problem (1). It is also applicable for nonlinear programming problems having only one nonlinear constraint, provided that the objective function and constraints fulfil the conditions listed below.

The convergence of the algorithm can certainly be proved under less restrictive assumptions, these assumptions were selected, because in our practical problems they were fulfilled and on the other hand, they allow a short and elegant convergence proof. For the sake of completeness our list also contains some already verified properties of $\mathcal{G}(\mathbf{x})$. The assumptions are as follows:

(a) The objective function $f(\mathbf{x})$ is concave, differentiable, with a Lipschitz-continuous gradient, i.e. there exists a constant $L>0$ such that for every $\mathbf{x}\in R^n$, $\mathbf{y}\in R^n$

$$\|\nabla f(\mathbf{x}) - \nabla f(\mathbf{y})\| \leqq L \cdot \|\mathbf{y}-\mathbf{x}\|$$

is fulfilled.

(b) $\mathcal{G}(\mathbf{x})$ is a logarithmic concave differentiable function, $\nabla\mathcal{G}(\mathbf{x})$ is Lipschitz-continuous and bounded on R^n.

(c) There exists a feasible point $\mathbf{y}$ such that $\mathcal{G}(\mathbf{y})>p$ holds.

(d) The set of feasible solutions of problem (1) is bounded.

(e) At any point $\mathbf{x}$ of the feasible domain there exists a base B of A such that for all $i\in I_B$, $x_i>0$ holds, where I_B is the set of basic indices. In other words we exclude degeneracy. Assume further, that $r(A)=m$, matrix A is of full rank.

The algorithm starts from a feasible point $\mathbf{x}^{(1)}$, furthermore $\varepsilon_1>0$ and the index set I_1 are selected in such a way that the corresponding columns of A form a base, and

132

for all $i \in I_1$, $x_i^{(1)} > \varepsilon_1$ is fulfilled. Let $\vartheta > 0$, $R > 0$ be fixed numbers during the procedure. The algorithm works in an iterative manner, at each iteration new feasible solution $\mathbf{x}^{(k)}$, new tolerance ε_k and new index set I_k are determined, $k = 1, 2, \ldots$, and each iteration is divided into steps. For the sake of simplicity the base corresponding to I_k is denoted by B, the linear part of the constraints is handled by storing and updating B^{-1}.

Let us assume that $\mathbf{x}^{(k)}$, ε_k, I_k, having the same properties as for $k = 1$, are determined, $k \geq 1$. For notational simplicity assume that the base consists of the first m columns of A, i.e. $I_k = \{1, \ldots, m\}$, denote the base by B. All vectors and the matrix A are partitioned into basic and nonbasic part: $\mathbf{x}^{(k)} = (\mathbf{y}^{(k)}, \mathbf{z}^{(k)})$, $A = (B, C)$, similarly for the direction $\mathbf{w}^{(k)} = (\mathbf{u}^{(k)}, \mathbf{v}^{(k)})$, where $\mathbf{y}^{(k)} \in R^m$, $\mathbf{u}^{(k)} \in R^m$ and B is $m \times m$.

The next iteration consists of the following steps:

Step 1. Compute the reduced gradient of $f(\mathbf{x})$ and $\mathcal{G}(\mathbf{x})$ with respect to the linear equality constraints:

$$\mathbf{r}^T = \nabla_{\mathbf{z}}^T f(\mathbf{x}^{(k)}) - \nabla_{\mathbf{y}}^T f(\mathbf{x}^{(k)}) B^{-1} C$$

(2)

$$\mathbf{s}^T = \nabla_{\mathbf{z}}^T \mathcal{G}(\mathbf{x}^{(k)}) - \nabla_{\mathbf{y}}^T \mathcal{G}(\mathbf{x}^{(k)}) B^{-1} C.$$

Step 2. Solve the following direction-finding subproblem:

(3)
$$\begin{aligned}
&\max \zeta \\
&\mathbf{r}^T \mathbf{v} \geq \zeta \\
&\mathbf{s}^T \mathbf{v} \geq \vartheta \zeta, \quad \text{if} \quad \mathcal{G}(\mathbf{x}^{(k)}) \leq p + \varepsilon_k \\
&v_i \geq 0, \quad \text{if} \quad z_i^{(k)} \leq \varepsilon_k, \quad k = 1, \ldots, n - m \\
&\|\mathbf{v}\| \leq 1.
\end{aligned}$$

Let the optimal solution of this problem be denoted by $(\hat{\mathbf{v}}, \hat{\zeta})$.

Step 3. If $\hat{\zeta} > \varepsilon_k$ continue at Step 4. Otherwise we consider two cases:

Case 1. If $\hat{\zeta} \leq \varepsilon_k$ and $\hat{\zeta} \neq 0$, take $\varepsilon_{k+1} = \frac{1}{2} \varepsilon_k$, $\mathbf{x}^{(k+1)} = \mathbf{x}^{(k)}$, $I_{k+1} = I_k$, this iteration is terminated, the next iteration starts at Step 2.

Case 2. If $\hat{\zeta} = 0$, solve the following subproblem:

(4)
$$\begin{aligned}
&\max \zeta \\
&\mathbf{r}^T \mathbf{v} \geq \zeta \\
&\mathbf{s}^T \mathbf{v} \geq \vartheta \zeta, \quad \text{if} \quad \mathcal{G}(\mathbf{x}^{(k)}) = p \\
&v_i \geq 0, \quad \text{if} \quad z_i^{(k)} = 0, \quad i = 1, \ldots, n - m \\
&\|\mathbf{v}\| \leq 1.
\end{aligned}$$

Let the optimal solution be denoted by $(\mathbf{v}_0, \zeta_0)$. Here again we distinguish two cases: if $\zeta_0 = 0$ the algorithm terminates, $\mathbf{x}^{(k)}$ is an optimal solution of problem (1); if $\zeta_0 > 0$ the iteration terminates, $\varepsilon_{k+1} = \frac{1}{2} \varepsilon_k$, $\mathbf{x}^{(k+1)} = \mathbf{x}^{(k)}$, $I_{k+1} = I_k$ the next iteration starts at Step 2.

Step 4. Let $\mathbf{v}^{(k)}=\hat{\mathbf{v}}$, $\mathbf{u}^{(k)}=-B^{-1}C\mathbf{v}^{(k)}$ and the direction: $\mathbf{w}^{(k)}=(\mathbf{u}^{(k)},\ \mathbf{v}^{(k)})$.

Step 5. Compute a starting step length:

$$\alpha_k=\begin{cases}\min\limits_{w_i^{(k)}<0}\left\{-\dfrac{x_i^{(k)}}{w_i^{(k)}}\right\}, & \text{if there exists } i,\ 1\leq i\leq n, \text{ for which } w_i^{(k)}<0 \text{ holds}\\[2ex] R, & \text{otherwise.}\end{cases}$$

Step 6. Determine the step length λ_k of the iteration as follows:

$\lambda_k=\dfrac{1}{2^{l_0}}\,\alpha_k$ where, $l=l_0$ is the first integer, for which

$$(5) \qquad\qquad f\left(\mathbf{x}^{(k)}+\frac{\alpha_k}{2^l}\,\mathbf{w}^{(k)}\right)\geqq f(\mathbf{x}^{(k)})+\frac{\alpha_k}{2^{l+1}}\,\hat{\zeta}$$

$$\mathcal{G}\left(\mathbf{x}^{(k)}+\frac{\alpha_k}{2^l}\,\mathbf{w}^{(k)}\right)\geqq p$$

$$l\geqq 0, \quad \text{if at Step 5 the } \alpha_k \text{ was computed according}$$
$$\text{to the first row}$$

holds, i.e. either at Step 5 α_k is computed according to the first row and $l=l_0=0$ holds, or system (5) is not fulfilled for $l=l_0-1$.

Step 7. $\mathbf{x}^{(k+1)}=\mathbf{x}^{(k)}+\lambda_k\mathbf{w}^{(k)}$.

Step 8. If at $\mathbf{x}^{(k+1)}$ for all $i\in I_k$ $x_i^{(k+1)}>\varepsilon_k$ holds, then this iteration terminates, $\varepsilon_{k+1}=\varepsilon_k$, $I_{k+1}=I_k$, the next iteration starts at Step 1.

Otherwise starting from the present base a new base is determined, such that for the corresponding index set I_{k+1} for every $i\in I_{k+1}$ $x_i^{(k+1)}>\varepsilon_{k+1}$ holds, where $\varepsilon_{k+1}=$ $=\dfrac{1}{2^q}\,\varepsilon_k$ and either $q=0$ or q is the first positive integer for which a base having the above property can be found. Assumption (e) implies that this is always possible having ε_{k+1} small enough, for a practical procedure the interested reader may consult [4].

For I_{k+1} the matrix B^{-1} is updated accordingly, the iteration terminates and a new iteration starts at Step 1.

In the next part of this section we analyze the algorithm defined above.

In the case of finite termination the current point is an optimal solution of our problem. In fact:

Lemma 2. *If the algorithm terminates at the k-th iteration at Step 3 with $\zeta_0=0$, then* $\mathbf{x}^{(k)}$ *is an optimal solution of problem* (1).

134

Proof. $\zeta_0=0$ implies that the following problem

$$\max \zeta$$

(6)

$$\nabla_{\mathbf{x}}^{T} f(\mathbf{x}^{(k)})\mathbf{w} \geqq \zeta$$
$$\nabla_{\mathbf{x}}^{T} \mathcal{G}(\mathbf{x}^{(k)})\mathbf{w} \geqq \vartheta\zeta \quad \text{if} \quad \mathcal{G}(\mathbf{x}^{(k)})=p$$
$$A\mathbf{w}=\mathbf{0}$$
$$v_i \geqq 0 \quad \text{if} \quad z_i^{(k)}=0, \quad i=1,\ldots,n-m.$$

also has a finite optimum, and at the optimal solution the value of its objective function is 0. Now the same argument works as in [9]. Using the Farkas Theorem and the fact that assumption (c) guarantees a strictly positive multiplier for the objective function row, from (6) the Kuhn–Tucker optimality conditions can be derived. According to Lemma 1 these are sufficient conditions of the optimality.

In the case of an infinite sequence of iterations the sequence of objective function values increases, furthermore the algorithm continually improves the value of the objective function:

Lemma 3. *Either the algorithm terminates at an optimal solution of problem* (1), *or for an infinite subsequence of the sequence of iterations it reaches Step 4, i.e. at this iterations the value of the objective function is strictly increased.*

Proof. Let us observe that the iteration can be terminated at Step 3 only a finite times in sequence. From this fact our assertion follows easily.

Before proceeding let us mention that there exists a constant $Q>0$, such that for every iteration $\|\mathbf{w}^{(k)}\| \leqq Q$ holds. In fact, because there exist only a finite number of possible bases there exists a $Q>0$ such that:

$$\|\mathbf{w}^{(k)}\|=\|(-B^{-1}C\mathbf{v}^{(k)}, \mathbf{v}^{(k)})\| \leqq Q\|\mathbf{v}^{(k)}\|=Q.$$

There is one step in the algorithm which requires some discussion, we have namely to prove that at Step 6 it is always possible to select λ_k as required there. This is a trivial consequence of the following statement:

Lemma 4. *If* $\lambda \leqq \dfrac{1}{2}\dfrac{\hat{\zeta}}{Q^2 L}$, *then* $f(\mathbf{x}^{(k)}+\lambda\mathbf{w}^{(k)}) \geqq f(\mathbf{x}^{(k)})+\dfrac{1}{2}\lambda\hat{\zeta}.$

Proof. Applying the Taylor Theorem and using the Lipschitz-continuity of $\nabla f(\mathbf{x})$ we get:

$$f(\mathbf{x}^{(k)}+\lambda\mathbf{w}^{(k)})=f(\mathbf{x}^{(k)})+\lambda\nabla_{\mathbf{x}}^{T} f(\mathbf{x}^{(k)})\mathbf{w}^{(k)}+\lambda[\nabla_{\mathbf{x}}^{T} f(\mathbf{x}^{(k)}+\Theta\lambda\mathbf{w}^{(k)})-\nabla_{\mathbf{x}}^{T} f(\mathbf{x}^{(k)})]\mathbf{w}^{(k)} \geqq$$

$$\geqq f(\mathbf{x}^{(k)})+\lambda\hat{\zeta}-\lambda^2 Q^2 L \geqq f(\mathbf{x}^{(k)})+\frac{1}{2}\lambda\hat{\zeta}, \text{ if } \lambda \leqq \frac{1}{2}\frac{\hat{\zeta}}{Q^2 L} \text{ where } 0<\Theta<1.$$

In the remainder of this section we give a convergence proof for the algorithm. The crucial point of convergence proofs for algorithms which work with feasible directions and use the same anti zig-zagging device is to prove that $\varepsilon_k \to 0$ as $k \to \infty$ [11]. This holds also in our case:

Theorem 2. *If the algorithm generates an infinite sequence of points* $\mathbf{x}^{(k)}$, *then* $\varepsilon_k \to 0$ *as* $k \to \infty$.

Proof. Using the same argument as in Lemma 4 if $\mathcal{G}(\mathbf{x}^{(k)}) \leqq p + \varepsilon_k$ and a more simple reasoning also based on Taylor's theorem if $\mathcal{G}(\mathbf{x}^{(k)}) > p + \varepsilon_k$ it can be derived that $\mathcal{G}(\mathbf{x}^{(k)} + \lambda \mathbf{w}^{(k)}) \geqq p$ provided that $\lambda \leqq \min \left\{ \dfrac{\vartheta \hat{\zeta}}{Q^2 L}, \dfrac{\varepsilon_k}{K \cdot Q} \right\}$, where K is an upper bound for $\| \nabla \mathcal{G}(\mathbf{x}) \|$. It is now clear that for $\lambda \leqq \min \left\{ \dfrac{1}{2} \dfrac{\hat{\zeta}}{Q^2 L}, \dfrac{\vartheta \hat{\zeta}}{Q^2 L}, \dfrac{\varepsilon_k}{K \cdot Q}, \dfrac{\varepsilon_k}{Q} \right\}$ the point $\mathbf{x}^{(k)} + \lambda \mathbf{w}^{(k)}$ is a feasible solution of problem (1), and in addition

$$(7) \qquad f(\mathbf{x}^{(k)} + \lambda \mathbf{w}^{(k)}) \geqq f(\mathbf{x}^{(k)}) + \tfrac{1}{2} \lambda \hat{\xi} \quad \text{holds.}$$

According to Lemma 3 there exists a subsequence for which $\hat{\zeta}_k > \varepsilon_k$ is fulfilled, where $\hat{\zeta}_k$ denotes $\hat{\zeta}$ at the k-th iteration. Let us denote the set of indices in such a subsequence by J. For $k \in J$ the point $\mathbf{x}^{(k)} + \lambda \mathbf{w}^{(k)}$ is a feasible solution of problem (1) and (7) holds, if $\lambda \leqq \varepsilon_k \cdot \varepsilon_0$, where $\varepsilon_0 = \min \left\{ \dfrac{1}{2 Q^2 L}, \dfrac{\vartheta}{Q^2 L}, \dfrac{1}{KQ}, \dfrac{1}{Q} \right\}$.

At Step 6 of the algorithm l_0 was the first integer, for which (5) holds, so necessarily $\lambda_k > \tfrac{1}{2} \varepsilon_k \cdot \varepsilon_0$, for $k \in J$. Substituting into (5) we get:

$$f(\mathbf{x}^{(k+1)}) \geqq f(\mathbf{x}^{(k)}) + \frac{1}{4} \varepsilon_0 \cdot \varepsilon_k^2, \quad k \in J.$$

Since $f(\mathbf{x}^{(k)})$ is a monotone increasing sequence bounded from above, it follows that $\varepsilon_k \to 0$, $(k \to \infty)$, $k \in J$. However ε_k is a monotone decreasing sequence so $\varepsilon_k \to 0$, $(k \to \infty)$ holds.

Based on this theorem, the convergence of the algorithm can be proved.

Theorem 3. *Either the algorithm terminates in a finite number of iterations at an optimal solution of* (1), *or every accumulation point of the sequence* $\mathbf{x}^{(k)}$, $k = 1, 2, \ldots$ *is an optimal solution of problem* (1).

Proof. Because of the fact that $f(\mathbf{x})$ is a concave function and $f(\mathbf{x}^{(k)})$, $k = 1, 2, \ldots$ a monotone sequence, it is sufficient to prove that in the case of an infinite sequence there exists a convergent subsequence of the sequence $\mathbf{x}^{(k)}$, $k = 1, 2, \ldots$ with a limit point being an optimal solution of (1). To select such a subsequence first of all we mention that assumption (e) implies that ε_k is reduced at Step 8 only a finite times. Using this remark, from Theorem 2 it follows that there exists a subsequence of the sequence of iterations, for which $\hat{\zeta}_k \leqq \varepsilon_k$, $k \in J$ holds, where J denotes the set of indices

136

in this subsequence. According to assumption (d) we may assume that the sequence $\mathbf{x}^{(k)}$, $k \in J$ converges to some feasible point $\mathbf{x}^*$, i.e. $\mathbf{x}^{(k)} \to \mathbf{x}^*$, $(k \to \infty)$, $k \in J$. We consider the following problem:

$$
\begin{aligned}
&\max \zeta \\
&\nabla_{\mathbf{x}}^T f(\mathbf{x}^*)\mathbf{w} \geqq \zeta \\
&\nabla_{\mathbf{x}}^T \mathcal{G}(\mathbf{x}^*)\mathbf{w} \geqq \vartheta \zeta \quad \text{if} \quad \mathcal{G}(\mathbf{x}^*)=p \\
&A\mathbf{w}=0 \\
&\qquad w_i \geqq 0 \quad \text{if} \quad x_i^*=0, \quad i=1,\ldots,n \\
&\|\mathbf{w}\| \leqq 1.
\end{aligned}
$$

(8)

Let us denote the optimal solution of this problem by $(\mathbf{w}^*, \zeta^*)$.

If here $\zeta^*=0$ holds, then using the same technique as in the proof of Lemma 2 it follows easily that $\mathbf{x}^*$ is an optimal solution of problem (1).

Let us assume that $\zeta^*>0$, we shall get a contradiction. If $\mathcal{G}(\mathbf{x}^*)>p$, then for k large enough $\mathcal{G}(\mathbf{x}^{(k)})>p+\varepsilon_k$, $k \in J$. Similarly, if for some i, $1 \leqq i \leqq n$, $x_i^*>0$ holds, then for k large enough $x_i^{(k)}>\varepsilon_k$, $k \in J$ holds. On the other hand, because of the continuity of $\nabla f(\mathbf{x})$ and $\nabla \mathcal{G}(\mathbf{x})$ for k sufficiently large

$$
\nabla_{\mathbf{x}}^T f(\mathbf{x}^{(k)})\mathbf{w}^* \geqq \frac{1}{2}\zeta^*
$$

$$
\nabla_{\mathbf{x}}^T \mathcal{G}(\mathbf{x}^{(k)})\mathbf{w}^* \geqq \frac{1}{2}\vartheta \zeta^* \quad \text{if} \quad \mathcal{G}(\mathbf{x}^*)=p
$$

holds. Let us further consider all bases of A, and denote by t the maximum of the norms of the corresponding non-basic parts of $\mathbf{w}^*$. From our considerations it follows that the non-basic parts of $\left(\frac{1}{t}\mathbf{w}^*, \frac{1}{2t}\zeta^*\right)$ form a feasible solution to the direction-finding subproblems (3), for $k \geqq k_0$, $k \in J$ where k_0 is sufficiently large. This implies that $\hat{\zeta}_k \geqq \frac{1}{2t}\zeta^*$ for $k \geqq k_0$, $k \in J$ which contradicts the fact that because $\hat{\zeta}_k \leqq \varepsilon_k$ for $k \in J$, the sequence $\hat{\zeta}_k$ tends to 0 for $k \to \infty$, $k \in J$. This completes the proof.

Some final comments:

If we choose the norm $\|\mathbf{v}\| = \max_{1 \leqq i \leqq n-m} |v_i|$, then the subproblems (3), (4) are linear programming problems with only two rows, they can easily be solved. The method can be extended for more than one nonlinear constraint, if we have l nonlinear constraints, the above-mentioned LP sub-problems contain $l+1$ rows. This suggests that the method might work well with only a few nonlinear constraints.

The method reduces to the reduced gradient method if there are only linear constraints, and it gives Zoutendijk's P1 method if except of bounds for the variables there are no linear constraints.

The first phase of the method, that is the determination of a starting feasible point $\mathbf{x}^{(1)}$ and the corresponding base is carried out by maximizing $\mathcal{G}(\mathbf{x})$ subject to the linear constraints, like in [10].

The idea of modifying the reduced gradient method in order to work with feasible points was first proposed by Kleinmichel and Sadowski [3], [12], but they use a different approach and get a different algorithm. Kleinmichel also gave a nice and general framework for proving the convergence of methods which operate with feasible directions [3].

5. COMPUTATIONAL EXPERIENCES WITH THE METHOD

The numerical difficulties, which arise in the implementation of the reduced gradient method for problem (1) are already explained in Section 2. Here we only mention that the method described in the previous section proved to be superior to the specialized version of the reduced gradient method for all of our stochastic programming test problems.

First the algorithm was tested on a series of smallsize problems, with satisfactory results. For instance the computational results are in accordance with a theorem of Slepian, which states that the two dimensional normal probability distribution function $\Phi(x_1, x_2, r)$ is a monotone increasing function of the correlations coefficient r, at any fixed (x_1, x_2) point.

Having thoroughly tested the algorithm we also solved a version of a model of the electric energy sector of the Hungarian Economy, a real-life problem. The detailed description of this model can be found in the paper of A. Prékopa, S. Ganczer, I. Deák, K. Patyi [10], together with a thorough analysis of Zoutendijk's P2 method for this case, and some very interesting computational results. The model is a stochastic programming problem as described in Section 2, with 49 linear constraints and 72 variables after introducing slack variables for some inequality constraints. The stochastic constraint contained 2 rows. We solved the problem with $p=0.90$ and $p=0.95$, and have got essentially the same optimum as reported in [10].

The code was developed on the framework given in the previous section, in FORTRAN language, for a CDC 3300 computer. Some computational results of the above-mentioned two runs are summarized in Table 1.

Table 1

	Phase I	Phase II	
		$p=0.90$	$p=0.95$
Starting value of the objective	0.49993	3875.00	3875.00
Optimal value of the objective	0.99643	4371.68	4370.73
Value of $\mathcal{G}(\mathbf{x})$ at the optimum	0.99643	0.89999	0.95001
Number of iterations	19	18	16
Number of the calculations of $\mathcal{G}(\mathbf{x})$	37	68	61
Computing time in seconds	51.24	45.71	40.82

Although we are aware of its insufficiency, just to give some comparison we mention that CDC's LP package REX needed 51 seconds to solve in two phases the deterministic version of the problem on the CDC 3300.

Acknowledgement. I wish to express my thanks to Professor A. Prékopa for his advices and for suggesting this research.

REFERENCES

[1] Abadie, J., Carpentier, J.: Généralisation de la méthode du gradient réduit de Wolfe au cas des contraintes non-linéaires, in: *Proc. IFORS Conf.,* eds. D. B. Hertz and J. Melese, Wiley, New York, 1966, 1041–1053.

[2] Deák, I.: Monte-Carlo evaluation of the multidimensional normal distribution function by the ellipsoid method, *Problems of Control and Information Theory,* 7 (3), 1978, 203–212.

[3] Kleinmichel, H.: On methods of feasible directions, Lecture at III. Conf. on Math. Progr., Mátrafüred, 1975.

[4] Kleinmichel, H., Sadowski, H.: Der verallgemeinerte RG-Algorithmus bei linearen Restriktionen, die Behandlung des Entartungsfalls und die Konvergenz des Verfahrens, *Beiträge zur Num. Math.,* 3 (1975), 37–55.

[5] Mayer, J.: Computational experiences with the reduced gradient method, in: *Progress in Op. Res.,* ed. A. Prékopa, North-Holland, 1974, 613–624.

[6] Prékopa, A.: On probabilistic constrained programming, in: *Proc. of the Princeton Symposium on Mathematical Programming,* Princeton Univ. Press, Princeton, New York, 1970, 113–138.

[7] Prékopa, A.: Logarithmic concave measures with application to stochastic programming, *Acta Sci. Math., Szeged,* 32 (1971), 301–316.

[8] Prékopa, A.: Contributions to the theory of stochastic programming, *Mathematical Programming,* 4 (1973), 202–221.

[9] Prékopa, A.: Eine Erweiterung der sogenannten Methode der Zulässigen Richtungen der nichtlinearen Optimierung auf den Fall quasikonkaver Restriktionen, *Math. Operationsforsch. u. Statist.,* 5 (1974), 281–293.

[10] Prékopa, A., Ganczer, S., Deák, I. and Patyi, K.: The STABIL stochastic programming model and its experimental application to the electric energy sector of the Hungarian Economy, *Proc. of the Intern. Conf. on Stochastic Programming,* Oxford, 15–17 July, 1974, Academic Press, in print.

[11] Pshenichnyj, B. N. and Danilin, Ju. M.: *Computationa methods in optimization,* Izd. Nauka, 1975, in Russian.

[12] Sadowski, H.: Untersuchungen zu den Verfahren der reduzierten Gradienten in der nichtlinearen Optimierung, Ph. D. Thesis, University of Technology, Dresden, GDR, 1973.

[13] Wolfe, P.: Methods of nonlinear programming, in: *Recent advances in mathematical programming,* eds. R. L. Graves and P. Wolfe, McGraw-Hill, New York, 1963. 76–77.

[14] Zoutendijk, G.: *Methods of feasible directions,* Elsevier Publ. Comp., Amsterdam and New York, 1960.

A STUDY OF PIVOT PROBABILITIES IN LP TABLEAUS

A. ORDEN

(Chicago, USA)

1. INTRODUCTION

The average ratio of the number of iterations of the simplex method to the number of tableau rows is generally reported to be about 1.5 or 2 to 1. The uniformly low value of this figure serves to point out that the simplex method converges with high efficiency over a very wide range of tableau sizes and structures. It seems at the same time to suggest that there may be some underlying characteristic of the algorithm which, once identified, would explain its efficiency. However, whether on deterministic grounds, such as the maximum number of iterations as a function of tableau size, or on statistics, such as mean and variance, the behavior of the algorithm with respect to the number of iterations remains ill understood.

In principle the mean of the number of iterations (and thereby the above mean ratio) might be found analytically by assuming initial probability distributions for tableau coefficients and deriving subsequent probabilities and resulting expected values under the operating rules of the simplex algorithm. It appears, however, that to carry out such an analysis would take an enormous effort.

Conceivably, restrictive assumptions can be found which would simplify matters sufficiently to permit determination or close approximation of expected values of the number of iterations. This study is intended as a step in that direction. Certain probabilities are evaluated for an initial iteration and, in a very limited way (a very small tableau size), for a second iteration. The results are interpreted in ways which bear crudely on simplex convergence.

The analysis deals only with phase 1 of the simplex method—the process of generating a tableau which exhibits a basic feasible solution. Progress of a series of iterations toward completion of a phase 1 computation may be associated either with reduction of the value of an artificial objective function or with removal of artificial unit vectors from the basis. We will deal here essentially with the latter. The notions and terminology of "artificials" will, however, not be used. The tableaus of phase 1 will instead be characterized in terms of canonical and noncanonical rows; viz.:

A row of a tableau is "canonical" if there is a unit vector with a 1 in that row under a proper (nonartificial) variable; otherwise the row is "noncanonical". For example, in the following tableau rows 1 and 2 are canonical, and row 3 is noncanonical.

$$
\begin{array}{cccccc}
\underline{x_1} & \underline{x_2} & \underline{x_3} & \underline{x_4} & \underline{x_5} & \underline{\text{RHS}} \\
\end{array}
$$

$$
\begin{bmatrix}
1 & 0 & 3 & 4 & 2 & 8 \\
0 & 1 & -2 & 2 & 5 & 6 \\
0 & 0 & 6 & -1 & 3 & 9
\end{bmatrix}
$$

In these terms (avoiding any reference to artificials), each iteration of phase 1 either steps up the number of canonical rows or leaves it unchanged, and all rows canonical culminates phase 1.

Assume—due to initial slack variables—that the starting tableau may contain canonical rows and that the initial coefficients in the nonbasic columns are independent, uniformly distributed random variables. The paper deals with the following matters:

1. In the first iteration, what is the probability that the pivot is in a noncanonical row—which steps up the number of canonical rows, as opposed to a pivot in an already canonical row, which leaves the number of canonical rows unchanged?

2. If the first iteration leaves the number of canonical rows unchanged, what are the probabilities in the second iteration of a pivot which steps up the number of canonical rows vs. one which fails to do so?

3. What preliminary indication do such first and second iteration probabilities give of the convergence properties of phase 1 of the simplex method?

Initial constraints and tableaus. Let the set of LP problems under consideration have r nonnegative variables $x_1, \ldots, x_r$; and m functional constraints, of which h are inequalities (which put canonical rows into the starting tableau) and $(m-h)$ are equations. r, m, and h are to be parameters in the probability evaluations. It will be useful later on to distinguish between coefficients in the inequalities, a_{ij}, and those in the equations, b_{ij}. All of these coefficients are assumed to be independent random variables uniformly distributed in $(-1, 1)$. For simplicity, all RHS constants are 1. We have, then, the problem sets (with no objective function, since we will deal only with phase (1):

(1)
$$
\sum_{j=1}^{r} a_{ij} x_j \leq 1, \quad i = 1, \ldots, h
$$

$$
\sum_{j=1}^{r} b_{ij} x_j = 1, \quad i = (h+1), \ldots, m
$$

$$
x_1 \ldots x_s \geq 0.
$$

Independent Cum. Dist. function of each a_{ij}, b_{ij}:

$$F_0(x) = \text{Prob} (\cdot) \leqq x = \begin{cases} 0, & x < -1 \\ (x+1)/2, & -1 < x < 1 \\ 1, & x > 1 \end{cases}$$

and Prob. density of each a_{ij}, b_{ij}:

$$f_0(x) = \frac{1}{2}, \quad -1 < x < 1.$$

The starting tableaus which correspond to (1) are shown in Fig. 1. The h inequalities result in h slack variables and h canonical rows.

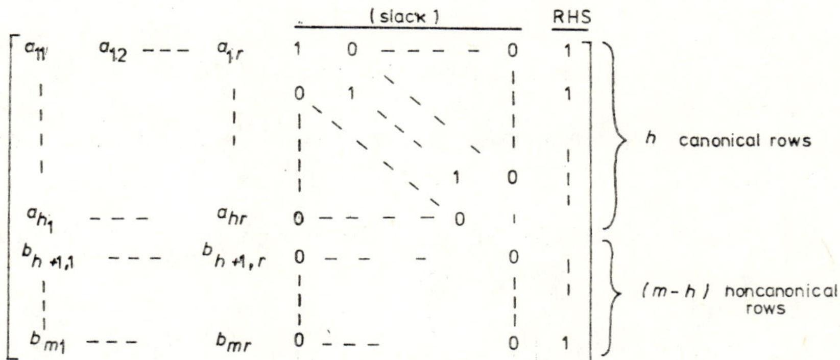

Fig. 1. Starting tableaus

Given such a set of tableaus, questions 1 and 2 above call for finding the following probabilities:

p_1^a, the probability that the first pivot is in a canonical row (an "a_{ij} row")

p_1^b, the probability that the first pivot is in a noncanonical (b_{ij}) row

p_2^a, the probability, when the first pivot is in a canonical row, that the second pivot is again in such a row

p_2^b, the probability, when the first pivot is in a canonical row, that the second pivot is in a noncanonical row.

A distraction. The possibility exists in each iteration of uncovering an infeasible system (all coefficients in a row $\leqq 0$, and RHS term > 0). Consequently, two more probabilities must be brought into consideration:

p_1^0 and p_2^0, the probabilities of encountering infeasibility in the first iteration and second iterations.

Fig. 2 shows the probabilities under consideration in the form of a probability tree.

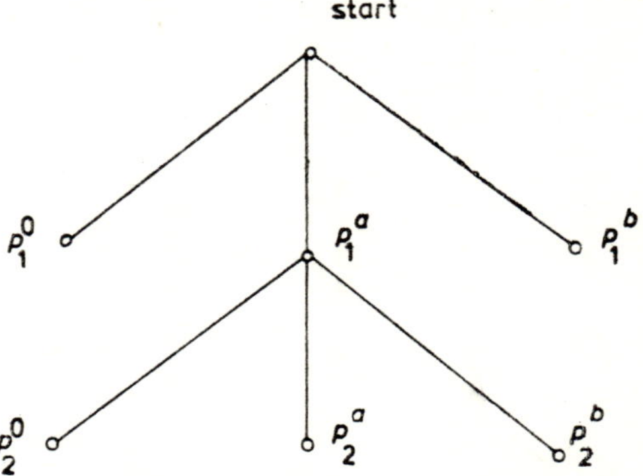

Fig. 2. First and second iteration probability tree

II. RESULTS AND OBSERVATIONS

This section gives the results of the study. The derivations of probabilities are given in Section III.

The first iteration probabilities, p_1^0, p_1^a, p_1^b can be expressed as functions of r, m, and h:

$$(2) \qquad p_1^0 = 1/2^r$$

$$p_1^b = \left(1 - \frac{h}{r+m-1}\right)\left(1 - \frac{1}{2^{r+m-1}}\right) - \frac{1}{2^r}\left(1 - \frac{h}{m-1}\right)\left(1 - \frac{1}{2^{m-1}}\right)$$

$$= 1 - p_1^0 - p_1^b.$$

While—as will be seen in Section III—derivation of the above expressions for the first iteration is fairly simple, work on the second iteration becomes complex and tedious; and going on to further iterations, even for very small tableaus, seems forbidding. Only the case $r=3$, $m=2$, $h=1$ has been worked through to completion for the second iteration. The results are

$$p_2^0 = .090, \quad p_2^a = .004, \quad p_2^b = .078.$$

The evaluated probability tree for 2 iterations, for the case $r=3$, $m=2$, $h=1$, is shown in Fig. 3.

Since the results consist only of probabilities of transitions of the number of canonical rows, $h \to h$ and $h \to h+1$, in the first two iterations, and in the second iteration only for a very small tableau size, they do not go far enough to yield any significant conclusions on expected values of the number of phase 1 iterations. Two observations will, however, be made.

Observation 1. Suppose in (1) that $h = m - 1$; i.e., the constraint systems contain $(m-1)$ inequalities and one equation. Then one successful step, $h \to h+1$, completes

144

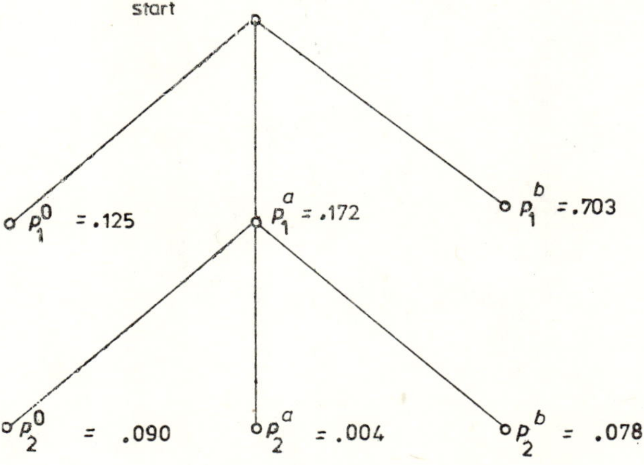

Fig. 3. The Prob. tree for $r=3$, $m=2$, $h=1$

phase 1. Inserting $h=m-1$ into (2) gives the probability for this case of phase 1 being completed in just one iteration:

$$(3) \qquad p_1^b = \left(1 - \frac{m-1}{r+m-1}\right)\left(1 - \frac{1}{2^{r+m-1}}\right)$$

$$= \frac{r}{r+m-1}\left(1 - \frac{1}{2^{r+m-1}}\right)$$

and for tableaus of any appreciable size,

$$p_1^b \cong \frac{r}{r+m-1}.$$

Thus the probability in the case $h=m-1$ of completing phase 1 in one iteration is $\sim \frac{1}{2}$ when $m=r$ (number of constraints equal to the number of variables, not including slacks) and tends toward 1 as m/r decreases.

Observation 2. A hypothesis follows which, if it could be substantiated, would fairly well explain the expected value behavior of phase 1 of the simplex method:

Let there be some initial probability distribution of the initial coefficients, such as is given in (1). Let g be the number of canonical rows to which the tableau arrives in iteration $k-1$, and let $p_k(g, g+1), p_{k+1}(g, g+1), p_{k+2}(g, g+1)$, etc. be the probabilities of the transition from g canonical rows to $g+1$ occurring in iterations k, $k+1$, $k+2$, ... Hypothesis:

$$(4) \qquad p_k(g, g+1) \leq p_{k+1}(g, g+1) \leq p_{k+2}(g, g+1)\ldots;$$

i.e., whenever a transition $g \to g+1$ does not occur, the prob. that it will occur in the next iteration is at least as high as in an iteration which has just occurred.

It is easy to show that if this hypothesis is valid, the expected value of the ratio of the number of iterations in phase 1 to the number of rows, m, is $\leq \log m$. The results shown in Fig. 3 serve as a bit of evidence for the hypothesis. Discarding the probabilities p_k^0 of encountering infeasibility, we have

$$p_1(1, 2) = \frac{.703}{.703 + .172} = .80$$

$$p_2(1, 2) = \frac{.078}{.078 + .004} = .95$$

which is in line with (4).

III. FIRST ITERATION PROBABILITIES

The probability evaluations will be based on the following simplex phase 1 iterative procedure:

Choose a noncanonical row as the criterion row for selection of the pivot column (in lieu of an artificial objective function)[1],[2] If the highest coefficient in that row is ≤ 0, stop—the system is infeasible. If the highest coefficient is > 0, use the column which contains that term as the pivot column.

Find the pivot row by the usual simplex minimum ratio rule involving the RHS column and the pivot column.

Carry out the pivot transformation of the tableau.

If the pivot is in an already canonical row, repeat with the same criterion row. If it is in a noncanonical row, that row becomes canonical.

Let row m in Fig. 1 be the criterion row. p_1^0 is the probability that $b_{m1} \ldots b_{mr}$, each independently having probability distribution $F_0(x)$, are all ≤ 0. (Assume that in each iteration only the criterion row is inspected for infeasibility.) Hence

$$p_1^0 = F_0^r(0) = (\tfrac{1}{2})^r.$$

p_1^b will be evaluated below as a function of r, m, and h. The results for p_1^0 and p_1^b give $p_1^a = 1 - p_1^0 - p_1^b$.

Let $\mu = \max\limits_{j=1\ldots r} b_{mj}$, and let $F_\mu(x) = \mathrm{Prob}\,(\mu < x)$ be its probability distribution. We have

$$F_\mu(x) = \mathrm{Prob}\,(\max_j b_{mj} < x) = \mathrm{Prob}\,(b_{mj} < x, j = 1 \ldots r) = F_0^r(x).$$

$\mu \leq 0$ is covered by p_1^0. We are interested now in $\mu > 0$, hence in the Prob distribution

$$(5) \qquad F_\mu(0, x) = \mathrm{Prob}\,(0 < \mu < x)$$
$$= F_\mu(x) - F_\mu(0)$$
$$= F_0^r(x) - F_0^r(0).$$

Assuming $\mu > 0$, there is a pivot column. For notational simplicity call it column 1, in which case we have $\mu = b_{m1} > 0$. The smallest positive value among $\{1/a_{i1}; 1/b_{i1}\}$ determines the pivot row. (At least one of these, namely $1/b_{m1}$, is positive.) We are interested in whether the pivot is in a canonical or a noncanonical row (an "a row" or a "b row").

146

Let

$$\alpha = \max_{i=1\ldots h} a_{i1}$$

and

$$\beta = \max_{i=(h+1)\ldots m} b_{i1} \quad \text{(given that } b_{m1}>0\text{).}$$

If $\beta>\alpha$ the pivot row is noncanonical, and if $\alpha>\beta$ it is canonical.

We will need the probability distributions:

$$F_\alpha(x) = \text{Prob } (\alpha < x)$$
$$F_\beta(x) = \text{Prob } (\beta < x \mid b_{m1}>0),$$

and

$$f_\beta(x) = F'_\beta(x) = \text{density function of } \beta.$$
$$F_\alpha(x) = \text{Prob } (\max_i a_{i1} < x)$$
$$= \text{Prob } (a_{i1} < x, \, i=1\ldots h).$$

Hence,

(6)
$$F_\alpha(x) = F_0^h(x)$$
$$F_\beta(x) = \text{Prob } [\max_i b_{i1} < x \mid 0 < b_{m1} < x], \quad (x>0)$$

$$= \text{Prob } [b_{i1} < x, \, i=(h+1)\ldots(m-1); \, 0 < b_{m1} < x]$$
$$= F_0^{m-1-h}(x)F_\mu(0, x), \quad \text{where } F_\mu(0, x)$$

is given by (5) The initial coefficients b_{i1} in rows $(h+1)$ to $(m-1)$ are independent of the process of finding the highest term in row m.)

$$= F_0^{m-1-h}(x)[F_0^r(x) - F_0^r(0)]$$
$$F_\beta(x) = F_0^{r+m-1-h}(x) - F_0^r(0)F_0^{m-1-h}(x).$$

Hence,

(7) $\quad f_\beta(x) = [(r+m-1-h)F_0^{r+m-2-h}(x) - F_0^r(0)(m-1-h)F_0^{m-2-h}(x)]f_0(x).$

The probability of the pivot being in a noncanonical row,

$$p_1^b = \text{Prob } [\beta>\alpha \mid \beta>0]$$

can now be evaluated by integrating the probability densities of α and β over the region $[\beta>\alpha, \, \beta>0]$, as indicated in Fig. 4. We get

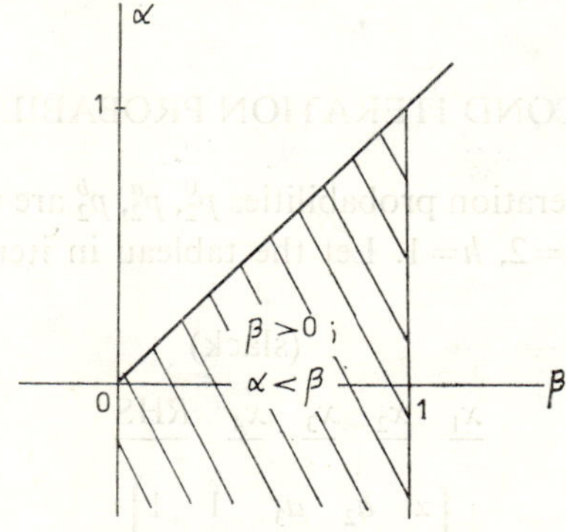

Fig. 4. Integration on α, β for pb_1

$$p_1^b = \int\limits_{v=0}^{\infty} \int\limits_{t=-\infty}^{v} f_\alpha(t) f_\beta(v) \, \mathrm{d}t \, \mathrm{d}v$$

$$= \int\limits_{0}^{\infty} F_\alpha(v) f_\beta(v) \, \mathrm{d}v.$$

Inserting (6) and (7) gives

$$p_1^b = \int\limits_{0}^{\infty} F_0^h(v)[(r+m-1-h)F_0^{r+m-2-h}(v) - F_0^r(0)(m-1-h)F_0^{m-2-h}(v)] f_0(v) \, \mathrm{d}v$$

$$= \int\limits_{0}^{\infty} [(r+m-1-h)F_0^{r+m-2}(v) - F_0^r(0)(m-1-h)F_0^{m-2}(v)] f_0(v) \, \mathrm{d}v$$

$$= \left\{ \frac{r+m-1-h}{r+m-1} F_0^{r+m-1}(v) - F_0^r(0) \frac{m-1-h}{m-1} F_0^{m-1}(v) \right\} \Bigg|_{v=0}^{\infty}.$$

Let $c = F_0(0)$

$$(8) \qquad p_1^b = \frac{r+m-1-h}{r+m-1}(1-c^{r+m-1}) - c^r \frac{m-1-h}{m-1}(1-c^{m-1}),$$

or

$$p_1^b = \left(1 - \frac{h}{r+m-1}\right)(1-c^{s+m-1}) - \left(1 - \frac{h}{m-1}\right)(c^r)(1-c^{m-1}).$$

Thus p_1^b has been evaluated in terms of r, m, h, and the value of $c = F_0(0)$, for any distribution, F_0, of the initial coefficients.

In the case of the uniform distribution on $(-1, 1)$ we have $c = \frac{1}{2}$, and as given earlier in (2):

$$p_1^b = \left(1 - \frac{h}{r+m-1}\right)\left(1 - \frac{1}{2^{r+m-1}}\right) - \left(1 - \frac{h}{m-1}\right)\left(\frac{1}{2^r}\right)\left(1 - \frac{1}{2^{m-1}}\right).$$

IV. SECOND ITERATION PROBABILITIES

As noted earlier, second iteration probabilities p_2^0, p_2^a, p_2^b are worked out in this paper only for the case $r=3$, $m=2$, $h=1$. Let the tableau in iteration 1 for this case be written:

$$
\begin{array}{ccccc}
 & & & \text{(slack)} & \\
 & & & \downarrow & \\
x_1 & x_2 & x_3 & x_4 & \text{RHS} \\
\end{array}
$$

$$
\begin{bmatrix}
\alpha & a_2 & a_3 & 1 & 1 \\
\beta & b_2 & b_3 & 0 & 1
\end{bmatrix},
$$

148

where α and β appear in column 1 in lieu of a_1 and b_1 in order to indicate that the pivot column of every tableau is considered to have been moved to solumn 1. (Since there is only one noncanonical row, no distinction can be made here between the μ and β of the previous section, and we call it β.)

Several probability distributions will be formulated below, conditional on α and β, which are ultimately to be integrated over α and β in order to obtain the desired probabilities p_2^0, p_2^a, p_2^b. The probability densities of α and β, which will be needed in the final integrations, are:

$$(9) \qquad f_\alpha(x) = \tfrac{1}{2}, \quad -1 < x < 1$$

(α is a term in row 1 of the starting tableau. Its density is unaffected by the initial selection of the highest term in row 2.)

$$f_\beta(x) = \frac{3}{8}(x+1)^2, \quad -1 < x < 1 \qquad \left[\frac{d}{dx} F_0^3(x) = \frac{d}{dx}\left(\frac{x+1}{2}\right)^3 = \frac{3}{8}(x+1)^2 \right].$$

Entering iteration 2 implies:

$\beta > 0$, otherwise infeasibility is encountered in iteration 1; and $\alpha > \beta$, otherwise the first pivot is in the noncanonical row, which in the case $h = 1$, $m = 2$, terminates phase 1.

Since all initial coefficients are in $(-1, 1)$, we have $\alpha < 1$; so we proceed into iteration 2 under the conditions:

$$(10.1) \qquad 0 < \beta < \alpha < 1.$$

When convenient we will use $\eta = \beta/\alpha$. It follows from (10.1) that

$$(10.2) \qquad \beta < \eta < 1.$$

Subject to these conditions, the tableaus which enter iteration 2—resulting from pivoting on α—may be written:

$$(11) \qquad \begin{bmatrix} 1 & a_2/\alpha & a_3/\alpha & 1/\alpha & 1/\alpha \\ 0 & (b_2 - \eta a_2) & (b_3 - \eta a_3) & (-\eta) & (1-\eta) \end{bmatrix}.$$

Row 2 is noncanonical.

The first step of iteration 2 is to find the highest among $(b_2 - \eta a_2)$, $(b_3 - \eta a_3)$, and $(-\eta)$. Since $(-\eta) < 0$, it is sufficient to find

$$\gamma = \max\{b_2 - \eta a_2,\ b_3 - \eta a_3\}.$$

We will need the probability distribution of γ, given α and β. First of all, note the densities of a_2, a_3, b_2, b_3. Since $\beta = \max\{b_1, b_2, b_3\}$, fixing β implies that the associated, conditional values of b_2 and b_3 are less than β, and that b_2 and b_3 are uniformly distributed, independently of one another, on $(-1, \beta)$; hence have constant probability densities $1/(\beta+1)$ on that interval. a_2 and a_3—which have no role in prior steps—have the original distributions. Hence,

(12)   For $j=2$, 3 the Prob. densities of a_j and b_j are:

$$f_{a_j}(x\,|\,\alpha,\beta)=\text{constant}=\tfrac{1}{2}, \quad -1<x<1$$
$$f_{b_j}(x\,|\,\alpha,\beta)=\text{constant}=1/(\beta+1), \quad -1<x<\beta.$$

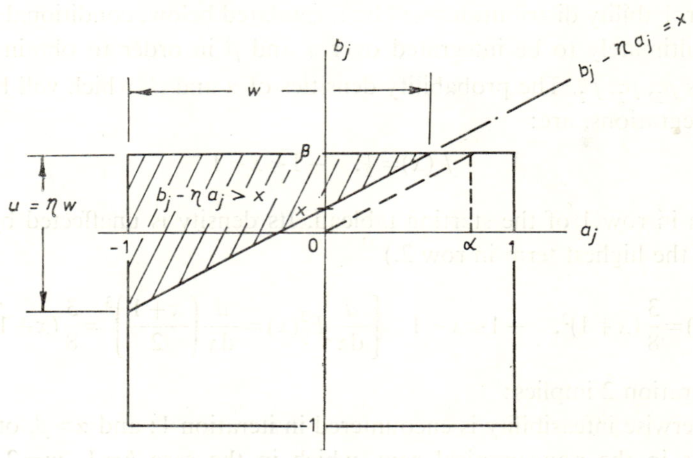

Fig. 5. Area representation of Prob $(b_j - \eta a_j > x)$

Fig. 5 represents a pair of random variables a_j, b_j ($j=2$ or 3) on $-1<a_j<1$, $-1<b_j<\beta$. Let x be a value which $b_j-\eta a_j$ may take on. The triangular area in the upper left corner of the rectangle corresponds to occurrence of $b_j-\eta a_j>x$, and the remainder of the rectangle to $b_j-\eta a_j<x$. Let w denote the length of the horizontal side and u the length of the vertical side of the triangle. Then

$$\frac{u}{w}=\eta=\beta/\alpha \quad \text{and} \quad \frac{w-1}{\beta-x}=\frac{\alpha}{\beta}, \quad \text{hence}$$

(13)
$$u=\beta w/\alpha$$
$$w=1+\alpha-\alpha x/\beta=1+(\beta-x)/\eta.$$

Setting $w=0$ gives $x=\beta+\eta$. At this x the line $b_j-\eta a_j=x$ goes through the upper left hand corner of the rectangle. $\beta+\eta$ is the highest value which $b_j-\eta a_j$ can have (conditional on α, β).

Setting $w=2$ gives $x=\beta-\eta$. In this case $b_j-\eta a_j=x$ goes through the upper right hand corner of the rectangle. By (10.2) $\beta-\eta$ is negative. It will suffice below to deal with $\beta-\eta<x<\beta+\eta$.

Based on areas in Fig. 5 and the densities of a_j and b_j given in (12), we have

$$\text{Prob } (b_j-\eta a_j>x)=\frac{uw}{2}f_{a_j}f_{b_j}=\frac{\beta w^2}{2\alpha}\cdot\frac{1}{2}\cdot\frac{1}{\beta+1}=\frac{\beta w^2}{4\alpha(\beta+1)}$$

(14)
$$\text{Prob } (b_j-\eta a_j<x)=1-\frac{\beta w^2}{4\alpha(\beta+1)}.$$

150

Since $\gamma=\max\{b_j-\eta a_j, j=2, 3\}$, squaring the right side of (14) gives the distribution of γ:

$$(15) \qquad F_\gamma(x \mid \alpha, \beta)=\mathrm{Prob}\,(\gamma<x \mid \alpha, \beta)=\left[1-\frac{\beta w^2}{4\alpha(\beta+1)}\right]^2$$

for $\beta-\eta<x<\beta+\eta$, where $w(x, \alpha, \beta)$ is given by (13). The density function is

$$(16) \qquad f_\gamma(x \mid \alpha, \beta)=\frac{d}{dx}\,F_\gamma=\frac{w}{\beta+1}\left[1-\frac{\beta w^2}{4\alpha(\beta+1)}\right].$$

Evaluation of p_2^0

Infeasibility is encountered at the start of the second iteration if $\gamma\leq 0$. To find the probability p_2^0 set $x=0$ in (13), giving $w=1+\alpha$ for use in (15), and integrate $F_\gamma(a \mid \alpha, \beta)$ over α and β, using the densities given in (9), and conditions (10.1). Figure 6 shows the domain of integration. We have

$$p_2^0=\int_{\alpha=0}^{1}\int_{\beta=0}^{\alpha}\left[1-\frac{\beta(1+\alpha)^2}{4\alpha(\beta+1)}\right]^2\cdot\frac{1}{2}\cdot\frac{3}{8}(\beta+1)^2\,d\alpha\,d\beta$$

$$=\frac{3}{256}\int_0^1\int_0^\alpha (2\beta+4-\alpha\beta-\beta/\alpha)^2\,d\alpha\,d\beta$$

$$=.090.$$

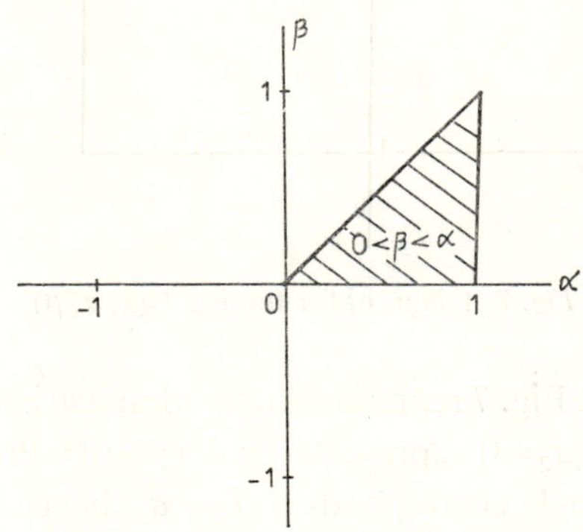

Fig. 6. Integration on α, β for p_2^0

Evaluation of p_2^a and p_2^b

$\gamma>0$ calls for carrying out an iteration pivot. Let it be assumed that the column which contains the higher of the quantities $b_j-\eta a_j$ is always put in column 2 so that column 2 can be called the pivot column. Then we write

$$(17) \qquad \begin{aligned}\gamma&=b_2-\eta a_2>b_3-\eta a_3\\ \gamma&=b_2-\eta a_2>0.\end{aligned}$$

The terms in the RHS column along with those in column 2 determine the pivot row. Of these four terms only a_2/α may be either negative or positive, the other three being positive. Under these conditions the ratio procedure for finding the pivot row is equivalent to:

If $\dfrac{1}{\alpha}(b_2 - \eta a_2) < \dfrac{a_2}{\alpha}(1-\eta)$, the pivot is in row 1.

If $\dfrac{1}{\alpha}(b_2 - \eta a_2) > \dfrac{a_2}{\alpha}(1-\eta)$, the pivot is in row 2.

Since, by (10.1), α is positive, these inequalities reduce to

(18)   If $b_2 < a_2$, the pivot is in row 1.
       If $b_2 > a_2$, the pivot is in row 2.

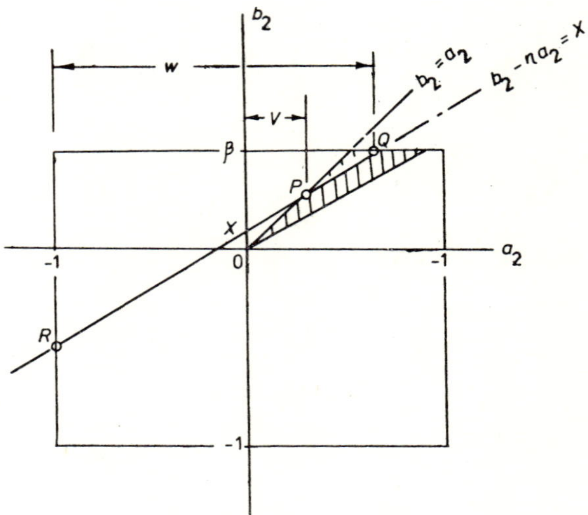

Fig. 7. Diagram for finding $\Theta(x, \alpha, \beta)$

Now (with α, β fixed) let Fig. 7 represent the random variables a_2, b_2. The triangular area above the line $b_2 - \eta a_2 = 0$ represents Prob $(\gamma > 0) = $ Prob $(b_2 - \eta a_2 > 0)$. Within that area, the shaded triangle corresponds to $b_2 < a_2$, hence to a pivot in row 1, and the remainder of the larger triangle to $b_2 > a_2$, hence a pivot in row 2.

For a fixed x consider a line $b_2 - \eta a_2 = x$ which passes through the shaded triangle. The highest such x passes through the point $b_2 = a_2 = \beta$; hence the highest value of $b_2 - \eta a_2$ at which there is a pivot in row 1 is $(\beta - \eta\beta)$. We will deal with a range of values of $\gamma = b_2 - \eta a_2$ indicated by $0 < x < \beta - \eta\beta$. The line $b_2 - \eta a_2 = x$ represents a value of γ in this interval. Let P be the point at which this line intersects the line $b_2 = a_2$. For a given value of x, the probability that the pivot is in row 1 corresponds to the ratio of the length of the line segment PQ to that of the line segment RQ. From the geometry, this ratio is

152

$$\Theta(x) = \frac{w - (v+1)}{w},$$

where v is the horizontal coordinate of P. It is the value of a_2 at the point of intersection of the lines $b_2 = a_2$ and $b_2 - \eta a_2 = x$; hence $v = x/(1-\eta)$. Hence

$$\Theta = \frac{1}{w}\left(w - 1 - \frac{x}{1-\eta}\right).$$

Integrating Θ (conditional probability of the 2-nd pivot being in row 1, given γ, α, and β) over γ between the limits $0 < \gamma < \beta - \beta\eta$ found above gives the probability of a pivot in row 1, given α, β:

$$\Pi_1(\alpha, \beta) = \int_{x=0}^{\beta - \eta\beta} \Theta(x, \alpha, \beta) f_\gamma(x \mid \alpha, \beta)\, dx,$$

where f_γ is given in (16)

$$= \int_0^{\beta - \eta\beta} \frac{1}{w}\left(w - 1 - \frac{x}{1-\eta}\right) \frac{w}{\beta + 1}\left[1 - \frac{\beta w^2}{4\alpha(\beta + 1)}\right] dx.$$

Use (13) to insert $\dfrac{\beta - x}{\eta}$ for $(w - 1)$ and $\left(\dfrac{\eta + \beta - x}{\eta}\right)^2$ for w^2. Also, to simplify the form of the integral, let $\lambda_1 = \eta + \beta$ and $\lambda_2 = \beta - \eta\beta$. Then

$$\Pi_1(\alpha, \beta) = \frac{1}{4\eta(1-\eta)(\beta+1)^2} \int_0^{\lambda_2} (\lambda_2 - x)[4\eta\beta + 4\eta - (\lambda_1 - x)^2]\, dx.$$

Routine integration gives

$$\text{(19)} \qquad \Pi_1(\alpha, \beta) = \frac{\beta^2(1-\eta)}{48\eta^2(\beta+1)^2}(24\beta\eta + 24\eta - \lambda_2^2 + 4\lambda_1\lambda_2 - 6\lambda_2^2).$$

Integration of Π_1 over the shaded area in Fig. 6, using the densities of α and β, takes the form

$$p_2^a = \int_{\alpha=0}^{1} \int_{\beta=0}^{\alpha} \Pi_1(\alpha, \beta) \cdot \frac{1}{2} \cdot \frac{3}{8}(\beta + 1)^2\, d\alpha\, d\beta.$$

Inserting (19) and carrying out another routine integration gives

$$p_2^a = .0040.$$

For p_2^b (see Fig. 2):

$$p_2^b = p_1^a - p_2^0 - p_2^a = .172 - .090 - .004 = .078.$$

REFERENCES

[1] Wolfe, P.: The composite simple algorithm, *SIAM REVIEW* 7 (1965), 42–54.
[2] Orden, A.: On the solution of linear equation/inequality systems, *Mathematical Programming* 1 (1971), 137–152.
[3] Halász, S.: On a stochastic programming problem with random coefficients, Colloquia Mathematica Societatis Janos Bolyai, 12, *Progress in Operations Research*, Vol. I, 493–498.
[4] Orden, A.: Computational investigation and analysis of probabilistic parameters of convergence of a simplex algorithm, Colloquia Mathematica Societatis Janos Bolyai, 12, *Progress in Operations Research*, Vol. II, 705–715.

NONLINEAR PROGRAMMING METHODS IN THE PRESENCE OF NOISE

B. T. POLJAK

(Moscow, USSR)

1. INTRODUCTION

Let us consider a nonlinear programming problem where the values of cost and constraint functions and their gradients have to be computed in the presence of noise. This may be, above all, the case when actual plants are optimized—the values of process output characteristics are obtained through measurements and these are inevitably noise—corrupted. Such problems, especially of unconstrained minimization, have been discussed in papers (e.g. by V. V. Kazakevich, A. A. Feldbaum, A. A. Pervozvansky and others [1]) on theory of extremal control. Another field are simplest stochastic programming problems where cost and constraint functions are of the form

$$f(x) = M\Phi(x, \omega) = \int \Phi(x, \omega) \, dP(\omega),$$

where the function $\Phi(x, \omega)$ is specified, the distribution $P(\omega)$ is unknown and only a sample $\omega_1, \ldots, \omega_n$ of that distribution is available. Then computing the accurate value of the regression function $f(x)$ and its gradient $\nabla f(x)$ in a subsequent point x_n is out of the question, but they can be approximated by $\Phi(x_n, \omega_n)$ and $\nabla_x \Phi(x_n, \omega_n)$. Recurrent minimization algorithms where this type of data is used have been widely used by Ya. Z. Tsypkin in theory of adaptive systems [2]. Finally, in a number of cases the cost function $f(x)$ has the same form $f(x) = \int \Phi(x, \omega) \, dP(\omega)$, where the distribution $P(\omega)$ is known, but computation of associated integrals would be too cumbersome. Then algorithms using $\Phi(x_n, \omega_n)$ and $\nabla_x \Phi(x_n, \omega_n)$ can be applied that can be regarded as versions of the Monte Carlo technique.

Recurrent stochastic algorithms were originally used instatistics for parameter estimation type problems. Developed in early fifties by Robbins and Monro, Kiefer and Wolfowitz, Dvoretzky, Blum and others, they are known as methods of stochastic approximation [3]. These can be regarded as gradient methods of unconstrained

155

minimization in the presence of random noise. Curiously enough, these studies were quite independent of work on deterministic gradient methods and in many respect ahead of them.

One of the first papers in which the method of stochastic approximation was extended to problems with constraints was that of V. Fabian [4]. A systematic study of stochastic methods for solution of nonlinear programming problems was undertaken by Yu. M. Yermol'ev and his followers [5]–[7] and later and independently by H. J. Kushner [8]–[9].

This paper will be concerned with certain new results in this field. Solution techniques will be presented for problems of nonconvex programming (e.g. problems with equality constraints) in the presence of noise. In the convex case algorithms will be described which converge without the assumptions that there is a unique solution and that the cost function is strictly convex. The basic purpose is to show in what way conventional deterministic methods of nonlinear programming can be modified so as to remain working even in the presence of noise. The resultant modified algorithms are "implementable" versions of associated "conceptual", using the terminology of [10], methods. It should be noted that these algorithms are of interest also for the deterministic case (i.e. for ordinary nonlinear programming problems).

True, stability of the gradient method in the presence of rounding off errors has been treated in a number of papers, see e.g. [11]. These, however, assumed that the errors are not random and their magnitude decreases when the number of iterations increases. In what follows random noise of a constant level, obviously, a more typical case, will be treated.

2. PROBLEMS WITH EQUALITY CONSTRAINTS

For the constrained extremum problem

(A)
$$\min f(x), \quad x \in R^N$$
$$g_i(x) = 0, \quad i = 1, \ldots, m$$

a number of iterative methods for finding a solution are known. Philosophicaly the most important of these are:

1. Lagrange multipliers method [12] (or primal-dual method)

$$L(x, y) = f(x) + (y, g(x)), \quad y \in R^m$$

(1)
$$x_{n+1} = \arg \min_x L(x, y_n)$$
$$y_{n+1} = y_n + \gamma g(x_{n+1}).$$

2. Penalty functions method [13]

$$\Phi_n(x) = \varepsilon_n f(x) + \|g(x)\|^2, \quad \varepsilon_n > 0, \quad \varepsilon_n \to 0$$

(2)
$$x_n = \arg \min_x \Phi_n(x).$$

156

3. Augmented Lagrangean method [14]–[16] (or multiplier method)

$$M(x, y) = L(x, y) + K/2 \|g(x)\|^2, \quad K > 0$$

(3)
$$x_{n+1} = \arg \min_x M(x, y_n)$$

$$y_{n+1} = y_n + Kg(x_{n+1}).$$

The substantiation and convergence studies for these methods have been treated, in particular, in papers [17]–[19], [33]. There is, first of all, one difficulty in extending the methods (1)–(3) to problems with noise. In E. Polak's terms, these methods are "conceptual" in that at each iteration an auxiliary problem of unconstrained minimization need be solved, which cannot be done accurately because of the noise. The well-known [17]–[19] criteria for approximate solution are not acceptable either. Therefore in further discussion we will deal with those modifications of the methods (1)–(3) whereby at each iteration only one step of the gradient method is made for the associated problem of unconstrained minimization.

About the problem (A) let us make the same assumptions that the functions are smooth and the solution is unique and regular which are necessary to prove convergence in the deterministic case.

A) There exists a point x^*, a local solution of (A); in the neighbourhood of that point $f(x)$, $g_i(x)$ are twice differentiable, $\nabla^2 f(x)$ and $\nabla^2 g_i(x)$ satisfy the Lipschitz condition, the vectors $\nabla g_i(x^*)$, $i = 1, \ldots, m$ are linearly independent and the second-order sufficiency condition holds. Under these assumptions Lagrange multipliers rule is applicable: there exist $y^* \in R^m$ such that $\nabla_x L(x^*, y^*) = \nabla_y L(x^*, y^*) = 0$ and the sufficiency condition is of the form $(\nabla_{xx}^2 L(x^*, y^*)x, x) > 0$ for $x \neq 0$, $\nabla_{xy}^2 L(x^*, y^*)x = 0$.

Assume that observable values of the functions and their gragients are noise-corrupted and denote by $F(x)$, $G_i(x)$ and $\nabla F(x)$, $\nabla G_i(x)$ the results of observations $f(x)$, $g_i(x)$ and $\nabla f(x)$, $\nabla g_i(x)$ respectively. About the noises $F(x) - f(x)$, $G_i(x) - g_i(x)$ and $\nabla F(x) - \nabla f(x)$, $\nabla G_i(x) - \nabla g_i(x)$ assume the following:

B) The noises depend only on the point x and are random, independent of each other, zero-mean and their variances are limited in the neighbourhood of x^* (i.e., for instance $MF(x) = f(x)$, $M(F(x) - f(x))^2 \leq \sigma^2$, etc. where M denotes the mean value).

With this notation the proposed methods are of the form:

1. Lagrange multipliers method

(4)
$$x_{n+1} = x_n - \gamma_n \left(\nabla F(x_n) + \nabla^T G(x_n)y_n \right)$$

$$y_{n+1} = y_n + \gamma_n G(x_n).$$

2. Penalty function method

(5)
$$x_{n+1} = x_n - \gamma_n \left(\varepsilon_n \nabla F(x_n) + \nabla^T G(x_n)G(x_n) \right).$$

3. Augmented Lagrangean method

$$(6) \qquad \begin{aligned} x_{n+1} &= x_n - \gamma_n\big(\nabla F(x_n) + \nabla^T G(x_n)(y_n + KG(x_n))\big) \\ y_{n+1} &= y_n + \gamma_n G(x_n). \end{aligned}$$

In the absence of noise (or with $F(x)=f(x)$, $\nabla F(x)=\nabla f(x)$, etc.) the method (4) and a continuous analog of the method (6) were proposed in [12]; the discrete method (6) was studied in [20]. For problems with noise the methods (4)–(6) seem never to have been studied.

Impose a constraint on the parameters γ_n.

C) $$\sum_{n=0}^{\infty} \gamma_n = \infty, \qquad \sum_{n=0}^{\infty} \gamma_n^2 < \infty.$$

This condition is ordinary for convergence of the stochastic approximation methods. If there are no noises ($\sigma^2=0$), the condition $\Sigma\gamma_n^2 < \infty$ may be eliminated.

In substantiating convergence of the methods (4)–(6) the following fact is significant. In the deterministic case convergence of any iterative method for solving nonconvex problems (such as nonlinear problems (A) with nonlinear $g_i(x)$) is of local nature. For instance, the method (1) converges only if an initial approximation x_0 is close to the solution. With random noise there is a (generally nonzero) probability that the iterative process will leave the convergence domain. Consequently, there is no convergence in the conventional probabilistic sense (e.g. mean square or with probability 1). Therefore in the Theorems below it is convergence with a probability $1-\delta$ that is stated where δ decreases when the performance of the initial approximation is better, the level of noise lower and the step-sizes γ_n shorter.

Theorem 1. *Assume that conditions A, B, and C hold and $\nabla^2_{xx}L(x^*,y^*)>0$ (i.e. this matrix is positive definite). Then for the method (4)*

$$P(x_n \to x^*, y_n \to y^*) \geq 1-\delta, \quad \delta = c_1 M(\|x_0 - x^*\|^2 + \|y_0 - y^*\|^2) + c_2\sigma^2 \sum_{n=0}^{\infty} \gamma_n^2,$$

where c_1 and c_2 are constants which depend on the deterministic part of the problem. The method converges at a rate $O(n^{-1})$ in the following sense. With $\gamma_n = \gamma/(n+a)$ for any $\delta_0>0$ one can choose γ, a, ϱ, c so that with $M(\|x_0 - x^*\|^2 + \|y_0 - y^*\|^2) \leq \varrho$ we will have

$$P\left(\|x_n - x^*\|^2 + \|y_n - y^*\|^2 \leq \frac{c}{n+a}\right) \geq 1-\delta_0.$$

Theorem 2. *Let the conditions A, B and C hold, the random values $G_i(x)$ and $\nabla G_i(x)$ are independent (or the observation noises of the functions and their gradients are independent) and*

$$\sum_{n=0}^{\infty} \gamma_n \varepsilon_n = \infty, \qquad \sum_{n=0}^{\infty} \gamma_n \varepsilon_n^2 < \infty.$$

Then for the method (5)

$$P(x_n \to x^*) \geq 1 - \delta, \quad \delta = c_1 M \|x_0 - x^*\|^2 + c_2 \sigma^2 \sum_{n=0}^{\infty} \gamma_n \varepsilon_n^2.$$

Sequences γ_n and ε_n which satisfy these conditions are, for instance, $\gamma_n = \gamma n^{-\frac{3}{4}}$, $\varepsilon_n = \varepsilon n^{-\frac{1}{4}}$ (this choice is, in a sense, optimal in the class of exponential dependencies), then the method converges at a rate of $O(n^{-\frac{1}{2}})$.

Theorem 3. *Let the conditions* A, B *and* C *hold and* $G_i(x)$ *and* $\nabla G_i(x)$ *are independent of each other. Then there is* K_0 *such that with* $K \geq K_0$ *for the method* (6)

$$P(x_n \to x^*, y_n \to y^*) \geq 1 - \delta, \quad \delta = c_1 M(\|x_0 - x^*\|^2 + \|y_0 - y^*\|^2) + c_2 \sigma^2 \sum_{n=0}^{\infty} \gamma_n^2.$$

With $\gamma_n = \gamma n^{-1}$ *the method converges at a rate of* $O(n^{-1})$.

Theorems 1–3 are proved by using the results of Refs [17–19] for deterministic methods and a general theorem [21] on local convergence of iterative stochastic processes with probability $1 - \delta$.

If there are no noises and the condition A holds, the methods (4) and (6) (the latter with K large enough) locally converge if $0 \leq \gamma_n \leq \gamma$, γ being small enough and

$$\sum_{n=0}^{\infty} \gamma_n = \infty.$$

A geometric convergence rate hold if $\lim_{n \to \infty} \gamma_n > 0$. In the same situation the method

(5) locally converges if $\varepsilon_n \to 0$, $\gamma_n \leq K \varepsilon_n$ (K is a certain constant), $\sum_{n=0}^{\infty} \gamma_n \varepsilon_n = \infty$.

Now compare the convergence of the methods. The Lagrange multiplier method (4) converges only with a very stringent assumption that $\nabla^2_{xx} L(x^*, y^*) > 0$ (which may be shown to be significant). The method of penalty functions (5) does not require that condition, but its convergence is as low as $O(n^{-\frac{1}{2}})$ whereas that of the method (4) is $O(n^{-1})$. The augmented Lagrangean method (6) has no such disadvantages in that it does not require strong convexity of $L(x, y^*)$ with respect to x and insures a convergence of about $O(n^{-1})$. Consequently, the effectivenesses of these methods are related as the deterministic methods (1)–(3) are [17]–[19].

Note also that the requirement that the noises $G_i(x) - g_i(x)$ and $\nabla G_i(x) - \nabla g_i(x)$ are independent is not too stringent. Thus if estimates of the gradients are obtained by finite difference approximation (which means that only values of the functions are observable) then by using symmetrical difference approximation

$$\nabla G_i(x) = (2\alpha)^{-1} \sum_{j=1}^{m} (G_i(x + \alpha e_j) - G_i(x - \alpha e_j)) e_j$$

the condition that $G_i(x)$ and $\nabla G_i(x)$ are independent will hold.

3. MINIMIZATION ON A CONVEX SET

Consider the problem

(B)
$$\min_{x\in\Omega} f(x), \quad x\in R^N$$

where $f(x)$ is a convex continuous function and Ω is a convex closed set such that a projection P_Ω onto it can be readily obtained (e.g. $\Omega=\{x: a\leqq x\leqq b\}$).

A major method of solving the problem (B) is subgradient projection method

(7)
$$x_{n+1}=P_\Omega(x_n-\gamma_n\partial f(x_n)),$$

where $\partial f(x)$ is any subgradient of $f(x)$ at the point x. For the smooth case (i.e. with $\partial f(x)=\nabla f(x)$) this method was proposed in Refs [22, 23]. For nonsmooth problems of unconstrained minimization the idea of the method was suggested by N. Z. Shor [24] and the method was substantiated by Yu. M. Yermol'yev [25]. In the general form the method (7) with various ways of choice γ_n was studied by the author of this paper [26, 27]. A survey of the results related to the method (7) is to be found in [28].

If the subgradient is noise-corrupted then an analog of (7) is the stochastic subgradient projection method

(8)
$$x_{n+1}=P_\Omega(x_n-\gamma_n\partial F(x_n)),$$

where $\partial F(x)$ is the observed value of $\partial f(x)$. This method was proposed and studied by Yu. M. Yermol'yev [5, 6] who showed, in particular, that under natural assumptions $\left(\text{such as that the noise variance and } \Omega \text{ are limited, } \gamma_n \text{ satisfies the conditions } \sum_{n=\gamma}^{\infty} \gamma_n=\infty, \ \sum_{n=0}^{\infty} \gamma_n^2<\infty\right)$ the method (8) converges to a set of minimum points of the problem (B) with a probability 1. Then if $\gamma_n=\gamma n^{-1}$, γ is large enough and $f(x)$ satisfies the condition $f(x)\geqq f(x^*)+l\|x-x^*\|^2$, $l>0$, for all $x\in\Omega$, then $M\|x_n-x^*\|^2==O(n^{-1})$.

Let us show one more interesting result on the convergence of the method (8). Let $f(x)$ satisfy the condition

(9)
$$f(x)\geqq f(x^*)+l\|x-x^*\|, \quad l>0, \quad x\in\Omega$$

(e.g. $f(x)$ is piecewise linear and has a unique minimum and $\Omega=R^N$ or (B) is a linear programming problem with a unique solution). Let the noise be not necessarily random, but satisfy the condition

(10)
$$\|\partial F(x)-\partial f(x)\|\leqq c<l, \quad x\in\Omega.$$

Let, finally $\|\partial f(x)\|\leqq K$, $x\in\Omega$.

Theorem 4. *If with the above assumptions we select*

$$\gamma_n = a\|x_0 - x^*\| q^n, \quad 0 < a < \frac{2(l-c)}{K^2 - c^2}, \quad 1 > q > \sqrt{1 - 2al + a^2 K^2} + ac,$$

then for the method (8) $\|x_n - x^*\| \leq \|x_0 - x^*\| q^n$.

Consequently, here despite the noise, linear convergence proceeds, a result opposite to conventional assertions on convergence of stochastic methods. Note also that adjustment of the step-size $\gamma_n = \gamma_0 q^n$ was proposed (for problems without noise) by N. Z. Shor [29].

Convergence rate of the method (8) with the condition (10) not holding remains an open question.

Now consider the case where there exists more than one solution of the problem (B), which is typical, for instance, in pattern recognition. The behaviour of the methods (7) and (8) may be quite irregular. In a number of cases they can be shown to converge to one of the solutions, but this solution may be rather arbitrary (e.g. be different for different realizations of the stochastic method (8)). Therefore, it is desirable to make the methods stable. In particular, they could converge to an extremum point featuring certain properties (e.g. to a normal solution i.e. a solution with a minimal norm). Thus we arrive at the following statement of the problem. Among all solutions X^* of the problem (B) it is required to find that for which the function $f_1(x)$ is minimal. Note that this statement is frequent in multi-criterion problems.

To solve this problem in the deterministic case the regularization method is used. An approximate solution x_n of the regularized problem is found

$$x_n = \arg \min_{x \in \Omega} \left(f(x) + \alpha_n f_1(x) \right)$$

and then the regularization parameter $\alpha_n > 0$ is made to go to 0. Convergence of this method was proved in [30].

This approach does not work in the stochastic case (the auxiliary minimization problem cannot be solved) but the approach mentioned above whereby only one step of the gradient method is made for the auxiliary problem, can be used here. In this way we obtain the method

$$(11) \qquad x_{n+1} = P_\Omega \left[x_n - \gamma_n \left(\partial F(x_n) + \alpha_n \partial F_1(x_n) \right) \right].$$

Here, as before, $\partial F(x)$, $\partial F_1(x)$ are values of the subgradients $\partial f(x)$, $\partial f_1(x)$ in the presence of noise. The method (11) for the deterministic case was proposed in [31]. (See also the paper by A. B. Bakushinsky of this Symposium.)

Suppose $f(x)$, $f_1(x)$ are convex continuous functions, Ω is a closed bounded set, $f_1(x)$ is a strongly convex function (e.g. $f_1(x) = \|x\|^2$),

$$X^* = \operatorname*{Arg\,min}_{x \in \Omega} f(x), \quad x^* = \operatorname*{arg\,min}_{x \in X^*} f_1(x),$$

the noises are zero-mean and their variances are limited:

$$M\partial F(x)=\partial f(x), \quad M\partial F_1(x)=\partial f_1(x), \quad M\|\partial F(x)-\partial f(x)\|^2\leqq\sigma^2,$$
$$M\|\partial F_1(x)-\partial f_1(x)\|^2\leqq\sigma^2.$$

Theorem 5. *If with the above assumptions*

$$\sum_{n=0}^{\infty}\gamma_n\alpha_n=\infty, \quad \gamma_n/\alpha_n\rightarrow0, \quad \alpha_n\rightarrow0$$

then in the method (11) $M\|x_n-x^*\|^2\rightarrow0$.

The sequences γ_n and α_n may be taken, for instance as

$$\gamma_n=\gamma n^{-1}, \quad \alpha_n=\frac{\alpha}{\log n} \quad \text{or} \quad \gamma_n=\gamma n^{r-1}, \quad \alpha_n=\alpha n^{-r}, \quad 0<r<\frac{1}{2}.$$

The condition that Ω is bounded may be weakened.

4. PROBLEMS OF CONVEX PROGRAMMING

For general convex programming problem

(C)
$$\min f(x), \quad x\in R^N$$
$$g_i(x)\leqq0, \quad i=1,\ldots,m, \quad x\in\Omega$$

(where $f(x)$ and $g_i(x)$ are convex functions, Ω is a convex set) the methods of solution in the presence of random noise were studied, as already mentioned, by V. Fabian, Yu. M. Yermol'yev and his followers, J. Kushner, V. Ya. Katkovnik and others [4, 5–9, 32]. We will discuss only the essentially most important, Lagrange multipliers method. For the deterministic case it is of the form

(12)
$$x_{n+1}=P_\Omega\big[x_n-\gamma_n\big(\partial f(x_n)+\partial^T g(x_n)y_n\big)\big].$$
$$y_{n+1}=[y_n+\gamma_n g(x_n)]_+.$$

This method (with $\Omega=R^N$, $f(x)$ and $g_i(x)$ smooth) was proposed in [12]. But its justification by Uzawa had an error. Various authors have proved a weaker result on convergence of the method (12) viz. that there is a subsequence x_{n_i} which converges to the solution. Convergence of the entire sequence x_n (and the behaviour of y_n) have been open questions until very recently when G. D. Maistrovsky obtained a solution.

The method (12) converges only under a very stringent assumption that $f(x)$ is strictly convex. This assumption is essential; for instance, one can show that for problems of linear programming the method (12) is certain to diverge no matter how γ_n is selected.

A stochastic analog of the method (12) was proposed by Yu. M. Yermol'yev and Z. V. Nekrylova [5] who proved convergence of the subsequence x_{n_i} with probability 1 provided that $f(x)$ is strictly convex

In what follows a modification of the Lagrange multipliers method is proposed which converges (with respect to x and y) without the assumption that $f(x)$ is strictly convex. That method was initially introduced (for deterministic problems) by A. B. Bakushinsky and the author [31]. Its framework is the same idea of recurrent regularization as in the above modification of the subgradient projection method. Specifically, a regularized Lagrange function is considered

$$L_n(x, y) = f(x) + (y, g(x)) + \frac{\alpha_n}{2} \|x\|^2 - \frac{\alpha_n}{2} \|y\|^2$$

and one step of the Arrow—Hurwicz—Uzawa process is made to find its saddle point. Consequently, we obtain the method

(13)
$$\begin{aligned}
x_{n+1} &= P_\Omega\big[x_n - \gamma_n(\partial F(x_n) + \partial^T G(x_n)y_n + \alpha_n x_n)\big] \\
y_{n+1} &= \big[y_n + \gamma_n(G(x_n) - \alpha_n y_n)\big]_+,
\end{aligned}$$

where as before $\partial F(x)$, $\partial G(x)$ and $G(x)$ denote "disturbed" values of $\partial f(x)$, $\partial g(x)$ and $g(x)$ respectively.

Let us assume that:

A) $f(x)$, $g_i(x)$ are convex continuous functions.
B) Ω is a convex closed bounded set.
C) The Slater condition holds.
D) All noises are zero-mean, have a limited variance and are independent in different points.

Theorem 6. Let the above assumptions hold and the parameters γ_n and α_n satisfy the conditions

(14)
$$\sum_{n=0}^{\infty} \gamma_n \alpha_n = \infty, \quad \gamma_n/\alpha_n \to 0, \quad \alpha_n \to 0, \quad \frac{\alpha_{n-1} - \alpha_n}{\gamma_n \alpha_n^2} \to 0.$$

Then in the method (13) $M\|x_n - x^*\|^2 \to 0$, $M\|y_n - y^*\|^2 \to 0$, where x^* is the solution of the problem (C) with the least norm, y^* is the Lagrange multiplier with the least norm.

To γ_n and α_n satisfying the conditions (14) belong $\gamma_n = \gamma n^{-r}$, $\alpha_n = \alpha n^{-s}$, $0 < s < r$, $s + r < 1$.

The method (13) may be applied, in particular, to the deterministic linear programming problem

$$\min (c, x)$$
$$Ax \leqq b, \quad x \geqq 0.$$

In this case it is of the form

$$(15) \qquad \begin{aligned} x_{n+1} &= [x_n - \gamma_n(c + A^T y_n + \alpha_n x_n)]_+ \\ y_{n+1} &= [y_n + \gamma_n(Ax_n - b - \alpha_n y_n)]_+ \end{aligned}$$

and converges (with the above conditions for γ_n and α_n) to the normal solution of the primal and dual problems.

REFERENCES

[1] Rastrigin, L. A.: *Extremal control systems,* Moscow, "Nauka", 1974 (in Russian).

[2] Tsypkin, Ja. Z.: *Adaptation and learning in automatic systems,* Acad. Press. 1971.

[3] Nevel'son, M. B., Hasminskiy, R. Z.: *Stochastic approximation and recurrent estimation,* Moscow, "Nauka", 1972 (in Russian).

[4] Fabian, V.: Stochastic approximation of constrained minima, *Trans. 4th Prague Conf. Inf. Th. Stat. Decis. Funct. Random Proc.,* 1965, Prague, 1967.

[5] Ermol'ev, Yu. M., Nekrylova, Z. V.: Method of stochastic gradients and its applications, in *Theory of optimal decisions,* No. 1, Kiev, Institute of cybernetics, 1967 (in Russian).

[6] Ermol'ev, Yu. M.: On the method of generalized stochastic gradients and stochastic quasi-Feuer sequences, *Kibernetika* (Kiev), No. 2, 1969.

[7] Ermol'ev, Yu. M.: *Stochastic programming methods,* Moscow, "Nauka", 1976 (in Russian).

[8] Kushner, H. J.: Stochastic approximation algorithms for constrained optimization problems, *Ann. Stat.* 2, 1974, No. 4.

[9] Kushner, H. J., Gavin, T.: Stochastic approximation type methods for constrained systems: algorithms and numerical results, *IEEE Trans. Autom. Control,* 19, 1974, No. 4.

[10] Polak, E.: *Computational methods in optimization. A unified approach.* Acad. Press. 1971.

[11] Poljak, B. T.: Convergence of feasible directions methods in extremal problems. *USSR Comput. Math. Math. Phys.* 11, 1971, No. 4.

[12] Arrow, K. J., Hurwicz, L., Uzawa, H.: *Studies in linear and non-linear programming,* Stanford Univ. Press., 1958.

[13] Courant, R.: Variational methods for the solution of problems of equilibrium and vibration, *Bull. Amer. Math. Soc.,* 49, 1943, No. 1.

[14] Hestenes, M. R.: Multiplier and gradient methods, *Journ. Optimiz. Theory Appl.* 4, 1969, No. 5.

[15] Powell, M. J. D.: A method for nonlinear constraints in minimization problems, in *Optimization,* London, Acad. Press., 1969.

[16] Haarhoff, P. C., Buys, J. D.: A new method of optimization of a nonlinear function subject to nonlinear constraints, *Comput. Journ.,* 13, (1970), No 2

[17] Poljak, B. T.: Iterative methods using Lagrange multipliers for solving extremal problems with equality constraints. USSR Comput. *Math. Math. Phys.* 10, (1970), No. 5.

[18] Poljak, B. T.: On the rate of convergence of penalty function method, *USSR Comput. Math. Math. Phys.* 11, (1971), No. 1.

[19] Poljak, B. T., Tretjakov, N. V.: Method of penalty prices for conditional extremal problems, *USSR Comput. Math. Math. Phys.,* 13, (1973), No. 1.

[20] Belen'kiy, V. Z., Volkonskiy, A. V., Ivankov, S. A., Pomanskiy, A. B., Shapiro, A. D.: Iterative methods in theory of games and programming, Moscow, *Nauka,* 1974 (in Russian).

[21] Poljak, B. T.: Convergence and rate of convergence of iterative stochastic algorithms. I. General case, *Automation and Remote Control,* (1976), No. 12.

[22] Goldstein, A. A.: Convex programming in Hilbert space, *Bull. Amer. Math. Soc.,* 70, (1964), No. 5.

164

[23] Levitin, E. S., Poljak, B. T.: Minimization methods under constraints, *USSR Comput. Math. Math. Phys.*, 6, (1966), No. 5.

[24] Shor, N. Z.: Application of gradient descent method to the solution of network transportation problem, in *Proc. sci. semin. theor. and appl. probl. of cybernet. and operat. research,* No. 1, Kiev, 1962, Sci. council Ac. Sci. Ukr. SSR on Cybernet. (in Russian).

[25] Ermol'ev, Yu. M.: Methods of solving nonlinear extremum problems, *Kibernetika* (Kiev), No. 4, 1966.

[26] Poljak, B. T.: A general method for solving extremum problems, Soviet Math. Dokl., 174, (1967), No. 1.

[27] Poljak, B. T.: Minimization of unsmooth functionals, *USSR Comput Math. Math. Phys.*, 9, (1969), No. 3.

[28] Shor, N. Z.: Minimization of nondifferentiable functions, *Ekonomica i Matem. Meth.*, (1976), No. 2, (in Russian).

[29] Shor, N. Z., Gamburd, P. R.: Some problems concerning convergence of generalized gradient descent, *Kibernetika* (1971), No. 6, (Kiev).

[30] Levitin, E. S., Poljak, B. T.: Convergence of minimizing sequences in conditional extremum problems, *Soviet. Math. Dokl.*, 168, (1966), No. 5.

[31] Bakushinskiy, A. B., Poljak, B. T.: On a solution of variational inequalities, *Soviet Math. Dokl.*, 219, (1974), No. 5.

[32] Katkovnik, V. Ja.: Linear estimation and stochastic optimization problems, Moscow, *Nauka* 1976 (in Russian).

[33] Bertsekas, D. P.: Multiplier methods: a survey. *Automatica,* 12, (1976), No. 2.

RELIABILITY TYPE INVENTORY MODELS BASED ON STOCHASTIC PROGRAMMING

A. PRÉKOPA and P. KELLE

(Budapest, Hungary)

1. INTRODUCTION

The models discussed in the present paper are generalizations of the models introduced previously by A. Prékopa [6] and M. Ziermann [13]. In the mentioned papers the initial stock level of one basic material is determined where the delivery and demand process allow certain homogeneity (in time) assumptions if they are random. Here we are dealing with more than one basic material and drop the time homogenity assumption. Only the delivery processes will be assumed to be random. They will be supposed to be stochastically independent. Out of the models discussed in this paper the first one was already introduced in [9]. All these models are stochastic programming models where algorithms serve for the determination of the initial stock levels instead of simple formulas. We have to solve nonlinear programming problems where one of the constraints is probabilistic. The function and gradient values of the corresponding constraining function are determined by simulation.

The numerical evaluation of the models discussed in the present paper is more sophisticated than those of the earlier models of Prékopa and Ziermann. However, if the delivery process is inhomogeneous, then with the present methodology we can get closer to reality and can handle many delivery processes simultaneously.

The most general model introduced in [6] is the following. Let M denote the initial stock level, $(0, T)$ the investigated time interval, α_t the amount of the basic material delivered up to time t and β_t the cumulative demand which occurred up to t, where $0 \leq t \leq T$. The initial stock level is to be determined in such a way that this be smallest M satisfying

$$(1.1) \qquad P\left(\inf_{0 \leq t \leq T} (M + \alpha_t - \beta_t) > 0 \right) \geq 1 - \varepsilon,$$

where ε is a previously prescribed, in practice, low value, e.g. $\varepsilon = 0.05$. Under the assumptions introduced in [6] in connection with the random processes α_t, β_t. Relation (1.1) holds with equality in case of the optimal initial stock. Thus an equation serves for the determination of M. This is called the Reliability Equation.

For the easier understanding of the generalizations presented in this paper we need to repeat the modelling of the random processes α_t, β_t introduced in [6]. Since the model for β_t is the same as that of α_t only its parameters are different, it will be enough to deal with α_t only.

Let λ be a real number satisfying $0 \leq \lambda \leq 1$ and $t_1, \ldots, t_n$ further $\tau_1, \ldots, \tau_{n-1}$ be independent samples taken from the population uniformly distributed in (0.1). Let $\tau_1^* \leq \tau_2^* \leq \ldots \leq \tau_{n-1}^*$ be the ordered sample corresponding to τ_i, $i = 1, \ldots, n-1$ and put $\tau_0^* = 0$, $\tau_n^* = 1$. Now we define α_t in the following manner:

$$(1.2) \qquad \alpha_t = c\lambda v/n + c(1-\lambda)\tau_v^*, \quad 0 \leq t \leq T,$$

where v is the number of those t_i which are smaller than t, c is a positive constant, cT equals the total demand occuring in the time interval $(0, T)$ and this is supposed to be equal the total amount of basic material delivered in the same time interval. If $\lambda = 1$, then α_t is the empirical probability distribution function belonging to the sample $t_1, \ldots, t_n$. In connection with β_t we use m instead of n and μ instead of λ.

In [6] it is mentioned that the following limit relations hold:

$$\lim_{\substack{m \to \infty \\ n \to \infty}} P\left[\left(\frac{mn}{m+n+m(1-\lambda)^2+n(1-\mu)^2}\right)^{\frac{1}{2}} \sup_{0 \leq t \leq 1} (\alpha_t - \beta_t) < y\right] =$$

$$(1.3) \qquad = \lim_{\substack{m \to \infty \\ n \to \infty}} P\left[\left(\frac{mn}{m+n+m(1-\lambda)^2+n(1-\mu)^2}\right)^{\frac{1}{2}} \sup_{0 \leq t \leq 1} (\beta_t - \alpha_t) < y\right] =$$

$$= \begin{cases} 1 - \exp(-2y^2), & \text{if } y > 0 \\ 0, & \text{if } y \leq 0. \end{cases}$$

Here we have fixed $T = 1$ for the sake of simplicity. This choice does not restrict the generality.

If we assume the left hand sides of (1.3) approximately equal to the right hand side value, then for a given ε the following $M = M_{\lambda,\mu}$ value turns out to be the approximate solution of the Reliability Equation:

$$(1.4) \qquad M_{\lambda,\mu} = c\left[\frac{1}{2}\left(\frac{1+(1-\lambda)^2}{n} + \frac{1+(1-\mu)^2}{m}\right) \log \frac{1}{\varepsilon}\right]^{\frac{1}{2}}.$$

If α_t is a deterministic process and $\alpha_t = ct$ $(0 \leq t \leq 1)$, then the corresponding M value can be obtained from (1.4) if we take the limit $n \to \infty$. We proceed similarly, if β_t is deterministic. We remark that the minimal amount δ delivered at one delivery time and λ are in the following relation: $\lambda = n\delta/c$. Similar relation holds for the parameters of the process β_t.

168

2. GENERALIZATION OF THE DELIVERY AND DEMAND PROCESSES

In this section we repeat the generalization of the delivery process as it is given in [9].

In Section 1 we mentioned the following assumptions in connection with the delivery process: (a) the number of delivery times is fixed, this was denoted by n; (b) the n delivery time points are so distributed in the interval $(0, 1)$ as the elements of a sample of size n taken from a population uniformly distributed in the same interval;

(c) the total delivered amount is constant and is equal to c which is at the same time equal to the total demand;

(d) the random vector the components of which are the random delivered amounts is stochastically independent of the random vector of the delivery time points; (e) denoting by δ the smallest amount to be delivered if one delivery occurs, the model for the distribution of the remaining amount among the n delivery times is the following: divide the interval $(0, c-n\delta)$ into n parts by choosing $n-1$ independent and uniformly distributed random points and assign the quantities equal to the lengths of the subintervals to the n delivery times. In what follows we maintain the assumptions (a), (c), (d), and modify the assumptions (b), (e).

For the modelling of the delivery process we choose L uniformly distributed independent random points in the interval $(0, c-n\delta)$, where $L>n-1$. Let $y_1^*, \ldots, y_L^*$ denote the ordered sample formed from the L random points. Out of this ordered sample we select those which have subscripts $k_1 < k_2 < \ldots < k_{n-1}$ and add to the fixed delivery amounts the following

$$(2.1) \qquad \eta_1 = y_{k_1}^*, \quad \eta_2 = y_{k_2}^* - y_{k_1}^*, \quad \ldots, \quad \eta_n = c - n\delta - y_{k_{n-1}}^*.$$

Thus the amounts delivered at the delivery times will be

$$(2.2) \qquad \delta + \eta_1, \quad \delta + \eta_2, \quad \ldots, \quad \delta + \eta_n.$$

Similar model is used for the delivery time points. To the fixed amount to be delivered at one occasion there corresponds a fixed time γ as the minimal distance between two consecutive delivery time points ($0 \leq \gamma \leq 1/n$). The delivery time points are selected from an ordered sample $x_1^* \leq x_2^* \leq \ldots \leq x_N^*$ of a sample of size N taken from a population uniformly distributed in $(0, 1-n\gamma)$, so that we select those elements which have subscripts $j_1 < j_2 < \ldots < j_n$, form the random variables

$$(2.3) \qquad \xi_1 = x_{j_1}^*, \quad \xi_2 = x_{j_2}^* - x_{j_1}^*, \quad \ldots, \quad \xi_n = x_{j_n}^* - x_{j_{n-1}}^*.$$

and finally take the partial sums of the random variables

$$(2.4) \qquad \gamma + \xi_1, \quad \gamma + \xi_2, \quad \ldots, \quad \gamma + \xi_n.$$

These partial sums represent the n delivery time points.

Let $s(z_1, \ldots, z_{n-1})$ denote the joint probability density function of the random variables $\eta_1, \ldots, \eta_{n-1}$. It is easy to see that this function has the following form

$$s(z_1, \ldots, z_{n-1})=$$

$$(2.5) \qquad =\left(\frac{1}{c-n\delta}\right)^n \frac{\Gamma(L+1)}{\Gamma(k_1)\Gamma(k_2-k_1)\ldots\Gamma(k_{n-1}-k_{n-2})\Gamma(L+1-k_{n-1})} \cdot$$

$$\cdot \left(\frac{z_1}{c-n\delta}\right)^{k_1-1} \left(\frac{z_2}{c-n\delta}\right)^{k_2-k_1-1} \cdots \left(\frac{z_{n-1}}{c-n\delta}\right)^{k_{n-1}-k_{n-2}-1} \left(1-\frac{z_1+\ldots+z_{n-1}}{c-n\delta}\right)^{L-k_{n-1}},$$

if $z_i>0$, $i=1, \ldots, n-1$; $z_1+\ldots+z_{n-1}<c-n\delta$ and $s(z_1, \ldots, z_{n-1})=0$

otherwise.

Similar formula gives the joint probability density function of the random variables $\xi_1, \ldots, \xi_n$ which we denote by $r(z_1, \ldots, z_n)$.

$$r(z_1, \ldots, z_n)=$$

$$(2.6) \qquad =\left(\frac{1}{1-n\gamma}\right)^{n+1} \frac{\Gamma(N+1)}{\Gamma(j_1)\Gamma(j_2-j_1)\ldots\Gamma(j_n-j_{n-1})\Gamma(N+1-j_n)} \cdot$$

$$\cdot \left(\frac{z_1}{1-n\gamma}\right)^{j_1-1} \left(\frac{z_2}{1-n\gamma}\right)^{j_2-j_1-1} \cdots \left(\frac{z_n}{1-n\gamma}\right)^{j_n-j_{n-1}-1} \left(1-\frac{z_1+\ldots+z_n}{1-n\gamma}\right)^{N-j_n},$$

if $z_i>0$, $i=1, \ldots, n$; $z_1+\ldots+z_n<1-n\gamma$ and $r(z_1, \ldots, z_n)=0$

otherwise.

Thus the random vectors $(\eta_1, \ldots, \eta_{n-1})$ and $(\xi_1, \ldots, \xi_n)$ have Dirichlet distributions. For properties of this multivariate distribution the reader is referred to [12].

3. THE INVENTORY MODELS

Model I [9]. The model for the delivery process is that one discussed in Section 2. The demand is assumed to have constant intensity i.e. the demand occurring in the interval $(0, t)$ is equal to ct where c is a constant. M denotes the initial stock level. The demand will be met continuously in the whole interval $(0, 1)$ if and only if the following relations hold

$$M \geqq \gamma+\xi_1,$$
$$M+\delta+\eta_1 \geqq 2\gamma+\xi_1+\xi_2,$$
$$M+2\delta+\eta_1+\eta_2 \geqq 3\gamma+\xi_1+\xi_2+\xi_3,$$

$$(3.1) \qquad \qquad \cdot$$
$$\cdot$$
$$\cdot$$

$$M+(n-1)\delta+\eta_1+\ldots+\eta_{n-1} \geqq n\gamma+\xi_1+\xi_2+\ldots+\xi_n.$$

170

Let us introduce the notations:

$$\zeta_1 = \xi_1,$$
$$\zeta_2 = \xi_1 + \xi_2 - \eta_1,$$

(3.2)

$$\vdots$$

$$\zeta_n = \xi_1 + \ldots + \xi_n - \eta_1 - \ldots - \eta_{n-1}.$$

The random vectors $\boldsymbol{\eta} = (\eta_1, \ldots, \eta_{n-1})$ and $\boldsymbol{\xi} = (\xi_1, \ldots, \xi_n)$ are independent and their probability density functions are logconcave functions in R^{n-1} resp. R^n. It follows that the altogether $2n-1$ components have a logconcave joint density in R^{2n-1}.

The notation of a logconcave probability measure is introduced in [7]. A probability measure P defined on the measurable subsets of R^m is said to be logconcave if for every pair A, B of convex subsets, of R^m and every $0 < \lambda < 1$ the following inequality holds:

(3.3)
$$P(\lambda A + (1-\lambda)B) \geqq [P(A)]^{\lambda} [P(B)]^{1-\lambda}.$$

The main theorem of [7] says that if a probability measure is generated by a logconcave probability density, then it is a logconcave measure. On the other hand any linear transform of a random vector having logconcave distribution has again logconcave distribution [9, Theorem 3]. Thus the random vector $\boldsymbol{\zeta} = (\zeta_1, \ldots, \zeta_n)$ has logconcave probability distribution.

We can write the Reliability Equation in our case, by taking into account only one basic material, as follows:

(3.4)
$$h(\mathbf{M}) = P(\zeta_i \leqq M + (i-1)\delta - i\gamma, \quad i = 1, \ldots, n) = p,$$

where $0 < p < 1$ and $p \approx 1$ in the practice.

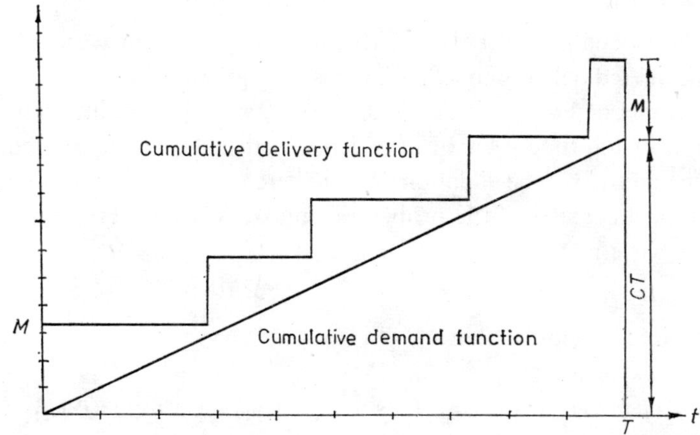

Fig. 1. The cumulative delivery function schould be above the cumulative demand function with a prescribed high probability (for each material)

The function $h(\mathbf{M})$ is logconcave on the halfline $[0, \infty)$ because the joint probability distribution function of a random vector having logconcave probability distribution is a logconcave point function [7]. Model I for more than one basic material consists of the following stochastic programming problem:

$$d^{(1)}M^{(1)}+\ldots+d^{(l)}M^{(l)}$$

is to be minimized supposing that

$$(3.5) \qquad h(\mathbf{M})=h_1(M^{(1)})\ldots h_l(M^{(l)})\geqq p$$

$$\mathbf{M}\geqq 0, \quad \mathbf{M}\in D,$$

where $\mathbf{M}=(M^{(1)}, \ldots, M^{(l)})$ and D is a subset of R^l determined by some constraints such as we prescribe that the components of $\mathbf{M}$ be smaller than or equal to certain upper bounds or that the initial stock amounts do not take more room than a certain upper limit and do not require more financial investment than a further upper limit, etc. The numbers $d^{(1)}, \ldots, d^{(l)}$ are nonnegative and they are some valuations of units of goods to be determined on the basis of local knowledge. Sometimes nonlinear objective function may turn out.

In the above discussion we assumed the demand function to be linear. Of course we can drope this assumption and use that model for the demand processes what was introduced in Section 2.

Stochastic programming models with independent joint constraints were considered first by Miller and Wagner [4].

Model II. This model differs from the previous one in that further constraints containing conditional expectation appear. With this we prescribe not only the rarity of the occurrence of unsatisfied demand but also prescribe upper bound for the average magnitude of the unsatisfied demand. Thus upper bound may depend on the basic material. We assume that unsatisfied demand will not be lost. Thus the model works with backorders.

If one of the inequalities (3.1) is violated, then it means that there was a lack just before the considered delivery time. As large is the violation as great is the unsatisfied demand. Assuming deterministic demand process with constant intensity, this means that the length of the time interval in which unsatisfied demand existed, is proportional to the magnitude of the violation. We did here take into account that no demand is lost. Our model consists of the problem formulated below. The superscripts refer to the various basic materials:

$$d^{(1)}M^{(1)}+\ldots+d^{(l)}M^{(l)}$$

is to be minimized supposing that

$$h(\mathbf{M})\geqq p,$$

$$(3.6) \quad E(\zeta_i^{(j)}-M^{(j)}-(i-1)\delta^{(j)}+i\gamma^{(j)} \mid \zeta_i^{(j)}-M^{(j)}-(i-1)\delta^{(j)}+i\gamma^{(j)}>0)\leqq g_i^{(j)},$$

$$i=1,\ldots,n; \quad j=1,\ldots,l,$$

$$\mathbf{M}\geqq 0, \quad \mathbf{M}\in D,$$

where the $g_i^{(j)}$ are constants and E is the symbol of expectation. The conditional expectation type constraints may even substitute the probabilistic constraint. For every i, j the random variable $\zeta_i^{(j)}$ has a logconcave probability density. It follows from this (see [8]) that the constraining functions in the conditional expectation constraints are monotonically decreasing functions of the variables $M^{(j)}$ and every such constraint is simply equivalent to a lower bound for the variable $M^{(j)}$ appearing in the constraint. We return to this question at the end of the section.

Model III. Again we assume that no demand will be lost. The difference between this model and (3.6) consists in a penalty term what we introduce now. Let us introduce the random variables

$$(3.7) \qquad k_i^{(j)} = \begin{cases} q_i^{(j)} \cdot \left(\zeta_i^{(j)} - M^{(j)} - (i-1)\delta^{(j)} + i\gamma^{(j)}\right) \\ \text{if } \zeta_i^{(j)} - M^{(j)} - (i-1)\delta^{(j)} + i\gamma^{(j)} > 0 \\ 0 \quad \text{otherwise} \end{cases}$$

$i = 1, \ldots, n; j = 1, \ldots, l$ where $q_i^{(j)} \geq 0$ for every i and j. It is easy to show that $E(k_i^{(j)})$ is a convex function of the variable $M^{(j)}$. To this it is enough to know that $\zeta_i^{(j)}$ has a continuous probability distribution. Since Model III has only a new objective function as compared to the model given by (3.6), it will be enough to formulate the new objective function. This is the following:

$$(3.8) \qquad \sum_{j=1}^{l} d^{(j)} M^{(j)} + \sum_{j=1}^{l} \sum_{i=1}^{n} E(k_i^{(j)}).$$

The construction of the above three models are in correspondence with the three general model constructions given in [8].

Model III contains Model I and Model II as special cases. We obtain Model II by putting $q_i^{(j)} = 0$, and Model I by putting $g_i^{(j)} = \infty$, $i = 1, \ldots, n; j = 1, \ldots, l$.

Now for a while we return to the conditional expectation contained in Problem (3.6) and the expectations $E(k_i^{(j)})$. Let $f_i^{(j)}$ resp. $F_i^{(j)}$ denote the probability density and the probability distribution functions of the random variable $\zeta_i^{(j)}$. For the sake of simplicity the superscripts will be ommitted in the sequel. It is easy to see that if ζ is a continuously distributed random variable and a is a constant, then the following equality holds

$$(3.9) \qquad E(\zeta - a \mid \zeta - a > 0) = \int_a^\infty [1 - F(x)] \, dx / [1 - F(a)] =$$

$$= \int_a^\infty x f(x) \, dx / [1 - F(a)] - a$$

where F is the probability distribution function of ζ. In view of this we can write

$$E(\zeta_i - M - (i-1)\delta + i\gamma \mid \zeta_i - M - (i-1)\delta + i\gamma > 0) =$$

(3.10)

$$= \frac{1}{1 - F_i(M + (i-1)\delta - i\gamma)} \int_{M+(i-1)\delta-i\gamma}^{1-n\gamma} x f_i(x)\,dx - M - (i-1)\delta + i\gamma.$$

Similarly we obtain

(3.11)
$$E(k_i) = q_i \int_{M+(i-1)\delta-i\gamma}^{1-n\gamma} [1 - F_i(x)]\,dx.$$

A simple argument shows that

(3.12)
$$f_i(x) = \frac{\Gamma(N+1)\Gamma(L+1)}{\Gamma(j_i)\Gamma(N-j_i+1)\Gamma(k_{i-1})\Gamma(L-k_{i-1}+1)} \frac{1}{(c-n\delta)(1-n\gamma)} \cdot$$

$$\cdot \int_0^b \left(\frac{x+u}{1-n\gamma}\right)^{j_i-1} \left(1 - \frac{x+u}{1-n\gamma}\right)^{N-j_i} \left(\frac{u}{c-n\delta}\right)^{k_i-1-1} \left(1 - \frac{u}{c-n\delta}\right)^{L-k_i-1}$$

(where $b = \min\{1-n\gamma-x, c-n\delta\}$), if $0 < x < 1-n\gamma$ and $f_i(x) = 0$ otherwise, $i = 2, 3,$
$\ldots, n$, further

(3.13) $\quad f_1(x) = \dfrac{\Gamma(N+1)}{\Gamma(j_1)\Gamma(N-j_1+1)} \dfrac{1}{1-n\gamma} \left(\dfrac{x}{1-n\gamma}\right)^{j_1-1} \left(1 - \dfrac{x}{1-n\gamma}\right)^{N-j_1},$

if $0 < x < 1-n\gamma$ and $f_1(x) = 0$ otherwise.

As we already remarked, the i-th conditional expectation-type constraint in Problem (3.6) can be converted into the simple inequality $M^{(j)} \geq M_i^{(j)}$ where $M_i^{(j)}$ is that value of $M^{(j)}$ for which the constraint holds with equality. This value can be determined by numerical intergration of the function f_j.

4. SOLUTION OF THE PROBLEMS

In this section we present a solution method to the problems discussed in the previous section. We restrict ourselves to the problem of Model I, since the solution of the two further problems requires only slight modification.

For the sake of simplicity let us agree that the constraint $M \in D$ be specialized so that it consist in the system of inequalities $M^{(j)} \leq 1, j = 1, \ldots, l$. These are, on the other hand, no real restrictions, because the equalities

(4.1)
$$h_j(1) = 1, \quad j = 1, \ldots, l$$

174

hold trivially and these imply that the optimal $M^{(1)}, \ldots, M^{(l)}$ values are automatically smaller than or equal to 1. The upper bounding of the $M^{(1)}, \ldots M^{(l)}$ values has the only significance that we shall be able to refer to well-known convergence theorem concerning the SUMT method what we are going to use here [2].

We apply the interior point version of the SUMT. Consider the following penalty function

$$G(r, \mathbf{M}) = \sum_{j=1}^{l} d^{(j)} M^{(j)} - r \left\{ \log \left[\prod_{j=1}^{l} h_j(M^{(j)}) - p \right] + \right.$$

(4.2)

$$\left. + \sum_{j=1}^{l} \log M^{(j)} (1 - M^{(j)}) \right\},$$

where r is a fixed positive number. It is easy to see that for every fixed $p > 0$ the function $h_1(M^{(1)}) \ldots h_l(M^{(l)}) - p$ is also logconcave which implies that for every fixed $r > 0$ the function $G(r, \mathbf{M})$ is convex on the set $\{\mathbf{M} \mid \mathbf{M} \geq 0\}$. From this we only need the fact that $G(r, \mathbf{M})$ is convex on the l-dimensional unit cube $\{\mathbf{M} \mid 0 \leq M^{(j)} \leq 1, j = 1, \ldots, l\}$. The SUMT interior point method works so that we take a sequence $r_1 > r_2 > \ldots$ consisting of positive numbers, tending to 0 and minimize $G(r_k, \mathbf{M})$ with respect to $\mathbf{M}$ (in principle) for every r_k. If $\mathbf{M}_k$ is the minimizing vector then $G(r_k, \mathbf{M}_k)$ tends to the minimum value of the objective function in Model I. Thus $\mathbf{M}_k$ is an approximate optimal solution to the problem if k is large enough.

The mentioned convergence is ensured if the set of feasible solutions is bounded, the constraining functions as well as the objective function are continuous on the set of feasible solutions, further there exists interior point of this set and all inequalities hold as strict inequalities at every interior point. As regards Model I here we have the constraints $0 \leq M^{(j)} \leq 1, j = 1, \ldots, t$ and the probabilistic constraint which restricts further the unit cube. The former ones hold strictly at every interior point of the unit cube thus it will be enough to consider the probabilistic constraint. Let $\mathbf{M}_1$ be an interior point in the set of feasible solutions. We shall show that $h(\mathbf{M}_1) > p$ where p is fixed and $0 < p < 1$ the line section connecting the points $\mathbf{1}$ and $\mathbf{M}_1$ is entirely feasible, because the set of feasible solutions is convex. Let $\mathbf{M}_0$ be a feasible point on the line connecting $\mathbf{1}$ and $\mathbf{M}_1$ lying outside the section so that $\mathbf{M}_1$ be between $\mathbf{1}$ and $\mathbf{M}_0$. Then there exists a $0 < \lambda < 1$ so that

$$\mathbf{M}_1 = \lambda \mathbf{1} + (1 - \lambda) \mathbf{M}_0$$

from which, using the logconcavity of the function $h(\mathbf{M})$, it follows that

(4.3) $$h(\mathbf{M}_1) \geq [h(\mathbf{1})]^{\lambda} [h(\mathbf{M}_0)]^{1-\lambda} \geq p^{1-\lambda} > p.$$

Thus we have shown that the SUMT interior point method is convergent in case of Model I.

Many general unconstrained minimization technique can be applied for the function (4.2). Some of them use only function values, some use gradient values too. In order to facilitate the application of methods belonging to the latter category, we present a methodology to compute the gradient values. Since $h(\mathbf{M})$ is the product of the functions $h_j(M^{(j)})$, $j=1, \ldots, l$ it will be enough to consider the derivatives of the functions $h_j(M^{(j)})$. Let us ommit the j, for the sake of simplicity. The function (3.4) is the joint probability distribution function of the random variables $\zeta_1, \ldots, \zeta_n$ at the point with coordinates $M+(i-1)\delta-i\gamma$, $i=1, \ldots, n$.

We remark that if $F(\mathbf{z})$ is the probability distribution function corresponding to a continuous probability distribution, then the following relation holds:

$$(4.4) \qquad \frac{\partial F(z)}{\partial z_i}=F(z_j, j\neq i|z_i)f_i(z_i), \quad i=1, \ldots,n,$$

where $f_1, \ldots, f_n$ are the probability density functions of the one-dimensional marginal distributions and $F(.\,|z_i)$ is the $n-1$-dimensional conditional probability distribution function given that the ith random variable equals z_i.

We assume that $n\delta<c$, $n\gamma<1$ (if one of the equalities $n\delta=c$, $n\gamma=1$ holds, our procedure can essentially be simplified). To compute the derivative of the function (3.4), first we take the partial derivatives with respect to all $z_1, \ldots, z_n$ the function

$$(4.5) \qquad P(\zeta_i\leq z_i+(i-1)\delta-i\gamma, i=1, \ldots, n)$$

and put $z_1= \ldots =z_n=M$. The sum of these equals the derivative of $h(\mathbf{M})$. The partial derivative of the function (4.5) with respect to z_i can be obtained by using the formula (4.4). Putting $z_1= \ldots z_n=M$ we obtain

$$P(\zeta_j\leq M+(j-1)\delta-j\gamma, j\neq i \mid \zeta_i=M+(i-1)\delta-i\gamma) \cdot f_i(M+(i-1)\delta-i\gamma)=$$

$$=f_i(M+(i-1)\delta-i\gamma) \cdot \int_0^v P(\zeta_j\leq M+(j-1)\delta-j\gamma, j\neq i \mid \xi_1+ \ldots +\xi_i=$$

$$(4.6) \qquad =M+(i-1)\delta-i\gamma+x,$$

$$\eta_1+ \ldots +\eta_{i-1}=x)\frac{\Gamma(N+1)}{\Gamma(j_i)\Gamma(N+1-j_i)} \frac{1}{1-n\gamma}\left(\frac{M+(i-1)\delta-i\gamma+x}{1-n\gamma}\right)^{j_i-1} \cdot$$

$$\cdot\left(1-\frac{M+(i-1)\delta-i\gamma+x}{1-n\gamma}\right)^{N-j_i} \frac{\Gamma(L+1)}{\Gamma(k_{i-1})\Gamma(L+1-k_{i-1})} \cdot$$

$$\cdot\frac{1}{c-n\delta}\left(\frac{x}{c-n\delta}\right)^{k_{i-1}-1}\left(1-\frac{x}{c-n\delta}\right)^{L-k_{i-1}} dx,$$

$$v=\min \{1-M-(i-1)\delta-(n-i)\gamma, c-n\delta\},$$

where $f_i(z)$ is the probability density function of the random variable ζ_i. The probability in the second row of (4.6) can be expressed as an absolute probability and thus we obtain an expression similar to (3.4).

176

We remind that the random variables $\xi_1, \ldots, \xi_n$ arise from a sample of size N, taken from a population uniformly distributed in the interval $(0, 1)$, in a way described in Section 2. The joint distribution of $\xi_1, \ldots, \xi_u$ given that $\xi_1 + \ldots + \xi_i = u = M + (i-1)\delta - i\gamma + x$ coincide with the joint distribution of two independent random vectors. These vectors consists of $j_i - 1$ resp. $N - j_i$ components and in both the joint densities are given by expressions of the type (2.6). In case of the first vector N, n, $1 - n\gamma$ should be replaced by $j_i - 1$, $i - 1$, u and in case of the second vector, by $N - j_i$, $n - i$, $1 - n\gamma - u$ respectively. Similar is the situation concerning the random variables $\eta_1, \ldots, \eta_{n-1}$.

We apply simulation for the computation of the probability $h(\mathbf{M})$. The computation of the gradient values is more sophisticated, because beyond simulation numerical integration is also needed. Hence it seems to be more economic to apply gradient free minimization technique when carrying out the unconstrained minimization of the penalty function.

5. SIMULATION TECHNIQUE FOR THE COMPUTATION OF THE VALUES OF THE FUNCTION $h(\mathbf{M})$

Two methods are proposed. The first one follows the modelling of the delivery processes. We take many samples of size N resp. L, order them and select the required elements. This method has the great disadvantage that the ordering of the sample elements requires much computer time. It is known that the ordering time of N elements increases in the order of magnitude of $N \cdot \log_2 N$.

The second method is more effective than the just mentioned former one. It is based on the fact that any Dirichlet distribution can be represented as the joint distribution of random variables $y_1, \ldots, y_n$ by

$$(5.1) \qquad y_i = x_i / (x_1 + \ldots + x_{n+1}), \quad i = 1, 2, \ldots, n$$

where $x_1, \ldots, x_{n+1}$ are independent, standard gamma distributed random variables with parameters $\vartheta_1 > 0, \ldots, \vartheta_{n+1} > 0$ i.e. x_i has the following probability density:

$$(5.2) \qquad (z^{\vartheta_i - 1} e^{-z}) / \Gamma(\vartheta_i).$$

In fact, the joint density of the random variables (5.1) is given by

$$(5.3) \qquad \frac{\Gamma(\vartheta_1 + \ldots + \vartheta_{n+1})}{\Gamma(\vartheta_1) \ldots \Gamma(\vartheta_{n+1})} z_1^{\vartheta_1 - 1} \ldots z_n^{\vartheta_n - 1} (1 - z_1 - \ldots - z_n)^{\vartheta_{n+1} - 1},$$

if $z_i > 0$, $i = 1, \ldots, n$, $z_1 + \ldots + z_n < 1$ and is 0 otherwise. Thus by a suitable choice of $\vartheta_1, \ldots, \vartheta_{n+1}$, the required Dirichlet density turns out.

Ahrens and Dieter [1] gave effective simulation technique for the simulation of the gamma distribution. Their method is particularly effective when the ϑ parameter is large or is not an integer.

The probability density functions, (2.5) and (2.6) slightly differ from the density function (5.3). The simulation technique described above requires only very simple modification in both cases. Let us consider the gamma probability density function

$$(5.4) \qquad (\lambda^\vartheta z^{\vartheta-1} e^{-\lambda z})/\Gamma(\vartheta), \quad z>0.$$

If $x_1, \ldots, x_{n+1}$ are independent and gamma distributed random variables with parameter pairs $\lambda, \vartheta_1; \ldots; \lambda, \vartheta_{n+1}$; where $\lambda=1-n\gamma$, $\vartheta_1=j_1$, $\vartheta_2=j_2-j_1$, $\ldots$, $\vartheta_n=j_n-j_{n-1}$, $\vartheta_{n+1}=N-j_{n+1}+1$, then the random variables defined by (5.1) have the same joint probability distribution as $\xi_1, \ldots, \xi_n$ do. On the other hand, x_i can be represented as the sum of ϑ_i independent and exponentially distributed random variables with the same parameter λ, for every $i=1, \ldots, n+1$. Finally the exponentially distributed random variables can be represented as negative logarithms of random variables uniformly distributed in the interval $(0, 1)$. The simulation of the joint distribution of the random variables can be carried out in a similar way.

In case of this second simulation technique we only take the logarithms of the $N+1$ sample elements but do not order them. The required computer time is much less then in the first case.

The probabilities are approximated by relative frequencies. The sample size ensuring prescribed precision can be determined by the inequality of Bernstein. This is the following. If v_m denotes the frequency of an event of probability p, in the course of m independent experiments and ε is a given positive number, then

$$(5.5) \qquad P\left(\left|\frac{v_m}{m}-p\right|\geqq\varepsilon\right)\leqq 2\exp\left[-\frac{m\varepsilon^2}{2p(1-p)(1+\varepsilon/(2p(1-p))^2)}-\right],$$

for

$$0<\varepsilon\leqq p(1-p).$$

If the probability on the left hand side equals δ, then for m we obtain the inequality

$$(5.6) \qquad m\geqq 1/\varepsilon^2 \cdot 2p(1-p)(1+\varepsilon/[2p(1-p)]^2) \cdot \log 2/\delta$$

if $0<\varepsilon\leqq p(1-p)$. For fixed δ and ε the largest value of the right hand side of (5.6) corresponds to $p=\frac{1}{2}$, and it is a monotonically decreasing function of p for $\frac{1}{2}<p<1$ (and monotonically increasing for $0<p<\frac{1}{2}$), provided $\varepsilon\leqq p(1-p)$. In such a way we can get a lower bound for m which is good for every p. This is important because our aim is to approximate the probability p. Sometimes we have certain bounds for p. This is the case in connection with such stochastic programming problems where we have probabilistic constraint i.e. lower bound for the probability.

In our models we use at least 0.8 as lower bound for the function $h(\mathbf{M})$. In practice this means that the factors are greater than or equal to 0.9. Using this information, the required sample size is much smaller then would be the case without any previous information. Table 1 well illustrates the variation of the lower bound for m as a function of ε and p when δ is fixed at 0.1.

178

Table 1

p	0.09	0.045	0.025	0.01
0.5	258	879	2,645	15,606
0.8	195	616	1,783	10,209
0.9	150	417	1,120	6,016
0.95	—	306	727	3,481

6. NUMERICAL EXAMPLE

As an example we consider a certain kind of product for the production of which two basic materials are needed and we want to determine the initial stock levels of the two basic materials ensuring the continuous production. The shortage in each of them stops the production and the cost of such an event is relatively high so that one of our main objectives is to avoid shortage by a prescribed probability near unity.

We assume that the demands for both basic materials are uniform in time and the unsatisfied demand remains i.e. the production plan has to be fulfilled. We assume that the basic materials are delivered from two different sources so that the two delivery processes can be supposed stochastically independent. The (0, 1) time interval is now a quarter of a year, 90 days in other terms. According to long term statistics deliveries occur 4 resp. 5 times concerning the first resp. second basic material during one period (90 day). Table 2 shows actual delivery days for six past periods concerning the first basic material.

Table 2

Periods	Deliveries			
	1	2	3	4
1	23	41	61	82
2	27	48	73	88
3	30	39	60	90
4	19	48	68	89
5	24	50	65	78
6	28	42	71	82
Column averages	25.17	44.66	66.33	84.83

The minimum distance between two cunsecutive deliveries is 9 days. Since 90 days form a time interval of lenght 1, this means that the mentioned minimum distance is $\gamma^{(1)} = 0.1$. For the average delivery times we get in the same way

$$z_1 = 0.28, \quad z_2 = 0.49, \quad z_3 = 0.73, \quad z_4 = 0.94.$$

Using our modeling of the delivery time process we can write

$$\bar{z}_i = i\gamma^{(1)} + E(x^{*}_{j_i}), \quad i = 1, \ldots, n,$$

where $x^{*}_{j_i}$ denotes the j_i-th element of the ordered sample of size N taken from the population uniformly distributed in the interval $(0.1 - n\gamma^{(1)})$. We have to find integers $N, j_1, \ldots, j_n$ for which the following equalities hold at least in good approximation:

$$E(x^{*}_{j_i}) = j_i(1 - n\gamma^{(1)})/N = \bar{z}_i - i\gamma^{(1)},$$

$$j_i = \frac{\bar{z}_i - i\gamma^{(1)}}{1 - n\gamma^{1}} N, \quad i = 1, \ldots, n.$$

Since the $\bar{z}_i$, $i = 1, \ldots, n$ and $\gamma^{(1)}$ are rationals in parctice, such integers, $N, j_1, \ldots, j_n$ always exist. It is not worth always to require that the above equalities hold exactly. In fact, if we work with large numbers, then the computer time will considerably be increased. In the above example the values of

$$(\bar{z}_i - i\gamma^{(1)})/(1 - n\gamma^{(1)}), \quad i = 1, 2, 3, 4$$

are $0.298; 0.493; 0.728; 0.903$ and choosing $N = 10, j_1 = 3, j_2 = 5, j_3 = 7, j_4 = 9$, the above equalities are well approximated.

Table 3 shows the delivered amounts of the first basic material in the same past six periods.

Table 3

Periods	Deliveries				
	1	2	3	4	Totals
1	630	400	670	800	2500
2	700	500	600	900	2700
3	730	580	550	740	2600
4	720	620	650	1010	3000
5	760	580	760	1100	3200
6	750	650	780	920	3100

Dividing the rows by the sums of the rows we get Table 4.

Table 4

	0.252	0.16	0.268	0.32
	0.259	0.185	0.222	0.333
	0.28	0.223	0.211	0.248
	0.24	0.206	0.216	0.336
	0.237	0.181	0.238	0.344
	0.242	0.21	0.252	0.297
Column averages	0.252	0.194	0.234	0.319

From the table we see that the minimal delivered amount is $\delta^{(1)} = 0.16$.

180

If the column averages are denoted by $\bar{u}_1, \ldots, \bar{u}_n$ and we introduce the further notation $v_i = \bar{u}_1 + \ldots + \bar{u}_i$, $i = 1, \ldots, n$, then similarly to the case of the delivery times we write the equalities

$$v_i = i\delta^{(1)} + E(\overset{*}{y}_{k_i}), \quad i = 1, \ldots, n-1,$$

where $\overset{*}{y}_{k_i}$ denotes the k_i-th element of a sample of size L taken from a population uniformly distributed in the interval $(0, 1 - n\delta^{(1)})$. We want to determine integers $L, k_1, \ldots, k_{n-1}$ so that the following equalities hold at least in good approximation. In our case the values of

$$(v_i - i\delta^{(1)})/(1 - n\delta^{(1)}), \quad i = 1, 2, 3$$

are 0.255; 0.35; 0.555. Thus the choices $L = 20$, $k_1 = 5$, $k_2 = 7$, $k_3 = 11$ provide good approximations.

We can proceed in a similar way concerning the second basic material. Assume that we obtained the following values: $n = 5$, $N = 10$, $j_1 = 2$, $j_2 = 3$, $j_3 = 5$, $j_4 = 7$, $j_5 = 9$, $\gamma^{(2)} = 0.15$. $L = 10$, $k_1 = 2$, $k_2 = 5$, $k_3 = 7$, $k_4 = 8$, $\delta^{(2)} = 0.12$. As regards the objective function, we have choosen $d^{(2)} = 3d^{(1)}$.

The SUMT interior point method started with values between 0.6 and 0.8. The $r_1, r_2, \ldots$ sequence was choosen to be 1, 1/5, 1/25, $\ldots$ and the initial values of the $k+1$-st unconstrained minimization were the optimal values of the k-th unconstrained minimization. The method was stopped when the change in the optimal values of the penalty function was less than 0.01. The method of Hooke and Jeeves [3] was ap-

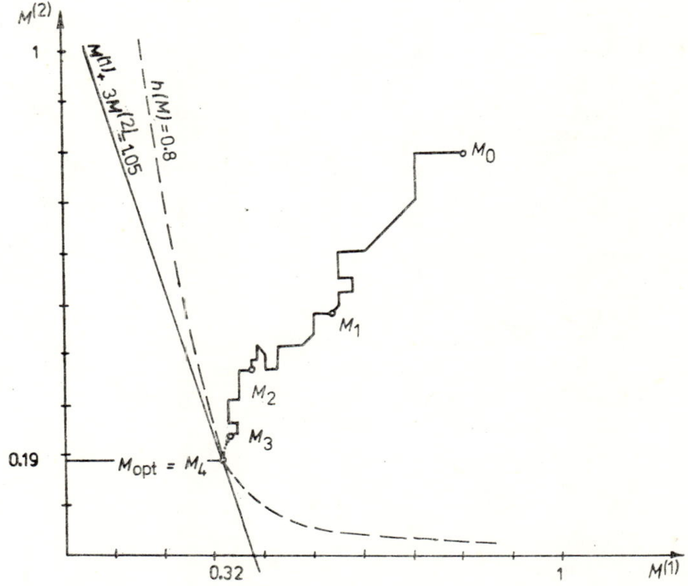

Fig. 2. The steps of the optimization procedure, where $\mathbf{M}_k$ is the optimal value of the k-th unconst- rained optimization

plied for the minimization of the penalty function. In our numerical example the minimizing $M^{(1)}$ and $M^{(2)}$ belonging to $r=1,125$ were accepted as optimal solutions of the problem. These are $M^{(1)}=0.32$ and $M^{(2)}=0.19$. This means that 32% of the total demand of the first material and 19% of the total demand of the second material will serve the production without shortage with probability $p=0.8$ and the cost will be minimum among all feasible alternatives.

The test programs written in FORTRAN run between 1.5 and 2.5 minutes on a CDC 3,300 computer. Further unconstrained optimization methods were also tested such as the method of Rosenbrock [11] and Powell [5]. The best computer time was produced by the method of Hooke and Jeeves, however. This method was successfully applied also in other stochastic programming problems where function values were determined by simulation.

REFERENCES

[1] Ahrens, J. H. and Dieter, K.: Computer Methods for Sampling from Gamma, Beta, Poisson and Binomial Distributions, *Computing,* 12 (1974), 223–246.

[2] Fiacco, A. V. and McCormick, G. P.: *Nonlinear Programming: Sequential Unconstrained Minimization Technique,* Wiley, New York 1968.

[3] Hooke, R. and Jeeves, T. A.: Direct Search Solution of Numerical and Statistical Problems, *Journal of the A. C. M.* 8 (1959), 215–229.

[4] Miller, B. L. and Wagner, H. M.: Chance Constrained Programming with Joint Constraints, *Operations Research,* 13 (1965), 930–945.

[5] Powell, M. J. D.: An Iterative Method for Finding the Minimum of a Function of Several Variables Without Calculating Derivatives, *Computer Journal,* 7 (1964), 155–162.

[6] Prékopa, A.: Reliability Equation for an Inventory Problem and its Asymptotic Solution, *Coll. on Appl. of Math. to Economics,* Akadémiai Kiadó, Budapest 1965, pp. 317–327.

[7] Prékopa, A.: On Logarithmic Concave Measures with Application to Stochastic Programming, *Acta Universitatis Szegediensis,* 32 (1971), 301–316.

[8] Prékopa, A.: Contributions to the Theory of Stochastic Programming, *Mathematical Programming,* 4 (1973), 202–221.

[9] Prékopa, A.: Stochastic Programming Models for Inventory Control and Water Storage Problems, Colloquia Mathematica Societatis János Bolyai, 7. *Inventory Control and Water Storage, Győr, 1971.* Bolyai János Math. Soc. and North-Holland Publ. Comp., Budapest 1973.

[10] Prékopa, A.: Generalizations of the Theorems of Smirnov with Applications to a Reliability Type Inventory Problem, *Mathematische Operationsforschung und Statistik,* 4 (1973), 293–297.

[11] Rosenbrock, H. H.: An Automatic Method for Finding the Greatest or Least Value of a Function, *The Computer Journal,* 3 (1960), 175–184.

[12] Wilks, S. S.: *Mathematical Statistics,* Wiley, New York 1962.

[13] Ziermann, M.: Anwendung des Srmirnov'schen Sätzen auf Lagerhaltungsproblem, *Publications of the Math. Inst. of the HAS.,* 8 (1965), Series B, 509–518.

ON OPTIMAL REGULATION OF A STORAGE LEVEL WITH APPLICATION TO THE WATER LEVEL REGULATION OF A LAKE

A. PRÉKOPA and T. SZÁNTAI

(Budapest, Hungary)

1. INTRODUCTION

In the paper [10] a system of stochastic programming models was introduced for the optimal control of a storage level. Each model in this system serves to determine the optimal policy for only one period ahead though the time horizon consists of many future periods. The optimal control thus obtained can be considered an open loop control methodology. The main purpose of this paper is to present an application by giving an optimal control method for the regulation of the water level of Lake Balaton in Hungary. By solving almost 600 stochastic programming problems we analyze what would have happened if we had controlled the water level using our method between 1922 and 1970, where one decision period is one month. The numerical results show that the proposed control methodology works quite well in this case.

Our water input stochastic process will be assumed to be Gaussian. No time homogeneity or independence, Markovian character or whatsoever will be supposed. Also the Gaussian nature of the input process is not an essential feature of our control methodology. We refer to the paper [11] where we used a multivariate gamma distribution introduced by the authors of this paper in connection with certain reservoir system operation model. It is possible to use also here the same multigamma or some other multivariate probability distribution. The case of the multivariate normal distribution—the application of which is supported by statistics in connection with Lake Balaton—is relatively simple because of the special properties of this multivariate probability density.

2. SHORT DESCRIPTION OF THE DYNAMIC MODEL SYSTEM USED FOR THE CONTROL OF THE STORAGE LEVEL

Taking into account that application of our control methodology which we are going to present in the further sections of this paper, we shall use the terms corresponding to the water level regulation of a lake.

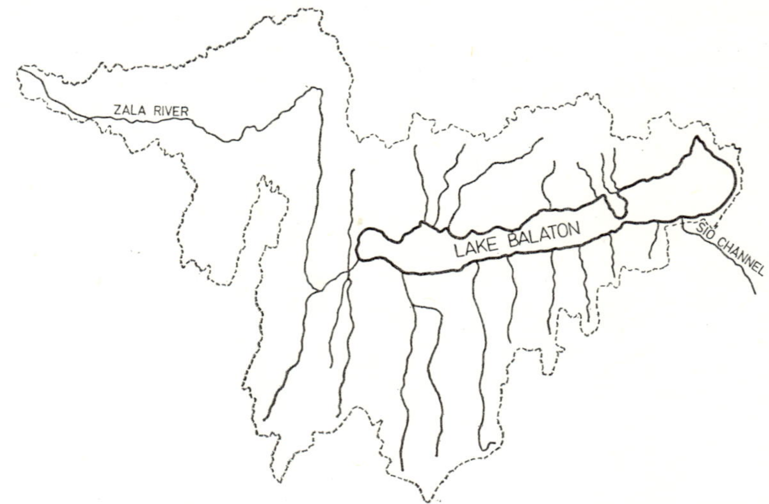

Fig. 1. Lake Balaton and catchment area in the western part of Hungary. Water is released through Sió channel into Danube river

Fig. 1 illustrates Lake Balaton in the western part of Hungary. Small rivers and rainfall represent positive inputs whereas evaporation represents negative input. The sum of these with positive resp. negative signs will be considered the water input. Thus the variables which denote water inputs in the subsequent periods can take on negative values too. Sometimes this in fact occurs.

Water can be released through the Sió channel into the Danube river. The monthly water quantities which can be released are limited by the capacity of the Sió channel. This capacity will be denoted symbolically by K in this section.

Instead of water levels we shall speak about water quantities. The connection relative to Lake Balaton between these two notions will be clarified later on. Let ζ_0 be the initial water content of the lake and $\xi_1, \xi_2, \ldots$ the monthly random water inputs. ζ_0 will be assumed to be nonrandom in our models. Let further $z_1, z_2, \ldots$ be decision variables belonging to the subsequent periods. These are the water quantities to be released through the channel in the subsequent periods. We decide on z_1 in the beginning of the first period, on z_2 in the beginning of the second period, etc. Introduce the notations

$$\zeta_k = \zeta_0 + \xi_1 + \ldots + \xi_k,$$
$$k = 1, 2, \ldots$$
$$Z_k = z_1 + \ldots + z_k,$$

The random process $\xi_1, \xi_2, \ldots$ will be assumed to be Gaussian. We prescribe further lower bounds $a_1, a_2, \ldots$ resp. upper bounds $b_1, b_2, \ldots$ for the water quantities being in the lake at the end of the subsequent periods. We consider the situation favourable if the inequalities

184

(2.1) $$a_k \leqq \zeta_k - Z_k \leqq b_k, \quad k=1, 2, \ldots$$

are satisfied, where the water quantites to be released z_1, z_2, ... are subject to the inequalities

(2.2) $$0 \leqq z_k \leqq K, \quad k=1, 2, \ldots$$

Since ξ_1, ξ_2, ... are random variables, the fulfilment of the inequalities (2.1) cannot be guaranteed with probability 1. Before formulating our decision principle we make a remark concerning stochastic programming model construction.

Stochastic programming problems are formulated in such a way that first we formulate a deterministic mathematical programming problem, which is called underlying deterministic problem, then observe that some of the parameters in this problem are random in reality; in view of our problem looses its original meaning, hence we formulate another decision principle by taking into account the probability distribution of the random variables involved. Underlying mathematical programming problems can be either minimization (resp. maximization) problems or problems, where we only wish to find at least one vector satisfying certain constraints. In this latter case the advised stochastic programming decision principle is to find that vector which maximizes the probability of the fulfilment of the random constraints subject to those constraints which do not contain random variables.

Inequalities (2.1) and (2.2) represent an underlying deterministic problem where we want to find z_1, z_2, ... such that the referred inequalities be satisfied. The number of the considered periods should be finite. Having observed that ξ_1, ξ_2, ... are random variables, we formulate the stochastic programming decision principle, in accordance with the above remark, so that we maximize the probability of the fulfilment of the inequalities (2.1) subject to the constraints (2.2).

Since we have the possibility to use a dynamic type decision methodology i.e. we have the possibility to decide in every period, the above-mentioned principle will be turned into a sequence of problems and conditional probabilities will be maximized where in the condition there stand the already realised values of the stochastic process ξ_1, ξ_2, ... The first problem in this sequence is the following

maximize

$$P(a_k \leqq \zeta_k - Z_k \leqq b_k, \quad k=1, \ldots, N)$$

(2.3) subject to

$$0 \leqq z_k \leqq K, \quad k=1, \ldots, N.$$

Out of the optimal solution z_1^*, $\ldots$, z_N^* we only accept z_1^* and formulate the next problem. For the sake of simplicity the asterisk will be omitted except for Section 5. Assume that we already fixed z_1, $\ldots$, z_n. Then in order to fix z_{n+1}, we formulate the following nonlinear programming problem maximize

$$P(a_k \leqq \zeta_k - Z_k \leqq b_k, \quad k=n+1, \ldots, n+N \mid \xi_1, \ldots, \xi_n)$$

(2.4)   subject to

$$0 \leqq z_k \leqq K, \quad k=n+1, \ldots, n+N.$$

Here $z_{n+1}, \ldots, z_{n+N}$ are the decision variables. Having computed the optimal solution, we only accept z_{n+1} as a final value. Thus our control methodology is fixed.

It should be mentioned that a positive lower bound for K may be required. Mathematically this does not present any difficulty. In fact if K_0 is a positive lower bound for the z_k, then using the new variables $y_k = z_k - K_0$, we can transform our problem into the already introduced form (2.3), (2.4).

3. MATHEMATICAL PROPERTIES OF THE MODEL SYSTEM INTRODUCED IN SECTION 2.

Before starting to discuss the subject mentioned in the title of this section, we recall some facts concerning logarithmic concave measures.

A nonnegative function $f(\mathbf{x})$, $\mathbf{x} \in R^m$ is said to be a logarithmic concave (point) function if for every $\mathbf{x}_1, \mathbf{x}_2 \in R^m$ and $0 < \lambda < 1$, we have

$$f(\lambda \mathbf{x}_1 + (1-\lambda)\mathbf{x}_2) \geqq [f(\mathbf{x}_1)]^{\lambda}[f(\mathbf{x}_2)]^{1-\lambda}.$$

A measure P defined on the measurable subset of the space R^m is said to be logarithmic concave if for every pair A, B of convex subsets of R^m and $0 < \lambda < 1$, we have

$$P(\lambda A + (1-\lambda)B) \geqq [P(A)]^{\lambda}[P(B)]^{1-\lambda}.$$

Here the sign $+$ denotes Minkowski addition, i.e. in connection with two sets D, G, $D + G = \{\mathbf{d} + \mathbf{g} \mid \mathbf{d} \in D, \mathbf{g} \in G\}$, further the constant multiple λG of the set G is defined by the equality $\lambda G = \{\lambda \mathbf{g} \mid \mathbf{g} \in G\}$.

In [8], [9] the following theorem was proved.

Theorem 1. *If a probability measure P is generated by a logarithmic concave probability density i.e. for every measurable set $C \subset R^m$ we have*

$$P(C) = \int_C f(\mathbf{x}) \, d\mathbf{x},$$

then P is a logarithmic concave measure.

Theorem 1 implies that if A is a convex subset of R^m, then

$$\int_{A+x} f(\mathbf{t}) dt$$

186

is a logarithmic concave function of the variable **x**. This implies further that if G is an $n \times N$ matrix and **z** is an N-component vector, then

$$\int_{A+Gz} f(\mathbf{t}) \, d\mathbf{t}$$

is a logarithmic concave function of the variable **z**.

Consider now the random vector of components $\xi_1, \ldots, \xi_{n+N}$, denote by **e** its expectation vector and by C its covariance matrix. By assumption this random vector has a normal distribution. Assume also that this distribution is nondegenerated. Then the probability density of this random vector exists and it is given by

$$f(\mathbf{x}) = \left(\frac{\det C^{-1}}{(2\pi)^n} \right)^{\frac{1}{2}} e^{-\frac{1}{2}(x-e)'C^{-1}(x-e)}, \quad \mathbf{x} \in R^{n+N}.$$

Out of the random vector components $\xi_1, \ldots, \xi_{n+N}$ we form two random vectors ξ^P, ξ^F which are the following

$$(3.1) \qquad \xi^P = \begin{pmatrix} \xi_1 \\ \vdots \\ \xi_n \end{pmatrix}, \quad \xi^F = \begin{pmatrix} \xi_{n+1} \\ \vdots \\ \xi_{n+N} \end{pmatrix}$$

and partition **e**, **x** accordingly. The obtained parts will be denoted by $\mathbf{e}^P$, $\mathbf{e}^F$ resp. $\mathbf{x}^P$, $\mathbf{x}^F$. Thus we have

$$(3.2) \qquad \mathbf{e}^P = E(\xi^P) = \begin{pmatrix} e_1 \\ \vdots \\ e_n \end{pmatrix}, \quad \mathbf{e}^F = E(\xi^F) = \begin{pmatrix} e_{n+1} \\ \vdots \\ e_{n+N} \end{pmatrix}.$$

The superscripts P and F are initials of the words "Past" resp. "Future". Let us rearrange and then partition the covariance matrix C so that we obtain the following

$$
\begin{pmatrix}
c_{n+1,n+1} \cdots c_{n+1,n+N} & c_{n+1,1} \cdots c_{n+1,n} \\
\cdots\cdots\cdots\cdots\cdots\cdots\cdots\cdots\cdots \\
c_{n+N,n+1} \cdots c_{n+N,n+N} & c_{n+N,1} \cdots c_{n+N,n} \\
c_{1,n+1} \quad \cdots c_{1,n+N} & c_{11} \quad \cdots c_{1n} \\
\cdots\cdots\cdots\cdots\cdots\cdots\cdots\cdots\cdots \\
c_{n,n+1} \quad \cdots c_{n,n+N} & c_{n1} \quad \cdots c_{nn}
\end{pmatrix} =
$$

$$(3.3)$$

$$
= \begin{pmatrix}
E[(\xi^F - e^F)(\xi^F - e^F)'] & E[(\xi^F - e^F)(\xi^P - e^P)'] \\
E[(\xi^P - e^P)(\xi^F - e^F)'] & E[(\xi^P - e^P)(\xi^P - e^P)']
\end{pmatrix} =
$$

$$
= \begin{pmatrix} S & U \\ U' & T \end{pmatrix}.
$$

It is known that the probability distribution of the random vector ξ^F given $\xi^P=\mathbf{x}^P$ is a normal distribution with expectation vector

(3.4)
$$\mathbf{e}^C=\mathbf{e}^F+UT^{-1}(\mathbf{x}^P-\mathbf{e}^P)$$

and covariance matrix

(3.5)
$$S-UT^{-1}U',$$

where the superscript in $\mathbf{e}^C$ refer to the word "Conditional". Thus the conditional probability density of ξ^F given $\xi^P=\mathbf{x}^P$ is the following

$$f(\mathbf{x}^F\mid\mathbf{x}^P)=\left[\frac{\det(S-UT^{+1}U')}{(2\pi)^N}\right]^{\frac{1}{2}}e^{-\frac{1}{2}(x-e^c)'(S-UT^{-1}U')(x^F-e^c)}.$$

The function $f(\mathbf{x}^F\mid\mathbf{x}^P)$ is logarithmic concave as a function of all variables in $\mathbf{x}^F$ and $\mathbf{x}^P$. Now we only need the fact it is logarithmic concave in $\mathbf{x}^F$ for every fixed $\mathbf{x}^P$. Consider the following set in the space of the vectors $\mathbf{x}^F$:

$$A=\{\mathbf{x}^F\mid a_k-\zeta_0-x_1-\ldots-x_n+z_1+\ldots+z_n\leqq x_{n+1}+\ldots+x_{n+k}\leqq$$
$$\leqq b_k-\zeta_0-x_1-\ldots-x_n+z_1+\ldots+z_n,\quad k=1,\ldots,N\}.$$

Then probability in the objective function of Problem (2.4) can be expressed in the following manner:

(3.8) $$P(a_k\leqq\zeta_k-Z_k\leqq b_k, k=n+1,\ldots,n+N\mid\xi_1=x_1,\ldots,\xi_n=x_n)=$$

$$=\int\limits_{A(z_{n+1},\ldots,z_{n+N})}f(\mathbf{x}\mid\mathbf{x}^P)\,d\mathbf{x},$$

where

(3.9) $$A(z_{n+1},\ldots,z_{n+N})=A+\begin{pmatrix}z_{n+1}\\z_{n+1}+z_{n+2}\\\vdots\\z_{n+1}+z_{n+2}+\ldots+z_{n+N}\end{pmatrix}.$$

Theorem 1 and Relation (3.8) jointly imply

Theorem 2. *The probability standing in* (3.8) *is a logarithmic concave function of the variables* $z_{n+1},\ldots,z_{n+N}$.

In [1] the following theorem is proved.

Theorem 3. *Let* **A** *be a convex set in* R^m *symmetric about the origin. Let f be a quasiconcave probability density in* R^m *with the property that $f(-\mathbf{x})=f(\mathbf{x})$ for every* $\mathbf{x}\in R^m$. *Then for every* $\mathbf{y}\in R^m$ *and* $0\leqq k\leqq 1$ *we have the inequality*

$$\int\limits_{A+k\mathbf{y}}f(\mathbf{x})\,d\mathbf{x}\geqq\int\limits_{A+\mathbf{y}}f(\mathbf{x})\,d\mathbf{x}.$$

188

In other words, this theorem states that the probability of the set $A+t\mathbf{y}$ is a monotonically decreasing function in $[0, \infty)$ of the variable t for every fixed $\mathbf{y}$.

Theorem 3 implies that if we take the unconstrained maximum of the probability (3.8) where only $z_{n+1}, \ldots, z_{n+N}$ are the variables, then the maximizing $z_{n+1}, \ldots, z_{n+N}$ satisfy the equalities

$$(3.10) \qquad \frac{a_k+b_k}{2} - \zeta_{\text{initial}} - e^C_{n+1} - \ldots - e^C_{n+k} + z_{n+1} + \ldots + z_{n+k} = 0, \quad k=1, \ldots, N,$$

where

$$(3.11) \qquad \zeta_{\text{initial}} = \zeta_0 + x_1 + \ldots + x_n - z_1 - \ldots - z_n$$

is the water content of the lake beginning of Period $n+1$. This is that period for which we want to find an optimal policy.

4. SOLUTION OF THE PROBLEMS FORMULATED IN SECTION 2

We shall consider Problem (2.4). Problem (2.3) is very similar and does not need a separate treatment. First we show that the optimization of the function (3.8) on the cube $0 \leq z_k \leq K$, $k=n+1, \ldots, n+N$ can be reduced to maximizations of the same functions on at most N faces of this cube. Sometimes the constrained optimal solution can be obtained directly without any computation. In fact first we solve the system of equations (3.10) with respect to $z_{n+1}, \ldots, z_{n+N}$. If we have $0 \leq z_k \leq K$ for $k=n+1, \ldots, n+N$, then this is the optimal solution also to the constrained problem (2.4). On the other hand, if for some i we have $z_i > K$ or for some k we have $z_k < 0$, then by Theorem 3 the optimum is attained on one of those faces of the cube which can be "seen" from the point with coordinates $z_{n+1}, \ldots, z_{n+N}$. These faces can be generated as follows. If $z_i > K$, then we adopt the face

$$z_i < K, \quad 0 \leq z_j \leq K, \quad j=n+1, \ldots, i-1, i+1, \ldots, n+N;$$

if $z_k < 0$, then we adopt the face

$$z_k = 0, \quad 0 \leq z_j \leq K, \quad j=n+1, \ldots, k-1, k+1, \ldots, n+N.$$

Obviously the number of such faces is at most N.

Example. Let $n=2$, $N=4$ and $z_3 > K$, $0 \leq z_4 \leq K$, $z_5 < 0$, $z_6 > K$. Then our face collection consits of the following three faces

$$\{z_3, z_4, z_5, z_6 \mid z_3 = K, \quad 0 \leq z_4, z_5, z_6 \leq K\},$$
$$\{z_3, z_4, z_5, z_6 \mid z_5 = 0, \quad 0 \leq z_3, z_4, z_6 \leq K\},$$
$$\{z_3, z_4, z_5, z_6 \mid z_6 = K, \quad 0 \leq z_3, z_4, z_5 \leq K\}.$$

When optimizing in the faces we can apply various nonlinear programming methods. We have tested on this and similar stochastic programming problems the method of feasible directions, SUMT, GRG the flexible tolerance method and the cutting plane method.

It is worth to describe shortly the application of the SUMT interior point method. Let us assume that we want to maximize the function (3.8) on the following face

$$\{z_{n+1}, \ldots, z_{n+N} \mid z_{n+1}=K, \quad 0 \leqq z_i \leqq K, \quad i=n+2, \ldots, n+N\}.$$

If instead of the function (3.8) we work with its logarithm, then the penalty function is given by the following formula

$$(4.1) \quad \log P(a_k \leqq \zeta_k - Z_k \leqq b_k, \quad k=n+1, \ldots, n+N \mid \xi_1, \ldots, \xi_n) - r \sum_{k=n+2}^{n+N} \log z_k(1-z_k),$$

where r is a fixed positive number and $z_{n+1}=K$ in the sums $Z_k=z_1+\ldots+z_k$, $k=n+1$, $\ldots, n+N$. The function (4.1) is convex and this fact makes the solutions of the unconstrained minimization problems relatively comfortable. Several unconstrained optimization method can be applied here [7] and we have tested a number of them. The use of a gradient free method seems to be advisable. The gradient of the function (4.1) is expressed in [10], but the formula is sophisticated.

Function values are computed by simulation at every step using a fast random number generation technique written in COMPASS for the CDC 3300 computer [3].

If we take a decreasing sequence $r_1, r_2, \ldots$ tending to zero, then the SUMT interior point method convergens (the conditions are trivially satisfied in our case) in the sense that (4.1) converges to the negative logarithm of the optimum value of Problem (2.4).

If $N=2$, then the original problem (2.4) is two-dimensional, but since the faces of the rectangle

$$\{z_{n+1}, z_{n+2} \mid 0 \leqq z_{n+1}, z_{n+2} \leqq K\}$$

are lines we have to optimize on at most two lines. This is done by the use of the Fibonacci search [14].

5. METHOD FOR THE REGULATION OF THE WATER LEVEL OF LAKE BALATON

Having performed a large number of computations it turned out that using only two conditioning random variables (instead of the whole past history) and optimizing for two steps ahead i.e. choosing $N=2$, a satisfactory water level control methodology can be obtained.

The lake is represented by a prism the surface of which is 600 km². We choose as water quantity unit that quantity which increases the water level by exactly 1 mm. (This quantity equals 600,000 m³.) All date will be given in this unit.

According to what is said above, four random variables will be involved in every optimization problem. They belong to four consecutive months and will be denoted

190

by ξ_1, ξ_2, ξ_3, ξ_4 in agreement with the earlier notations. To the earliest month corresponds ξ_1, then comes ξ_2, etc.

The prescribed lower resp. upper bounds are as follows:

	Lower bounds	Upper bounds
February–June	3100 mm	3400 mm
July–January	3000 mm	3300 mm

The original by prescribed levels communicated to us were 2900 mm resp. 3400 mm for every month. We observed that our control methodology allowed to keep the water level between the narrower limits 3000 mm resp. 3300 mm (with a satisfactory probability). However, due to large input water quantites in the first half of the year, the corresponding limits were increased by 100 mm and this improved the controllability for the most important summer months. Thus in all cases we have to solve the following type of problem:

$$\text{maximize} \quad P\left(\begin{matrix} a_3 \leq \zeta_{\text{initial}} + \xi_3 \qquad - z_3 \qquad \leq b_3 \\ a_4 \leq \zeta_{\text{initial}} + \xi_3 + \xi_4 - z_3 - z_4 \leq b_4 \end{matrix} \middle| \xi_1, \xi_2\right)$$

5.1)

$$\text{subject to}$$

$$0 \leq z_3 \leq 200,$$
$$0 \leq z_4 \leq 200,$$

where a_3, b_3, a_4, b_4 are chosen according to the above table of the lower resp. upper bounds. The subscripts of a_3, b_3, z_3 and a_4, b_4, z_4 are chosen in accordance with the subscripts of ξ_3 and ξ_4.

It will be still more comfortable to operate with the following transformed random variables

$$(5.2) \qquad \zeta_1 = \xi_1, \quad \zeta_2 = \xi_2, \quad \zeta_3 = \xi_3, \quad \zeta_4 = \xi_3 + \xi_4.$$

The covariance matrix D of the random variables ζ_1, ζ_2, ζ_3, ζ_4 can be obtained from the covariance matrix

$$C = \begin{pmatrix} c_{11} & c_{12} & c_{13} & c_{14} \\ c_{12} & c_{22} & c_{23} & c_{24} \\ c_{13} & c_{23} & c_{33} & c_{34} \\ c_{14} & c_{24} & c_{34} & c_{44} \end{pmatrix}$$

of the random variables ξ_1, ξ_2, ξ_3, ξ_4. We have

$$(5.3) \qquad D = \begin{pmatrix} c_{11} & c_{12} & c_{13} & c_{13}+c_{14} \\ c_{12} & c_{22} & c_{23} & c_{23}+c_{24} \\ c_{13} & c_{23} & c_{33} & c_{33}+c_{34} \\ c_{13}+c_{14} & c_{23}+c_{24} & c_{33}+c_{34} & c_{33}+c_{44}+2c_{34} \end{pmatrix}.$$

Table I. Natural Water Content Changes of Lake Balaton (Rainfall+Inflow-Evaporation)

	January	February	March	April	May	June	July	August	September	October	November	December
1921	62	133	16	26	43	−35	−102	−143	−68	4	77	58
1922	102	119	128	192	−3	−52	−103	−18	118	198	96	72
1923	85	122	239	94	−17	9	−52	−75	−30	60	119	154
1924	38	94	262	171	104	26	−50	−7	−4	−26	10	56
1925	48	75	86	48	54	−18	−22	−56	87	−8	232	106
1926	130	165	77	23	−7	74	101	97	−11	89	173	139
1927	121	79	137	62	−25	−48	−54	32	45	−1	19	36
1928	99	126	111	31	90	−51	−97	−59	81	49	76	73
1929	95	74	170	208	38	−24	−90	−101	−76	39	159	16
1930	79	131	106	172	−51	−95	−124	−27	−21	182	178	235
1931	178	208	302	166	37	−59	−156	−32	43	16	92	37
1932	101	35	135	65	59	−95	−97	−65	−59	74	18	48
1933	51	73	92	21	41	−1	−119	−41	−8	48	240	136
1934	137	69	77	−37	−62	−19	−42	−46	27	9	103	72
1935	63	147	85	23	−21	−85	−136	−57	−13	18	31	172
1936	162	199	107	63	85	4	−90	−107	27	126	97	94
1937	105	146	343	260	−4	40	8	10	67	99	222	331
1938	214	118	71	18	79	−39	−98	58	−11	34	10	87
1939	114	77	73	−61	74	9	−137	−37	2	97	92	77
1940	68	67	339	111	96	76	22	159	247	174	243	90
1941	127	231	204	216	129	−39	−83	−35	−57	70	200	148
1942	115	215	367	307	178	−66	−57	−99	−56	−27	23	61
1943	112	179	−12	−5	−13	108	32	−106	−2	−2	148	129
1944	57	113	241	11	52	96	−32	−72	−50	146	225	254
1945	191	309	215	35	−37	−79	−68	−99	−28	19	99	97
1946	68	118	60	−41	−75	−51	−63	−93	−52	7	86	84
1947	87	182	448	117	−17	−54	−77	−148	−85	−5	22	88
1948	110	106	8	85	−19	17	136	−41	−30	46	47	37
1949	85	9	−16	4	41	−73	−77	−75	−84	−2	175	52
1950	115	188	63	63	−33	−155	−90	−58	−1	67	216	190

1951	139	154	206	−27	112	167	39	−35	18	−36	21	85
1952	114	155	166	37	−49	−38	−153	−128	−26	133	97	155
1953	110	71	18	−5	40	22	−93	−51	−59	8	−22	17
1954	73	56	192	28	175	16	22	−54	−31	21	76	124
1955	121	134	152	66	10	−46	39	81	26	140	185	77
1956	88	108	160	114	63	14	−3	−75	−85	22	54	114
1957	46	293	75	17	12	−56	19	−68	14	−26	78	36
1958	99	89	47	24	−60	89	−29	−83	−5	−19	32	85
1959	98	40	11	59	9	82	63	−55	−69	−40	46	140
1960	127	147	67	63	33	−80	40	−57	−7	141	154	207
1961	121	113	39	13	50	8	−80	−110	−70	−11	90	51
1962	116	70	185	95	−20	−50	21	−115	−42	−1	220	149
1963	191	130	368	174	9	−17	−127	7	96	75	71	109
1964	42	73	250	123	77	−2	−30	−31	−2	170	82	190
1965	164	111	131	190	193	244	145	131	72	15	187	448
1966	122	202	151	146	46	38	89	87	48	19	240	203
1967	204	184	143	141	−11	96	−55	−107	60	17	33	40
1968	104	85	50	10	−65	−100	−140	70	21	40	167	69
1969	139	313	234	48	14	84	−58	−30	11	20	72	110
1970	111	182	382	223	35	−18	−64	49	−5	12	62	85
Expectations	108.96	132.34	151.22	79.74	29.78	−4.52	−43.44	−38.30	−0.74	46.00	109.46	114.46
Dispersions	41.52	67.04	112.84	83.51	63.11	73.98	73.96	69.58	61.95	62.51	75.15	80.60

Table II. Correlations of Natural Water Content Changes of Lake Balaton

	January	February	March	April	May	June	July	August	September	October	November	December
January	1.000	.356	.069	.105	−.012	.125	−.019	.148	.159	−.054	−.167	−.016
February	.356	1.000	.284	.199	−.064	−.046	−.018	−.097	.043	−.070	−.037	.045
March	.069	.284	1.000	.630	.244	.036	−.090	.139	.220	.128	−.023	.124
April	.105	.199	.630	1.000	.284	−.017	.047	.193	.167	.159	.106	.247
May	−.012	−.064	.244	.284	1.000	.333	.198	.201	.074	−.054	−.028	.141
June	.125	−.046	.036	−.017	.333	1.000	.579	.263	.223	−.152	.016	.338
July	−.019	−.018	−.090	.047	.198	.579	1.000	.352	.120	−.122	.217	.355
August	.148	−.097	.139	.193	.201	.263	.352	1.000	.597	.260	.301	.287
September	.159	.043	.220	.167	.074	.223	.120	.597	1.000	.345	.276	.132
October	−.054	−.070	.128	.159	−.054	−.152	−.122	.260	.345	1.000	.387	.308
November	−.167	−.037	−.023	.106	−.028	.016	.217	.301	.276	.387	.1000	.513
December	−.016	.045	.124	.247	.141	.338	.355	.287	.132	.308	.513	1.000
January	−.180	.020	.080	.158	.006	.028	.172	.194	.093	.204	.526	.559
February		−.003	.085	.079	.114	.041	.054	.201	.017	.112	.393	.349
March			−.021	−.117	−.034	.016	−.190	.076	.100	.255	.262	.108
April				.227	.228	.106	−.031	.097	.154	.270	.216	.101
May					.140	−.071	−.091	.311	.165	.321	−.039	.187
June						−.083	.019	.084	.152	.127	.035	.195
July							.174	−.034	−.078	−.094	−.053	.163
August								.150	.206	.180	.136	.257
September									−.046	−.020	−.000	.115
October										−.093	−.114	−.174
November											−.113	−.100
December												.037

	January	February	March	April	May	June	July	August	September	October	November	December
January	−.180											
February	.020	−.003										
March	.080	.085	−.021									
April	.158	.079	−.117	.227								
May	.006	.114	−.034	.228	.140							
June	.028	.041	.016	.106	−.071	−.083						
July	.172	.054	−.190	−.031	−.091	.019	.174					
August	.194	.201	.076	.097	.311	.084	−.034	.150				
September	.093	.017	.100	.154	.165	.152	−.078	.206	−.046			
October	.204	.112	.255	.270	.321	.127	−.094	.180	−.020	−.093		
November	.526	.393	.262	.216	−.039	.035	−.053	.136	−.000	−.114	−.113	
December	.559	.349	.108	.101	.187	.195	.163	.257	.115	−.174	−.100	.037
January	1.000	.356	.069	.105	−.012	.125	−.019	.148	.159	−.054	−.167	−.016
February	.356	1.000	.284	.199	−.064	−.046	−.018	−.097	.043	−.070	−.037	.045
March	.069	.284	1.000	.630	.244	.036	−.090	.139	.220	.128	−.028	.124
April	.105	.199	.630	1.000	.284	−.017	.047	.193	.167	.159	.106	.247
May	−.012	−.064	.244	.284	1.000	.333	.198	.201	.074	−.054	−.028	.141
June	.125	−.046	.036	−.017	.333	1.000	.579	.263	.223	−.152	.016	.338
July	−.019	−.018	−.090	.047	.198	.579	1.000	.352	.120	−.122	.217	.355
August	.148	−.097	.139	.193	.201	.263	.352	1.000	.597	.260	.301	.287
September	.159	.043	.220	.167	.074	.223	.120	.597	1.000	.345	.276	.132
Oktober	−.054	−.070	.128	.159	−.054	−.152	−.122	.260	.345	1.000	.387	.308
November	−.167	−.037	−.028	.106	−.028	.016	.217	.301	.276	.387	1.000	.513
December	−.016	.045	.124	.247	.141	.338	.355	.287	.132	.308	.513	1.000

With the aid of ζ_1, ζ_2, ζ_3, ζ_4, Problem (5.1) can be written in the following manner

$$\text{maximize} \qquad P\left(\begin{matrix} a_3 \leq \zeta_{\text{initial}} + \zeta_3 - z_3 & \leq b_3 \\ a_4 \leq \zeta_{\text{initial}} + \zeta_4 - z_3 - z_4 \leq b_4 \end{matrix} \middle| \zeta_1, \zeta_2 \right)$$

(5.4)

subject to

$$0 \leq z_3 \leq 200,$$
$$0 \leq z_4 \leq 200.$$

Let us rearrange and then partition the covariance matrix D in a way indicated here below

$$
\begin{array}{cccc}
\zeta_3 & \zeta_4 & \zeta_1 & \zeta_2
\end{array}
$$

S_1	U_1	ζ_3 ζ_4
U_1'	T_1	ζ_1 ζ_2

Since $\zeta_1 = \xi_1$, $\zeta_2 = \xi_2$ it follows that $T_1 = T$ where T is taken out from that special case of (3.3) in which $n = 2$, $N = 2$. Then we have

$$
(5.6) \qquad E\left[\begin{pmatrix} \zeta_3 \\ \zeta_4 \end{pmatrix} \middle| \begin{pmatrix} \zeta_1 \\ \zeta_2 \end{pmatrix}\right] = \begin{pmatrix} e_3 \\ e_3 + e_4 \end{pmatrix} + U_1 T_1^{-1}\left[\begin{pmatrix} \zeta_1 \\ \zeta_2 \end{pmatrix} - \begin{pmatrix} e_1 \\ e_2 \end{pmatrix}\right].
$$

On the next pages we present the input data (Table 1) for the 50 years between 1921 and 1970. Taking into account a longer (less reliable but improved by hydrological considerations) time series, the Institute of Water Management of the Technical University of Budapest advised the use of a Gaussian process as the mathematical model of the water input process, after a careful statistical analysis [2].

Together with the realized values of the input process we present the corresponding expectations and variances.

Then we present a part of a 24×24 correlation matrix (Table 2). Assuming that the random input of one month is stochastically independent of the inputs of such months which are farther then one year then it turns out that all nonzero correlations will be contained in a 24×24 correlation matrix. In practice, however, only correlations very near the diagonal will be needed because the others are very small. That part of the correlation matrix what we present here is larger than the necessary part but it well illustrates that dependencies exist only between very near months.

As it was mentioned in the Introduction we carried out the monthly optimizations between 1922 and 1970.

First we computed twelwe D matrices and to each D we also computed the corresponding two matrices $U_1 T_1^{-1}$, $S_1 - U_1 T_1^{-1} U_1'$. These matrices are fixed, they do not depend on actual values of the input time series. Then using the actual values of ζ_1 and ζ_2 we computed all conditional expectations (5.6). Finally came the 588 optimizations (one for every month in the years 1922-1970) out of which a large number were trivial i.e., the solution of the equation (3.10) specialized to our case ($n=2$, $N=2$) produced such z_3 and z_4 for which $0 \leq z_3, z_4 \leq 200$.

As an example we consider the problem of finding the optimal water quantity to be released in July, 1953. In this case the random variables ξ_1, ξ_2, ξ_3, ξ_4 have the following meanings

ξ_1 input water quantity in May, 1953
ξ_2 input water quantity in June, 1953
ξ_3 input water quantity in July, 1953
ξ_4 input water quantity in August, 1953.

The expectations, dispersions and the correlation matrix can be obtained from the presented tables. They are reproduced here:

$$E(\xi_1)=29.78, \quad E(\xi_2)=-4.52, \quad E(\xi_3)=-43.44, \quad E(\xi_4)=-38.30,$$
$$D(\xi_1)=63.11, \quad D(\xi_2)=73.98, \quad D(\xi_3)=73.96, \quad D(\xi_4)=69.58,$$

$$R = \begin{matrix} & \xi_1 & \xi_2 & \xi_3 & \xi_4 & \\ & 1.000 & 0.333 & 0.198 & 0.201 & \xi_1 \\ & 0.333 & 1.000 & 0.579 & 0.263 & \xi_2 \\ & 0.198 & 0.579 & 1.000 & 0.352 & \xi_3 \\ & 0.201 & 0.263 & 0.352 & 1.000 & \xi_4 \end{matrix}$$

The transformed variables (5.2) have the following expectations and covariance matrix

$$E(\zeta_1)=27.98, \quad E(\zeta_2)=-4.52, \quad E(\zeta_3)=-43.44, \quad E(\zeta_4)=-81.74,$$

$$D = \begin{matrix} & \zeta_1 & \zeta_2 & \zeta_3 & \zeta_4 & \\ & 3982.87210011 & 1554.73630744 & 924.18788880 & 1806.81784260 & \zeta_1 \\ & 1554.73630744 & 5473.04040002 & 3168.03370320 & 4521.83367240 & \zeta_2 \\ & 924.18788880 & 3168.03370320 & 5470.08160018 & 7281.52175378 & \zeta_3 \\ & 1806.81784260 & 4521.83367240 & 7281.52175378 & 13934.33830761 & \zeta_4 \end{matrix}$$

From here we obtain

$$S_1 = \begin{pmatrix} 5470.08160018 & 7281.52175378 \\ 7281.52175378 & 13934.33830761 \end{pmatrix},$$

$$U_1 = \begin{pmatrix} 924.18788880 & 3168.03370320 \\ 1806.81784260 & 4521.83367240 \end{pmatrix},$$

$$T_1 = \begin{pmatrix} 3982.87210011 & 1554.73630744 \\ 1554.73630744 & 5473.04040002 \end{pmatrix},$$

$$T_1^{-1} = \begin{pmatrix} 0.00028239 & -0.00008022 \\ -0.00008022 & 0.00020550 \end{pmatrix},$$

$$U_1 T_1^{-1} = \begin{pmatrix} 0.00684480 & 0.57689906 \\ 0.14748962 & 0.78430377 \end{pmatrix}.$$

The realised water input data are the following

$$\xi_1 = \zeta_1 = 40, \quad \xi_2 = \zeta_2 = 22,$$

hence the conditional expectation equals

$$(5.7) \quad E\left[\binom{\zeta_3}{\zeta_4}\,\middle|\,\zeta_1 = 40, \zeta_2 = 22\right] = \binom{-43.44}{-8174} + U_1 T_1^{-1}\left[\binom{40}{22} - \binom{29.78}{-4.52}\right] = \binom{-28.07}{-59.43}.$$

The covariance matrix of the conditional distribution of ζ_3, ζ_4 given that $\zeta_1 = 40$, $\zeta_2 = 22$ is the following

$$(5.8) \qquad S_1 - U_1 T_1^{-1} U_1' = \begin{pmatrix} 3636.12006366 & 4660.51286423 \\ 4660.51286423 & 10121.36024427 \end{pmatrix}.$$

Since $\zeta_{\text{initial}} = 3205$, the optimization problem (5.4) can be written in the following manner

$$\text{maximize } P\left(\begin{array}{l} -205 \le \zeta_3 - z_3 \quad\;\; \le 95 \\ -205 \le \zeta_4 - z_3 - z_4 \le 95 \end{array}\,\middle|\,\zeta_1 = 40, \zeta_2 = 22\right)$$

subject to

$$0 \le z_3 \le 200,$$
$$0 \le z_4 \le 200,$$

and the above probability distribution is two-dimensional normal with expectation vector (5.7) and covariance matrix (5.8).

If we compute z_3, z_4 according to (3.10), we obtain the values

$$z_3 = 27, \quad z_4 = -31.$$

It follows that the optimal z_4 to Problem (5.9) equals

$$z_4^* = 0.$$

and z_3^* is the optimal solution of the following one-dimensional problem

$$(5.10) \qquad \text{maximize } P\left(\begin{array}{l} -205 \le \zeta_3 - z_3 \le 95 \\ -205 \le \zeta_4 - z_3 \le 95 \end{array}\,\middle|\,\zeta_1 = 40, \zeta_2 = 22\right)$$

subject to

$$0 \le z_3 \le 200.$$

The Fibonacci search gives in 15 steps the result

$$z_3^* = 2.$$

In higher dimensional cases the values of the objective function are determined by simulation. In the two-dimensional case numerical integration is satisfactorily effective. We use a reduction formula and then one-dimensional numerical integration.

The reduction formula states that if $f(x, y; r)$ is the two-dimensional normal probability density function with standard marginal distributions and r is the correlation coefficient ($|r| < 1$), then we have

$$\int\limits_a^b \int\limits_c^d f(x, y, r)\, dy\, dx = \int\limits_a^b \left[\Phi\left(\frac{d - rx}{\sqrt{1 - r^2}}\right) - \Phi\left(\frac{c - rx}{\sqrt{1 - r^2}}\right) \right] \varphi(x)\, dx$$

where φ and Φ denote the one-dimensional standard normal density resp. distribution function. The one-dimensional numerical integration is done by the Romberg–Havie procedure [4]. The computational precision is 10^{-3}.

In Table III. we summarize the results of the 588 optimizations. The + resp. − signs mean that the water level is higher resp. lower than desired. We see that only a few such signs occur. Moreover the first month of 1922 do not count because we need a few periods for the running-in of this control methodology. In 1946 the lock was repaired which caused a high water level. The variation of the controlled water level is illustrated on Fig. 2.

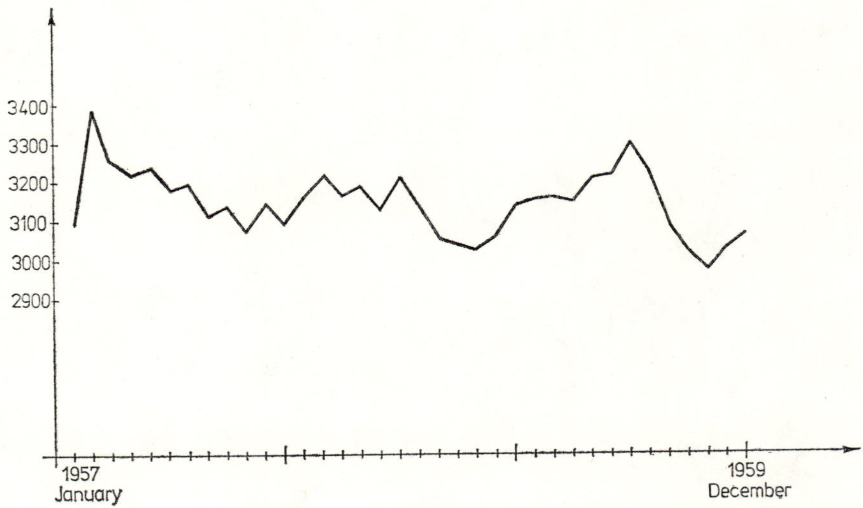

Fig. 2. Variation of the controlled water level of Lake Balaton (Illustration for three years)

Table III. Numerical Results.
Application of the Proposed Control Methodology for Lake Balaton between the years 1922–1970. The x_3 are the Final Accepted Values. Water Levels are Computed with These.
+ Sign (resp. − Sign) Means Water Level Above 3400 mm (resp. Below 2900 mm)

		x_3	x_4	WATER LEVEL	PROBABILITY
1922	JANUARY	0.	0.	2314.(—)	0.00
1922	FEBRUARY	0.	0.	2433.(—)	0.00
1922	MARCH	0.	0.	2561.(—)	0.00
1922	APRIL	0.	0.	2753.(—)	0.00

Table III (cont.)

		x_3	x_4	WATER LEVEL	PROBABILITY
1922	MAY	0.	0.	2750.(—)	0.00
1922	JUNE	0.	0.	2698.(—)	0.00
1922	JULY	0.	0.	2595.(—)	0.00
1922	AUGUST	0.	0.	2577.(—)	0.00
1922	SEPTEMBER	0.	0.	2695.(—)	0.00
1922	OCTOBER	0.	0.	2893.(—)	0.00
1922	NOVEMBER	0.	129.	2989.	74.10
1922	DECEMBER	0.	84.	3061.	90.06
1923	JANUARY	9.	22.	3137.	96.20
1923	FEBRUARY	3.	144.	3256.	70.79
1923	MARCH	155.	75.	3341.	61.29
1923	APRIL	200.	54.	3235.	88.62
1923	MAY	21.	0.	3196.	82.43
1923	JUNE	0.	0.	3205.	71.23
1923	JULY	0.	0.	3153.	84.81
1923	AUGUST	0.	0.	3078.	79.77
1923	SEPTEMBER	0.	0.	3048.	80.39
1923	OCTOBER	0.	40.	3108.	80.80
1923	NOVEMBER	67.	119.	3160.	77.34
1923	DECEMBER	131.	112.	3182.	90.72
1924	JANUARY	151.	41.	3069.	96.20
1924	FEBRUARY	0.	99.	3163.	67.16
1924	MARCH	52.	65.	3373.	61.29
1924	APRIL	200.	95.	3344.	85.70
1924	MAY	141.	0.	3307.	82.43
1924	JUNE	74.	73.	3258.	76.86
1924	JULY	66.	0.	3143.	85.98
1924	AUGUST	0.	0.	3136.	79.03
1924	SEPTEMBER	3.	57.	3129.	92.30
1924	OCTOBER	26.	116.	3077.	84.40
1924	NOVEMBER	7.	87.	3080.	77.34
1924	DECEMBER	0.	67.	3136.	90.50
1925	JANUARY	62.	0.	3122.	96.20
1925	FEBRUARY	0.	115.	3197.	70.34
1925	MARCH	76.	62.	3207.	61.29
1925	APRIL	5.	27.	3250.	88.72
1925	MAY	16.	0.	3288.	82.42
1925	JUNE	47.	63.	3223.	76.86
1925	JULY	0.	0.	3201.	85.30
1925	AUGUST	14.	0.	3131.	81.95
1925	SEPTEMBER	0.	6.	3218.	92.04
1925	OCTOBER	138.	121.	3071.	84.40
1925	NOVEMBER	26.	97.	3277.	77.34
1925	DECEMBER	200.	195.	3183.	77.45
1926	JANUARY	162.	63.	3151.	96.20
1926	FEBRUARY	40.	151.	3276.	70.79
1926	MARCH	192.	89.	3162.	61.29
1926	APRIL	0.	0.	3185.	86.00
1926	MAY	0.	0.	3178.	76.80
1926	JUNE	0.	0.	3252.	69.81
1926	JULY	81.	0.	3271.	85.83
1926	AUGUST	130.	13.	3238.	81.95
1926	SEPTEMBER	151.	58.	3076.	92.30
1926	OCTOBER	0.	119.	3165.	84.23
1926	NOVEMBER	139.	131.	3198.	77.34
1926	DECEMBER	200.	128.	3137.	90.72
1927	JANUARY	113.	53.	3146.	96.20
1927	FEBRUARY	37.	155.	3188.	70.79

Table III (cont.)

		x_3	x_4	WATER LEVEL	PROBABILITY
1927	MARCH	61.	68.	3264.	61.29
1927	APRIL	85.	35.	3240.	88.72
1927	MAY	12.	0.	3203.	82.43
1927	JUNE	0.	0.	3155.	72.52
1927	JULY	0.	0.	3101.	61.25
1927	AUGUST	0.	0.	3133.	63.79
1927	SEPTEMBER	23.	70.	3155.	92.30
1927	OCTOBER	70.	134.	3084.	84.40
1927	NOVEMBER	34.	98.	3069.	77.34
1927	DECEMBER	0.	67.	3105.	90.44
1928	JANUARY	28.	0.	3176.	96.20
1928	FEBRUARY	40.	140.	3262.	70.79
1928	MARCH	161.	77.	3212.	61.29
1928	APRIL	23.	24.	3220.	88.72
1928	MAY	0.	0.	3310.	82.05
1928	JUNE	87.	71.	3172.	76.86
1928	JULY	0.	0.	3075.	72.15
1928	AUGUST	0.	0.	3016.	48.87
1928	SEPTEMBER	0.	0.	3097.	52.93
1928	OCTOBER	16.	120.	3130.	84.40
1928	NOVEMBER	107.	119.	3099.	77.34
1928	DECEMBER	48.	99.	3125.	90.72
1929	JANUARY	69.	17.	3150.	96.20
1929	FEBRUARY	20.	145.	3204.	70.79
1929	MARCH	78.	65.	3297.	61.29
1929	APRIL	133.	41.	3461.(+)	88.72
1929	MAY	200.	45.	3299.	76.92
1929	JUNE	25.	57.	3250.	76.86
1929	JULY	15.	0.	3146.	85.27
1929	AUGUST	0.	0.	3045.	75.32
1929	SEPTEMBER	0.	0.	2969.	54.15
1929	OCTOBER	0.	0.	3008.	38.95
1929	NOVEMBER	0.	69.	3167.	75.62
1929	DECEMBER	155.	123.	3028.	90.72
1930	JANUARY	0.	5.	3107.	96.19
1930	FEBRUARY	0.	101.	3238.	69.87
1930	MARCH	142.	77.	3202.	61.29
1930	APRIL	11.	23.	3363.	88.72
1930	MAY	145.	0.	3167.	82.38
1930	JUNE	0.	0.	3072.	53.06
1930	JULY	0.	0.	2948.	15.01
1930	AUGUST	0.	0.	2921.	2.97
1930	SEPTEMBER	0.	0.	2900.	8.62
1930	OCTOBER	0.	24.	3082.	14.85
1930	NOVEMBER	92.	166.	3168.	77.34
1930	DECEMBER	189.	129.	3214.	90.72
1931	JANUARY	200.	80.	3192.	96.20
1931	FEBRUARY	122.	171.	3278.	70.79
1931	MARCH	200.	108.	3380.	61.21
1931	APRIL	200.	103.	3346.	79.80
1931	MAY	147.	0.	3235.	82.43
1931	JUNE	0.	36.	3176.	76.62
1931	JULY	0.	0.	3020.	69.09
1931	AUGUST	0.	0.	2988.	17.73
1931	SEPTEMBER	0.	0.	3031.	48.74
1931	OCTOBER	0.	69.	3047.	81.97
1931	NOVEMBER	4.	105.	3136.	77.34
1931	DECEMBER	87.	104.	3086.	90.72

Table III (cont.)

		x_3	x_4	WATER LEVEL	PROBABILITY
1932	JANUARY	26.	15.	3161.	96.20
1932	FEBRUARY	26.	140.	3170.	70.79
1932	MARCH	23.	57.	3282.	61.29
1932	APRIL	101.	40.	3245.	88.72
1932	MAY	18.	0.	3287.	82.43
1932	JUNE	47.	64.	3145.	76.86
1932	JULY	0.	0.	3048.	47.34
1932	AUGUST	0.	0.	2983.	33.58
1932	SEPTEMBER	0.	0.	2924.	26.62
1932	OCTOBER	0.	8.	2998.	19.06
1932	NOVEMBER	0.	88.	3016.	76.22
1932	DECEMBER	0.	44.	3064.	86.50
1933	JANUARY	0.	0.	3115.	96.05
1933	FEBRUARY	0.	108.	3188.	70.05
1933	MARCH	66.	61.	3215.	61.29
1933	APRIL	15.	28.	3220.	88.72
1933	MAY	0.	0.	3261.	81.88
1933	JUNE	18.	60.	3242.	76.86
1933	JULY	23.	0.	3101.	85.52
1933	AUGUST	0.	0.	3060.	60.82
1933	SEPTEMBER	0.	0.	3052.	85.17
1933	OCTOBER	0.	60.	3100.	82.45
1933	NOVEMBER	58.	115.	3281.	77.34
1933	DECEMBER	200.	200.	3217.	70.14
1934	JANUARY	200.	74.	3154.	96.20
1934	FEBRUARY	51.	155.	3172.	70.79
1934	MARCH	39.	67.	3210.	61.29
1934	APRIL	4.	27.	3169.	88.72
1934	MAY	0.	0.	3107.	71.05
1934	JUNE	0.	0.	3088.	33.21
1934	JULY	0.	0.	3046.	40.89
1934	AUGUST	0.	0.	3000.	45.18
1934	SEPTEMBER	0.	0.	3027.	42.11
1934	OCTOBER	0.	56.	3036.	80.52
1934	NOVEMBER	0.	90.	3139.	77.25
1934	DECEMBER	94.	107.	3117.	90.72
1935	JANUARY	66.	24.	3114.	96.20
1935	FEBRUARY	0.	117.	3261.	70.35
1935	MARCH	174.	80.	3172.	61.29
1935	APRIL	0.	0.	3195.	87.87
1935	MAY	0.	0.	3174.	78.89
1935	JUNE	0.	0.	3089.	67.04
1935	JULY	0.	0.	2953.	23.66
1935	AUGUST	0.	0.	2896.(—)	3.21
1935	SEPTEMBER	0.	0.	2883.(—)	1.42
1935	OCTOBER	0.	20.	2901.	9.20
1935	NOVEMBER	0.	35.	2932.	43.31
1935	DECEMBER	0.	24.	3104.	50.99
1936	JANUARY	60.	22.	3205.	96.20
1936	FEBRUARY	118.	161.	3286.	70.79
1936	MARCH	200.	108.	3193.	61.11
1936	APRIL	4.	14.	3252.	88.72
1936	MAY	21.	0.	3315.	82.42
1936	JUNE	86.	70.	3233.	76.86
1936	JULY	22.	0.	3121.	85.78
1936	AUGUST	0.	0.	3014.	70.77
1936	SEPTEMBER	0.	0.	3041.	29.59
1936	OCTOBER	0.	48.	3167.	81.86

Table III (cont.)

		x_3	x_4	WATER LEVEL	PROBABILITY
1936	NOVEMBER	164.	146.	3100.	77.34
1936	DECEMBER	72.	105.	3122.	90.72
1937	JANUARY	75.	26.	3152.	96.20
1937	FEBRUARY	30.	148.	3269.	70.79
1937	MARCH	177.	83.	3435.	61.29
1937	APRIL	200.	147.	3495.	46.19
1937	MAY	200.	70.	3291.	64.33
1937	JUNE	2.	47.	3328.	76.86
1937	JULY	134.	0.	3203.	85.61
1937	AUGUST	32.	7.	3180.	81.95
1937	SEPTEMBER	52.	50.	3195.	92.30
1937	OCTOBER	115.	133.	3179.	84.40
1937	NOVEMBER	173.	137.	3228.	77.34
1937	DECEMBER	200.	180.	3359.	86.97
1938	JANUARY	200.	195.	3373.	15.74
1938	FEBRUARY	200.	200.	3291.	42.93
1938	MARCH	175.	85.	3187.	61.29
1938	APRIL	0.	2.	3205.	88.36
1938	MAY	0.	0.	3284.	80.17
1938	JUNE	58.	69.	3188.	76.86
1938	JULY	0.	0.	3090.	78.52
1938	AUGUST	0.	0.	3148.	55.80
1938	SEPTEMBER	56.	87.	3081.	92.30
1938	OCTOBER	0.	112.	3115.	84.26
1938	NOVEMBER	67.	109.	3057.	77.34
1938	DECEMBER	0.	58.	3144.	89.97
1939	JANUARY	80.	2.	3178.	96.20
1939	FEBRUARY	58.	148.	3198.	70.79
1939	MARCH	71.	67.	3200.	61.29
1939	APRIL	0.	18.	3139.	88.67
1939	MAY	0.	0.	3213.	58.54
1939	JUNE	0.	62.	3222.	76.84
1939	JULY	13.	0.	3072.	85.77
1939	AUGUST	0.	0.	3035.	46.82
1939	SEPTEMBER	0.	0.	3037.	76.64
1939	OCTOBER	0.	55.	3134.	80.96
1939	NOVEMBER	114.	134.	3112.	77.34
1939	DECEMBER	76.	104.	3113.	90.72
1940	JANUARY	61.	22.	3120.	96.20
1940	FEBRUARY	0.	126.	3187.	70.59
1940	MARCH	59.	61.	3466.(+)	61.29
1940	APRIL	200.	163.	3377.	30.64
1940	MAY	174.	3.	3300.	82.43
1940	JUNE	70.	72.	3305.	76.86
1940	JULY	149.	0.	3179.	86.13
1940	AUGUST	15.	14.	3322.	81.95
1940	SEPTEMBER	200.	160.	3369.	87.77
1940	OCTOBER	200.	200.	3343.	36.85
1940	NOVEMBER	200.	200.	3386.	20.00
1940	DECEMBER	200.	198.	3276.	9.88
1941	JANUARY	200.	117.	3203.	96.10
1941	FEBRUARY	88.	148.	3346.	70.79
1941	MARCH	200.	137.	3350.	53.40
1941	APRIL	200.	31.	3366.	88.68
1941	MAY	163.	0.	3332.	82.40
1941	JUNE	106.	79.	3187.	76.86
1941	JULY	0.	0.	3104.	80.08
1941	AUGUST	0.	0.	3069.	62.99

Table III (cont.)

		x_3	x_4	WATER LEVEL	PROBABILITY
1941	SEPTEMBER	0.	0.	3012.	87.72
1941	OCTOBER	0.	24.	3082.	69.05
1941	NOVEMBER	40.	122.	3242.	77.34
1941	DECEMBER	200.	172.	3190.	87.28
1942	JANUARY	172.	61.	3133.	96.20
1942	FEBRUARY	24.	156.	3324.	70.79
1942	MARCH	200.	127.	3491.(+)	57.66
1942	APRIL	200.	154.	3598.(+)	12.74
1942	MAY	200.	67.	3576.(+)	9.21
1942	JUNE	200.	149.	3310.	40.00
1942	JULY	59.	0.	3194.	85.74
1942	AUGUST	0.	0.	3095.	81.53
1942	SEPTEMBER	0.	0.	3039.	81.06
1942	OCTOBER	0.	19.	3012.	76.90
1942	NOVEMBER	0.	34.	3035.	73.46
1942	DECEMBER	0.	47.	3096.	87.32
1943	JANUARY	29.	1.	3179.	96.20
1943	FEBRUARY	53.	144.	3305.	70.79
1943	MARCH	200.	106.	3093.	60.62
1943	APRIL	0.	0.	3088.	38.15
1943	MAY	0.	0.	3075.	34.71
1943	JUNE	0.	0.	3183.	26.88
1943	JULY	37.	0.	3178.	85.99
1943	AUGUST	20.	20.	3052.	81.95
1943	SEPTEMBER	0.	0.	3050.	44.39
1943	OCTOBER	0.	42.	3048.	81.86
1943	NOVEMBER	0.	85.	3196.	77.27
1943	DECEMBER	171.	120.	3153.	90.72
1944	JANUARY	122.	45.	3088.	96.20
1944	FEBRUARY	0.	111.	3201.	69.45
1944	MARCH	98.	71.	3344.	61.29
1944	APRIL	200.	60.	3155.	88.53
1944	MAY	0.	0.	3207.	73.24
1944	JUNE	0.	36.	3303.	76.49
1944	JULY	156.	0.	3115.	86.10
1944	AUGUST	0.	0.	3043.	77.91
1944	SEPTEMBER	0.	0.	2993.	60.29
1944	OCTOBER	0.	12.	3139.	59.20
1944	NOVEMBER	129.	151.	3235.	77.34
1944	DECEMBER	200.	188.	3289.	83.48
1945	JANUARY	200.	177.	3280.	91.76
1945	FEBRUARY	200.	191.	3389.	70.61
1945	MARCH	200.	160.	3404.(+)	35.50
1945	APRIL	200.	69.	3239.	82.96
1945	MAY	16.	1.	3187.	82.43
1945	JUNE	0.	0.	3108.	68.17
1945	JULY	0.	0.	3040.	31.79
1945	AUGUST	0.	0.	2941.	34.46
1945	SEPTEMBER	0.	0.	2913.	2.81
1945	OCTOBER	0.	8.	2932.	16.88
1945	NOVEMBER	0.	36.	3031.	56.57
1945	DECEMBER	0.	94.	3128.	90.56
1946	JANUARY	9.	8.	3187.	96.13
1946	FEBRUARY	8.	9.	3297.	51.19
1946	MARCH	9.	9.	3348.	54.57
1946	APRIL	9.	9.	3298.	21.09
1946	MAY	9.	6.	3214.	8.79
1946	JUNE	6.	0.	3157.	5.07

		x_3	x_4	WATER LEVEL	PROBABILITY
1946	JULY	0.	0.	3094.	11.64
1946	AUGUST	0.	0.	3001.	33.03
1946	SEPTEMBER	0.	0.	2949.	45.03
1946	OCTOBER	0.	0.	2956.	74.23
1946	NOVEMBER	0.	0.	3042.	69.54
1946	DECEMBER	0.	0.	3126.	75.57
1947	JANUARY	0.	0.	3213.	96.19
1947	FEBRUARY	0.	0.	3395.	53.64
1947	MARCH	0.	0.	3843.(+)	54.73
1947	APRIL	0.	0.	3960.(+)	79.00
1947	MAY	0.	0.	3943.(+)	32.30
1947	JUNE	0.	0.	3889.(+)	7.49
1947	JULY	0.	0.	3812.(+)	12.31
1947	AUGUST	0.	65.	3664.(+)	18.64
1947	SEPTEMBER	200.	120.	3379.	1.73
1947	OCTOBER	200.	106.	3174.	83.50
1947	NOVEMBER	96.	92.	3099.	77.34
1947	DECEMBER	12.	84.	3175.	90.72
1948	JANUARY	113.	5.	3172.	96.20
1948	FEBRUARY	50.	148.	3228.	70.79
1948	MARCH	116.	74.	3120.	61.29
1948	APRIL	0.	0.	3205.	58.97
1948	MAY	0.	0.	3186.	80.22
1948	JUNE	0.	0.	3203.	68.84
1948	JULY	0.	0.	3339.	85.07
1948	AUGUST	200.	5.	3098.	81.95
1948	SEPTEMBER	0.	0.	3068.	83.20
1948	OCTOBER	0.	63.	3114.	83.28
1948	NOVEMBER	68.	113.	3093.	77.34
1948	DECEMBER	27.	91.	3104.	90.72
1949	JANUARY	36.	3.	3153.	96.20
1949	FEBRUARY	12.	139.	3150.	70.79
1949	MARCH	0.	45.	3134.	61.24
1949	APRIL	0.	0.	3138.	60.86
1949	MAY	0.	0.	3179.	58.30
1949	JUNE	0.	12.	3106.	74.20
1949	JULY	0.	0.	3029.	36.44
1949	AUGUST	0.	0.	2954.	29.03
1949	SEPTEMBER	0.	0.	2870.(—)	8.73
1949	OCTOBER	0.	12.	2868.(—)	2.69
1949	NOVEMBER	0.	30.	3043.	18.30
1949	DECEMBER	32.	128.	3063.	90.72
1950	JANUARY	21.	39.	3157.	96.20
1950	FEBRUARY	30.	143.	3315.	70.79
1950	MARCH	200.	113.	3178.	59.82
1950	APRIL	0.	0.	3241.	87.05
1950	MAY	6.	0.	3202.	82.41
1950	JUNE	0.	0.	3047.	71.63
1950	JULY	0.	0.	2957.	3.19
1950	AUGUST	0.	0.	2899.(—)	4.42
1950	SEPTEMBER	0.	0.	2898.(—)	1.31
1950	OCTOBER	0.	21.	2965.	15.50
1950	NOVEMBER	0.	72.	3181.	73.61
1950	DECEMBER	200.	141.	3171.	90.72
1951	JANUARY	164.	73.	3146.	96.20
1951	FEBRUARY	53.	163.	3247.	70.79
1951	MARCH	156.	87.	3297.	61.29
1951	APRIL	153.	36.	3117.	88.72

Table III (cont.)

		x_3	x_4	WATER LEVEL	PROBABILITY
1951	MAY	0.	0.	3229.	55.99
1951	JUNE	22.	77.	3375.	76.86
1951	JULY	200.	72.	3214.	79.08
1951	AUGUST	63.	31.	3116.	81.95
1951	SEPTEMBER	0.	0.	3134.	91.64
1951	OCTOBER	35.	114.	3062.	84.40
1951	NOVEMBER	0.	77.	3083.	77.31
1951	DECEMBER	0.	75.	3168.	90.65
1952	JANUARY	105.	5.	3177.	96.20
1952	FEBRUARY	56.	147.	3276.	70.79
1952	MARCH	188.	85.	3254.	61.29
1952	APRIL	91.	29.	3200.	88.72
1952	MAY	0.	0.	3151.	80.94
1952	JUNE	0.	0.	3113.	52.77
1952	JULY	0.	0.	2960.	45.98
1952	AUGUST	0.	0.	2832.(—)	3.99
1952	SEPTEMBER	0.	0.	2806.(—)	0.00
1952	OCTOBER	0.	0.	2939.	0.00
1952	NOVEMBER	0.	94.	3036.	72.80
1952	DECEMBER	9.	105.	3182.	90.72
1953	JANUARY	147.	36.	3145.	96.20
1953	FEBRUARY	35.	157.	3181.	70.79
1953	MARCH	51.	66.	3148.	61.29
1953	APRIL	0.	0.	3143.	74.44
1953	MAY	0.	0.	3183.	61.06
1953	JUNE	0.	15.	3205.	74.70
1953	JULY	2.	0.	3110.	85.70
1953	AUGUST	0.	0.	3059.	68.30
1953	SEPTEMBER	0.	0.	3000.	81.39
1953	OCTOBER	0.	16.	3008.	62.19
1953	NOVEMBER	0.	52.	2986.	74.67
1953	DECEMBER	0.	12.	3003.	65.02
1954	JANUARY	0.	0.	3076.	75.22
1954	FEBRUARY	0.	78.	3132.	66.33
1954	MARCH	0.	59.	3324.	61.29
1954	APRIL	170.	47.	3182.	88.72
1954	MAY	0.	0.	3357.	78.57
1954	JUNE	170.	91.	3202.	76.86
1954	JULY	10.	0.	3214	86.11
1954	AUGUST	46.	2.	3114.	81.95
1954	SEPTEMBER	0.	0.	3083.	90.38
1954	OCTOBER	0.	72.	3104.	83.96
1954	NOVEMBER	48.	104.	3132.	77.34
1954	DECEMBER	76.	99.	3180.	90.72
1955	JANUARY	135.	25.	3166.	96.20
1955	FEBRUARY	55.	153.	3245.	70.79
1955	MARCH	146.	81.	3251.	61.29
1955	APRIL	81.	30.	3236.	88.72
1955	MAY	10.	0.	3236.	82.43
1955	JUNE	0.	30.	3190.	76.60
1955	JULY	0.	0.	3229.	75.02
1955	AUGUST	51.	0.	3259.	81.93
1955	SEPTEMBER	168.	66.	3117.	92.30
1955	OCTOBER	30.	142.	3227.	84.40
1955	NOVEMBER	200.	177.	3212.	76.88
1955	DECEMBER	200.	153.	3089.	89.97
1956	JANUARY	54.	46.	3123.	96.20
1956	FEBRUARY	0.	137.	3231.	70.75
1956	MARCH	122.	72.	3269.	61.29

 (cont.)

		x_3	x_4	WATER LEVEL	PROBABILITY
1956	APRIL	102.	35.	3281.	88.72
1956	MAY	62.	0.	3282.	82.42
1956	JUNE	38.	64.	3258.	76.86
1956	JULY	52.	0.	3203.	85.75
1956	AUGUST	28.	2.	3100.	81.95
1956	SEPTEMBER	0.	0.	3015.	85.18
1956	OCTOBER	0.	8.	3037.	65.20
1956	NOVEMBER	0.	77.	3091.	76.88
1956	DECEMBER	24.	93.	3181.	90.72
1957	JANUARY	130.	18.	3097.	96.20
1957	FEBRUARY	0.	110.	3390.	69.59
1957	MARCH	200.	145.	3265.	33.55
1957	APRIL	60.	0.	3222.	88.70
1957	MAY	0.	0.	3234.	81.82
1957	JUNE	0.	34.	3178.	76.69
1957	JULY	0.	0.	3197.	69.06
1957	AUGUST	12.	0.	3117.	81.93
1957	SEPTEMBER	0.	0.	3131.	89.26
1957	OCTOBER	29.	105.	3076.	84.40
1957	NOVEMBER	10.	87.	3144.	77.34
1957	DECEMBER	81.	100.	3099.	90.72
1958	JANUARY	37.	11.	3162.	96.20
1958	FEBRUARY	26.	140.	3225.	70.79
1958	MARCH	105.	69.	3166.	61.29
1958	APRIL	0.	0.	3190.	84.50
1958	MAY	0.	0.	3130.	77.41
1958	JUNE	0.	0.	3219.	41.55
1958	JULY	54.	0.	3136.	85.69
1958	AUGUST	0.	0.	3053.	80.72
1958	SEPTEMBER	0.	0.	3048.	61.90
1958	OCTOBER	0.	47.	3029.	81.78
1958	NOVEMBER	0.	58.	3061.	76.47
1958	DECEMBER	0.	68.	3146.	90.26
1959	JANUARY	85.	7.	3159.	96.20
1959	FEBRUARY	32.	147.	3167.	70.79
1959	MARCH	23.	58.	3155.	61.29
1959	APRIL	0.	0.	3214.	76.17
1959	MAY	0.	0.	3223.	80.95
1959	JUNE	0.	19.	3305.	76.18
1959	JULY	142.	0.	3225.	85.94
1959	AUGUST	74.	15.	3096.	81.95
1959	SEPTEMBER	0.	0.	3027.	84.26
1959	OCTOBER	0.	21.	2987.	72.79
1959	NOVEMBER	0.	16.	3033.	65.57
1959	DECEMBER	0.	57.	3173.	88.28
1960	JANUARY	126.	20.	3174.	96.20
1960	FEBRUARY	68.	155.	3253.	70.79
1960	MARCH	160.	85.	3160.	61.29
1960	APRIL	0.	0.	3223.	84.94
1960	MAY	0.	0.	3256.	82.15
1960	JUNE	5.	57.	3171.	76.86
1960	JULY	0.	0.	3211.	61.20
1960	AUGUST	25.	0.	3129.	81.90
1960	SEPTEMBER	0.	0.	3122.	91.38
1960	OCTOBER	15.	104.	3248.	84.40
1960	NOVEMBER	200.	187.	3202.	76.09
1960	DECEMBER	200.	126.	3209.	90.70
1961	JANUARY	195.	59.	3135.	96.20
1961	FEBRUARY	39.	165.	3209.	70.79

Table III (cont.)

		x_3	x_4	WATER LEVEL	PROBABILITY
1961	MARCH	100.	76.	3149.	61.29
1961	APRIL	0.	0.	3162.	78.30
1961	MAY	0.	0.	3212.	69.47
1961	JUNE	0.	38.	3220.	76.58
1961	JULY	8.	0.	3132.	85.64
1961	AUGUST	0.	0.	3022.	74.58
1961	SEPTEMBER	0.	0.	2952.	32.90
1961	OCTOBER	0.	0.	2941.	29.68
1961	NOVEMBER	0.	21.	3031.	51.68
1961	DECEMBER	0.	85.	3082.	90.27
1962	JANUARY	25.	17.	3173.	96.20
1962	FEBRUARY	47.	142.	3197.	70.79
1962	MARCH	66.	66.	3316.	61.29
1962	APRIL	159.	44.	3251.	88.72
1962	MAY	33.	0.	3199.	82.43
1962	JUNE	0.	0.	3149.	71.48
1962	JULY	0.	0.	3170.	58.56
1962	AUGUST	0.	0.	3055.	81.58
1962	SEPTEMBER	0.	0.	3013.	43.63
1962	OCTOBER	0.	9.	3012.	68.51
1962	NOVEMBER	0.	51.	3232.	75.10
1962	DECEMBER	200.	170.	3181.	88.90
1963	JANUARY	167.	67.	3205.	96.20
1963	FEBRUARY	125.	159.	3210.	70.79
1963	MARCH	102.	86.	3476.(+)	61.29
1963	APRIL	200.	163.	3450.(+)	18.92
1963	MAY	200.	51.	3259.	80.07
1963	JUNE	0.	38.	3242.	76.79
1963	JULY	8.	0.	3107.	85.13
1963	AUGUST	0.	0.	3114.	61.20
1963	SEPTEMBER	0.	73.	3210.	92.30
1963	OCTOBER	138.	137.	3147.	84.40
1963	NOVEMBER	137.	129.	3081.	77.34
1963	DECEMBER	31.	98.	3159.	90.72
1964	JANUARY	110.	21.	3091.	96.20
1964	FEBRUARY	0.	103.	3164.	68.93
1964	MARCH	42.	61.	3372.	61.29
1964	APRIL	200.	91.	3295.	86.60
1964	MAY	85.	0.	3286.	82.43
1964	JUNE	48.	67.	3237.	76.86
1964	JULY	20.	5.	3187.	85.71
1964	AUGUST	2.	0.	3154.	81.95
1964	SEPTEMBER	6.	45.	3146.	92.30
1964	OCTOBER	42.	111.	3274.	84.40
1964	NOVEMBER	200.	200.	3156.	70.11
1964	DECEMBER	127.	101.	3219.	90.72
1965	JANUARY	188.	38.	3195.	96.20
1965	FEBRUARY	112.	164.	3194.	70.79
1965	MARCH	79.	79.	3246.	61.29
1965	APRIL	66.	29.	3370.	88.72
1965	MAY	156.	0.	3407.(+)	82.38
1965	JUNE	200.	103.	3451.(+)	76.82
1965	JULY	200.	127.	3396.	18.60
1965	AUGUST	200.	102.	3327.	76.21
1965	SEPTEMBER	200.	110.	3199.	90.88
1965	OCTOBER	129.	161.	3085.	84.40
1965	NOVEMBER	46.	105.	3226.	77.34
1965	DECEMBER	200.	150.	3474.(+)	90.28
1966	JANUARY	200.	200.	3396.	0.00

Table III (cont.)

		x_3	x_4	WATER LEVEL	PROBABILITY
1966	FEBRUARY	200.	200.	3398.	37.48
1966	MARCH	200.	133.	3349.	45.48
1966	APRIL	180.	20.	3314.	88.72
1966	MAY	98.	0.	3263.	82.41
1966	JUNE	8.	60.	3293.	76.86
1966	JULY	102.	0.	3280.	85.83
1966	AUGUST	132.	6.	3234.	81.95
1966	SEPTEMBER	143.	58.	3110.	92.30
1966	OCTOBER	59.	147.	3099.	84.40
1966	NOVEMBER	57.	106.	3282.	77.34
1966	DECEMBER	200.	200.	3285.	72.05
1967	JANUARY	200.	163.	3289.	94.38
1967	FEBRUARY	200.	188.	3273.	70.52
1967	MARCH	190.	99.	3225.	61.29
1967	APRIL	53.	21.	3314.	88.72
1967	MAY	96.	0.	3207.	82.41
1967	JUNE	0.	0.	3303.	72.86
1967	JULY	148.	0.	3100.	85.94
1967	AUGUST	0.	0.	2993.	72.95
1967	SEPTEMBER	0.	0.	3053.	15.55
1967	OCTOBER	0.	70.	3070.	83.66
1967	NOVEMBER	30.	106.	3073.	77.34
1967	DECEMBER	0.	83.	3113.	90.70
1968	JANUARY	43.	0.	3174.	9620.
1968	FEBRUARY	41.	141.	3218.	70.79
1968	MARCH	96.	68.	3172.	61.29
1968	APRIL	0.	0.	3182.	85.85
1968	MAY	0.	0.	3117.	75.28
1968	JUNE	0.	0.	3017.	34.75
1968	JULY	0.	0.	2877.(—)	3.64
1968	AUGUST	0.	0.	2947.	0.00
1968	SEPTEMBER	0.	6.	2968.	60.31
1968	OCTOBER	0.	59.	3008.	62.78
1968	NOVEMBER	0.	87.	3175.	76.82
1968	DECEMBER	167.	126.	3077.	90.72
1969	JANUARY	37.	40.	3179.	96.20
1969	FEBRUARY	64.	146.	3428.(+)	70.79
1969	MARCH	200.	152.	3462.(+)	23.48
1969	APRIL	200.	101.	3310.	56.84
1969	MAY	89.	2.	3234.	82.43
1969	JUNE	0.	33.	3318.	76.66
1969	JULY	158.	0.	3103.	85.96
1969	AUGUST	0.	0.	3073.	72.91
1969	SEPTEMBER	0.	0.	3084.	88.47
1969	OCTOBER	0.	98.	3104.	84.30
1969	NOVEMBER	55.	105.	3121.	77.34
1969	DECEMBER	62.	98.	3168.	90.72
1970	JANUARY	120.	22.	3160.	96.20
1970	FEBRUARY	42.	151.	3300.	70.79
1970	MARCH	200.	105.	3482.(+)	60.79
1970	APRIL	200.	158.	3505.(+)	13.90
1970	MAY	200.	76.	3340.	61.28
1970	JUNE	72.	57.	3250.	76.86
1970	JULY	18.	0.	3168.	85.32
1970	AUGUST	0.	0.	3217.	80.32
1970	SEPTEMBER	117.	77.	3095.	92.30
1970	OCTOBER	0.	125.	3107.	84.40
1970	NOVEMBER	52.	101.	3117.	77.34
1970	DECEMBER	52.	95.	3150.	90.72

REFERENCES

[1] Anderson, T. W.: The Integral of a Symmetric Unimodal Function over a Symmetric Convex Set and Some Probability Inequalities, *Proc. Amer. Math. Soc.* 6 (1955), 1970–1976.

[2] The Balaton (Institute for Water Management of the Technical University of Budapest, Budapest, 1970, in Hungarian).

[3] Deák, I.: A Fast Monte Carlo Method for Computing Probabilities of Sets in Higher Dimensional Spaces in Case of Normal Distribution, *Water Resources Research* to appear.

[4] Havie, T.: On a modification of Romberg's algorithm, *BIT* 6 (1966), 24–30.

[5] Himmelblau, D. M.: *Applied Nonlinear Programming* (McGraw-Hill, New York, 1972).

[6] Fiacco, A. W. and McCormick, G. P.: *Nonlinear Programming Sequential Unconstrained Minimization Techniques* (Wiley, New York, London, 1968).

[7] Powell, M. J. D.: Recent Advances in Unconstrained Optimization, *Mathematical Programming* 1 (1971), 26–57.

[8] Prékopa, A.: On Optimization of Stochastic Systems, Academic Doctorate Theses, Budapest, 1970.

[9] Prékopa, A.: Logarithmic Concave Measures with Application to Stochastic Programming, *Acta Scientiarum Mathematicarum* 32 (1971), 301–316.

[10] Prékopa, A.: Optimal Control of a Storage Level Using Stochastic Programming, *Problems of Control and Information Theory* 4 (1975), 193–204.

[11] Prékopa, A. and Szántai, T.: On Multi-Stage Stochastic Programming (with Application to Optimal Control of Water Supply), in: Colloquia Mathematica Societatis János Bolyai 12. *Progress in Operations Research* Ed. A. Prékopa (János Bolyai Mathematical Society, Budapest and North-Holland Publishing Company, Amsterdam, London, 1976) 733–755.

[12] Szesztai, K.: *The Hydrology of Lake Balaton* (Institute for Water Management, Budapest, 1962, in Hungarian).

[13] Wilks, S. S.: *Mathematical Statistics* (Wiley, New York, London, 1962).

[14] Wilde, D. J.: *Optimum Seeking Methods* (Prentice Hall Inc., Englewood Cliffs, N. J., 1964).

OPTIMAL CONTROL OF DISCRETE PROCESSES WITH INCOMPLETE INFORMATION

H.-J. SEBASTIAN

(Leipzig, GDR)

We consider a model of discrete processes with parameters and develop a theory of optimal control for the case in which the information about parameters is incomplete. We obtain new functional equations in the form of extended Bellman-equations of dynamic programming.

Model. At first we give some notations and definitions used throughout the paper.

We consider a finite stage process with discrete time t,

$$t = t_i = i \cdot \Delta, \quad i = 0, 1, 2, \ldots, N, \quad \Delta > 0.$$

We define a state vector x_i

$$x_i = (x_i^1, x_i^2, \ldots, x_i^a) \in R^a \quad (a\text{-dimensional Euclidean space-}R^a),$$

a control vector $v_{i'}$

$$v_{i'} = (v_{i'}^1, v_{i'}^2, \ldots, v_{i'}^b) \in R^b$$

and a parameter vector $m_{i'}$

$$m_{i'} = (m_{i'}^1, m_{i'}^2, \ldots, m_{i'}^c) \in R^c, \quad i \in I = \{0, 1, 2, \ldots, N\},$$
$$i' \in I' = \{0, 1, 2, \ldots, N-1\}.$$

The parameter vector contains selected influence quantities being important for the process description.

Furthermore, we have a state space X_i, a parameter space $M_{i'}$ and a control space $V_{i'}(x_{i'}, m_{i'})$ for each fixed integer, $i \in I$, $i' \in I'$, which are subspaces from R^a, R^c, R^b:

$$x_i \in X_i \subseteq R^a, \quad m_{i'} \in M_{i'} \subseteq R^c$$
$$v_{i'} \in V_{i'}(x_{i'}, m_{i'}) \subseteq R^b.$$

$V_{i'}$ denotes the union of sets $V_{i'}(x_{i'}, m_{i'})$, $V_{i'} = \bigcup_{x_{i'} \in X_{i'} \wedge m_{i'} \in M_{i'}} V_{i'}(x_{i'}, m_{i'})$. A function

$T_i: X_i \times V_i \times M_i \to X_{i+1}$ is called i-th state transformation. We denote the state transformations by the equations

$$(1) \qquad x_{i+1} = T_i(x_i, v_i, m_i),$$

x_0 is a known vector, $i \in \mathcal{J}'$.

Now we consider two cases with respect to the informations.

(1) At first we have the case "complete information". In this case, in each time t_i, all components of z_i are known,

$$z_i = (x_i, m_i, m_{i-1}, \ldots, m_{N-1}), \quad z_i \in X_i \times M_i \times M_{i+1} \times \ldots \times M_{N-1}.$$

(2) If, in each time t_i, all components of z_{in}

$$z_{in} = (x_i, m_i, m_{i+1}, \ldots, m_{i+n-1}), \quad z_{in} \in X_i \times M_i \times M_{i+1} \times \ldots \times M_{i+n-1}$$

$n \geq 0$, integer, are known, but the vectors

$$m_{i+n}, \quad m_{i+n+1}, \ldots, m_{N-1}$$

are unknown, we have an important special case of incomplete information about process parameters.

n denotes a fixed integer with $n \ll N$.

At last we define the objective function Z_i for the subprocess from t_i to t_N. A function

$$(2) \qquad Z_i(v_i, v_{i+1}, \ldots, v_{N-1}; x_i, \ldots, x_N) = \sum_{s=i}^{N-1} W_r(x_r, v_r) + W_N(x_N) \Rightarrow \text{Max.}$$

is called objective function if

$$W_r: X_r \times V_r \to R^1 \text{ and } W_N: X_N \to R^1$$

are real-valued functions. (We have also studied the cases, where Z_i is not an additive function or Z_i is a vector-valued function.) Now we have to maximize Z_i.

THE OPTIMIZATION PROBLEM IN THE CASE OF INCOMPLETE INFORMATION

Definition 1. (Feasibility)

(a) A function $h_i: X_i \times M_i \times \ldots \times M_{i+n-1} \to V_i(x_i, m_i)$ is called feasible control function. Each special value $v_i = h_i(\mathring{z}_{in}) \in V_i(\mathring{x}_i, \mathring{m}_i)$ is called feasible control.

(b) A $(N-i)$-tuple $h^i = (h_i, h_{i+1}, \ldots, h_{N-1})$ is called feasible control policy for the process from t_i to t_N if and only if each h_j, $j \in \{i, i+1, \ldots, N-1\}$, is a feasible control function. Each special value

$$v^i = (v_i, v_{i+1}, \ldots, v_{N-1}) = \left(h_i(\mathring{z}_{in}), h_{i+1}(\mathring{z}_{i+1,n}), \ldots, h_{N-1}(\mathring{z}_{N-1,n})\right)$$

of a feasible control policy h^i is called feasible solution if, for $j=i, i+1, \ldots, N-1$

$$\mathring{x}_{i+1}=T_j(\mathring{x}_j, \mathring{v}_j, \mathring{m}_j)$$

is valid.

Now we define a modified objective function. At first we replace in the objective function, $Z_i(v_i, v_{i+1}, \ldots, v_{N-1}, x_i, \ldots, x_N)$, the $(N-i)$-tupel $(v_i, v_{i+1}, \ldots, v_{N-1})$ by a feasible control policy h^i and all the state vectors x_j with $j>i$ by the state transformation,

$$x_j=T_{j-1}(x_{j-1}, h_{j-1}(z_{j-1, n}), m_{j-1}).$$

As a result of this procedure we obtain the modified objective function Z_i' in the following form:

$$Z_i'(h^i; z_i)=Z_i(h_i(z_{in}), h_{i+1}(z_{i+1,n}), \ldots, h_{N-1}(z_{N-1,n});$$
$$x_i, T_i(x_i, h_i(z_{in}), m_i), \ldots, T_{N-1}(\ldots)).$$

However, this form of the objective function is not suitable for the definition of the optimal control problem, because the function depends on the variables m_{i+n}, $m_{i+n+1}, \ldots, m_{N-1}$. This parameter vectors are unknown in the moment t_i.

In the following we shall study a modified function, which is defined by an operator.

Definition 2. $\tilde{Z}_i(h^i, z_{in})$ is called modified objective function if $\tilde{Z}_i$ is defined by the equation

$$(3) \qquad \tilde{Z}_i(h^i; z_{in})= \mathcal{L}_n^{(N-(i+n))} Z_i'(h^i; z_i)$$

and if the operator has the properties:

1. The range $\mathcal{D}(\mathcal{L}_n^{(r-n+1)})$ of the operator $\mathcal{L}_n^{(r-n+1)}$ is a function space. We denote the elements of $\mathcal{D}$ by $f(y, m_i, m_{i+1}, \ldots, m_{i+r})$ with: $y \in Y \subseteq R^\rho$; $m_i, m_{i+1}, \ldots,$ $m_{i+r} \in M \subseteq R^c$. In the following we assume $M_0=M_1=\ldots=M_{N-1}=M$.[1]

2. For all elements $f \in \mathcal{D}(\mathcal{L}_n^{(r-n+1)})$ we have

$$(4) \quad \mathcal{L}_n^{(r-n+1)}f(y, m_i, m_{i+1}, \ldots, m_{i+r})=\begin{cases} f(y, m_i, m_{i+1}, \ldots, m_{i+r}), & \text{for} \quad r \leq n-1 \\ \tilde{f}(y, m_i, m_{i+1}, \ldots, m_{i+n-1}), & \text{for} \quad r \geq n. \end{cases}$$

3. If $f_1, f_2 \in \mathcal{D}(\mathcal{L}_n^{(r_1-n+1)})$ and $r_1 \geq n$, $r_2 \leq n-1$, we have the equation

$$\mathcal{L}_n^{(r_1-n+1)}(\lambda f_1(y, m_i, \ldots, m_{i+r_1})+f_2(y, m_i, \ldots, m_{i+r_2}))=$$
$$(5)$$
$$=\lambda \mathcal{L}_n^{(r_1-n+1)}f_1(y, m_i, \ldots, m_{i+r_1})+f_2(y, m_i, \ldots, m_{i+r_2}).$$

(For $r=N-i-1$ is $\mathcal{L}_n^{(r-n+1)}= \mathcal{L}_n^{(N-i-1-n+1)}= \mathcal{L}_n^{(N-(i+n))}$.) Now we consider some importand examples

$$(6) \quad 1. \quad \mathcal{L}_n^{(N-(i+n))} Z_i'(h^i; z_i)= \min_{m_{i+n}, \ldots, m_{N-1} \in M'} Z_i'(h^i; z_i), \quad M' \subseteq M$$

[1] For our theory this assumption is not necessary.

$\mathcal{L}$ is a minimum operator. We have no information about the parameter vectors $m_{i+n}, \ldots, m_{N-1}$.

(7) 2. $\mathcal{L}_n^{(N-(i+n))} Z_i'(h^i; z_i) = \alpha \cdot \underset{m_{i+n}, \ldots, m_{N-1} \in M'}{\text{Min}} Z_i'(h^i; z_i) +$

$$+ (1-\alpha) \cdot \underset{m_{i+n}, \ldots, m_{N-1} \in M'}{\text{Max}} Z_i'(h^i; z_i)$$

$0 \leqq \alpha \leqq 1$, Hurwicz Criterion.

(8) 3. $\mathcal{L}_n^{(N-(i+n))} Z_i'(h^i; z_i) = \int_{M^{N-(i+n)}} Z_i'(h^i; z_i)\, dF(m_{i+n}) \ldots dF(m_{N-1}).$

In this case $\tilde{Z}_i(h^i; z_{in})$ is the expected value from $Z_i'(h^i; z_i)$ (a special stochastic case). Now we can define the optimal control problem.

Definition 3. (Optimality). The feasible control policy $\hat{h}^i$ is an optimal control policy if and only if $\hat{h}^i$ satisfies the inequality

(9) $$\tilde{Z}_i(\hat{h}^i; z_{in}) \geqq \tilde{Z}_i(h^i; z_{in})$$

for all feasible control policies h^i and for all $z_{in} \in X_i \times M^n$. "The problem" determine an optimal policy $\hat{h}^i$ is denoted by $(OP)_i$, the set of solutions of $(OP)_i$ is denoted by $L_{(OP)_i}$.

If the operator $\mathcal{L}_n^{(N-(i+n))}$ has the properties "multiplicativity" and "monotonicity", we can prove the main Theorem 1. Theorem 1 gives a decomposition of the problem $(OP)_i$ into optimization problems in Euclidean spaces.

Definition 4. (a) We consider the class of operators $\mathcal{L}_n^{(r-n+1)}$ with the property

(10) $\mathcal{L}_n^{(r-n+1)} f(y, m_i, \ldots, m_{i+r}) = \mathcal{L}_n^{(1)} \Big(\mathcal{L}_{n+1}^{(1)} (\ldots (\mathcal{L}_r^{(1)} f(y, m_i, \ldots, m_{i+r}) \ldots)) \Big)$

for all $f \in \mathcal{D}(\mathcal{L}_n^{(r-n+1)})$ and for each $(y, m_i, \ldots, m_{i+n-1}) \in Y \times M^n$. An operator $\mathcal{L}_n^{(r-n+1)}$ with the property (10) is called a "multiplicative operator".

(b) Operator $\mathcal{L}_k^{(1)}$, $k \in \{n+1, \ldots, r\}$ is equivalent to operator $\mathcal{L}_n^{(1)}$ (we write in this case $\mathcal{L}_k^{(1)} =: \mathcal{L}_n$ for $k = n, n+1, \ldots, r$, if and only if equation

(11) $$\mathcal{L}_k^{(1)} f^*(y, m_i, \ldots, m_{i+k}) = \mathcal{L}_n^{(1)} f^*(y_k', m_{i+k-n}, \ldots, m_{i+k})$$

with $y_k' = (y, m_i, \ldots, m_{i+k-n-1})$ is valid. We write in this case

$$\mathcal{L}_n^{(r-n+1)} = (\mathcal{L}_n)^{r-n+1}.$$

(c) An operator $\mathcal{L}_n$ is called monotonous if an only if:

$$\text{If} \qquad f(y, m_i, \ldots, m_{i+n}) \geqq g(y, m_i, \ldots, m_{i+n})$$
$$\text{for all } (y, m_i, \ldots, m_{i+n}) \in Y \times M^{n+1}$$

(12)

$$\text{than} \;\rightarrow\; \mathcal{L}_n f(y, m_i, \ldots, m_{i+n}) \geqq \mathcal{L}_n g(y, m_i, \ldots, m_{i+n})$$
$$\text{for all } (y, m_i, \ldots, m_{i+n-1}) \in Y \times M^n.$$

We call an operator $\mathcal{L}_n^{(r-n+1)}$ having the properties (a), (b), (c) "multiplicative operator with generating monotonous operator $\mathcal{L}_n$".

Let us now consider functional equations

(13)
$$f_0(z_N) = W_N(x_N)$$

$$f_{N-j}(z_j) = \operatorname*{Max}_{v_j \in V_j(x_j, m_j)} \big[W_j(x_j, v_j) + f_{N-j-1}(T_j(x_j, v_j, m_j), m_{j+1}, \ldots, m_{N-1}) \big]$$

(14)
$$z_j = z_{jn} \quad \text{for} \quad j = N-1, N-2, \ldots, N-n$$

$$f_{N-j, \mathcal{L}_n}(z_{jn}) = \operatorname*{Max}_{v_j \in V_j(x_j, m_j)} \big[W_j(x_j, v_j) + \mathcal{L}_n f_{N-j-1, \mathcal{L}_n}(T_j(x_j, v_j, m_j), m_{j+1}, \ldots, m_{j+n}) \big]$$

(15)
$$\text{for } j = N-n-1, N-n-2, \ldots, i$$

and

$$f_{n, \mathcal{L}_n}(z_{N-n\,n}) = f_n(z_{N-n}),$$

[i is a fixed integer, $i \in \{N-n-1, \ldots, 0\}$.]

The functional equation (14) is Bellman's functional equation for processes with parameters, and equation (15) is an extended functional equation of dynamic optimization. It is easily shown that equations (13), (14), (15) define a feasible control policy h^i, if all maximum-operators exist. In the following we assume that such a solution h^i of (13), (14), (15) exists. The set of solutions of equations (13), (14), (15) is denoted by $L_{[FG]_i}$. Now we can show that the following theorem is valid.

We consider $(OP)_i$ with a multiplicative operator $\mathcal{L}_n^{N-(i+n)} = (\mathcal{L}_n)^{N-(i+n)}$, $i \in \{N-1, N-2, \ldots, 0\}$.

Theorem 1. *If $\mathcal{L}_n$ is monotonous, then we have*
1. *the relation*

$$L_{(OP)_i} \geqq L_{(FG)_i} \quad \text{and}$$
$$\text{if } \tilde{h}_i \in L_{(FG)_i}, \quad \text{then} \quad f_{N-i, \mathcal{L}_n}(z_{in}) = \tilde{Z}_i(\tilde{h}_i; z_{in}).$$

2. *If $L_{(FG)_i} \neq \Phi$ and $L_{(OP)_i}$ has only one element, then $L_{(OP)_i} = L_{(FG)i}$.*

You can find the proof of the theorem in [4], [6], [9]. Our result (Theorem 1) is stated as follows. Each solution $\tilde{h}^i$ of the functional equations (13), (14), (15) is an optimal control policy in the sense of definition 3. The equation $L_{(OP)_i} = L_{(FG)_i}$ is very often not true. There are examples for $L_{(OP)_i} \supset L_{(FG)_i}$. However, we also have sufficient conditions for the simple property 2. [9].

For examples of technological processes in the building industry, we have found analytic solutions of Bellman's equation (14) and of the extended Bellman equation (15). These solutions are published in [2], [3], [5].

Furthermore, we have proved some properties of the optimal return functions, $f_{N-i,\,\mathcal{L}_n}(z_{in})$, such as continuity, monotony and concavity, [4], [7]. At last we have found approximate methods for the numerical solution of functional equations (13), (14), (15) with a computer [8], [10]. Computer programs show the effectiveness of methods.

REFERENCES

[1] Sieber, N., Sebastian, H.-J.: Ein mathematisches Modell zur Optimalregelung von Versorgungssystemen, *Matematische Operationsforschung und Statistik*. Berlin: Akademie-Verlag (1974), H. 9.

[2] Sebastian, H.-J.: Ein nichtlineares, mehrdimensionales Regelungsmodell für den Teilprozeß Betondeckenbau beim Autobahnbau. Teil 1: Determinierte Transportzeiten, *Wiss. Zeitschrift der Hochschule für Bauwesen Leipzig*, (1971), H. 3.

[3] Gnaudschun, D.: Ein Regelungsmodell zur objektiven Leitung der Deckenherstellung im Straßenbau, *Wiss. Zeitschrift der Hochschule für Bauwesen Leipzig*, (1972), H. 3.

[4] Sebastian, H.-J.: Erweiterungen der dynamischen Optimierung. Dissertation B. Hochschule für Bauwesen, Leipzig, 1975.

[5] Sieber, N., Sebastian, H.-J., Gnaudschun, D.: Über geschlossene Lösungen zur Optimalregelung eines speziellen Versorgungssystems mittels dynamischer Optimierung, *Mathematische Operationsforschung und Statistik*, (1976), H. 3.

[6] Sebastian, H.-J.: Dynamische Optimierung unter Einbeziehung zeitabhängiger Einflußparameter, *Mathematische Operationsforschung und Statistik*, (1976), H. 3.

[7] Себастиан, Х.-Ю.: Динамическое программирование для задач с параметралии зависящих от времени, *в ж. Дифференциальные уравнения*. Том 2, Минск, 1978.

[8] Sieber, N., Sebastian, H.-J., Wittig, W. S.: Probleme der Numerik bei speziellen dynamischen Systemen, *Wiss. Zeitschrift der Hochschule für Bauwesen Leipzig*, (1976), H. 4.

[9] Sebastian, H.-J., Schäfer, W.: Eine erweiterte Bellmansche Funktionalgleichung als notwendige und hinreichende Optimalitätsbedingung für ein Optimalsteuerungsproblem mit unvollständiger Information, *Mathematische Operationsforschung und Statistik*, Ser. Optimization 9, 1978, No. 1.

[10] Sebastian, H.-J.: Theorie zur numerischen Behandlung einer erweiterten Bellmanschen Funktionalgleichung der diskreten dynamischen Optimierung, *Mathematische Operationsforschung und Statistik*, Ser. Optimization 9, 1978, No. 2.

STRUCTURAL PROBLEMS IN THE THEORY OF BOUNDED CONTROL FOR LINEAR DISCRETE TIME PARTIALLY OBSERVED SYSTEMS UNDER NONSTOCHASTIC BOUNDED DISTURBANCES

GY. SONNEVEND

(Budapest, Hungary)

1. INTRODUCTION

We consider here the discrete time linear controlled system under disturbances, in the case of incomplete state observation, given by the equation

$$(1.1) \qquad x(k+1)=Fx(k)+Gu(k)+Dv(k),$$

$$(1.2) \qquad y(k)=Hx(k),$$

$$(1.3) \qquad z(k)=Zx(k), \quad k=0, 1, 2, \ldots,$$

where at moment k, $x(k)\in R^n$ is the state vector, $u(k)\in R^m$ the control vector, $v(k)\in R^d$ the disturbance vector, $y(k)\in R^p$ the vector of the measured output, $z(k)\in R^l$ the vector of the output to be regulated, $F\in R^{n\times n}$, $G\in R^{n\times m}$, $H\in R^{p\times n}$, $D\in R^{d\times s}$, $Z\in R^{l\times n}$ are constant matrices. The vector $u(k)$ can be chosen by us, yet its value can depend only on the available information about the system, that is the exact knowledge of the matrices F, G, D, H, Z and the observed values $y(0), y(1), \ldots, y(k)$. The vector $v(k)$ is assumed to be unknown and arbitrarily chosen by some opponent, or by nature; no probabilistic assumptions about the realisations $\vartheta(k)$, $k=0, 1, \ldots$, are made. Here we deal mainly with the case of bounded controls, when the vectors Gu and $-D\vartheta$ can be chosen only from (should belong to) compact sets P and Q respectively. In this case the admissible controls $u(k)=U(y(0), \ldots, y(k))$, may depend on the knowledge of the sets P and Q. In the case of bounded controls we will write (1.1) simply as

$$x(k+1)=Fx(k)+u(k)-v(k), \quad u\in P, \quad v\in Q.$$

The vector $z(k)$ represents that part of the system (those linear combinations of the state variables) which we would like to control, i.e. regulate.

A continuous time system is defined, when we replace k by $t>0$, and $x(k+1)$ by $dx/dt=\mathring{x}(t)$.

$$(1.4) \qquad \mathring{x}(t)=Fx(t)+Gu(t)+Dv(t)$$

$$(1.5) \qquad\qquad y(t) = Hx(t)$$

$$(1.6) \qquad\qquad z(t) = Zx(t), \quad t > 0.$$

Our aim here is to find some conditions (necessary and sufficient, if possible) for the existence, and methods for the construction, of *invariant sets*—into which the state vector can be confined for all time—with some prescribed properties. The simplest of them is whether there exists a (compact) invariant set inside a given set, L, or one which contains a given set, M, (and how to connstruct the largest, resp. smallest such set).

In the applications one frequently is faced with the problem of the existence and construction of invariant sets whose projection into the orthogonal complement of the linear subspace, $\mathrm{Ker}\, Z = \{x \mid Zx = 0\}$, is small (compact) — *the problem of surveillance or following* —, or is far from (do not contain) the origin — *the problem of evasion*, see [1], [2] for background, and [9] for some examples.

1.7 **Definition.** A set $K \subset R^n$ is said to be invariant if whenever $x(0) \in K$, then there exists a control $u(0) = U(y(0)) \in P$, such that $x(1) \in K$, for all unknown values $v(0) \in Q$.

Notice that by the time invariant character of the dynamics of the system, the invariance of a set K means that the controller can keep the state of the system always in K, $x(k) \in K$, $k = 0, 1, 2, \ldots$, whenever $x(0) \in K$. The control $u(k) = U(y(k))$ is assumed to be "memory less", yet the memory is here embodied in the knowledge that $x(0) \in K$, $(x(k) \in K)$, see below. Specially, if no controls are present, i.e. $G = 0$, the invariance of a set K means that the original information, $x(0) \in K$, allows us to conclude that $x(k) \in K$, for all $k = 1, 2, \ldots$. We say that a set K is *conditionally invariant*, if the same is true at least in the cases, when $y(k) = 0$, for all $k = 0, 1, 2, \ldots$, here $\bar{y}$ is the result of the best state observation procedure (see below).

1.8 **Remark.** The notion of an invariant set in a continuous time system is defined analogously, changing k into t. Yet in the case of continuous time systems the class of allowed control (disturbance) functions $u(t)$, $v(t)$ is not uniquely determined in the case of unbounded controls in the sense that the requirement of their measurability, as functions of time, may be too restrictive, i.e. for some purposes generalized functions: impulse controls should be allowed as inputs.

Moreover one of the simplest ways to construct invariant sets in the case of continuous time systems seems to be the method of periodisation, see [9], i.e. their construction from an invariant set in the discrete time system corresponding to the original continuous time system via discretized measurements $y(k)$, $k = 0, 1, 2$ for some positive δ. Obviously a bounded control continuous time system generates a bounded control discrete time system

$$F \to e^{\delta F}, \quad P \to \int_0^\delta e^{\tau F} \cdot P \, d\tau, \quad Q \to \int_0^\delta e^{\tau F} \cdot Q \, d\tau,$$

218

Notice, however that the discrete time systems arising in this way always have a matrix F, which is invertible.

It is well known, and can be seen already on the simplest examples, that optimal strategies $U(x(t))$, (even any "closed-loop" i.e. dynamic strategy which is better than the best "open-loop" (i.e. passive) strategy) are essentially "discontinuous", see [13]. The discretisation in time alleviates these difficulties and at the same time allows to use fully the linear structure inherent in the dynamics of the problem, see [6], where. the principal tools for that methods of analysis of bounded control systems, which is followed here, in sections 2–4, were introduced.

Our main interest is here in the structural properties of bounded control systems. Before investigating the problems of "optimal" control—for example the problem of the best stabilisation of the output $z(k)$ in the system (1.1)–(1.3)—it is desireable to have more knowledge about the existence and about the family of all possible invariant sets, and to find a simple "parametrisation" for this family (at least in the cases when P and Q, together with F, G, H, D, Z are "simple". In section 4 we show that for a special class of sets P and Q one can find a decomposition technique, which allows the construction of invariant set (investigated in section 3) in relatively simple way. The elements of this class of sets are integrals (sums) of line segments and inherit some of the nice structural properties of the class of all linear subspaces of R^n, which plays a fundamental role in linear system theory. The importance of special kinds of "invariant subspaces" (and methods to construct them in the theory of linear control system was exhibited in many recent works. See [4], [5], [3]. It was our aim to investigate the analogues of these methods and results in the case of bounded control systems.

Regulation in systems with unbounded controls

Let us introduce the following notations. The range, kernel and total inverse of a linear operator $A : E_1 \rightarrow E_2$, are denoted by $R(A)$, Ker A, A^{-1}.

$$A^{-1}e_2 = \{e_1 \mid e_1 \in E_1 \ Ae_1 = e_2\}.$$

The space, dual to X, X^*, of continuous linear functionals over X, is for $X = R^n$ also an n-dimensional vector space, and in (1.2) the components of the vector y are interpreted as elements of X^*. The value of a linear functional φ over X, on the element x is denoted by $\langle \varphi, x \rangle$ and is interpreted as a scalar product, whenever the elements of the space X^* are identified with elements of the space X with the help of the symmetrical bilinear form $\langle x, y \rangle = \Sigma x^i y^i$ arising from a positive definite quadratic form $\langle x, x \rangle = \|x\|^2 = \Sigma x_i^2$. It should be noted however that, for the most parts the theory developed here do not relies on, or is equally true for, any such special quadratic form, it is purely affine. (For a more complete analysis of structural properties of linear control systems, for the duality between observation and control, as well as for the problem considered in this point (in the case of continuous time and unbounded controls), see [3]). We notice only that the dual system to the system (F, G, D, H, Z) given by (1.1)–(1.3) is the system $(-F^*, H^*, Z^*, G^*, D^*)$, where A^* is used to denote the dual operator to A, defined

$$\langle A^* \varphi_2, e_1 \rangle = \langle \varphi_2, Ae_1 \rangle, \quad \text{for all} \quad \varphi_2 \in E_2^*, e_1 \in E_1.$$

A special quadratic form would have meaning if, (in the case of unbounded controls), we were interested in quadratic optimisation, for example

$$\inf_u \int_0^N \langle Rx, x \rangle + \langle Su, u \rangle \, \mathrm{d}t, \quad \text{when} \quad \int_0^N \langle Tv, v \rangle \, \mathrm{d}t \leq \varrho = \text{known},$$

or when, in the case of bounded controls, we are interested to find that invariant set which has the smallest distance from a given set M, for example $M = \mathrm{Ker}\, Z$.

The fixed time soft regulation problem consist in finding whether it is possible, (and how to do it), to quarantee with some admissible control, U, $z(N+k)=0$, for some fixed, given, value of N, for all $k=0, 1, 2, \ldots$, and for all initial conditions, $x(0)$ (known modulo $\mathrm{Ker}\, H$), see [1], [14], [15].

1.9. **Theorem.** *The fixed time regulation problem is solvable for all initial state $x(0)$ known* $\mathrm{mod}\, (\mathrm{Ker}\, H)$, *for some* $N = N(y(0))$ *if and only if it is solvable for* $N = 2n - 1$, *where n is the dimension of the state vector, irrespective of the value of $y(0)$, and a necessary and sufficient condition for this is given by the two conditions*

(1.10) $$W_* = mp(F, \mathrm{Ker}\, H, R(D)) \subseteq M(F, R(G), \mathrm{Ker}\, Z) = V^*,$$

(1.11) $$V^* + R(F, G) = X = R^n.$$

Here we have used the following notations:

(1.12) $$R(F, G) = R(G) + FR(G) + \ldots + F^{n-1}R(G),$$

for the subspace of "reacheable" states in the system without disturbances, $D = 0$.

The subspace $W_* \cap \mathrm{Ker}\, H$ is exactly the subspace of "unobservable" states in the partially observed system (1.1)–(1.2), for known $u(k)$, or what is the same, for $G=0$. This is easily verified from the following recursion

(1.13) $$W_* = W(n-1), \quad W(i+1) = F(W(i) \cap \mathrm{Ker}\, H) + R(D), \quad W(0) = X = R^n.$$

Notice that W_* is the largest conditionally invariant subspace, (set), contained in $F\, \mathrm{Ker}\, H + R(D)$, (when $G=0$).

It can be shown that these relations assure that the value of $x(k)$ modulo W_* can

220

be computed from the observations $y(k-n)$, $y(k-n+1)$, ..., $y(k-1)$, for $k \geqq n$, whatever be the unknown initial state $x(0)$. One has

$$(1.14) \qquad F(W_* \cap \text{Ker } H) + R(D) = W_*, \quad (F+LH)W_* \subseteqq W_*, \quad \text{for some} \quad L \in R^{np},$$

the pair $(F+LH, H)$ on the factor space X/W_* is completely observable.

The subspace V^* is the largest subspace in Ker Z which can be made invariant using appropriate controls $u(k) = U(x(k))$ in the completely observed system (1.1), (1.2) without disturbances, $(D=0, H$ invertible). V^* is computed by the recursion

$$(1.15) \qquad V^* = V(n-1), \quad V(i+1) = F^{-1}(V(i) + R(G)) \cap \text{Ker } Z, \quad V(0) = \text{Ker } Z.$$

The meaning of the condition (1.10) is that the disturbance $Dv(k)$ can be "localized" into the subspace V^* by a linear "feedback control $u(k) = K\tilde{H}x(k) = K\tilde{y}(k)$.

$$(1.16) \qquad\qquad\qquad (F+GK\tilde{H})V^* \subseteqq V^*, \quad R(D) \subseteqq V^*$$

where $\tilde{H}x(k)$ is computable from $y(j)$, $j=k, k-1, \ldots, k-n$, i.e. Ker $\tilde{H} = $ Ker $H \cap W_*$, and K is some constant matrix, again K can be chosen so that $d(I\lambda - (F+GK)R^n/V^*) = = \lambda^{n^1}$, $n^1 = \dim R^n/V^*$.)

The second condition, (1.11) allows to control an arbitrary point $x(k)$ into the subspace V^* in no more then $(n-1)$ units of time. (In the factor system defined on X/V^* the whole state, x modulo V^*, is observable because Ker $\tilde{H} \subseteqq V^*$ and every state is controllable to zero, because of (1.11).) Moreover it can be shown that soft regulation can be achieved by a linear regulator of the following type

$$u(k)=0, \quad \text{if} \quad k<n, \quad u(k)=K_1 y(k)+K_2 \xi(k) \quad u(t)=\hat{K}_1 y(t)+\hat{K}_2 \xi(t)$$
$$\xi(k+1)=F\xi(k)+L[H\xi(k)-y(k)], \quad \xi(0)=0 \quad \xi(t)=(F+\hat{L}H)\xi(t)-Ly(t)$$

and the keeping of $x(N+k)$ in Ker Z is possible with the same regulator (in which the matrices K_1, K_2, L from (1.14), (1.16) are such that $K\tilde{H}x = K_1 y + K_2 H_0 x$, (Ker $H_0 = = W_*$) j_n. The case of continuous time $\hat{K}$ and $\hat{L}$ are chosen as to make, $(F+\hat{L}H)$ and $(F+G\hat{K})$ stable on R^n/W_*, resp. R^n/V^*). Thus we get asymptotic, exponentially fast regulation in the continuous time system (1.4)–(1.6): $\|z(t)\| \leqq \varrho(\tilde{y}(o))e^{-at}$, this last inequality can be satisfied with arbitrary large $\alpha > 0$ if and only if (1.10), (1.11) hold.

For more details see [3]. Specially we obtained that for the possibility of soft regulation it is necessary that the state of the system be observable modulo Ker Z. Notice that it is possible that for a given system, $x(N) \in \text{Ker } Z$ is solvable for all initial position $x(0)$ mod Ker H, for some moment $N \leqq N(y(0))$ (which might be fixable or not), yet it is not possible to guarante $x(N+k) \in \text{Ker } Z$ for all $k \geqq 0$, for any N — because the subspace Ker Z may not contain any invariant subspace.

The fixed time regulation problem, (without the softness condition), has been solved for the case of bounded controls and complete observation H invertible by L. S. Pontrjagin, see [1], who introduced a useful tool the "alternating integral". In the case of incomplete observability, Ker $H \cap W_* \neq 0$, the problem is much more difficult, see section 3.

2. THE BANACH SPACE OF CONVEX SETS

In control problems with bounded controls the operation of "*addition* of sets" plays a fundamental role.

The properties of *convexity* of the sets which occur here allow us to formulate the conditions of existence easier (and to construct (compute) the controls for assuring the invariance) of a set. Therefore we assume in the following that the sets P and Q are *convex* and compact.

Let us denote the family of convex compact subsets of the linear space $X = R^n$ by $\Omega_0(X)$. This family becomes a positive cone, where addition and scalar multiplication with positive numbers are defined as follows

$$(2.1) \qquad K_1 + K_2 = \{k_1 + k_2 \,|\, k_1 \in K_1, k_2 \in K_2\}, \quad \lambda K = \{\lambda k \,|\, k \in K\}, \quad \lambda > 0.$$

Now this cone (semigroup, commutative without nulldivisors) can be embedded into a linear space $\Omega^0(X)$ formed by pairs (A, B), $A, B \in \Omega_0(X)$, if we define addition and scalar multiplication componentwise

$$(A, B) + (C, D) = (A + C, B + D), \quad \lambda(A, B) = (\lambda A, \lambda B), \quad \lambda > 0, \quad (-1)(A, B) = (B, A)$$
$$(2.2) \qquad (A, B) = (C, D) \quad \text{iff} \quad A + D = B + C,$$

i.e. when we identify the pairs (A, B) and (C, D) whenever $A + D = B + C$, (this assures that the definition of scalar multiplication as given above is correct).

The *transitivity* of this relation between pairs holds because of the convexity of the sets in $\Omega_0(X)$, this easily can be seen by introducing the support functions, (see below), of the sets involved. It is easy to check that this equivalence relation is "linear" i.e. consistent with the linear structure of $\Omega^0(X)$. Moreover, we can define a norm on

$$\Omega^0(X) \quad \text{by} \quad \|(A, B)\| = d(A, B),$$

where $d(A, B)$ denotes the Hausdorff distance of the sets A and B in $\Omega_0 X$:

$$(2.3) \qquad d(A, B) = \min \{\varepsilon \,|\, A \subseteq B + \varepsilon B_1(X), B \subseteq A + \varepsilon S_1(X), \text{ where } S_1(X)$$

denotes the unit ball of radius 1 in X.}

In order to prove that this norm satisfy the "triangle inequality", let us introduce the support (Minkowski) functional $m(\varphi, K)$, defined over $\Omega_0(X)$, $\varphi \in X^*$.

$$m(\varphi, K) = \max \{\langle \varphi, x \rangle \,|\, x \in K, S_1(X^*), \|\varphi\| = 1\}.$$

It is easy to see that (denoting the unit sphere of X^* by $S_1(X^*)$)

$$(2.4) \qquad d(A, B) = \max \{m(\varphi, A) - m(\varphi, B) \,|\, \varphi \in S_1(X^*)\}$$

from this $d(A, B) + d(C, D) \geq d(A + C, B + D)$ follows by the additivity of the support functional

$$m(\varphi, K_1 + K_2) = m(\varphi, K_1) + m(\varphi, K_2).$$

222

The following norm turns $(\Omega_0 - \Omega_0)$ into a (noncomplete) Hilbert space

$$(2.5) \qquad \|(A, B)\|_H = \int_{S_1(X^*)} |m(\psi, A) - m(\psi, B)|^2 \, d\psi.$$

2.6. Theorem. *The normed linear space $\Omega^0(X)$ is dense in the space of continuous functions over $S_1(X^*)$.*

The Ascoli–Arzela theorem yields the following important compactness theorem of Blaschke: the elements of $\Omega_0(X)$ which belong to a compact set (i.e. a bounded and closed set in $X = R^n$ form a compact set in the topology induced on $\Omega_0(X)$ by the Hausdorff metrics. Now for a Cauchy sequence of sets (A_n, B_n), $n = 1, 2, \ldots$ the closure of the union of the sets $A_n + B_m$, is not always a compact set. The density of Ω^0 in $C(S_1(X^*)$ is proved using the lattice theoretic version of the theorem of Stone–Weierstrass, see [7], [12]. The separability of $\Omega^0(X)$ follows from the fact that the elements of $\Omega_0(X)$ can be approximated arbitrary closely (in the Hausdorff metrics) by polyhedrons, whose metrics have rational coordinates.

Notice that $\Omega^0(X) = \Omega_0(X) + (-1)\Omega_0(X)$, and $\Omega_0(X)$, consisting of elements $(A, 0)$, is closed in $\Omega^0(X)$, yet it contains no open set! Also notice that the two norms (2.4), (2.5) are *equivalent* over all bounded subsets of $\Omega_0(X)$.

Now we define integrals ("continuous" sums) of sets in $\Omega_0(X)$, taking these sets as elements of the Banach space $\Omega^0(X)$.

2.7. Definition. Let (S, Σ, μ) be a measure space. We say that the function

$$x(s) : S \to \Omega^0(X)$$

is finitely valued if $x(s)$ is equal to K_i on $S_i \in \Sigma$, where S_i are dispoint and $\mu(S_i) < \infty$ for $i = 1, \ldots, m$, and $x(s) = 0$ on $S - \Sigma S_i$

$$\int_S x(s)\mu\,(ds) = \sum_{i=1}^{m} \mu(S_i)K_i.$$

A function $x(s)$ is said to be (Bochner) μ—integrable if there exist a sequence of finitely valued functions, $x_n(s)$, which strongly converges to $x(s)$ almost everywhere (in μ)—i.e. $x(s)$ is "strongly measurable"—in such a way, that

$$\lim_{n \to \infty} \int_S \|x(s) - x_n(s)\|\mu\,(ds) = 0.$$

Then for any set $A \in \Sigma$ the (Bochner) integral of $x(s)$ over A is defined by

$$(2.8) \qquad \int_A x(s)\mu\,(ds) = \lim_{n \to \infty} \int_S C_A(s)X_n(S)\mu\,(ds),$$

where C_A is the defining function of the set A, and the limes in understood as strong convergence in the Banach space $\Omega^0(X)$. A theorem of Bochner asserts that a strongly measurable function $x(s)$ is integrable iff $\|x(s)\|$ is integrable, see [7], ch V. §5.

For example the integrals arising in the case of continuous time systems, see remark (1.8) can be interpreted as given here (over the real interval $S=[0, \delta]$, endowed with the Lebesque measure). For us important are the integrals introduced in section 4.

2.9. **Definition.** Let A and B elements of $\Omega_0(X)$, the maximal solution, W, of the inclusion $A+W\subseteq B$ is said to be the *geometric difference* of the set A and B,

$$(2.10) \qquad A\overset{*}{-}B=\{\omega \mid A\supseteq B+\omega,\ \omega\in R^n\}, \quad A\overset{*}{-}B+B\subseteq A.$$

The meaning of the set $(A\overset{*}{-}B)$ is that any vector $b\in B$, "perturbed" by a vector $\omega\in(A\overset{*}{-}B)$ belongs to A for all b and ω. It is easy to see that $A\overset{*}{-}B$ belongs to $\Omega_0(X)$ if is exist, i.e. is compact and convex (even if only A is convex).

We say that A contains B translationally iff $A\overset{*}{-}B\neq\emptyset$.

The following relations hold true

$$(2.11) \qquad A\overset{*}{-}(B+C)=(A\overset{*}{-}B)\overset{*}{-}C.$$

$$(A+B)\overset{*}{-}C\supseteq(A\overset{*}{-}C)+B, \quad \text{equality holds iff} \quad (A, C)=A\overset{*}{-}C, 0)$$
$$A\overset{*}{-}(B\overset{*}{-}C)\subseteq(A+C)\overset{*}{-}B, \quad \text{equality holds iff} \quad (B, C)=(B\overset{*}{-}C, 0)$$
$$A\overset{*}{-}(B\overset{*}{-}C)\supseteq(A\overset{*}{-}B)+C, \quad \text{equality holds iff} \quad (A, B)=(A\overset{*}{-}B, 0),$$
$$(B, C)=B\overset{*}{-}(C, 0).$$

We see that the condition

$$(2.12) \qquad A=B+(A\overset{*}{-}B) \quad \text{or} \quad (A, B)=(A\overset{*}{-}B, 0)$$

expresses an important special case, when the computations involving the operation $(\overset{*}{-})$ are easier to be implemented and possess nice algebraic properties. We say that the set B "sweeps out" the set A if (2.12) is true. (It is known that the alternating integral defined in [1] coincides with its "first approximation" if one has sweeping in the first approximations for all moments of time, this is a consequence of the second inclusion in (2.11), see [11], [12], and section 4.

If the set A and B are centralsymmetrical, i.e. $A=-A$, $B=-B$, then $(A\overset{*}{-}B)$ is also centralsymmetrical, (as well as $(A+B)$ and λA, for $\lambda>0$, and $A\overset{*}{-}B\neq\emptyset$, iff $A\supseteq B$.

Let $X=\text{Ker }\tilde{H}+H_0$, $H_0\cap\text{Ker }\tilde{H}=0$, i.e. H_0 is a linear complementary subspace to $\text{Ker }\tilde{H}$. Then if $A\overset{*}{-}B\neq 0$, then $A(\xi)\overset{*}{-}B(\xi)\neq\emptyset$ for all $\xi\in H_0$, where

$$(2.13) \qquad A(\xi)=\{x\mid\tilde{H}x=\tilde{H}\xi,\ x\in A\}.$$

Obviously, the converse of this is generally not true, yet in the case when A and B are centralsymmetrical $A\overset{*}{-}B\neq 0$ if and only if $A(\xi)\supseteq B(\xi)$ for all $\xi\in H_0$, (such that $B(\xi)\neq\emptyset$).

224

2.14. **Lemma.** *Let $A \in \Omega_0(X)$ be centralsymmetrical, then $A(\xi)$ is translationally included in $A(0)$ if for all ξ the set $A(\xi)$ has a centrum, i.e. $A(\xi)+l$ is centralsymmetrical for some vector l, or if the sets $A(\xi)$ are similar (homothetical) and similarly posed for all ξ. The sum of (two) sets, A_1, A_2, $\ldots$, which (both) have the property that $A_i(\xi)$ is translationally included in $A(0)$ for all ξ, has also this property, if A_i are symmetrical with respect to* Ker H_0.

2.15. **Definition.** For given sets A, B in $\Omega_0(X)$ the inclusion

$$A \subseteq V + B$$

has many solutions $V \in \Omega_0(X)$, which form a closed convex set in $\Omega_0(X)$. Let us denote by $V(A, B, M)$ that solution $V \in \Omega_0(X)$ which has the smallest (Hausdorff) distance from a given compact set M in $X = R^n$. (Notice that $V(A, B, M)$ may contain a non-convex set $\tilde{V}$, such that $A \subseteq \tilde{V} + B$ is satisfied.)

2.16. **Proposition.** *The set $V(A, B, M)$ can be constructed in the following way. Consider the convex hull of the set $(A \cup B)$. Let $e(a)$ any of the extremal points of the set* conv $(A \cup B)$ *which belongs to A, let $B(a)$ that unique point of the set $(e(a) - B)$ which has the smallest Hausdorff distance from the set M. Then $V(A, B, M)$ is the convex hull of the points $B(a)$.*

3. EXISTENCE AND CONSTRUCTION OF INVARIANT SETS

In this section we assume that the state of the system is observed modulo Ker $\tilde{H} =$ $=$ Ker $H \cap W_*$. We use the notation (2.13).

3.1. **Proposition.** *Let the sets K_0, $K_1 \in \Omega_0(X)$ be given. The necessary and sufficient condition for the existence of a control $u(0) = U(\tilde{H}x(0), K_0, K_1)$, which guarantees $x(1) \in K_1$ whenever $x(0) \in K_0$ is given by*

$$(3.2) \qquad\qquad -F\xi \subseteq [-K_1 \overset{*}{-} (-FK_0(\xi) + Q)] + P,$$

for all $\xi \in H_0$, such that $K_0(\xi) \neq \emptyset$. This follows from the definitions of the operations $+$, $(\overset{*}{-})$ and the fact that we have to guarantee $F\xi + FK_0(\xi) + u - v \in K_1$, for all $\xi \in H_0$, such that $K_0(\xi) \neq \emptyset$, where the vectors $v \in Q$ are unknown as well as the elements of $FK_0(\xi)$.

The fixed time regulation problem: to guarantee $x(N) \in K(N)$ for a given set $K(N)$ and a fixed value N, for all elements $x(0)$ belonging to a given set $K(0)$ is equivalent to the existence of sets $K(l)$, $l = 1, 2, \ldots, N-1$ such that the pairs $K_0 = K(j)$, $K_1 = K(j+1)$, satisfy (3.2) for all $j = 0, \ldots, N-1$, (a phase constraint $x(k) \in L$ would imply the requirement $K(k) \subseteq L$).

The family $R(K_1)$ of sets K_0 which satisfy (3.2) for a given convex set K_1, is convex, (and compact if K_1 is compact and F is invertible). Thus $R(.)$ defines a point to set mapping in the Banach space $\Omega^0(X)$ which is however not everywhere defined—even on the cone $\Omega_0(X)$—indeed $(-K_1)$ should contain translationally the set $FH(\infty)+Q$, see below. An invariant set K is an invariant point $K \in R(K)$ of this point to set mapping. We say that a point to set mapping $\sigma \to R(\sigma)$ defined over a metrical space is contractive if $R(\sigma)$ are compact and there exists an $\alpha < 1$, such that

$$D\big(R(\sigma_1),\, R(\sigma_2)\big) \leqq \alpha \mathrm{d}(\sigma_1,\, \sigma_2), \quad \text{for all} \quad \sigma_1, \quad \sigma_2 \in \mathrm{Dom}\,(R),$$

where Dom (R), is the domain of definition of R, and $D(.,\,.)$ denotes the Hausorff distance of compacts.

3.3. Lemma. *Suppose that the point to set mapping over a complete metrical space, $\sigma \to R(\sigma)$, has compact values and if it is defined at σ, then it is also defined at that (any) element of $R(\sigma)$, which is nearest to σ. Then for this mapping there exists a compact invariant set, if it is contractive.*

Indeed select an arbitrary point σ in the domain of definition of R. Suppose we have defined $\sigma_i \in \mathrm{Dom}\,(R)$, let σ_{i+1} that point of $R(\sigma_i)$ which is nearest to σ_i, (this point is unique if the sets $R(\sigma_i)$ are convex). It is easy to see that $\mathrm{d}(\sigma_N,\, \sigma_{N+1}) \leqq \leqq \alpha^N \mathrm{d}(\sigma_0,\, \sigma_1)$, thus by completeness $\sigma_N \to \sigma^*$, when $N \to \infty$. Then $\sigma^* \in R(\sigma^*)$, and the points which can be reached from σ^*, along the mapping $R(.)$, constitute a compact invariant set of diameter less then $\mathrm{d}(1-\alpha)^{-1}$, if the diameter of the set $R(\sigma^*)$ is less than 2d.

In the case of complete observability, Ker $\tilde{H}=0$ we have the condition (3.2) in the form

$$-FK_0 \subseteqq [-K_1 \overset{*}{-} Q]+P.$$

Let us notice here that in order to compute the vector $u(0)$ we have to project $-Fx(0)$ "along" P into a point $[-K_1 \overset{*}{-} Q]\,.\eta$. (for example take that point $.\eta.$ which is nearest to $-Fx(0)$ in the intersection of $(-K_1 \overset{*}{-} Q) \cap (-F(x_0)-P)$.

3.4. Theorem. *The largest invariant set K, with respect to inclusion, in a given set $L \in \Omega_0(X)$ exists and can be constructed as the limes $K_\infty \in \Omega_0(X)$ of the sets $L=K_0 \supseteqq \supseteqq K_1 \supseteqq K_2, \ldots$, whenever we have complete observability* Ker $\mathrm{H} \cap W_* = 0$, *and if*

$$(3.5) \qquad (K_j \overset{*}{-} Q) \neq 0, \quad \text{for all} \quad j=0, 1, 2, \ldots, \quad \text{where}$$

$$K_{j+1} = -F^{-1}\big((-K_j \overset{*}{-} Q)+P\big) \cap L, \quad K_0 = L.$$

Indeed, if such a sequence exists i.e. (3.5) is fulfilled, then we have $K_j \supseteqq K_{j+1}$, for all $j=0, 1, \ldots$. This follows inductively from $K_0 = L \supseteqq K_1$, by the monotonity of the operations involved. The limit of a decreasing sequence of compact always exists and by the continuity of the operations involved we have for $K = \lim_j K_j$.

226

$$K = -F^{-1}((-K \overset{*}{-} Q) + P) \cap L,$$

we see from this that $K \in \Omega_0(X)$. Notice that K_j is the set of all those points, $x(-j)$, starting from which we can guarantee that $x(k) \in L$ for all $k = -j+1, \ldots, -1, 0$. From this it follows immediately that K_j contains every invariant set, which lies in L, for all $j = 0, 1, 2, \ldots$.

3.6. *Example.* As a very simple example suppose that for some value of $j \geqq 1$ we have $F^j Q \subseteq P$, then $K = Q + FQ + \ldots + F^{j-1} Q$ is an invariant set, see [9] for applications of this example.

If there are no controls: $P = 0$, the smallest invariant set containing a given set L is obtained as the limit of $\tilde{K}_n$,

$$\tilde{K}_{n+1} = [(F\tilde{K}_n + Q) \cup L], \quad \tilde{K}_0 = L.$$

In general, for $P \neq 0$, there does not exist a smallest invariant set containing a given set M. This is, however, true if we consider conditionally invariant sets.

3.7. **Proposition.** *The largest (smallest) conditionally invariant set contained in $F \operatorname{Ker} H + Q$, (containing Q), — the analogue of the set W_*, $W(\infty)$, ($\tilde{W}(\infty)$ is obtained by the recursion.*

$$W(\infty) = FH(\infty) + Q, \quad H_{n+1} = (FH_n + Q) \cap \operatorname{Ker} H, \quad H_1 = \operatorname{Ker} H$$
$$\tilde{W}(\infty) = F\tilde{H}(\infty) + Q, \quad \tilde{H}_{n+1} = (F\tilde{H}_n + Q) \cap \operatorname{Ker} H, \quad \tilde{H}_1 = 0.$$

Notice that $H(\infty) \subseteq \operatorname{Ker} H \cap W_*$, and if Q is centralsymmetrical, $H(\infty)$ also is centralsymmetrical and compact if

$$\left| \sum_{m=0}^{\infty} F^m Q \right| \cap (\operatorname{Ker} H \cap W_*) \text{ is compact.}$$

The following theorem is the partial analogue of theorem 1.9.

3.8. **Theorem.** *Suppose, that Q is centralsymmetrical, and $Q(0)$ contains translationally $Q(\xi)$ for all $\xi \in H_0$. Then $W(\infty)$ has the same properties, and if it is compact, and if*

$$(3.9) \qquad W(\infty) = FH(\infty) + Q \subseteq F^{-1}P + H(\infty)$$

is valid, then the set $W(\infty) \in \Omega_0(X)$ is invariant. Moreover, if $W(\infty)\xi$ is centralsymmetrical for all ξ, then adding some compact subset K^0 in H_0 to $W(\infty)$, we obtain a more general condition for the invariance of the set $W(\infty) + K^0$:

$$(3.10) \qquad K^0 + W(\infty) \subseteq F^{-1}(P + K^0) + H(\infty).$$

If a set K, containing the origin, is invariant, then $\tilde{W}(\infty) \subseteq K$.

This is a consequence of Lemma (2.14) and the propositions (3.1), (3.7).

If the system is not completely observable, i.e. when Ker $\tilde{H}=$ Ker $H \cap W_{*} \neq 0$, then, generally there does not exist a largest invariant set in a given set L, because the union of two invariant sets is not necessarily invariant. Yet if there exists at least one, then we might try to find that one, K^{∞}, for which

$$h(K) = \max \{d(k, M) \mid k \in K\} \quad \text{is minimal, where} \quad d(k, M)$$

denotes the euklidean distance of the point from an arbitrarily given compact set, M, among those invariant sets K which are "extremal" in the sense that there does not exist any invariant set which contains K other than K. (Notice that on the other hand, it is not excluded that such a K contains an invariant set smaller than K.) Suppose that, P, Q and $L \in \Omega_0(X)$, are centralsymmetrial. Define the sequence of sets K_n as follows

(3.11)
$$K_0 = L, \quad K_{n+1} = \{\xi + H_{n+1}(\xi) \mid \xi \in H_0, H_{n+1}(\xi) \neq \emptyset\}$$

(3.12)
$$H_{n+1}(\xi) = -F^{-1}(-K_n \overset{*}{-} Q - F\xi + p_n(\xi)) \cap L(\xi).$$

Here the function $p_n(\xi)$ must be chosen so that the set K_{n+1} be convex, and then there does not exist any solutions $K_0 = K_1$, $K_1 = K_n$ of the inclusion (3.2) such that K_0 properly contains K_{n+1}. Moreover, if Q is centralsymmetrical, then K_{n+1} can be chosen to be central symmetrical, and $K_{n+1} \supseteq Q$ holds iff $K_{n+1}(\xi) = H_{n+1}(\xi) \supseteq Q(\xi)$ for all $\xi \in H_0$ such that $Q(\xi) \neq \emptyset$. Thus we chose $p_n(\xi)$ in such a way that $H_{n+1}(\xi) \supseteq Q(\xi)$ be satisfied (if possible) and $H_{n+1}(\xi)$ has least Hausdorff distance from the set M, see Proposition 2.16, and Lemma 2.14.

By these requirements $p_n(\xi)$ is uniquely defined, and exists only for a subset of elements ξ in H_0.

In the same way as in the proof of the Theorem 3.4, the sets K_n, inductively defined by (3.11)–(3.12), $n = 0, 1, 2, \ldots$ can be interpreted as the set of points $x(-n)$, starting from which it is possible to guarantee $x(k) \in L$, for all $k = -n, \ldots, -1, 0$, with the additional property that any other set $\bar{K}_n$, which has the same property of confinement of the $x(k)-s$ in L and which is not contained in any other set with the same property of confinement, such that for all possible controls $u(k) = U(\tilde{H}x(k))$, $k = -n, \ldots, -2$, for at least one vector $x(-n) \in \bar{K}_n$, among the points $x(k)$, $k = -n, \ldots, -1$, at least one point $x(k)$ will have a greater distance from the set M, then $d(K_k, M)$.

From this follows that $K_n \supseteq K_{n+1} \supset \supset K^{\infty}$ and $\lim K_n = K^{\infty}$, for $n \mapsto \infty$. Summarizing we have thus the following theorem.

3.13. Theorem. *Suppose that the sets P, Q, $L \in \Omega_0(X)$ are centralsymmetrical. For the existence of an invariant set in L it is necessary and sufficient that the sets $H_{n+1}(\xi)$ in (3.11) contain $Q(\xi)$ for all ξ, such that $Q(\xi) \neq 0$, for an appropriate choice of $p_n(\xi)$ P, for all $n = 0, 1, \ldots$, this choice then can be made in such a way that the monotonically decreasing sets K_n converge for $n \to \infty$ to a set $K^{\infty} \in \Omega_0(X)$ which is invariant not con-*

tained in any other invariant set, and has (under these conditions) the smallest Hausdorff distance from a given set M.

With the help of Proposition 2.16 we can construct for Ker $\tilde{H}=0$ that convex invariant set K, which contains a given set M yet do not contains any other convex invariant set, and has the smallest Hausdorff distance from M. Consider the sequence of sets K_n $n=0$, 1, $\ldots$, in $\Omega_0 X$ $K_0=\text{conv } M$, $FK_n+Q\subseteq K_{n+1}-P$, $K_{n+1}=V(FK_n+ +Q, -P, M)$ see definition 2.15. Notice that $K_n\subseteq K_{n+1}$, $\ldots$, and if $\lim K_n$ exists, then it may contain an invariant set K, only if K is not convex.

4. STRUCTURAL PROPERTIES OF SETS WHICH ARE INTEGRALS OF LINE SEGMENTS

The operations on convex compact sets involved in the construction of invariant sets, namely addition, geometric substraction, projection, taking the image under a linear map F, $\ldots$, are quite complicated and difficult to be implemented on "arbitrary" sets $P, Q, L, K, \ldots$. Here we introduce a special, "simple" class of convex compact sets (with a useful "representation" for them) which includes the sets usually arising in applications, and for which the above operations are easier to be implemented. On the other hand, the simple structure—*decomposability into a sum of independent elements*—inherits some of the nice structural (algebraic) properties of the set of all linear subspaces of the space $X=R^n$.

4.1. **Definition.** A convex compact set K is said to be *simple* if it is the integral (sum) of line segments:

$$(4.2) \qquad K=\frac{1}{2}\int_{S^{n-1}} W(\varphi)\mu_K(d\varphi), \quad W(\varphi)=\{\sigma\varphi \mid |\sigma|\leq 1\}, \|\varphi\|=1,$$

where S^{n-1} denotes the unit sphere in R^n and μ_K is a finite positive Borel measure on S^{n-1}, which is symmetrical with respect to the origine. The integral is defined as in (2.9). As it is well known, any Borel measure can be decomposed into two parts, one which is singular, the other which is absolutely continuous with respect to the usual homogen measure on S^{n-1}, (induced by the unique translationally invariant measure on R^n). The unicity of μ_K for a given K is a deep fact (see [12]).

4.3. **Proposition.** *A simple set is centralsymmetrical, projection sums (integrals), linear images, limits of simple set are simple. The decomposition (4.2) is unique (if it exists). If a simple set A can be swept out by a set B, then B and $A\overset{*}{-}B$ are simple, and we have $d\mu_A(\varphi)\leq d\mu_B(\varphi)$, for both singular and absolutely continuous part of these measures.*

We have already introduced a representation of elements K of $\Omega_0(X)$ by a function defined on $S^{n-1}(X^*)$, the support function $m(\psi, K)$. It is known that an arbitrary (positive homogenous) function (on R^n) $l(\psi)$, is the support function of some set $L \in \Omega_0(X)$ if and only if

$$\infty > \varphi(\psi_1) + \varphi(\psi_2) \geqq \varphi(\psi_1 + \psi_2), \quad \text{for all} \quad \psi_1, \ \psi_2.$$

On the other hand, *any* positive Borel regular measure $\mu(d\varphi)$ on $S^{n-1}(X)$ uniquely defines some element $K = K(\mu)$ of $\Omega_0(X)$.

As an example of *decomposition techniques* based on the representation (4.2) consider the case of control systems (1.1)–(1.3) where the set Q has a decomposition (4.2) where we separate the singular and absolutely continuous parts

$$\sum_k q_k + \int_{S^{n-1}} \mu'_Q(\varphi)\, d\varphi = C(Q),$$

$$\mu_Q(d\varphi) = u'_{Q(\varphi)} d\varphi + q_k \delta(\varphi - \varphi_k) d\varphi.$$

Suppose that if we replace the distrubance set Q by the one dimensional set $(\mu'_Q(\varphi) + + q_k \delta(\varphi, \varphi_k)) W(\varphi)$ $\varphi \in \operatorname{supp} \mu_Q$, then there exist an invariant set, $K(\varphi)$, in the system for P replaced by $P(\varphi) \subseteq R(G)$ for example: $P(\varphi) = r(\varphi) P$, $r(\varphi) > 0$. Notice that if such a value $r(\varphi)$ exists, (than there exists also a minimal one, $r_0(\varphi)$), then there exists an invariant set in the original system

$$K = \int_{S^{n-1}} K(\varphi) \mu_Q(d\varphi), \quad \text{iff} \quad \int_{S^{n-1}} P(\varphi) \mu_Q(d\varphi) \subseteq P.$$

Specially such a set exists iff the integral of the minimal function, $r_0(\varphi)$, is not greater then one.

If we introduce the partial order on $\Omega^0(X)$, generated by the "positive" cone, $\Omega_0(X): (A, B) \geqq (C, D)$ if $B = D$, $A = C + K$, for $K \in \Omega_0(X)$, then on the subset of $\Omega_0(X)$ consisting of simple sets this order is a *lattice:* max (min) (A, B) has a representation with the measure max (min) (μ_A, μ_B). On the whole $\Omega_0(X)$ this partial order is not a lattice. An absolutely continuous central symmetrical measure represents by (4.2) an element of Ω_0 iff a.e. hyperspace has a positive in it (see [12]).

4.4. **Theorem.** *In the two dimensional case, $n = 2$, every centralsymmetrical set K has a representation (4.2) and in arbitrary dimensions to each element, A of $\Omega_0(R^n)$ there exist an uniquely determined simple set, $K(A)$ such that $K(A) = A + \tilde{A}$, $\tilde{A} \in \Omega_0(X)$ and if $A + B$ is simple then $A + B = K(A) + C$, for some C simple, i.e. every element of $\Omega_0(X)$ has a smallest simple majorant.*

The first statement is a consequence of Chocquet's theorem on the representation of elements of a compact convex set, in a metrisable space, as integrals of the extremal points of this set with respect to some normed regular Borel measure, (over the G_δ set of extremal points of this set), see [8].

Indeed, for an arbitrary element A of $\Omega_0(R^2)$ and vector $x \in R^2$, the set $A \cap (A + x)$ sweeps out the set A therefore, if from $A = A_1 + A_2$ follows that $A_1 = \lambda A$, $A_2 = (1 - \lambda) A$ for some $0 \leqq \lambda \leqq 1$, then A must be a triangle or a line segment, which later is the only possibility if A is centralsymmetrical (We do not know which are the "extremal" sets of $\Omega_0(R^n)$ for $n \geq 3$.) The second part of the theorem is easily proved first for poly-ledrons convex hulls of a finite number of points. A polyhedron is simple if and only if each face of it is centralsymmetrical (simple). The polyhedron $K(A)$ is the sum of all the translationally "different" edges of A, (two edges are not "different" if one can be moved by some translation into the other).

Finally the statement follows from the approximability of the elements of $\Omega_0(R^n)$ by convex polyhedrons.

Ellipsoids are simple sets (as linear images of balls, whose representing measure is just the constant multiple of the homogenuous measure on S^{n-1}).

For specifying conditions, which ensure that the intersections, $A(\xi)$ of a simple set, A, with linear (parallel) manifolds of a given direction is simple, or centralsymmet-rical, see [12]. As a consequence of Lemma (2.14) the sets $A(\xi)$ are translationally included in $A(0)$, if the support, μ_A, is symmetrical with respect to Ker H_0 this has applications by Theorem (3.8).

To deal with optimisation problems concerning invariant sets, the knowledge of the dual space of $\Omega_0(X)$ is important.

For example a necessary condition for a set K to yield the extremal value in the problem

$$\inf\{f_0(K) \mid K \in A \subseteq \Omega_0(X), R(K)=0\},$$

where $f_0(K)$ is a continuous convex functional defined over a compact, convex subset, A, of $\Omega_0(X)$, and $R: A \to \Omega^0(X)$ is a linear continuous mapping — is, that there exists an element, γ, of the dual space, $[\Omega^0(X)]^*$, and a real number λ_0 (not both zero), such that K yields the global minimum of

$$\lambda_0 f_0(K) + \langle \gamma, R(K) \rangle, \quad \text{over} \quad K \in A.$$

The determination of the dual space, $[\Omega^0(R^n)]^*$, is easier than the problem of the enumeration of all "*extremal*" sets, (elements) of the cone $\Omega_0(R^n)$, (for $n \geq 3$, mentioned above) as a consequence of Riesz' representation theorem one obtains from Theorem 2.6, and from the *density* of the subspace generated by simple sets in the subcone of central symmetrical sets, see [12].

4.5. **Theorem.** *The dual space of the strongly closed subspace of $\Omega^0(X)$ generated by simple sets is the usually normed space of signed regular, Borel measures, μ_γ, symmetrical over $S_1(X^*)$; i.e. for all $\gamma \in (\Omega^0)^*$, K simple or K centralsymmetrical one has*

$$\gamma(K) = \int_{S_1(X^*)} m(\Psi, K)\mu_\gamma(d\Psi), \quad \|\gamma\| = \int_{S_1(X^*)} |\mu|(d\Psi)_l = \|\mu_\gamma\|.$$

An other useful parametrisation for a subclass of the elements of $\Omega_0(X)$ for the problem of contructing invariant sets consist to represent, (if possible), the sets P, K, L, in the form

$$K = \sum_{n}^{\infty} k_n F^n Q, \quad k_n \geqq 0, \quad K = \int^{\infty} k(\tau) e^{\tau F} Q \, d\tau,$$

for a given set Q. This is specially advanteguos if we are given Q and L, M and we would like to find the "smallest" set, P, (belonging to a given subspace $R(G)$), such that there exist a compact invariant subset contained in L, (or containing M), because this class of sets is *closed* under many of the operations used in the construction of invariant sets in section 3, see Example 3.6.

REFERENCES

[1] Pontrjagin, L. S.: On the evasion Process in Differential Games, *Applied Math. and Optimisation,* 1, (1975), No. 1.

[2] Krasovskij, N. N., Subbotin, A. I.: *Positional Differential Games* (In Russian), Nauka, Moscow, 1974.

[3] Sonnevend, G.: Output regulation in partially observed linear systems under distrubances, *Lecture Notes in Control and Information Sciences,* Vol. 6, (1978), 214—229. 1978.

[4] Wonham, W. M.: Linear Multivariable Control, *Lecture Notes in Economics and Math. Syst. Theory,* Vol. 104, Springer, Berlin, 1974.

[5] Basile, G., Marro, G.: On the synthesis of observers, 3rd *IFAC Symposium on Sensitivity,* Ischia, 1973.

[6] Pontrjagin, L. S.: On linear differential games, 2. *Dokl. Akad. Nauk,* 175, (1976), No. 4.

[7] Yosida, K.: *Functional Analysis,* 2nd edition, Springer, Berlin, 1966.

[8] Phelps, R. R.: *Lectures on Choquet's theorem,* Van Nostrand, Princeton, 1966.

[9] Sonnevend, G.: On constructing invariant sets in linear differential games, *Lecture Notes in Computer Science,* Vol. 27, ed. by G. I. Marchuk, Springer, 1975.

[10] Blaschke, W.: *Kreis und Kugel,* Walter de Gruyter, Berlin, 1956.

[11] Pshenitsnij, B. N., Sagaidak, M. I.: *Kibernetika* (Kiev), (1970), No. 2.

[12] Sonnevend, G.: Optimisation of adaptive numerical algorithms, to appear in *Analysis Mathematica* (1979).

[13] Barabanova, N. N., Subbotin, A. I.: On classes of strategies in differential games of evasion, *Prikladnaja Mathematika i Mechanika,* 35, (1971), No. 3.

[14] Davison, E. J.: The robust control of a servomechanism problem for linear time-invariant multivariable systems, *IEEE Transactions on Automatic Control,* AC-21, (1976), No. 1.

[15] Kalman, R. E., Falb, P. L., Arbib, M. A.: *Topics in Math. System Theory,* McGraw Hill, New York, 1969.

A DYNAMIC PROGRAMMING
MODEL FOR AN ASSEMBLING PROBLEM
AND A BUILDING GAME

J. WIJNGAARD

(Eindhoven, The Netherlands)

INTRODUCTION

In this paper the application of a dynamic programming model to an assembling problem with risk is considered. Let P be a product consisting of a number of components. Each time a connection is made between two subassemblies there is a certain risk that something happens which causes that one or both subassemblies fall apart into their basis components. In this case not only the connection which one was making has to be repaired, but also the connection in the subassemblies which are fallen apart. One can think for instance of an instrumentmaker with very unfirm hands. A problem with the same structure comes from the following game with dominoes. One builds a row of upright dominoes with interdistances which are just smaller than the height of the dominoes. Then one gives a push to the first domino. The fun of the game is to see how subsequently all dominoes fall down. The building of the row in this game is analogous to the assembling of P. Each time a domino is set up there is a certain probability that it will fall to the left or to the right. (The hand is not so firm as it should be) and this falling domino will take all its left or right neighbours with it.

The problem which will be investigated here is how to connect the components (in which order). Which assembling strategy minimizes the expected assembling time? Assume for instance that P consists of the components a, b, c, d, e which has to be connected in the following way

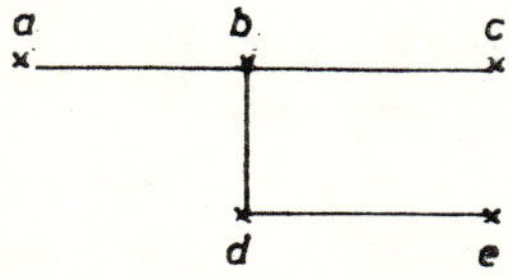

A method to make the product is by first connecting a with b, then $a-b$ with c, then $a-b-c$ with d and at last $a-b-c$ with e. In connecting $a-b-c$ with d it is

$$|$$
$$d$$

possible that the subassembly $a-b-c$ is destroyed. Another method is by first connecting a with b, then $a-b$ with c, then d with e and at last $a-b-c$ with $d-e$. Now, in connecting d with e, there is no risk for $a-b-c$. This strategy is better than the first one.

Dynamic programming will be used to find the optimal assembling strategy. As in most applications of dynamic programming the difficulty is how to reduce the state space. In this case this can be done by proving that the optimal strategy is of a certain type, namely a sequence. A sequence is a rank ordering of activities (connections). In each state one has to choose the activity with the smallest rank.

The intention of this paper is not to solve very important practical problems, but to show a problem which is interesting as dynamic programming problem. The structure of this problem is totally different from the more usual dynamic programming models in queueing-, inventory-, maintenance- and investment problems. Since the starting state and end state are given it can be interpreted as a stochastic shortest path problem, where one may decide to which of the next places one will go in each step, but where "nature" decides if one reaches this place or has to go back a number of steps.

STANDARD DYNAMIC PROGRAMMING

Assume that the product P consists of $n+1$ components, n connections have to be made. The time necessary for each connection is 1, also if an accident happens during this connection. Each time two subassemblies are connected there is a probability p for each of the two subassemblies to fall apart and a probability q that both subassemblies fall apart. It is clear that the problem can be formulated as a total costs dynamic programming problem. Each state can be represented by a subset of $\{1, 2, \ldots, n\}$, indicating the activities which are ready. That means that there are 2^n states. The possible actions in each state are the activities which are not yet ready. The transition probabilities are determined by p, q and the assembling strategy. It is clear that if n is not very small the number of states is too large to apply dynamic programming in a standard way. In the next section it will be shown how the structure of the optimal strategy can be used to make the problem manageable.

SEQUENCES

Definition. Let the numbers $1, 2, \ldots, n$ be assigned to the n activities. A strategy, which chooses in each state of all activities which have to be done yet that with the smallest number, is called a sequence.

In the first place it is important to notice that the class of all sequences is indeed smaller than the class of all stationary strategies. Assume for instance that P has the following structure $a-b-c-d-e-f$. Consider the strategy which chooses in state $a\,b\,c\,d\,e\,f$, (the starting state) the activity $a-b$, in state $a-b\,c\,d\,e\,f$ the activity $c-d$, in state $a-b\,c-d\,e\,f$ the activity $e-f$, in state $a-b\,c-d\,e-f$ the activity $d-e$ and in state $a-b\,c\,d\,e-f$ the activity $b-c$. It is easy to verify that this strategy is not a sequence. We shall prove now that there is an optimal strategy which is a sequence. Therefore we need the concept k-sequence.

Definition. Assign the numbers $1, \ldots, k$ to k of the n activities. A strategy is called a sequence if it chooses the activity with the smallest number in each state where (1) there are numbered activities which have to be done yet and (2) there are no unnumbered activities which are ready.

The proof of the optimality of sequences is now by induction to n, the number of activities and to k, k-sequences.

Theorem. *There exists an optimal strategy which is a sequence.*

Proof. We use induction to n, the number of activities. First it is true for $n=1$. Assume that it is true for all $n \le l$. We have to prove that it is also true for $n=l+1$. To do this we use k-sequences. It will be proved that an optimal strategy exists for $n=l+1$ which is a k-sequence for $k=1, \ldots, l+1$. Since an $(l+1)$-sequence is again a sequence this completes the induction argument. The proof of the existence of an optimal strategy which is a k-sequence, $k=1, 2, \ldots, l+1$, is also by induction. It is easy to see that each strategy is a l-sequence.

Now assume that v_j is an optimal strategy which is a j-sequence. We have to prove the existence of an optimal strategy which is a $(j+1)$-sequence.

Following strategy v_j one reaches at last a state s where all numbered activities are ready and no unnumbered activities. This state s consists of a number of subassemblies $s_1, \ldots, s_m$. Let a be the activity which has to be chosen in state s according to strategy v_j. This activity connects two subassemblies, say s_i and s_{i+1}. If an accident happens during activity a and one or both subassemblies fall apart, one has to repair first these subassemblies, since v_j is a j-sequence, and then to connect the two by activity a. The history can repeat some times, but at last one will succeed and the state s' is reached which differs from s in that the subassemblies s_i and s_{i+1} are connected now. This new subassembly is called s_i'.

Let $t_1, \ldots, t_{i-1}, t_i', t_{i+2}, \ldots, t_m$ be the minimal expected times to make the subassemblies $s_1, \ldots, s_{i-1}, s_i', s_{i+2}, \ldots, s_m$ let $r_{s'}'$ be the minimal expected time necessary to complete the project if it is now in state s' and let d be the minimal expected duration of the whole project.

Since v_j is optimal it follows that

$$(1) \qquad d = \sum_{h=1}^{i-1} t_h + t_i' + \sum_{h=i+2}^{m} t_h + r_{s'}.$$

Let $n_1, \ldots, n_m$ be the number of connections in the subassemblies $s_1, \ldots, s_m$ $\left(n_i \geqq 0, \right.$

$\left. \sum_{h=1}^{m} n_h = j \right)$. Since $n_h \leqq l$ there is an optimal sequence to make s_h (the induction assumption on n). Assign the numbers $1, \ldots, j$ to the j activities in $s_1, \ldots, s_m$ such that

1. in each subassembly one has only subsequent numbers,
2. the rankorder within each subassembly corresponds to the optimal sequence for that subassembly.

The number $j+1$ is assigned to activity a.

Define the following strategy v_{j+1}:

1. v_{j+1} prescribes the activity with the smallest number in each state where numbered activities have to be done yet and no unnumbered activities are ready.
2. v_{j+1} coincides with v_j in all other states. It is clear that this strategy v_{j+1} is a $(j+1)$-sequence. Let $(v_{j+1}v_j)$ be the strategy which applies v_{j+1} until the first moment, after that state s' is reached, that all unnumbered connections are destroyed again. The expected duration under this strategy is

$$\sum_{k=1}^{i-1} t_h + t'_i + \sum_{h=i+2}^{m} t_h + r_{s'} = d.$$

Hence, the strategy $(v_{j+1}v_j)$ is optimal. This implies by standard dynamic programming arguments that the strategy v_{j+1} is also optimal.

This completes the proof of the theorem.

Using the existence of an optimal sequence it is possible to compute this sequence.

Let b be some activity. Consider all sequences with the number n assigned to activity b. In each of these sequences this activity connects the two subassemblies s_{b_l} and s_{b_r}. Let v_b be the best of all these sequences, let t_{b_l} and t_{b_r} be the minimal expected assembling times necessary for s_{b_l} and s_{b_r} and let d_b be the expected duration of the whole project under sequence v_b. Then

$$(2) \qquad d_b = t_{b_l} + t_{b_r} + 1 + qd + p(d - t_{b_l}) + p(d - t_{b_r}).$$

The terms qd, $p(d - t_{b_l})$, $p(d - t_{b_r})$ are the contributions of the probability that both subassemblies fall apart, the subassembly s_{b_r} falls apart and the subassembly s_{b_l} falls apart.

The expression (2) implies

$$(3) \qquad d_b = \frac{(1-p)(t_{b_l} + t_{b_r}) + 1}{1 - 2p - q}.$$

For the minimal expected duration we have therefore

$$(4) \qquad d = \frac{1}{1 - 2p - q} + \frac{1-p}{1 - 2p - q} \min_{b} (t_{b_l} + t_{b_r}).$$

With aid of this expression it is possible to compute the optimal sequence for each of the subassemblies and for the whole product. The minimal expected assembling time for each of the subassemblies depends of course on the number of connections needed, but also on the structure of the subassembly. If the structure is as in Fig. 1a, it is only possible to add each time only one component. If the structure is as in Fig. 1b, it is possible to make first subassemblies. In the computing procedure it is not neces-

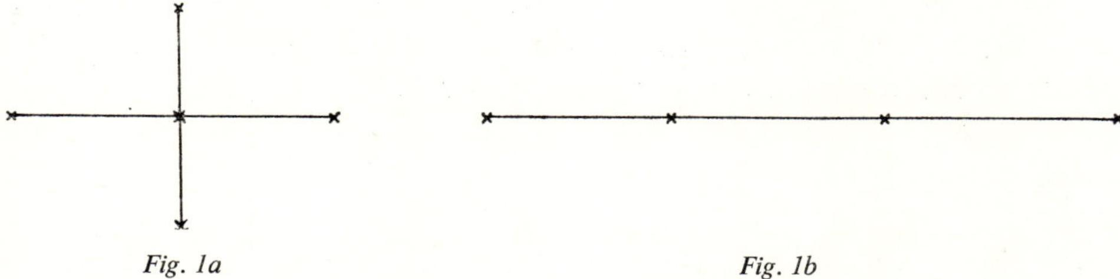

Fig. 1a   Fig. 1b

sary of course to distinguish between subassemblies with the same number of components and the same structure.

If the product has a linear structure (see Fig. 1b) there are only linear subassemblies. Notice that the domino problem has such a linear structure. In the linear case it is possible to give a very easy analytical solution. It is possible to prove, using relation (4), that the last activity in the optimal sequence is the middle connection or one of the middle connections. If there are 8 connections to be made the sequence as shown in Fig. 2 if optimal. But there are more optimal sequences of course, for instance all sequences with the number 8 assigned to the same activity and the other numbers such that the rank order of the numbers in the left and right subassembly stay the same.

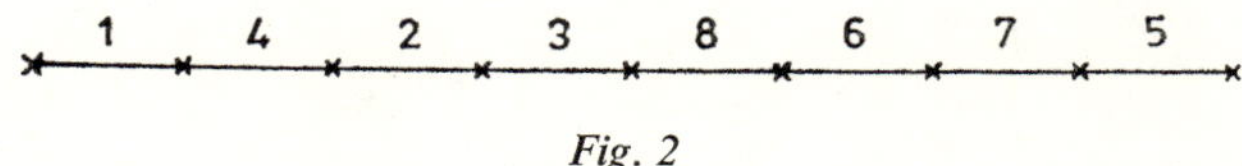

Fig. 2

Now we have to say something about the minimal expected duration of the project. Therefore we have to distinguish between first connections and repairments. We say that a connection is involved in another connection if it is part of one of the subassemblies which are connected by this last connection.

Let $a_1, \ldots, a_n$ be the n activities (connections) and let k_i, $i = 1, \ldots, n$ be the number of first connections in which the connection a_i is involved, using some sequence v. Notice that $0 \leq k_i \leq n-1$ for all i. Let

$$\alpha := \frac{1-p}{1-2p-q} \quad \text{and} \quad \beta := \frac{1}{1-2p-q}.$$

Then from (4) it follows for d_v, the expected duration under v,

(5)
$$d_v = \beta \sum_{i=1}^{n} \alpha^{k_i}.$$

This holds as well for linearly structured products as for products with a more general structure. For a linear product with 2^m-1 activities we get for d, the minimal expected duration.

$$(6) \qquad d=\beta(1+2+4\alpha^2+\ldots+(2\alpha)^{m-1})=\frac{\beta}{2\alpha-1}(2\alpha)^m.$$

In the derivation of this expression it has to be used that in the optimal sequence for a linear product the last activity is the middle one.

PART 6

OPTIMIZATION IN NETWORKS

ABOUT THE DETERMINATION OF MAXIMUM FLOW IN A NETWORK

C. DINESCU and B. SĂVULESCU

(Bucharest, Romania)

DEFINITIONS AND RESULTS

Considering a net with a flow of $(n+2)$ nodes, $R=(X, \mathcal{A}, c, x_0, x_{n+1})$, where

$X=\{x_0, x_1, x_2, \ldots, x_{n+1}\}$ is the set of nodes;
$\mathcal{A}=$ is the set of arcs,

$c(x_i, x_j)=c_{ij}$ is the capacity of the arc $(x_i, x_j)\in\mathcal{A}$;

x_0 and x_{n+1} are respectively, the input and output of the net (extremities).

Definition 1. We denote by *elementary flow** of the net R, the flow φ, such that any circuit within the net contains at least one arc with a null flow.

Theorem 1. *If one can build up a flow φ upon a net, having the total value Φ, then one can also build up an elementary flow φ', with a total value Φ.*

The *demonstration* of the theorem is based on the observation of the fact that by reducing flows on all arcs of a circuit, with the same quantity, one obtains a new flow, with the same total value.

We denote by R' the extended net, obtained from R by adding the arc (x_{n+1}, x_0) of infinite capacity.

It can be observed that to a flow φ on R, there corresponds a flow on R', defined as follows:

$$\varphi'_{ij}=\begin{cases} \varphi_{ij}, & \text{for} \quad (x_i, x_j)\not\equiv(x_{n+1}, x_0), \\ \Phi, & \text{for} \quad (x_i, x_j)\equiv(x_{n+1}, x_0). \end{cases}$$

For our convenience, we admit that the flow on arc (x_{n+1}, x_0) shall not be considered, either to define the total flow, or to define the elementary flow.

* Notions introduced by the authors.

Lemma 1. *If in a net R there exists a flow of total value:*

$$\Phi = \min \left\{ \sum_{(x_0, x_j) \in \mathcal{A}} c_{0j}; \quad \sum_{(x_i, x_{n+1}) \in \mathcal{A}} c_{i, n+1} \right\},$$

then that flow has a maximum total value in the net.

Lemma 2. *The maximal total value of the flow in a net R', does not change if the value of the capacity of any cut through the net is considered as the capacity of the arc (x_{n+1}, x_0).*

The *demonstrations* of both lemmas are immediate.

In the following we admit to take as value of the capacity of arc (x_{n+1}, x_0), the value:

$$c_{n+1, 0} = \min \left\{ \sum_{(x_0, x_j) \in \mathcal{A}} c_{0j}; \quad \sum_{(x_i, x_{n+1}) \in \mathcal{A}} c_{i, n+1} \right\}.$$

Definition 2. We call *semi-flow of the net R* (respectively R'), a function Ψ, defined on the set of arcs of the net, satisfying the conditions:

$$0 \leq \Psi_{ij} \leq c_{ij}, \quad \text{for} \quad (x_i, x_j) \in \mathcal{A},$$

where $\Psi_{ij} = \Psi(x_i, x_j)$, *A semi-flow is called normal if:* $\Psi_{n+1, 0} = c_{n+1, 0}$.

Definition 3. We call a *surplus related to the semi-flow Ψ*, a function defined on the set of nodes of the net, such that:

$$s_i = s(x_i) = \sum_{(x_i, x_j) \in \mathcal{A}} \Psi_{ij} - \sum_{(x_j, x_i) \in \mathcal{A}} \Psi_{ij}, \quad i = 0, 1, 2, \ldots, n+1.$$

Observations

1. If, for a semi-flow Ψ, defined on the net R, we have: $s_i = 0$, for $i = 1, 2, 3, \ldots, n$, the semi-flow represents a flow on the net R.

2. A semi-flow identically null, defined on the extended net R', is a flow on that net.

Lemma 3. *For any semi-flow Ψ, defined on a net, be it extended or not, we have:*

$$\sum_{x_i \in X} s_i = 0.$$

The *demonstration* is direct.

Definition 4. Upon the semi-flow that can be built-up in a net, we define the non-negative real function:

$$\Theta(\Psi) = \begin{cases} \displaystyle\sum_{s_i > 0} s_i, & \text{if there exists at least } s_i > 0, \\ 0, & \text{in the contrary case,} \end{cases}$$

called the *total positive surplus.*

240

Definition 5. We call *elementary semi-flow*, a semi-flow Ψ, such that in any circuit of the net, that does not contain arc (x_{n+1}, x_0), there is at least one arc having a null value by the function Ψ.

Theorem 2. *Given a semi-flow Ψ on a net. an elementary semi-flow Ψ' can be built up on the same net, such that: $\Theta(\Psi')=\Theta(\Psi)$.*
The *demonstration* is similar to the demonstration of Theorem 1.

Lemma 4. *Given a semi-flow Ψ, on an extended net, such that a surplus $s_i>0$ should correspond to the node $x_i\,(x_i\not\equiv x_{n+1})$, one can build up another semi-flow Ψ', on the same net, its corresponding surplus s'_i being null, the mentioned semi-flow keeping all its previous surpluses null, excepting the one corresponding to the output. Moreover, we have $\Theta(\Psi')\leqq\Theta(\Psi)$.*
The lemma is constructive, given also the method of finding Ψ'.

Lemma 5. *Given a semi-flow Ψ, on an extended net, such that to the node $x_i\,(x_i\not\equiv x_0)$, there corresponds a surplus $s_i<0$, we can build up another semi-flow Ψ', having as corresponding surplus s'_i null, the said semi-flow keeping all the previous null surpluses, excepting the one corresponding to the input. Moreover, we have $\Theta(\Psi')\leqq\Theta(\Psi)$.*
The *demonstration* is similar to that of Lemma 4.

Corollary. *Given a semi-flow Ψ, on an extended net, one can determine another semi-flow Ψ', to which only one strictly negative surplus will correspond for the input of the net, and only one positive surplus for the output, and $\Theta(\Psi')=s'_{n+1}\leqq\Theta(\Psi)$.*

Theorem 3. *If a normal semi-flow Ψ_n does exist on an extended net R', then there exists at least one flow φ on the net R', with a total value Φ, such that*

$$\Phi\geqq c_{n+1,0}-\Theta(\Psi_n).$$

The *proof* can be done by using Lemmas 1 and 2.

Theorem 4. *Given an extended net, and the set $\{\varphi\}$ of all the flows that can be built up on it. Between the set $\{\Phi\}$, of all the values of total flows, and the set $\{\Psi_n\}$ of the normal semi-flows that can be built up on the net, there exists the relation:*

$$\max\,\{\Phi\}=c_{n+1,0}-\min\,\{\Theta(\Psi_n)\}.$$

Proof. Let φ^* be with $\Phi^*(\varphi^*)=\max\,\{\Phi(\varphi)\}$. To φ^* corresponds Ψ_n^* where:
$$\Psi_{ij}^*=\begin{cases}c_{n+1,0}, & \text{for } (x_{n+1}, x_0),\\ \varphi_{i,j}^*, & \text{otherwise.}\end{cases}$$

According to Lemma 2, we can write: $\Theta(\Psi_n^*)=c_{n+1,0}-\Phi^*(\varphi^*)^{(1)}$. Let $\Theta(\bar\Psi_n)=$

$= \min_{\Psi_n} \{\Theta(\Psi_n)\}$. According to Theorem 1, there is a $\bar\varphi$ having the total value $\bar\Phi(\bar\varphi)$ such that:

$$\bar\Theta(\bar\varphi) \geq c_{n-1,0} - \Theta(\bar\Psi_n)^{(2)} \quad \text{or} \quad \min_{\Psi_n} \{\Theta(\Psi_n)\} \geq c_{n+1,0} - \bar\Phi(\bar\varphi)^{(3)}$$

As $\Phi^*(\varphi^*)$ is maximum, we have: $\Phi^*(\varphi^*) \geq \bar\Phi(\bar\varphi)^{(4)}$. From (1), (3) and (4) it results:

$$\min_{\Psi_n} \{\Theta(\Psi_n)\} \geq c_{n+1,0} - \bar\Phi(\bar\varphi) \geq c_{n+1,0} - \Phi^*(\varphi^*) = \Theta(\Psi_n^*)$$

and therefore, we have: $\min_{\Psi_n} \{\Theta(\Psi_n)\} = \Theta(\Psi_n^*)^{(5)}$.

From (1) and (5) it results:

$$\min_{\Psi_n} \{\Theta(\Psi_n)\} = c_{n+1,0} - \Phi^*(\varphi^*),$$

which means that the theorem is proved.

Algorithm

Step 1. One considers Ψ_n, for example, $\Psi_{ij} = c_{ij}$ for every (x_i, x_j) and then: s_i is calculated for all $x_i \in X$.

Step 2. For every $x_i \in X$ with $s_i > 0$, the following operations are made:

a) one transfers from x_i to x_j with $s_j < 0$, the value $\min \{s_i; \Psi_{ij}\}$ if there is $(x_i, x_j) \in \mathcal{A}$ and $\Psi_{ij} > 0$ (this transfer is made by changing Ψ_{ij} with $\Psi_{ij} - \min \{s_i, \Psi_{ij}\}$);

b) one transfers from x_i to x_j with $s_j < 0$, $\min \{s_i; c_{ji} - \Psi_{ji}\}$, if there is $(x_j, x_i) \in \mathcal{A}$ and $c_{ji} > \Psi_{ji}$ (this transfer is done by changing Ψ_{ji} with $\Psi_{ji} + \min \{s_i; c_{ji} - \Psi_{ji}\}$).

For the remaining arcs one keeps the preceeding semi-flows. After performing operations a) and b) we can obtain:

c) for every $x_i \in X$, $x_i \neq x_{n+1}$, $x_i \neq x_0$ it results $s_i = 0$; in that case, Ψ_n corresponds to a maximum flow and the algorithm ends;

d) there is at least one $x_i \in X$, $x_i \neq x_{n+1}$, $x_i \neq x_0$ with $s_i > 0$; in this case, Ψ_n does not correspond to a maximum flow, and we go to step 3.

Step 3. We label all vertices $x_i \in X$, $x_i \neq x_{n+1}$ according to the rule:
— label every $X_i \in x$ with $s_i > 0$ by $(+)$;
— if $x_i \in X$ is labelled, we label every $x_j \in X$ by (i), if
— there is $(x_i, x_j) \in \mathcal{A}$ and $\Psi_{ij} > 0$, or if there is $(x_j, x_i) \in \mathcal{A}$ with $\Psi_{ji} < c_{ji}$;

After performing the labelling operations, we can obtain:
(i) none of the vertices x_j for which $s_j < 0$, can be labelled; in this case, $\Theta(\Psi_n)$ have minimum value and go to step 4;
(ii) the vertices x_j for which $s_j < 0$ have been labelled; in this case, by using the labelling, one determines a chain from the vertex x_k (with $s_k > 0$) to vertex x_j (with $s_j < 0$) and then one transfers the semi-flow to the already built-up chain, taking into account a) and b) from step 2;

The labelling operation is continued until the situation (i) is reached than go to step 4.

Step 4. For all $x_i \in X$, with $s_i > 0$, one transfers the corresponding surplus to x_{n+1} and for all $x_i (x$, with $s_i < 0)$ one transfers the corresponding surplus to x_0.
Finally, the required maximum flow is obtained.

Theorem 5. *Given a normal semi-flow* Ψ_n, *on an extended net* R', *for which by means of the labelling process of the nodes, it is impossible to give a label to any node with negative surplus, then a minimal total positive surplus corresponds to the semi-flow* Ψ_n.
The demonstration is made using the min-cut theorem of Fulkerson.

Notes

1. The transfer rules for the positive surplus given above, can also be used for the cumulation of some positive surpluses within a smaller number of nodes.
2. Rules for the transfer of negative surpluses can be similarly deduced, as the transfer of positive ones, between the extremities of an arc, and can also be used in the case of the final transfer of the whole negative surplus into the input of the net.

Other advantages of the algorithm

1. The algorithm can be applied to those practical situations when in intermediary vertices we have accumulations or losses or flow. For example:
— if x_t is an accumulation vertex in which β_t units of flow are accumulated, we
   put

$$s_t = \sum_{(x_t, x_j) \in \mathcal{A}} \Psi_{tj} - \sum_{(x_i, x_t) \in \mathcal{A}} \Psi_{it} - \beta_t,$$

— if x_t is a vertex from which β_t units of flow are lost, we put:

$$s_t = \sum_{(x_t, x_j) \in \mathcal{A}} \Psi_{tj} - \sum_{(x_i, x_t) \in \mathcal{A}} \Psi_{it} - \beta_t.$$

2. The algorithm seems to be succeptible to improve in case of multiflow items.
3. The algorithm is efficient for those networks in which values corresponding to the maximum flow are very close to the capacities c_{ij}.
4. It can be easily adapted to modifications that could happen to the value of capacities of the arcs, as well as in the case of changing the source or sink of the net.

REFERENCES

[1] Dantzig, G. B. and Fulkerson, D. R.: On the Max-Flow Min-Cut Theorem of Networks, *Ann. of Math. Studies* no. 38, Princeton Univ. Press, Princeton, N. J., 1956.
[2] Elias, P. and others: A note on the maximum flow through a network, *IRE Trans. Inform. Theory,* IT-2, 1956.

[3] Ford, L. R. and Fulkerson, D. R.: Maximal flow through a network, *Canad. Jr. Math.*, (1956).
[4] Ford, L. R. and Fulkerson, D. R.: *Flows in Networks,* Princeton Univ. Press, 1962.
[5] Ford, R. L. and Fulkerson, D. R.: A simple algorithm for maximal network flows and an application to the Hitchcock problem, *Canad. Jr. Math.*, 9, (1957).
[6] Fulkerson, D. R.: An out-of-kilter method for minimal cost flow problems, *Jr. Soc. Ind. App. Math.*, 9, (1961).

NETWORK IMPROVEMENTS VIA MATHEMATICAL PROGRAMMING

HOANG HAI HOC

(Montreal, Canada)

I. FORMULATION OF THE PROBLEM

This paper addresses the following network improvement or design problem. Consider a directed network with N nodes, and assume that nodes $1, 2, \ldots, P, P \leq N$, are origins and destinations. Let D be a $P \times P$ matrix whose (s, t) entry denotes flow requirement from origin s to destination t. Let also be given K projects, each of which involves an improvement of one or more existing arcs or an addition of one or more new arcs. Assume that $\{1, 2, \ldots, M\}$ is the index set of arcs involved in network improvement projects, and that existing arcs numbered by $\{M+1, M+2, \ldots, L\}$ do not depend on any improvement project. Let us use the following notations:

$x_l^s =$ flow along arc l and associated with origin s

$x_l =$ total flow along arc l

$c_k =$ investment cost of improvement project k

$L(k) =$ set of arcs involved in project k

$$y_k = \begin{cases} 1, & \text{if project } k \text{ is implemented,} \\ 0, & \text{otherwise.} \end{cases}$$

$b =$ investment budget

$f_l(x_l) =$ cost of shipping x_l units of total flow along arc l,

$$M+1 \leq l \leq L$$

$f_l(x_l, y_k) =$ cost of shipping x_l units of total flow along arc l,

$$1 \leq l \leq M, \quad l \in L(k).$$

Assuming that $L(1), \ldots, L(K)$ constitute a partition of $\{1, 2, \ldots, M\}$, the network improvement/design problem addressed is formulated as follows

Problem (NIP):

$$(1) \qquad \text{minimize} \sum_{l=1}^{M} f_l(x_l, y_k) + \sum_{l=M+1}^{L} f_l(x_l)$$

subject to

$$(2) \qquad D(s,j) + \sum_{l \in O(j)} x_l^s = \sum_{l \in \mathcal{I}(j)} x_l^s,$$
$$j = 1, 2, \ldots, N; \quad s = 1, 2, \ldots, P; \quad j \neq s$$

$$(3) \qquad x_l = \sum_{s=1}^{P} x_l^s, \quad l = 1, 2, \ldots, L$$

$$(4) \qquad \sum_{k=1}^{K} c_k y_k \leq b$$

$$(5) \qquad y_k = 0 \text{ or } 1, \quad k = 1, 2, \ldots, K$$

$$(6) \qquad x_l^s \geq 0, \quad l = 1, 2, \ldots, L; \quad s = 1, 2, \ldots, P$$

where

$O(j)$ = set of arcs incident to node j and directed away from j,
$\mathcal{I}(j)$ = set of arcs incident to node j and directed toward j.

We call attention to the fact that there may exist several variants of the network improvement/design problem. If there exist for each arc l in $L(k)$ two cost functions f_l^1 and f_l^0 depending on whether $y_k = 1$ or $y_k = 0$, then a simple link capacity expansion problem results. Such a link capacity expansion problem can be generalized to permit several cost functions corresponding to several discrete capacity alternatives. If the cost function f_l^0 is such that

$$f_l^0(x_l) = \begin{cases} 0 & \text{for } x_l = 0 \\ \infty & \text{for } x_l > 0 \end{cases}$$

then we have a link addition problem. Moreover, the fact that all links are involved in some project of the link addition type gives rise to a network design problem. A generalized version of link addition or design problems also includes several capacity alternatives for every added link.

For simplicity of presentation we shall only pay attention to the simple link capacity expansion problem mentioned above. Before discussing solution approaches in Sections 3 and 4, we shall consider some applications within the context of transportation or computer communication networks and present related work in Section 2. Section 5 is devoted to computational performances of the solution approaches discussed.

246

II. APPLICATIONS AND RELATED WORK

Given an existing urban road network, with projected increases in demands for road travel between various pairs of nodes, congestion will generally exist if the capacity of the network is not expanded. The decision maker must then consider building new links into the network, or increasing the capacity of some existing links. He will consider the various types of possible improvements to the network, as well as several locations in the network as candidates for improvement. Then, given a list of candidates for improvement he must select that combination which will result in the best road network according to given criteria.

The question of selection necessarily implies the concepts of evaluation and evaluation criterion. A simple, relatively easy to quantify, criterion for road network evaluation relates to travel time required to cover each link of the network. If the travel time is specified as a function of total traffics in the links and link capacities, it is then possible using traffic assignment procedures to estimate traffic volume and travel time in the network for a given set of trips. Hence, a straightforward objective for designing a road network is to minimize total travel time subject to a limited investment budget. This objective is not uncommon in transportation planning, and is somewhat preferred to that of minimizing the sum of investment and travel costs because of the difficulties associated with the latter arising from the dependence of the optimal solution on the choice of a monetary equivalance of travel time.

Within the framework of traffic assignment problems two most common behavioral assumptions lead to the concepts of user optimal and system optimal flow patterns. On one hand, one assumes in system optimal models that flows are distributed over the arcs of the network in such a manner that the sum of travel time for all users is minimized. On the other hand, it is assumed in user optimal models that each user of the network seeks to minimize his own travel cost and that the aggregation of individual decisions results in a set of equilibrium flows satisfying the following two conditions for every origin-destination pair s, t: (i) if two or more paths between node s and note t are actually traveled, then the cost of each traveler between s and t must be the same for each of these paths; (ii) there does not exist an alternative unused path between nodes s and t with less cost than that of the paths that are traveled. The user optimal flows are generally distinct from, and also more attractive than, the system optimal flows since they seem consistent with assumed rational behavior of individual users of road networks. However, a detailed comparison of the output from the system and user optimal models indicated that the difference between the two sets of flows is small enough to justify the simplyfing assumption in network improvement/ design problems, namely that total user cost is minimized [10].

Another application naturally arises from the following context. Let us consider a packet-switching computer-communication network. In such a network messages are segmented into packets and each packet traveling from source s to destination t is stored in a queue at each intermediate node i while awaiting transmission, and is sent forward to the next node j in the path from s to t when channel $l=(i,j)$ is free. There are several queues at each node, one for each output channel. Packet flow or data traffic requirements between nodes arise at random and packets are of random length. Therefore, channel flows, queue lengths and packet delays are all random variables. Under appropriate assumptions [9] and for a given network topology consisting of N nodes and L links the average delay T of a packet traveling from source to destination as a function of average flows on channels can be expressed as follows:

$$T = \frac{1}{\gamma} \sum_{l=1}^{L} \frac{f_l}{C_l - f_l}$$

with

$$0 \leq f_l \leq C_l$$

$$\gamma = \sum_{s=1}^{N} \sum_{t=1}^{N} r_{st},$$

where

r_{st} = average packet rate from s to t
f_l = total bit rate on channel l
C_l = signaling rate of channel l.

One is interested in determining channel location and capacity in such a manner that the average delay T is minimized subject to a budget constraint. This constitutes a generalized network design problem mentioned in Section I.

Several authors have proposed various formulations of network improvement/design problems and algorithms for solving those models Ochoa–Rosso [15] formulated a multistage mixed integer linear programming model for addition of links into an urban transportation network of multicommodity flow type. He attempted to minimize total user travel cost while imposing construction cost via a budget constraint, and showed how Benders' partitioning and Dantzig–Wolfe decomposition may be used to devise efficient algorithms. He also examined branch and bound approach for solving network design problems via user optimal flow traffic assignment models. The problem of the choice of investments in a transportation network was expressed by Bruynooghe [2] in a non-linear program certain variables of which can only take the values of 0 or 1. He used a branch-and-bound type method to evaluate the optimum without considering all possible alternatives, and to obtain a good solution and an upper limit of the distance of the optimal solution. However, as in the work of Ochoa–Rosso, no computational results were reported in Bruynooghe's work.

More recently, Steenbrink and Leblanc have also discussed network design problems. Steenbrink [16] decomposed the problem of cost minimization for a given trip matrix into a number of subproblems yielding for every link the optimal dimension relative to any given traffic flow, and a master problem in which the traffic flows are chosen in such a way that the objective function is minimized and the constraints are met. This is accomplished by a stepwise assignment of the trip matrix to the network according to the routes with the least value for the marginal objective function. His technique does not always converge to the optimum network, but may be quite good for extremely large network where the computational requirements of integer programming are excessive. Leblanc [12] developed a branch-and-bound scheme permitting one to avoid Braess' paradox which arises when adding a link to a network increases the total congestion in the network. This phenomenon is due to the fact that the desirability of a network configuration is determined by equilibrium flows which are not optimal with respect to the total travel cost cirterion.

III. SOLUTION METHOD BY PARTITIONING

1. Generalized Benders Partitioning

The network improvement problem that we address falls into the class of problems of the following form;

Problem (P): Minimize $f(x, y)$ subject to $G(x, y)=0$, $x \in X$, $y \in Y$, where y is a vector of complicating variables in the sense that (P) is much easier optimization problem in x when y is temporarily held fixed. G is an m-vector of constraint functions defined on

$$X \times Y \subseteq R^{n_1} \times R^{n_2}.$$

In fact, for fixed y Problem (NIP) becomes a capacitated traffic assignment problem for which efficient dual-adequate algorithms do actually exist. For this reason we attempt in this section to solve the Problem (NIP) by generalized Benders' decomposition approach proposed by Geoffrion [3].

To fix ideas let us use the following notations to put Problem (NIP) into the form of Problem (P):

$$f(x, y)= \sum_{l=1}^{M} f_l(x_l, y_k)+ \sum_{l=M+1}^{L} f_l(x_l)$$

$$= \sum_{l=1}^{M} [y_k f_l^1(x_l)+(1-y_k)f_l^0(x_l)]+ \sum_{l=M+1}^{L} f_l(x_l)$$

$$X=\{(x_1, x_2, \ldots, x_L, x_1^1, x_1^2, \ldots, x_l^s, \ldots, x_L^P): x_l \geq 0, x_l^s \geq 0\}$$

$$Y=\left\{(y_1, y_2, \ldots, y_K): y_k=0 \text{ or } 1, \sum_{k=1}^{K} c_k y_k \leq b\right\}.$$

$G(x, y)=$ multicommodity flow conservation equations (2), and equations (3) defining total arc flows, written into the form

$$G(x_1, x_2, \ldots, x_L, x_1^1, x_1^2, \ldots, x_l^s, \ldots, x_L^P)=0.$$

For all arcs l, $l=1, 2, \ldots, M$, the functions f_l^0 and f_l^1 are assumed to be increasing, convex, and continously differentiable on $[0, \infty)$. The functions $f_l^0 - f_l^1$, $l=1, 2, \ldots, M$, are assumed to be increasing and convex on $[0, \infty)$. We also assume that f_l, $l=M+1$, $\ldots, L$, are increasing, convex and continously differentiable functions on $[0, \infty)$.

The projection of (P) onto Y is

$$(7) \qquad \text{minimize } v(y) \quad \text{subject to} \quad y \in Y \cap V,$$

where

$$v(y) = \text{minimum} f(x, y) \text{ subject to } G(x, y) = 0, \quad x \in X$$

$$V \triangleq \{y : G(x, y) = 0 \quad \text{for some} \quad x \in X\}.$$

In the particular case of problem (NIP), for every $y \in Y$ there exists a feasible solution in X. Hence, $V \triangleq Y$.

Since X is a non-empty closed convex set, f is continuous and convex on X for each fixed y, $G(x, y)$ is linear on X and independent of y, and for each $\bar{y} \in Y v(\bar{y})$ is finite and the subproblem $S(\bar{y})$

$$(8) \qquad \underset{x \in X}{\text{minimize}} f(x, \bar{y}) \text{ subject to } G(x, \bar{y}) = 0$$

possesses an optimal multiplier vector, the optimal value of $S(\bar{y})$ equals that of its dual on Y, that is

$$v(\bar{y}) = \underset{u}{\text{maximum}} \, [\underset{x \in X}{\text{minimum}} f(x, \bar{y}) - uG(x, \bar{y})], \quad \text{all} \quad \bar{y} \in Y.$$

Consequently, the problem (P) is equivalent to the following master problem:

Problem (MP)

$$(9) \qquad \underset{y \in Y, y_0}{\text{minimize}} y_0$$

subject to

$$(10) \qquad y_0 \geq \underset{x \in X}{\text{minimum}} \, [f(x, y) - uG(x, y)], \quad \text{all} \quad u.$$

The natural strategy for solving the problem (MP), since it has a very large number of constraints, is relaxation. One begins by solving a relaxed version of (MP) that ignores all but a few of the constraints (10). Suppose that $(\hat{y}, \hat{y}_0)$ is optimal in a relaxed version of (MP). Then, $(\hat{y}, \hat{y}_0)$ satisfies (10) if, and only if, $\hat{y}_0 \geq v(\hat{y})$. Consequently, the subproblem $S(\hat{y})$ is the natural means for testing $(\hat{y}, \hat{y}_0)$ for feasibility in (MP), and its optimal multiplier vector $\hat{u}$ such that

$$\hat{y}_0 < \underset{x \in X}{\text{minimum}} \, [f(x, \hat{y}) - \hat{u}G(x, \hat{y})]$$

can be used as an index of a violated constraint.

250

It will be convenient in stating the generalized Benders' partitioning procedure to define the following function

$$L^*(y, u) = \underset{x \in X}{\text{minimum}} \, [f(x, y) - uG(x, y)], \quad y \in Y.$$

The procedure is then represented by the following steps:

Step 1
Let a point $\bar{y} \in Y$ be known.
Solve the subproblem $S(\bar{y})$ and obtain an optimal multiplier vector $\bar{u}$ and the function $L^*(y; \bar{u})$.

Put $J = 1$, $u^1 = \bar{u}$, ubd $= v(\bar{y})$.

Select a convergence tolerance parameter $\varepsilon > 0$, and go to Step 2.

Step 2
Solve the current relaxed master problem

$$\underset{y \in Y, y_0}{\text{minimize}} \, y_0$$

subject to

$$y_0 \geq L^*(y; u^j), \quad j = 1, 2, \ldots, J$$

by any applicable algorithm.
Let $(\hat{y}, \hat{y}_0)$ be an optimal solution; $\hat{y}_0$ is a lower bound on the optimal value of Problem (NIP).
If ubd $\leq \hat{y}_0 + \varepsilon$, terminate. Otherwise go to Step 3.

Step 3
Solve the revised subproblem $S(\hat{y})$.
If $v(\hat{y}) \leq \hat{y}_0 + \varepsilon$, terminate. Otherwise determine an optimal multiplier vector $\hat{u}$ and the function $L^*(y; \hat{u})$. Increase J by 1 and put $u^J = \hat{u}$.
If $v(\hat{y}) < $ ubd, put ubd $= v(\hat{y})$ which is an upper bound on the optimal value of Problem (NIP). Return to Step 2.

2. Computational Considerations

It is now unescapable to pay attention to the prospect for efficient computational implementation of generalized Benders' partitioning algorithm.

a) *Solution of Subproblem $S(y)$*: For fixed y subproblem $S(y)$ is a convex cost multi-commodity flow problem without explicit capacity constraints. Amongst procedures proposed for solving this problem, we have implemented and tested three algorithms suggested by Leblanc et al [11], Leventhal et al [13], and Nguyen [14] respectively, and found that Nguyen's Algorithm is the most suited to our task of solving sub-problem $S(y)$. The interested reader may consult reference [8] for more details of implementing and testing three algorithms considered. However, we remark that since no explicit capacity constraints exist problem $S(y)$ can be solved by adjusting sequentially the flows originated at various nodes one by one. Nguyen's algorithm exploits this structure and is of the convex simplex type using spanning arborescence rooted at an origin node for basis representation in the decomposed flow problem associated with the origin considered. This permits the use of augmented predecessor index [5] or augmented threaded index [6] method for efficient computational imple-mentation. Furthermore, Nguyen's algorithm can easily be restarted while changing from one network configuration to another, and is dual-adequate in the sense that it yields for every subproblem $S(\hat{y})$, $\hat{y} \in Y$, an optimal multiplier vector $\hat{u}$ which always exists.

b) *Construction of the Functions $L^*(y; u)$*: The structure of the linear constraints

$$G(x) = G(x_1, x_2, \ldots, x_L, x_1^1, x_1^2, \ldots, x_l^s, \ldots, x_L^P) = 0$$

is very peculiar. It only consists of reduced incidence matrices, one reduced incidence matrix for each origin. For origin s the corresponding reduced incidence matrix is obtained from the node-arc incidence matrix of the network by eliminating the row corresponding to node s. Hence, the multiplier vector is composed of $(L + PN)$ compo-nents $(u_1, u_2, \ldots, u_L, v_1^1, v_2^1, \ldots, v_j^s, \ldots, v_N^P)$, P of which can arbitrarily be set to zero:

$$v_1^1 = v_2^2 = \ldots = v_P^P = 0.0$$

The dual problem of $S(y)$ is as follows:

$$\text{maximize}_u \Gamma(u) = \text{maximize}_u \left\{ \text{minimize}_{x \in X} \sum_{s,t} v_t^s D(s, t) + \right.$$
$$\left. = \sum_{l=1}^{L} \left[f_l \left(\sum_s x_l^s \right) - \sum_{s=1}^{P} (v_{f(l)}^s - v_{d(l)}^s) x_l^s \right] \right\},$$

where arc l is assumed to be directed from node $d(l)$ to node $f(l)$. We have defined for $\bar{y} \in Y$ and for an optimal multiplier vector $\bar{u}$ of $S(\bar{y})$ the function

$$L^*(y; \bar{u}) = \text{minimum}_{x \in X} [f(x, y) - \bar{u} G(x, y)]$$

$$= \text{minimum}_{x \in X} \left\{ \sum_{s,t} v_t^s D(s, t) + \sum_{l=1}^{L} \left[f_l \left(\sum_s x_l^s \right) - \sum_{s=1}^{P} (v_{f(l)}^s - v_{d(l)}^s) x_l^s \right] \right\}.$$

We now need closer examination of optimal multiplier vector $\bar{u}$ for obtaining $L^*(y; \bar{u})$ in explicit form. If $\bar{x} = (\bar{x}_l^s)$ is an optimal solution to subproblem $S(\bar{y})$, then it is not difficult to show that the optimal multiplier vector $\bar{u}$ is given by

$$\bar{u}_l=\begin{cases}\dfrac{df_l}{dx_l}(\sum_s \bar{x}_l^s), & l=M+1,\ldots,L \\[2mm] y_k\dfrac{df_l^1}{dx_l}(\sum_s \bar{x}_l^s(+(1-y_k)\dfrac{df_l^0}{dx_l}(\sum_s \bar{x}_l^s), & l=1,2,\ldots,M, \ \in L(k)\end{cases}$$

$\bar{v}^s=$ shortest distance from s to j, $s=1,2,\ldots,P$, $j=1,2,\ldots,N$, under arc length $\bar{u}_l$, $l=1,2,\ldots,L$.

Let $s(l)$ be an origin such that

$$\bar{v}_{f(l)}^{s(l)}-\bar{v}_{d(l)}^{s(l)}=\operatorname*{maximum}_{s=1,2,\ldots,P}(v_{f(l)}^s-v_{d(l)}^s)$$

then the minimum solution defining $L^*(y;\bar{u})$ is as follows

$$\tilde{x}_l^s=0, \quad s=1,2,\ldots,P, \quad s\neq s(l)$$

where $\quad\tilde{x}_l^{s(l)}=$ solution of the equation $\gamma(x_l)=\bar{v}_{f(l)}^{s(l)}-\bar{v}_{d(l)}^{s(l)}=\bar{u}_l$

$$\gamma(x_l)=\begin{cases}\dfrac{df_l}{dx_l}(x_l), & l=M+1,\ldots,L \\[2mm] y_k\dfrac{df_l^1}{dx_l}(x_l)+(1-y_k)\dfrac{df_l^0}{dx_l}(x_l), & l=1,2,\ldots,M, \ l\in L(k).\end{cases}$$

Consequently, defining

$$\Delta_l^i=f_l^i(\tilde{x}_l^{s(l)})-\bar{u}_l\tilde{x}_l^{s(l)}, \quad i=1,2, \quad l=1,2,\ldots,M$$

$$\Delta_l^0=f_l(\tilde{x}_l^{s(l)})-\bar{u}_l\tilde{x}_l^{s(l)}, \quad l=M+1,\ldots,L$$

we obtain

$$L^*(y;\bar{u})=\alpha+\sum_{k=1}^{K}\beta_k y_k$$

where

$$\alpha=\sum_{l=1}^{L}(\bar{u}_l\bar{x}_l+\Delta_l^0)$$

$$\beta_k=\sum_{l\in L(k)}(\Delta_l^1-\Delta_l^0), \quad k=1,2,\ldots,K.$$

c) *Solution of Relaxed Master Problem*: Each relaxed master problem to be solved is a linear mixed pseudo-boolean programming problem.

Problem (RM) $\qquad\qquad$ minimize y_0
$${\scriptstyle y_0,y}$$

subject to

$$b-\sum_{k=1}^{K}c_k y_k\geqq 0$$

$$y_0-\alpha^j-\sum_{k=1}^{K}\beta_k^j y_k\geqq 0, \quad j=1,2,\ldots,J$$

where α^j, β_k^j, $k=1,2,\ldots,K$ are defined as in Subsection b) above, relative to optimal multiplier vector u^j. We solve problem (RM) using Balas' additive algorithm [1].

IV. SOLUTION METHOD BY BRANCH-AND-BOUND

1. Branch-and-Bound Approach

As a mixed integer programming problem Problem (NIP) can be solved using the branch-and-bound approach. A branch-and-bound algorithm will iteratively build an arborescence structure, each node of which represents a partial solution to the mixed integer programming problem. For example, if there are 5 variables y_k a node z of the arborescence may represent partial solution $(0, 1, -, -, -)$. Only the first two components of the decision vector y have specified values of 0 or 1; the variables y_3, y_4, and y_5 are as yet unspecified. The index sets $I_0(z)$, $I_1(z)$, and $I(z)$ will be used to denote the components of the partial solution y represented by node z which are fixed at 0, fixed at 1, and unspecified, respectively. The set of successors of node z is the set $S(z)$ of all possible completions of the partial solution $(0, 1, -, -, -)$ represented by z in the arborescence, that is

$$S(z) = \{(y_1, y_2, \ldots, y_5) : y_1 = 0, y_2 = 1, y_i = 0 \text{ or } 1, i = 3, 4, 5\}.$$

Starting with all decision variables unspecified, a branch-and-bound algorithm iteratively selects a node z representing a partial solution and creates two new nodes, each connected to z by a directed branch, by partitioning $S(z)$ into two parts. Each time a new node z^1 is added to the arborescence, a lower bound on the objective function evaluated at all the solutions in $S(z^1)$ must be computed. Indeed, the primary concern in any branch-and-bound technique is the method of generating the lower bounds on the values of the successors of any given node. The most apparent method of computing such lower bounds suggested by Ochoa-Rosso [15], is to set all free or unspecified variables equal to one and find user optimal flows on the resulting network. There may be here a pitfall known as Braess' paradox [16]. However, this pitfall will not occur, if system optimal flows are used to evaluate network configurations. Furthermore, Leblanc [12] has shown that even when user optimal flows are used for network evaluation, a correct lower bound on the values of the successors of any node z can be computed by setting all unspecified or free variables equal to one and then choosing flows to minimize total travel cost in this resulting network. He also pointed out that whenever two new nodes are added to the arborescence, it is only necessary to compute one lower bound, since the other one has already been computed. In the next subsection, we propose another useful method of computing lower bounds; the bounding procedure thus obtained actually is more powerful than those suggested by Ochoa-Rosso and Leblanc.

2. New Bounding Procedure

We are now discussing a new bounding procedure based on the superadditive property of the objective function with respect to zero-one variables y_k, $k=1, 2, \ldots, K$, as formally expressed in the following propositions.

We first note that a network configuration can always be characterized by the characteristic vector y of its set of implemented projects E. Let us denote by $H^{(0)}$ the network configuration where all projects are implemented; that is

$$y_j^0 = 1, \quad j = 1, 2, \ldots, K$$

$$f_l(x_l, y_j^0) = f_l^1(x_l), \quad l = 1, 2, \ldots, M, \quad l \in L(j)$$

and by $H^{(k)}$, $k=1, 2, \ldots, K$ the network configuration where all projects, except project k, are implemented; that is

$$y_j^k = 1, \quad j = 1, 2, \ldots, K, \quad j \neq k$$

$$y_k^k = 0$$

$$f_l(x_l, y_j^k) = \begin{cases} f_l^1(x_l), & l = 1, 2, \ldots, M, \quad l \notin L(k) \\ f_l^0(x_l), & l \in L(k). \end{cases}$$

Let

$V^k =$ optimal value $v(y^k)$ of the network evaluation criterion as applied to $H^{(k)}$, $k = 0, 1, 2, \ldots, K$

$\bar{x}_l^k =$ total flow along arc l in the system optimal flows of $H^{(k)}$

$Q_k = V^k - V^0$

$$F_k = \sum_{l \in L(k)} f_l^0(\bar{x}_l^k) - f_l^1(\bar{x}_l^k)$$

Proposition 1. $V^k - V^0 \geqq F_k$, $\quad k = 1, 2, \ldots, K$.

Proof. Consider the transition from $H^{(0)}$ to $H^{(k)}$. During this transition the cost function of each arc l, $l \in L(k)$, changes from f_l^1 to f_l^0; then, because of the assumed properties of f_l^0 and f_l^1 mentioned in Subsection III.1, some of the flow $\bar{x}_l^0$ along arc l in the system optimal flows of $H^{(0)}$ must be deviated elsewhere to give rise to the flow value $\bar{x}_l^k$ along arc l in the system optimal flows of $H^{(k)}$; that is

$$\bar{x}_l^0 \geqq \bar{x}_l^k, \quad l \in L(k)$$

$$\bar{x}_l^0 \leqq \bar{x}_l^k, \quad l = 1, 2, \ldots, L, \, l \notin L(k).$$

Consequently, the changes from f_l^1 to f_l^0, $l \in L(k)$, cause an increase of at least

$$f_l^0(\bar{x}_l^k) - f_l^1(\bar{x}_l^k)$$

in the contribution of each arc l, $l \in L(k)$, to the objective function, while the contribution of other arc is not decreased. Hence

$$V^k - V^0 \geqq \sum_{l \in L(k)} [f_l^0(\bar{x}_l^k) - f_l^1(\bar{x}_l^k)] = F_k$$

for all $k = 1, 2, \ldots, K$.

Proposition 2. *If $E^1 = E \cup \{k\}$, $k \notin E$, then*

$$v(y) \geqq v(y^1) + F_k.$$

Proof. Let $\bar{x}_l$ be total flow on arc l in the system optimal flow pattern of network characterized by E. It follows from Proposition 1 that

$$v(y) - v(y^1) \geqq \sum_{l \in L(k)} [f_l^0(\bar{x}_l) - f_l^1(\bar{x}_l)].$$

Consider now the transition from network $H^{(k)}$ to network E. During this transition some projects are excluded, and this results in some deviation elsewhere in network E of flows along arcs involved in excluded pojects to achieve a new system optimal flow pattern. Since all arcs in $L(k)$ do not involve in excluded projects one must then have

$$\bar{x}_l \geqq \bar{x}_l^k, \quad l \in L(k)$$

Consequently, it follows from the assumed properties of f_l^0 and f_l^1 mentioned in Subsection III.1 that

$$v(y) - v(y^1) \geqq \sum_{l \in L(k)} [f_l^0(\bar{x}_l) - f_l^1(\bar{x}_l)] \geqq$$

$$\geqq \sum_{l \in L(k)} [f_l^0(\bar{x}_l^k) - f_l^1(\bar{x}_l^k)] = F_k.$$

Proposition 3. *The following inequalities hold for any network characterized by E, the characteristic vector of which is $y = (y_1, y_2, \ldots, y_K)$*

$$v(y) \geqq V^0 + \underset{\substack{j=1,\ldots,K \\ y_j=0}}{\text{maximum}} \left[Qj + \sum_{\substack{k=1 \\ k \neq j}}^{K} (1 - y_k)F_k \right] \geqq V^0 + \sum_{k=1}^{K} (1 - y_k)F_k.$$

Proof. This result follows from Proposition 2, inductively.

Proposition 3 above provides us with a lower bound on the total travel time for a particular selection of projects which is stronger than the bound used by Ochoa-Rosso and Leblanc. Here, one actually is looking more "downwardly" into the arborescence, trying to satisfy budget constraint by excluding somewhat less profitable projects, and evaluating the effect on the objective function of such an exclusion. For illustration purposes let us locate at the root of the arborescence where all variables y_k are

256

unspecified. Then, by Proposition 3 a lower bound on $v(y_1, y_2, \ldots, y_k)$ is obtained by solving the following knapsack problem

$$\text{minimize } V^0 + \sum_{k=1}^{K} F_k(1 - y_k)$$

subject to

$$\sum_{k=1}^{K} y_k c_k \leqq b$$

$$y_k = 0 \quad \text{or} \quad 1, \quad k = 1, 2, \ldots, K.$$

One also may use as lower bound an optimal fractional solution to the knapsack problem; that is

$$y_k = 0, \quad k < r$$
$$y_k = 1, \quad k > r$$

$$y_r = \frac{1}{c_r}\left(b - \sum_{k=r+1}^{K} c_k\right),$$

where the variables are renumbered so that

$$F_1/c_1 \leqq F_2/c_2 \leqq \ldots \leqq F_K/c_K$$

and r is the least integer, $0 < r \leqq K$, for which $\sum_{k=r+1}^{K} c_k \leqq b$.

This bounding process is a generalization of the result found in [7] to networks where congestion is not negligible, and applies of course to any node of the arborescence. It provides us with a lower bound greater than Ochoa-Rosso and Leblanc's bound by an amount equal to optimal value of knapsack problem defined on unspecified variables and remaining, uncommitted budget at the node considered. Furthermore, the first inequality of Proposition 3 does give an even more powerful lower bound

$$V^0 + \operatorname*{minimum}_{y:\, cy \leqq b} \left[\operatorname*{maximum}_{j:\, y_j = 0} \left\{ Q_j + \sum_{\substack{k=1 \\ k \neq j}}^{K} (1 - y_k) F_k \right\} \right]$$

which is unfortunately much more complicated to exploit. Hence, in the next subsection we will only pay attention to a manner of exploiting, within the framework of branch-and-bound approach, the lower bound obtained from the second inequality of Proposition 3 using optimal fractional solution of associated knapsack problem.

3. Branch and Backtrack Algorithm

Suppose that we have on hand a trial solution $y^0 = (y_1^0, \ldots, y_k^0)$, Z^0 is the corresponding value of the objective function and

$$\lambda^0 = \sum_{k=1}^{K} c_k - b - \sum_{k=1}^{K} c_k y_k^0.$$

Let us denote by $S(z)$ the evaluation of arborescence node z obtained from fractional optimal solution to associated knapsack problem, and by $\lambda(z)$ the difference between $\left(\sum_{k=1}^{K} c_k - b\right)$ and the sum of all c_k for which specified variable y_k has the value zero at node z. Formally, a branch-and-backtrack algorithm for network improvement problems may be stated as follows:

(1) Set $l=1$, $\lambda = \sum_k c_k - b$, and all components of y equal to 2, i.e., all variables are unspecified. Solve the associated knapsack problem for a fractional optimal solution and obtain $S(l)$. If $S(l) \geqq Z^0$, then the trial solution y^0 is optimal; stop. Otherwise, set the fractional component y_r of y to zero and $R(l) = r$. Go to Step 2.

(2) Solve associated knapsack problem with l specified components of y added as constraints and obtain $S(l+1)$.

(a) If $S(l+1) \geqq Z^0$, then go to Step 3. Otherwise go to Step 2 (b).

(b) Set $l=l+1$, $x_r = 0$, $R(l) = r$, $\lambda = \lambda - c_r$.

If $\lambda \leqq 0$, then go to Step 2 (c). Otherwise go to Step 2 (a).

(c) Compute $Z = v(\bar{y})$ with all unspecified components of y set to 1.

If (Z, λ) is lexicographically less than (Z^0, λ^0), then replace (Z^0, λ^0) by (Z, λ), and y^0 by $\bar{y}$. Go to Step 3.

(3) (a) If the $r = R(l)$ component of y is equal to zero, then change x_r to one, set $\lambda = \lambda + c_r$, and go to Step 2 (a). Otherwise, go to Step 3 (b).

(b) If $l=1$, then y^0 is an optimal solution; Stop.

Otherwise change the $R(l)$ component of y to 2, set $l=l-1$, and go to Step 3 (a).

V. COMPUTATIONAL RESULTS

In order to shed some insights into the computational performances of two approaches proposed to obtain an optimal solution to a network improvement problem, we have implemented and tested the generalized Benders' partition (GBP) and branch-and-backtrack (B & B) methods. These two programs were written in FORTRAN IV language using IBM's FORTRAN IV level G Compiler under control of the operating system OS/360-MVT. During solution processes using GBP or B & B method one very often needs to evaluate the objective function for a network configuration represented by a fixed investment decision vector y. The accomplishment of this computational task, which consists in solving a convex cost multicommodity flow problem without explicit capacity constraints, is materialized by means of an adapted convex simplex algorithm [4] for the reasons mentioned in Subsection III.2 (a). In the case of the GBP method it is also necessary to solve relaxed master problems

258

which are special mixed pseudo-boolean programs. Balas' additive algorithm[1] was used to find optimal solutions for such problems.

For testing purposes two networks have been considered: Sioux Falls network in [11] and Hull network in [14]. Computational results for all test problems run on an IBM System 360 Model 50 using Ampex ECS under OS/360-MVT are presented in Tables 1 and 2. Results in Table 1 permit somewhat rough estimates of computa-

Table 1. Computational Performance Comparison between GBP and B & B Methods

Number of projects K	Ratio b/S_c	GBP method		B&B method	
		Number of iterations	[Computing time	Number of evaluations	Computing time
5	.694	2	206	7	256
10	.694	4	238	17	458
15	.694	6	333	35	816

Table 2. Computational Performances of GBP Method

	Number of projects K	Ratio b/S_c	Number of iterations	Computing time	Distance to optimal solution
A 24 node, 76 arc and 528 O/D pair network	10	.694	4	238	0.55
	10	.520	7	320	0.51
	15	.694	6	333	1.00
	15	.520	5	375	0.02
A 155 node, 376 arc and 720 O/D pair network	10	.600	9	598	0.99
	15	.600	18	1649	1.76
	25	.600	5	3448	10.31

tional performances of the GBP and B & B methods for small size problems based on a 24 node, 76 arc and 528 origin-destination pair network, called Sioux Falls network. We note that all computing times are measured in CPU seconds, and that an iteration of the GBP method consists in solving a relaxed master problem for a potential y and computing system optimal flows for the network configuration represented by that y. On the one hand, the number of iterations needed to reduce the upper limit of the distance to the optimal solution (measured by ratio of the difference between the upper and lower bounds on the objective function by the lower bound itself) to an acceptable tolerance (say, one per cent) indicates the overall computational effort required by the GBP method. On the other hand, the computational effort required by the B & B method can be naturally measured by the number of network evaluations, i.e., the number of traffic assignment problems solved. From computational point of view two relevant parameters characterising a problem are the number of projects K and the ratio of budget b by sum of all project costs S_c. The results obtained show that at least for small size problems the GBP method seems very competitive to the B & B method since it requires a small number of iterations. However, results in Table 2 indicate that for problems of somewhat moderate size, computing time needed to solve relaxed master problems grows very rapidly with respect to the number of projects. Computing time quoted will then be greatly reduced if efficient algorithms for solving relaxed master problems are available. One possible candidate for purposefully replacing Balas' additive algorithm can be found in reference [4]. Moreover to reveal the overall convergence characteristic of the GBP method we have plotted in Fig. 1 the limit of the distance to the optimal solution as function of the number of iterations.

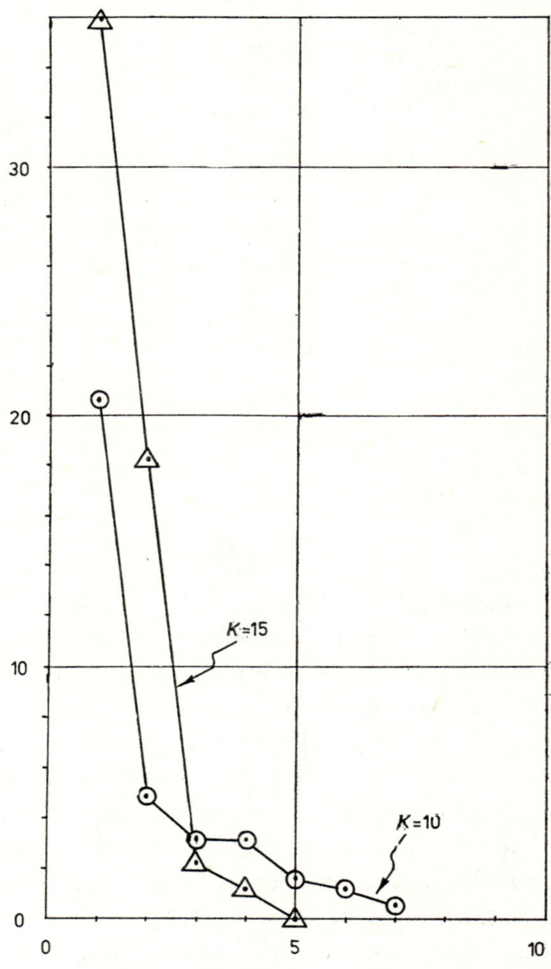

Fig. 1a. Convergence characteristic of GBP Method (Case of Sioux Falls Network with $b/S_c = 0.520$)

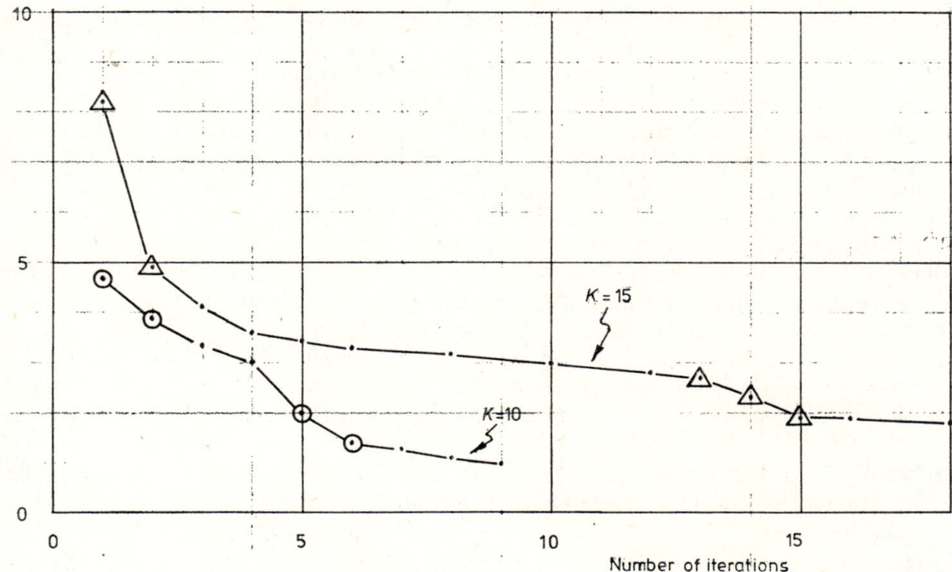

Fig. 1b. Convergence Characteristic of GBP Method (Case of Hull Network with $b/S_c=0.600$)

Last, but not least, it is worthy to note that early indications of efficiency discussed above must be regarded as suggestive rather than conclusive, due to the limited range and size of problems so far examined and to largely unexplored alternatives for implementing the GBP and B & B methods. Nevertheless, it seems reasonable to conclude that the GBP method may permit us to solve problems of relatively realistic size (i.e., problems of selecting an optimal combination amongst about 30 improvement projects defined on a network of a few thousand arcs and hundred nodes) without requiring very excessive computing time (i.e., not more than several hours on a medium size computing system).

REFERENCES

[1] Balas, E.: An Additive Algorithm for Solving Linear Programs with Zero-One Variables, *Operations Research,* 13 (1965), 517–545.

[2] Bruynooghe, M.: An Optimal Method of Choice of Investments in a Transport Network, Paper presented at the Planning & Transport Research & Computation Seminars on Urban Traffic Mode Research, London, May 8–12, 1972.

[3] Geoffrion, A. M.: Generalized Benders Decomposition, *J. of Optimization Theory and Applications,* 10 (1972), 237–260.

[4] Geoffrion, A. M.: An Improved Implicit Enumeration Approach for Integer Programming, *Operations Research,* 17 (1969), 437–454.

[5] Glover, F., Karney, D. and Klingman, D.: The Augmented Predecessor Index Method for Locating Stepping Stone Paths and Assigning Dual Prices in Distribution Problems, *Transportation Science,* 6 (1972), 171–180.

[6] Glover, F., Klingman, D. and Stutz, J.: Augmented Threaded Index Method for Network Optimization, *INFOR*, 12 (1974), 293–298.

[7] Hoang Hai Hoc: A Computational Approach to the Selection of an Optimal Network, *Management Science*, 19 (1973), 488–498.

[8] Hoang Hai Hoc: Implementation and Use of Nonlinear Cost Multicommodity Flow Subroutines, Paper Submitted to the ACM–SIGMAP Bicentennial Conference on Mathematical Programming, Gaithersburg, Maryland, Nov. 29–Dec. 1, 1976.

[9] Kleinrock, L.: Analytic and Simulation Methods for Computer Network Design, 1970 SJCC, AFIPS Conf. Proc., 36 (1970), 569–579.

[10] Leblanc, L. J. and Morlok, E. K.: An Analysis and Comparison of Behavioral Assumptions in Traffic Assignment, Paper presented at the International Symposium on Traffic Equilibrium Methods, Montreal, Canada, Nov. 21–23, 1974.

[11] Leblanc, L. J., Morlok, E. K. and Pierskalla, W. P.: An Efficient Approach to Solving the Road Network Equilibrium Traffic Assignment Problem, *Transportation Research*, 9 (1975), 309–318.

[12] Leblanc, L. J.: An Algorithm for the Discrete Network Design Problem, *Transportation Science*, 9 (1975), 183–199.

[13] Leventhal, T. L., Nemhauser, G. L. and Trotter, L. E.: Traffic Assignment: Computer Program II, DT–FHA Report No. 25, Department of Operations Research, Cornell University, Ithaca, N.Y., Nov. 1971.

[14] Nguyen, S.: An Algorithm for the Traffic Assignment Problem, *Transportation Science*, 8 (1974), 203–216.

[15] Ochoa-Rosso, F.: Applications of Discrete Optimization Techniques to Capital Investment and Network Synthesis Problems, Research Report No. R68–42, Department of Civil Engineering, M.I.T., Cambridge, Mass., Jan. 1968.

[16] Steenbrink, P. A.: *Optimization in Transport Networks*, Wiley, New York, N.Y., 1974.

THEORY OF FLOWS IN CONTINUA AS APPROXIMATION TO FLOWS IN NETWORKS

MASAO IRI

(Tokyo, Japan)

INTRODUCTION

The theory of flows in networks is now well established from the standpoints of theory as well as of application, and many effective practical algorithms are being developed for dealing with problems of flows in networks [3], [4], [5].

In the present paper, the author tries to mould a theory of flows in continua on the theory of flows in networks, and discusses the use of the theory in applications. Since everything should be coordinate-independent, the tensorial characters of the concepts appearing in the theory are important.

A continuum model of flow problems has been considered by R. E. Gomory and T. C. Hu, and according to the materials expounded in Chapter 12 of Hu's book [4], the continua they considered are isotropic and they pointed out that, in order to avoid pathological phenomena, fairly complicated continuity assumptions are needed. In the following, we shall assume sufficient "smoothness" (and "stability" in Rockafellar's sense [9]) for the functions and fields concerned, but we shall *not* assume "isotropy". In fact, "anisotropy" is a most interesting issue in the theory of flows in continua. The works of G. Monge and P. Appell on "déblai and remblai" [1], [7] and of L. Kantorovitch on "translocation of masses" [6], which have been frequently referred to in this context, also deal with minimum-cost flow problems of a special kind on an isotropic continuum.

1. DISCRETE APPROXIMATION OF CONTINUA AND CONTINUOUS APPROXIMATION OF DISCRETA

Present-day numerical analysts usually speak of "discretization" or "discrete approximation" of continuous field phenomena, such as stress and strain fields of elastic continua, governed by partial differential equations of some sorts. However, the concept of elastic continua itself is a kind of approximation to microscopically discrete lattice structures of constituent molecules. Thus, we are concerned with two-stage approximation: "discretum to continuum" and then "continuum to discretum".

The importance of this kind of two-stage approximation may be recognized with regard to flow theory.

We shall not mention the viewpoint that a flow network is an idealization of a highly lumped case of actually distributed flow field. However, if we consider a nation-wide network of roads, it is no doubt better to treat the network of inter-city arterial roads as an ordinary (discrete) network, whereas the intra-city roads are too fine and too dense to treat as a network, so that we are tempted to regard the latter kind of roads approximately as a continuum with some appropriate characteristics. In this way we are led to a compound system consisting partly of continua and partly of discrete networks.

If we succeed in establishing a way of dealing with flows in continua which is in complete analogy with the way of dealing with flows in networks, then it will be easy to combine two kinds of subsystems, one continuous and the other discrete, into a compound one, because we have a number of mathematical tools for that purpose (such as Schwartz' distributions [11], de Rham's currents [8], Sato's hyperfunctions, etc.). Thus, we shall devote ourselves to the formalization of flows in continua in the sequel.

2. MATHEMATICAL TOOLS

Basically, we need no novel mathematical theory or technique to establish the theory of flows in continua, but we have only to "assemble", in a suitable manner, mathematical tools ready to hand.

As we have already mentioned, the formulae for a compound system of continua and discreta may be formulated in a single expression by the aid of de Rham's currents or the like.

In order to guarantee the invariance of the expressions for continua under coordinate transformations, we may resort to the classical formalism of tensors [10], which is suitable also for treating the anisotropic character of the continuum.

The variational methods which are necessary for formulating the typical flow problems in continua may be found in classical works such as [2], and other more recent methodology concerning convexity and duality in [9].

3. BASIC CONCEPTS ON FLOWS IN CONTINUA
AND THEIR TENSORIAL EXPRESSIONS

The continuum to be considered may, in general, be an n-dimensional differentiable manifold, but, for the sake of simplicity, we shall confine ourselves to a simply connected subregion V of the n-dimensional Euclidean space with a sufficiently smooth

$(n-1)$-dimensional surface ∂V. (In practical applications, we ordinarily have $n=2$ or 3.)

As for the basic concepts regarding flows, tensions and characteristics of the continuum, we shall try to make natural extensions of those for any ordinary network, based on the formalisms delineated in the author's book [5].

3.1. Flow. A *flow* $\xi^{\varkappa}$ is a field of contravariant vector density of weight 1 in V with no divergence:

$$(1) \qquad \frac{\partial \xi^{\varkappa}}{\partial x^{\varkappa}} = 0,$$

which is the continuous counterpart of a flow in an arc (or a branch [5]). Here, as well as in the following, we denote by $x^{\varkappa}$ the coordinates of a point in V, and we adopt the summation convention, i.e. (1) would be written as

$$(1') \qquad \sum_{\varkappa=1}^{n} \frac{\partial \xi^{\varkappa}}{\partial \eta^{\varkappa}} = 0$$

in the ordinary expression without that convention.

3.2. Tension. A *tension* $\eta_{\varkappa}$ is a field of covariant vector in V, which is expressible as (the negative of) the gradient of a scalar field ζ called its *potential*:

$$(2) \qquad \eta_{\varkappa} = -\frac{\partial \zeta}{\partial x^{\varkappa}}\,.$$

A tension here is the continuous counterpart of a tension across an arc in a network, and a potential is that of a potential at a vertex (or a node [5]).

3.3. Cost. The cost characteristic with respect to flow, $\varphi(x, \xi)$, of the continuum, is a function of position $x^{\varkappa}$ and flow $\xi^{\varkappa}$ at x which is convex in ξ and whose value is a scalar density of weight 1.

The *cost characteristic with respect to tension, $\psi(\varkappa, \eta)$,* of the continuum is a function of position $x^{\varkappa}$ and tension $\eta_{\varkappa}$ at x which is convex in η and whose value is a scalar density of weight 1.

The two functions $\varphi(x, \xi)$ and $\psi(x, \eta)$ at the same position are conjugate to each other:

$$(3) \qquad \begin{aligned} \psi(x, \eta) &= \max_{\xi}\, [\eta_{\varkappa}\xi^{\varkappa} - \varphi(x, \xi)], \\ \varphi(x, \xi) &= \sup_{\eta}\, [\xi^{\varkappa}\eta_{\varkappa} - \psi(x, \eta)] \end{aligned}$$

or

$$(3') \qquad \frac{\partial \varphi}{\partial \xi^{\varkappa}} = \eta_{\varkappa}, \quad \frac{\partial \psi}{\partial \eta_{\varkappa}} = \xi^{\varkappa}$$

(where partial derivatives should be interpreted as subgradients if necessary). φ and ψ are the continuous counterparts of the cost functions associated with arcs of a network.

3.4. Capacity. When we speak of the capacities of an arc in a network, we essentially consider a (closed) interval within which the value of the flow in the arc should be confined. The *capacity convex* of the continuum at position x is a closed convex set $C(\varkappa)$ (containing the origin) of the vector space of ξ's at x. Physical considerations are necessary in order to claim that the set of feasible flows $C(x)$ at each position be convex, but they are omitted here.

Let us consider an $((n-1)$-dimensional) *surface element* at position x, which is represented by a covariant vector density of weight -1, $S_\varkappa$. For a given $S_\varkappa$, there is (or are) a vector (or vectors) $\xi^\varkappa$ on the boundary of $C(x)$ such that the surface element $S_\varkappa$ is (parallel to) a supporting hyperplane of $C(x)$ at x. We shall denote the $\xi^\varkappa$('s) by $\xi^\varkappa(S)$ in the following. The dependence of $\xi^\varkappa(S)$ on S is homogeneous of degree 0, i.e. $\xi^\varkappa(\alpha S)=\xi^\varkappa(S)$ for any $\alpha>0$. The flux of flow $\xi^\varkappa$ which crosses surface element $S_\varkappa$ is equal to $\xi^\varkappa S_\varkappa$, and the $\xi^\varkappa(S)$ is also characterized by the relation:

$$(4) \qquad \sup_{\xi \in C} \xi^\varkappa S_\varkappa = \xi^\varkappa(S)S_\varkappa.$$

The *polar* $C^*(x)$ of $C(x)$ is the convex set of the vector space of S's at x such that $\xi^\varkappa(S)S_\varkappa \leq 1$, i.e. $C^*(x)$ is the set of those surface elements across which at most one unit of flux can flow so long as the flow is in $C(x)$.

It should be noted that, if we define $\varphi_C(x, \xi)$ as the function such that

$$(5) \qquad \varphi_C(x,\xi)=\begin{cases} 0 & \text{for } \xi \in C(x), \\ \infty & \text{for } \xi \notin C(x), \end{cases}$$

then we have conjugate function $\psi_C(x, \eta)$ satisfying

$$(6) \qquad \psi_C(x, \eta)=\xi^\varkappa(\eta)\eta_\varkappa.$$

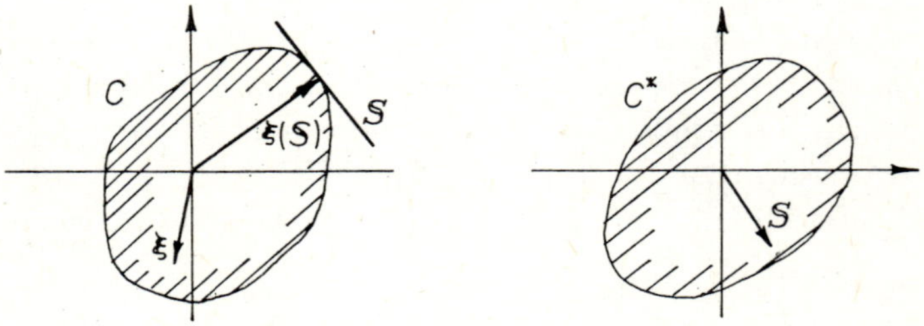

Fig. 1. Capacity convex and its polar

3.5. Velocity or distance. The concept of distance (or length) of an arc may be formulated in the case of continuum as follows. The *velocity convex* of the continuum at position x is a closed convex set $D(x)$ (containing the origin) of the vector space of certain contravariant vectors $v^\varkappa$, and is interpreted as the set of $v^\varkappa$'s such that a "particle" can move in the continuum with velocity $v^\varkappa$ at x. (The physical discussion about why $D(x)$ must be convex is omitted.) The polar $D^*(x)$ of $D(x)$ is the set of tensions η_x (at x) such that

$$(7) \qquad \sup_{v \in D} v^x \eta_x \leqq 1,$$

i.e. that it takes at least one unit of time for a particle with a velocity in $D(x)$ to go from one to the other of the pair of hyperplanes defined by η_x (see [10]). (In other words, an η_x on the boundary of $D^*(x)$ represents the pair of hyperplanes such that the second hyperplane is the wave front of the particles which were on the first a unit time before and which move with velocities in $D(x)$.) Let us put

$$(8) \qquad \sup_{\eta \in D^*} l^x \eta_x = l^x \eta_x(l),$$

where we obviously have

$$(9) \qquad \eta_x(\alpha l) = \eta_x(l)$$

for any $\alpha > 0$. Since $D(x)$ is the polar of $D^*(x)$, and since $v^x \eta(v) = 1$ if $v^x \in \partial D$, $l^x \eta_x(l)$ is equal to the shortest time in which a particle can pass through vector l^x.

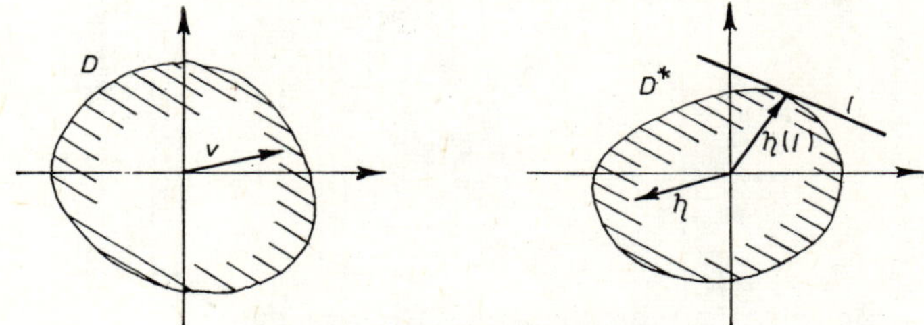

Fig. 2. Velocity convex and its polar

If $\psi_{D^*}(x, \eta)$ is defined as

$$(10) \qquad \psi_{D^*}(x, \eta) = \begin{cases} 0 & \text{for} \quad \eta \in D(x), \\ \infty & \text{for} \quad \eta \notin D(x), \end{cases}$$

then the conjugate $\varphi_{D^*}(x, \xi)$ satisfies

$$(11) \qquad \varphi_{D^*}(x, \xi) = \xi^x \eta_x(\xi).$$

4. FORMALISMS OF FLOW PROBLEMS IN CONTINUA

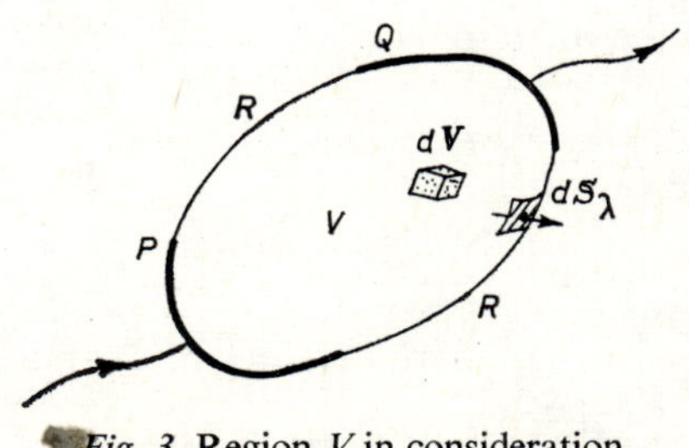

Fig. 3. Region V in consideration

For the sake of simplicity we shall take up the "two-terminal problems" for a region V with entrance (source) P and exit (sink) Q, where $P \cap Q = \emptyset$ and $P \cup Q \subset \partial V$. We put $R = \partial V - (P \cup Q)$. We shall denote the n-dimensional volume element (scalar density of weight -1) of V by dV and the $(n-1)$-dimensional surface element (covariant vector density of weight -1) of ∂V (oriented outward) by $dS_\varkappa$.

4.1. Minimum-cost flow/tension. Let us suppose that a given conjugate pair $\varphi(x, \xi)$ and $\psi(x, \eta)$ of cost characteristics satisfy certain smoothness and/or stability conditions [9]. The problem is to determine the flow ξ^K so as to

$$(12) \qquad \text{minimize } \Phi[\xi] \equiv \int_V \varphi(x, \xi)\, dV$$

subject to the constraints:

$$(13.1) \qquad \frac{\partial \xi^\varkappa}{\partial x^\varkappa} = 0 \quad \text{in } V,$$

$$(13.2) \qquad \Xi = - \int_P \xi^\varkappa\, dS_\varkappa,$$

$$(13.3) \qquad \Xi = \int_Q \xi^\varkappa\, dS_\varkappa,$$

$$(13.4) \qquad \xi^\varkappa dS_\varkappa = 0 \quad \text{on } R,$$

where Ξ is a given constant. Note that (13.3) is dependent on (13.1), (13.2) and (13.4).

The dual to the above problem is to determine the tension $\eta_\varkappa$, the potential ζ and the constants (i.e. not fields) Z_P and Z_Q so as to

$$(14) \qquad \text{minimize } \Psi[\eta, \zeta, Z_P, Z_Q] \equiv - \Xi(Z_P - Z_Q) + \int_V \psi(x, \eta)\, dV$$

subject to the constraints:

$$(15.1) \qquad \eta_\varkappa = - \frac{\partial \zeta}{\partial x^\varkappa} \quad \text{in } V,$$

$$(15.2) \qquad \zeta = Z_P \quad \text{on } P,$$
$$(15.3) \qquad \zeta = Z_Q \quad \text{on } Q.$$

The conditions to be satisfied by the solution $\xi^\varkappa$ of the primal problem and by the solution $\eta_\varkappa, \zeta, Z_P, Z_Q$ of the dual are (13.1)$\sim$(13.4) and (15.1)$\sim$(15.3) and

$$(16) \qquad \frac{\partial \varphi(x, \xi)}{\partial \xi^\varkappa} = \eta_\varkappa, \quad \text{or equivalently,} \quad \frac{\partial \psi(x, \eta)}{\partial \eta_\varkappa} = \xi^\varkappa.$$

Here, ζ, Z_P and Z_Q play the roles of the Lagrangean multipliers for the primal constraints (13.1), (13.2) and (13.3), respectively, and $\xi^\varkappa$ plays the role of that for the dual constraint (15.1). As for the infimum values of Φ and Ψ, we have

$$(17) \qquad \inf \Phi + \inf \Psi = 0.$$

If we delete the conditions (13.2) and (13.3) from the primal problem and add the term $Z \int_P \xi^\varkappa \, dS_\varkappa$ to the Φ with Z a given constant, then the term $-\Xi(Z_P - Z_Q)$ is deleted from the Ψ of the dual problem and the new condition $Z_P - Z_Q = Z$ is introduced.

4.2. Maximum flow. When a capacity convex $C(x)$ is given everywhere, the maximum-flow problem is to determine the flow $\xi^\varkappa$ so as to

$$(18) \qquad \text{maximize} -\int_P \xi^\varkappa \, dS_\varkappa = \int_Q \xi^\varkappa \, dS_\varkappa$$

subject to the constraints:

$$(19.1) \qquad \frac{\partial \xi^\varkappa}{\partial x^\varkappa} = 0 \quad \text{in } V,$$

$$(19.2) \qquad \xi^\varkappa dS_\varkappa = 0 \quad \text{on } R,$$

$$(19.3) \qquad \xi^\varkappa \in C(x) \quad \text{in } V.$$

By making use of the function φ_C defined by (5), the constraint (19.3) can be incorporated in the objective function to convert the original problem into that of

$$(20) \qquad \text{minimizing } \Phi[\xi] \equiv \int_P \xi^\varkappa \, dS_\varkappa + \int_V \varphi_C(x, \xi) \, dV$$

subject to the constraints (19.1) and (19.2).

The dual to this problem is then to

$$(21) \qquad \text{minimize } \Psi[\eta, \zeta] \equiv \int_V \psi_C(x, \eta) \, dV$$

subject to the constraints:

$$(22.1) \qquad \eta_\varkappa = -\frac{\partial \zeta}{\partial x^\varkappa} \quad \text{in } V,$$

$$(22.2) \qquad \zeta = 1 \quad \text{on } P,$$

$$(22.3) \qquad \zeta = 0 \quad \text{on } Q.$$

The conditions to be satisfied by the solution $\xi^\varkappa$ of the primal problem and by the solution $\eta_\varkappa$, ζ of the dual are (19.1)$\sim$(19.3) and (22.1)$\sim$(22.3) and

$$(23) \qquad \text{``if } \eta_\varkappa \neq 0, \quad \text{then } \xi^\varkappa = \xi^\varkappa(\eta)\text{'',}$$

where (23) is equivalent to (16) in this case. Furthermore, in the equation

$$(24) \qquad \inf \Phi + \min \Psi = 0$$

where $\inf \Phi$ is the supremum value of flux feasible from entrance P to exit Q through V, whereas $\min \Psi$ may be interpreted as follows.

For an arbitrary surface S in V which separates Q from P (oriented from P to Q,) we define the value $\Omega(S)$ by

$$(25) \qquad \Omega(S)= \int_S \xi^\varkappa (\mathrm{dS}) \, \mathrm{dS}_\varkappa,$$

where $\mathrm{dS}_\varkappa$ is the element of S. $\Omega(S)$ is the maximum value of flux which can cross S under the capacity constraint around S, and the infimum value of $\Omega(S)$ with respect to S is to be compared to the concept of "minimum cut" in a capacitated network, and by obvious calculation, we have

$$(26) \qquad \inf_S \Omega(S) \geqq \sup_\xi (-\Omega).$$

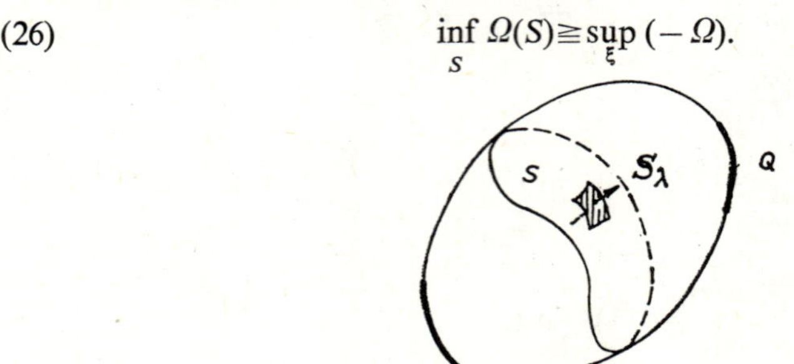

Fig. 4. Surface S separating entrance P from exit Q

On the other hand, for an arbitrary tension $\eta_\varkappa$ and its potential ζ satisfying $(22.1)\sim\sim(22.3)$, if we denote the surface $\zeta=$const. (oriented from P to Q) by S_ζ, Ψ may be expressed as follows (cf. (6) and Fig. 5):

$$(27) \qquad \Psi= \int_V \psi(\varkappa, \eta) \, \mathrm{dV} = \int_V \xi^\varkappa(\eta)\eta_\varkappa \, \mathrm{dV} = \int_{\zeta=0}^{\zeta=1} \mathrm{d}\zeta \int_{S_\zeta} \xi^\varkappa (\mathrm{dS}) \, \mathrm{dS}_\varkappa =$$

$$= \int_0^1 \Omega(S_\zeta) \, \mathrm{d}\zeta.$$

Fig. 5. Volume element with reference to constant-potential surfaces

$$\eta_\varkappa \, \mathrm{dV} = \mathrm{d}\zeta \, \mathrm{dS}_\varkappa$$

$$\left(\eta_\varkappa = -\frac{\partial \zeta}{\partial x^\varkappa} \right)$$

In fact, with (22.1), we have

$$(28) \qquad \eta_\varkappa d\mathbf{V} = d\zeta dS_\varkappa.$$

Therefore, we have

$$\Psi \geqq \inf_\zeta \Omega(S_\zeta) \geqq \inf_S \Omega(S)$$

and

$$(29) \qquad \inf_{\eta,\zeta} \Psi \geqq \inf_S \Omega(S).$$

Hence, from (24), (26) and (29) it follows that

$$(30) \qquad \inf_S \Omega(S) = \sup_\xi (-\varPhi)$$

which is the continuous counterpart of the "maximum-flow minimum-cut theorem".

The above arguments are intuitive but not quite rigorous since φ_C and ψ_C, and, consequently, the solution $\eta_\varkappa$, ζ are not sufficiently smooth or even continuous in general. However, we could get the same conclusion in a rigorous way by smoothing unsmooth functions and regions and then taking limits.

4.3. Minimum route (or shortest path). When a velocity convex $D(x)$ is given everywhere in V, the minimum-route problem (or, shortest-path problem) is to find a path from a point on P to a point on Q along which it takes the minimum time for a particle to go from P to Q with velocities in $D(x)$. The problem stated in this form is mathematically formulated in terms of $D^*(x)$ as that of finding the path L from P to Q which

$$(31) \qquad \text{minimizes } \varSigma(L) \equiv \int_L \eta_\varkappa (dl)\, dl^\varkappa,$$

where dl^K is the line element (contravariant vector) along L, and $\inf_L \varSigma(L)$ is the "shortest distance" from P to Q.

In analogy with the "maximum-separation minimum-route theorem" (as it is called in [5]) in a network, we may consider the variational problem of

$$(32) \qquad \text{minimizing } -Z$$

subject to the constraints:

$$(33.1) \qquad \eta_K = -\frac{\partial \zeta}{\partial x^\varkappa} \quad \text{in } V,$$

$$(33.2) \qquad \zeta = Z \quad \text{on } P,$$

$$(33.3) \qquad \zeta = 0 \quad \text{on } Q,$$

$$(33.4) \qquad \eta_\varkappa \in D^*(x) \quad \text{in } V,$$

where Z is a constant, or, using ψ_{D*} of (10), the problem of

$$(34) \qquad \text{minimizing} \quad \Psi[\eta, \zeta, Z] \equiv -Z + \int_V \psi_{D*}(x, \eta)\, dV$$

subject to the constraints (33.1), (33.2) and (33.3). We shall call the supremum value of $-\Psi(=Z)$ the "maximum separation".

The dual to the above problem is then to

$$(35) \qquad \text{minimize} \quad \Phi[\xi] \equiv \int_V \varphi_{D*}(x, \xi)\, dV$$

subject to the constraints:

$$(36.1) \qquad \frac{\partial \xi^\varkappa}{\partial x^\varkappa} = 0 \quad \text{in } V,$$

$$(36.2) \qquad \int_P \xi^\varkappa\, dS_\varkappa = -1,$$

$$(36.3) \qquad \xi^\varkappa dS_\varkappa = 0 \quad \text{on } R.$$

The conditions to be satisfied by the solution $\xi^\varkappa$ of the dual problem and by the solution $\eta_\varkappa$, ζ, Z of the primal are $(33.1) \sim (33.4)$ and $(36.1) \sim (36.3)$ and

$$(37) \qquad \text{``if} \quad \xi^\varkappa \neq 0 \quad \text{then} \quad \eta_\varkappa = \eta_\varkappa(\xi),\text{''}$$

where (37) is equivalent to (16) in this case.

We may transform the expression of $\Phi[\xi]$ by means of (11) as

$$(38) \qquad \Phi = \int_V \varphi_{D*}(x, \xi)\, dV = \int_V \eta_\varkappa(\xi)\xi^\varkappa\, dV = -\int_P \xi^\varkappa\, dS_\varkappa \int_{L_x} \eta_\varkappa(\,dl)\, dl^\varkappa =$$

$$= -\int_P \Sigma(L_x)\xi^\varkappa\, dS_\varkappa,$$

where L_x is the "stream line" of $\xi^\varkappa$ starting from a point x on P (cf. Fig. 6). By virtue of (36.2), we have from (38)

$$(39) \qquad \Phi \geq \inf_{x \in P} \Sigma(L_x) \geq \inf_L \Sigma(L)$$

and

$$(40) \qquad \inf_\xi \Phi \geq \inf_L \Sigma(L).$$

On the other hand, it follows from $(33.1) \sim (33.4)$ that, for any path L from P to Q

$$(41) \qquad Z = \int_L \eta_\varkappa\, dl^\varkappa \leq \int_L \eta_\varkappa(\,dl)\, dl^\varkappa = \Sigma(L),$$

272

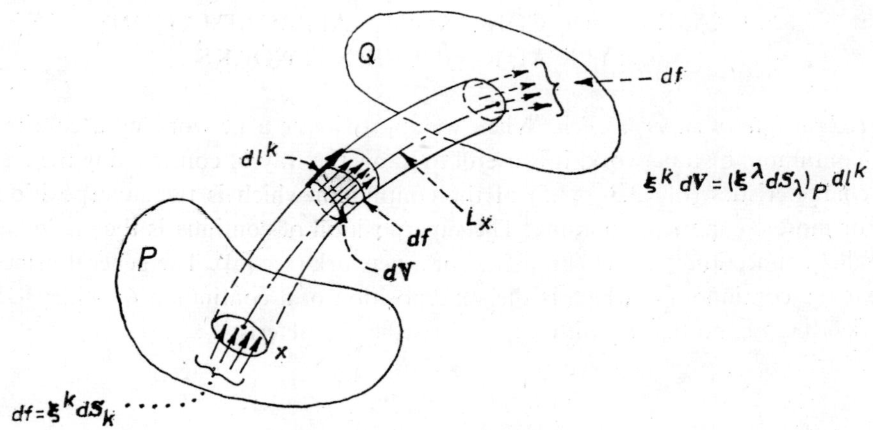

Fig. 6. Volume element with reference to stream lines

so that we have

(42)
$$-\inf \Psi = \sup_L Z \leqq \inf (L).$$

Combining (40) and (42) with equation $\min \Phi + \min \Psi = 0$, we have

(43)
$$\sup Z = \inf_L \Sigma(L).$$

The relation (43) is the continuous counterpart of "max-separation min-route theorem".

4.4. Other problems. We can formulate various problems in the same way by choosing appropriate cost characteristics of the continuum and boundary conditions of the fields. For example, Monge, Appell and Kantorovitch's problem of déblais and remblai [1], [6], [7] is the minimum-cost flow problem with $\varphi(x, \xi) = [(\xi^1)^2 + (\xi^2)^2 + (\xi^3)^2]^{\frac{1}{2}}$ and with distributed entrance and exit, where, instead of $(13.1) \sim (13.4)$ we put

$$\frac{\partial \xi^\varkappa}{\partial x^\varkappa} = \begin{cases} 0 & \text{in} & V - (V_0 \cup V_1), \\ \Xi_0 & \text{in} & V_0, \\ -\Xi_1 & \text{in} & V_1, \end{cases}$$

$$\xi^\varkappa\,dS_\varkappa = 0 \qquad \text{on} \quad \partial V$$

with given scalar fields Ξ_0 in V_0 and Ξ_1 in V_1 such that

$$\int_{V_0} \Xi_0\,dV = \int_{V_1} \Xi_1\,dV$$

$$(V_0 \cap V_1 = \emptyset).$$

5. EXAMPLES OF CONTINUOUS APPROXIMATIONS OF TYPICAL REGULAR NETWORKS

5.1. Principle of superposition. When we approximate a network by a continuum or a continuum by a network, it is useful to know the way of constructing the resulting characteristics (of §3.3~§3.5) of the continuum which is the superposition of two or more component continua. The superposition of continua is the analogue of "parallel connection" of arcs (branches) of a network (see [5]). The general principle is that the continuum $\mathcal{C}$ which is the superposition of a continuum $\mathcal{C}_1$ with characteristic $\psi_1(x, \eta)$ and that $\mathcal{C}_2$ with $\psi_2(x, \eta)$ has the characteristic

$$(44) \qquad \psi(x, \eta) = \psi_1(x, \eta) + \psi_2(x, \eta).$$

The $\varphi(x, \xi)$ of $\mathcal{C}$ is determined from those $\varphi_1(x, \xi)$ and $\varphi_2(x, \xi)$ of $\mathcal{C}_1$ and $\mathcal{C}_2$ by the formula:

$$(45) \qquad \varphi(x, \xi) = \min_{\xi = \xi_1 + \xi_2} [\varphi_1(x, \xi_1) + \varphi_2(x, \xi_2)].$$

In the case of capacity convexes, (44) is written as

$$(46) \qquad C(x) = C_1(x) + C_2(x),$$

where "$+$" in (46) means the sum of subsets of a vector space.

In the case of velocity convexes, (44) is written as

$$(47) \qquad D^*(x) = D_1(x) \cap D_2(x),$$

and (45) as

$$(48) \qquad D(x) = \text{convex combination of } D_1(x) \text{ and } D_2(x),$$

5.2. Examples of capacity convexes

network   approximate continuum

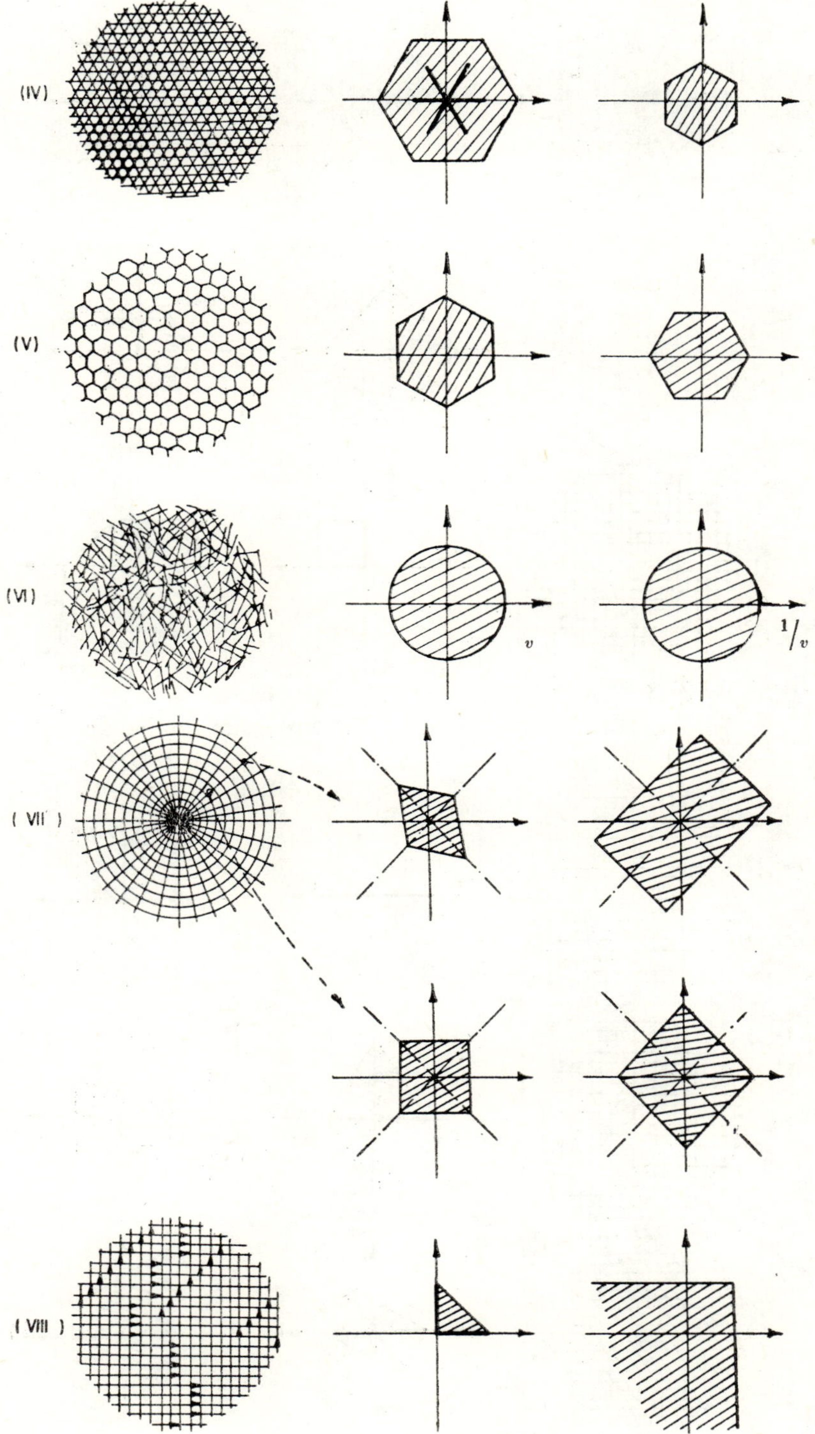

(IV)
(V)
(VI)
(VII)
(VIII)
v
1/v

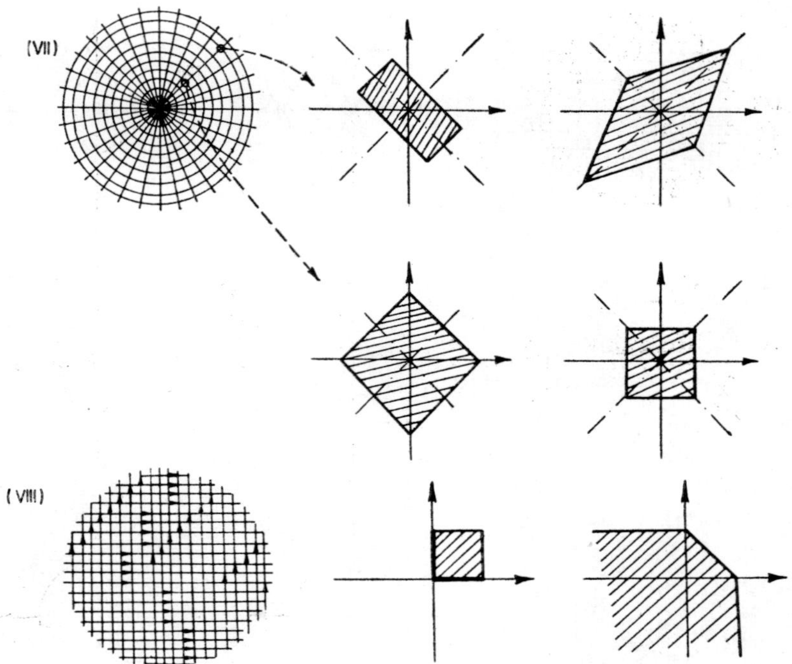

5.3. Examples of velocity convexes

network   approximate continuum

$D(x)$ $D^*(x)$

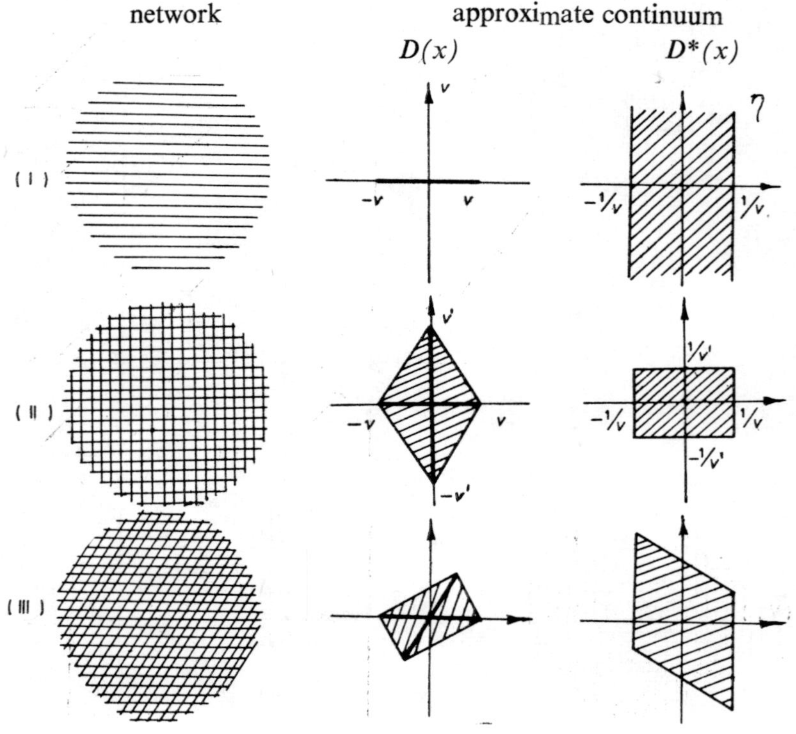

276

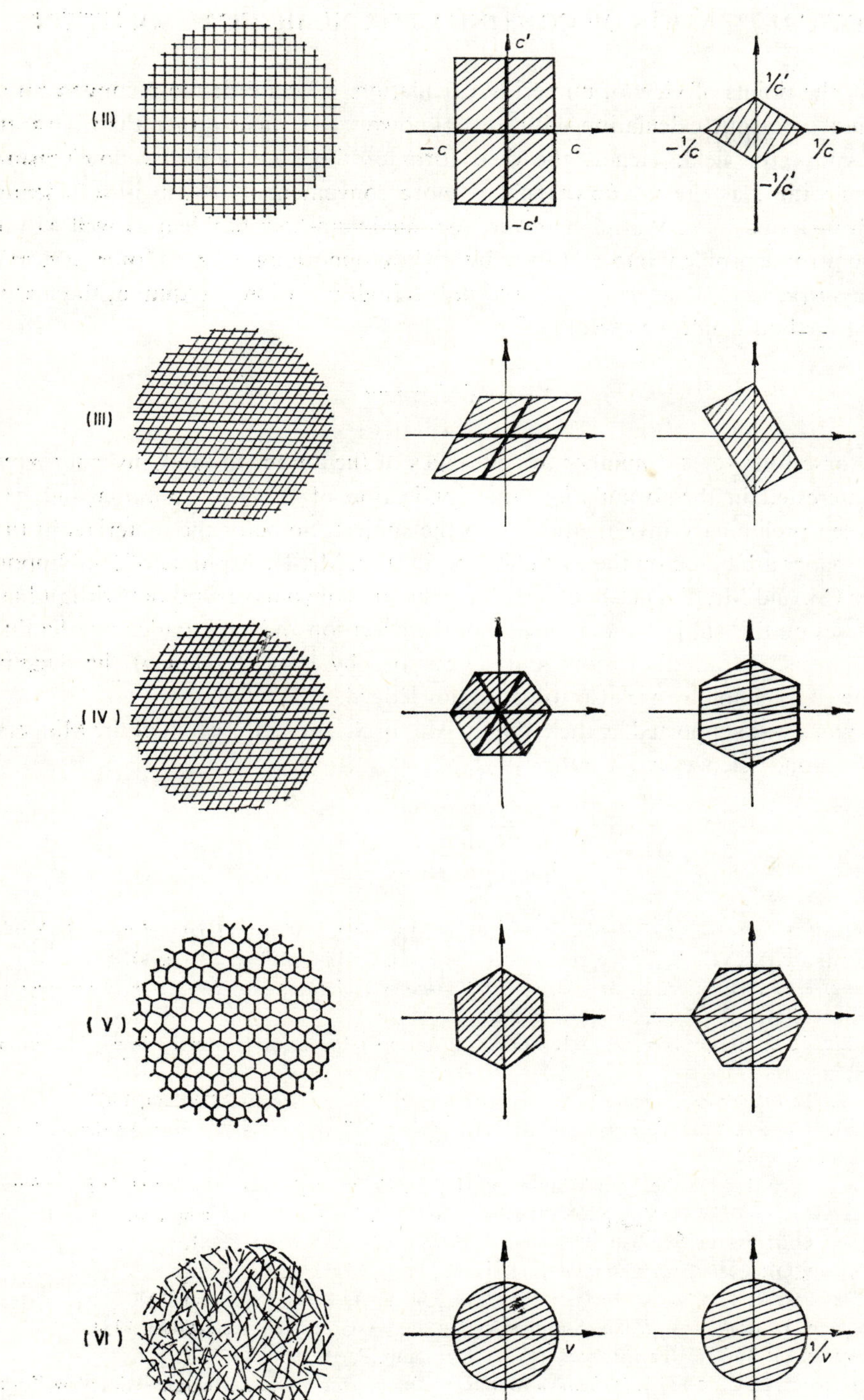
(II)
c'
-c
c
-c'
1/c'
-1/c
1/c
-1/c'
(III)
(IV)
(V)
(VI)
v
1/v

6. DISCRETIZATION OF CONTINUA FOR NUMERICAL SOLUTION

From the points of view of numerical calculation, it would be more convenient if we could carry out calculation with an unknown scalar field rather than with an unknown vector field. Hence, the dual form of the minimum-cost flow/tension problem with ζ as the unknown field is more convenient if $\psi(\varkappa, \eta)$ is sufficiently smooth in $\varkappa$ and in η. We may convert the maximum-flow problem as well as the minimum-route problem into that form by slightly smoothing $\psi_C(\varkappa, \eta)$ (or $\psi_{D^*}(\varkappa, \eta)$).

Unless $\psi(\varkappa, \eta)$ has a special form and unless high precision is required, the finite-element method would be useful [12].

Acknowledgement

For these four years a number of colleagues of the author in the University have been interested in the formulation and application of flows in continua and has performed preliminary investigations into the subject. Some of the materials in the present paper are based on their results. Specifically, Mr. K. Aoshima of the Nippon Electric Co. and Mr. Y. Ōhashi of the University of Tokyo have written their graduation theses on this subject under the author's supervision and have made considerable contributions. Useful discussions and suggestions by Prof. T. Han of the Sagami Institute of Technology are also to be acknowledged.

This work was supported by the Grant in Aid for Scientific Research of the Ministry of Education, Science and Culture of Japan.

REFERENCES

[1] Appel, P.: Le problème géométrique des déblais et remblais. *Mémorial des Sciences Mathématiques,* Fasc. XXVII, Académie des Sciences de Paris, Gauthier-Villars, Paris, 1928.

[2] Arrow, K. J., Hurwicz, L. and Uzawa, H.: *Studies in Linear and Non-linear Programming* Stanford University Press, Stanford, 1958.

[3] Ford, L. R. Jr. and Fulkerson, D. R.: *Flows in Networks.* Princeton University Press, Princeton, 1962.

[4] Hu, T. C.: *Integer Programming and Network Flows.* Addison-Wesley, Reading, 1969.

[5] Iri, M.: *Network Flow, Transportation and Scheduling—Theory and Algorithms.* Academic Press, New York, 1969.

[6] Kantorovitch, L.: On the Translocation of Masses. *Comptes Rendus (Doklady) de l'Académie des Sciences de l'URSS,* Vol. XXXVII (1942), No. 7-8, 199—201 (Mathematics).

[7] Monge, G.: Déblai et remblai. *Mémoires de l'Académie des Sciences,* 1781.

[8] de Rham, G.: *Variétés différentiables.* Hermann, Paris, 1955.

[9] Rockafellar, R. T.: *Convex Analysis.* Princeton University Press, Princeton, 1970.

[10] Schouten, J. A.: *Ricci-Calculus.* Springer Verlag, Berlin–Göttingen–Heidelberg, 1954.

[11] Schwartz, L.: *Théorie des distributions I, II.* Hermann, Paris, 1950, 1951.

[12] Zienkiewicz, O. C.: *The Finite Element Method in Engineering Science.* McGraw-Hill, New York, 1971.

ON A SPECIAL TYPE JOB SEQUENCING PROBLEM

P. KAS

(Budapest, Hungary)

1. INTRODUCTION

Let us consider a computer with $K \geq 2$ identical parallel processors. Our task is to execute a system of jobs by the processors. Assume that we have the following informations about the jobs:

—the system of jobs to be done consists of n jobs, denote this set by $\{J_1, \ldots, J_n\}$,

—we have d_i, $i=1, \ldots, n$ non-negative integers representing the execution times of jobs,

—we have k_i, $i=1, \ldots, n$ non-negative integers referring to the maximal number of processors that may be assigned to the corresponding job at the same minute,

—the schedule must not violate the precedence relations between the jobs $(<)$, the precedence relations are represented by a directed graph.

Suppose that the execution of jobs goes ahead in successive minutes. In every minute t we assign $o \leq x_{it} \leq \min \{K, k_i\}$ processors to the job J_i and at the end of every minute the executions are interrupted and the assignment changed. Our optimality criterion can be formulated as follows. Let us assume that we have the $[0, T]$ time interval in which all jobs must be completed. Determine the minimum number of processors needed by completing all jobs during the given interval.

Denote by x_{it} $i=1, \ldots, n$, $t=1, \ldots, T$ the number of processors assigned to the job J_i during the minute t, then one has to solve the following problem (P1):

$$\min K$$

s.t.

$$(1) \quad \begin{cases} \displaystyle\sum_{t=1}^{T} x_{it}=d_i & i=1, \ldots, n \\[2ex] \displaystyle\sum_{i=1}^{n} x_{it}\leq K & t=1, \ldots, T \\[2ex] 0\leq x_{it}\leq k_i & \text{non-negative integer} \end{cases}$$

$$(2) \quad \text{if } x_{it}>0 \text{ and } x_{jt'}>0, \text{ where } J_j<J_i, \text{ then } t'<t \text{ for all } i, j, t, t'.$$

First we shall deal with a more simple problem. We change the (2) type restrictions to more simple ones.

Instead of considering the precedence relations suppose that we have $[q_i, r_i]$ $i=1, \ldots, n$ time intervals during which the job J_i has to be completed. That is we must not assign any processor to the job J_i before $t=q_i$ and after $t=r_i$. Of course, if we have a feasible sequencing according to (1), (2) constraints that sequencing will automatically define the $[q_i, r_i]$ intervals. Let us consider the following problem (P2):

$$\min K$$

s.t.

(1)
$$\begin{cases} \sum_{t=1}^{T} x_{it}=d_i \quad i=1, \ldots, n \\[2ex] \sum_{i=1}^{n} x_{it} \leqq K \quad t=1, \ldots, T \\[1ex] 0 \leqq x_{it} \leqq k_i \quad \text{non-negative integer} \end{cases}$$

(2′)
$$x_{it}=0, \quad \text{if} \quad t<q_i \quad \text{or} \quad t>r_i.$$

Here $T=\max_{1 \leqq i \leqq n} r_i$ and we assume that the $[q_i, r_i]$ intervals are defined such that

$$(r_i-q_i)k_i \geqq d_i$$

hold for all i.

2. NETWORK FLOW FORMULATION

In this section we shall reformulate and solve the problem (P2) as a network flow problem of special type.

The connection between the problem treated before and the following network flow problem can easily be seen.

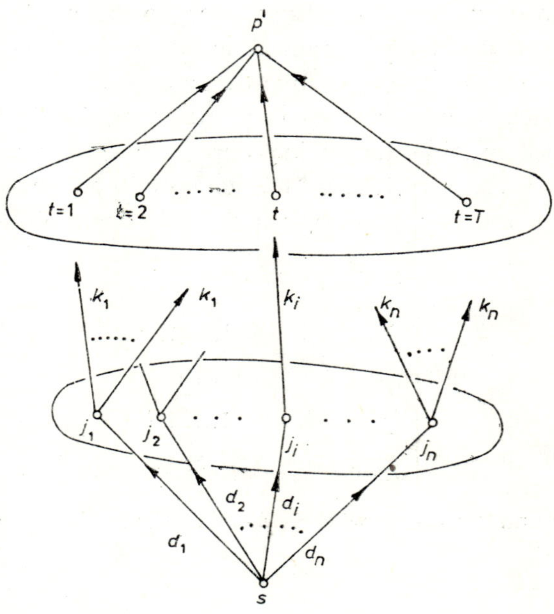

Denote this network by (N1).

Here the vertices—except for s, s'—can be partitioned into two subsets U, V

—vertices belonging to the set U represent the minutes $t=1, \ldots, T$
—vertices belonging to the set V represent the jobs J_i, $i=1, \ldots, n$.

The set of edges can be partitioned into three subsets:

—(s, J_i) type edges with capacity d_i, $i=1, \ldots, n$
—(J_i, t) type edges where $q_i \leq t \leq r_i$ with capacity k_i
—(t, s') type edges with no capacity constraints $t=1, \ldots, T$.

Since

$$(r_i - q_i)k_i \geq d_i \quad i=1, \ldots, n$$

relations are satisfied one can prove that there exists a flow in (N1) network with

value $v_0 = \sum_{i=1}^{n} d_i$.

Obviously to solve the problem (P2) it is sufficient to find a flow f_0 in (N1) with value v_0 such that

$$\max_{1 \leq t \leq T} f_0(t, s') \leq \max_{1 \leq t \leq T} g(t, s')$$

holds for all flows g with value v_0.

We shall deal with these problem more generally.

Suppose that we have an arbitrary network $[N, A, c]$ with s source and s' sink points. Let $H \subset A$ be an arbitrary subset of edges and the capacity function is defined to be $+\infty$ on the edges of H. That is

$$c(x, y) = \begin{cases} \text{non-negative integer if } (x, y) \notin H \\ +\infty \qquad\qquad\qquad\quad \text{if } (x, y) \in H. \end{cases}$$

Assume that there exists a flow with value v^* in $[N, A, c]$ network. We can define the following problem (P3).

We want to find a flow f^* in $[N, A, c]$ with value v^* such that

$$\max_{(x,y) \in H} f^*(x, y) \leq \max_{(x,y) \in H} f(x, y)$$

for all flows f with value v^*.

To solve the problem (P3) the following algorithm can be proposed:

Let $f_1(x, y)$ be a feasible flow in $[N, A, c]$ with value v^*.

Step 1. Denote by $K_1 = \max_{H} \{f_1(x, y)\} = f_1(x_1, y_1)$

and $H_1 = \{(x, y) : (x, y) \in H, f_1(x, y) \geq K_1 - 1\}$.

Obviously $(x_1, y_1) \in H_1$.

Change the capacities in the following way:

$$\tilde{c}(x,y)=\begin{cases} 0 & \text{if} \quad (x,y)\in H_1 \\ 0 & \text{if} \quad f_1(x,y)=c(x,y) \\ 1 & \text{if} \quad f_1(y,x)>0 \\ 1 & \text{if} \quad f_1(x,y)<c(x,y) \end{cases}$$

GO TO 2.

Step 2. Search for an unsaturated path in $[N,A,\tilde{c}]$ from x_1 to y_1. If there exists such a path denote the set of edges of that path by $\mathcal{P}$ and GO TO 3.
Otherwise

$$K_1 = \min_{\{f:v(f)=v^*\}} \max_H \{f(x,y)\} \text{ STOP.}$$

Step 3. Modify the arc flows as follows:

$$f_1(x,y):=\begin{cases} f_1(x,y)+1 & \text{if} \quad (x,y)\in\mathcal{P} \\ f_1(x,y)-1 & \text{if} \quad (y,x)\in\mathcal{P} \\ f_1(x,y)-1 & \text{if} \quad (x,y)=(x_1,y_1) \\ f_1(x,y) & \text{otherwise} \end{cases}$$

GO TO 1.

Theorem 1. *If the capacities are non-negative integers the proposed simple algorithm gives a flow f^* which is a solution of* (P3) *in finite number of steps.*

Proof. Denote by $H_{K_1}=\{(x,y):(x,y)\in H, f_1(x,y)=K_1\}$. The presented algorithm terminates after finite number of steps because either K_1 or $|H_{K_1}|$ will decrease in the successive steps.
After terminating we have a flow f_1 in $[N,A,c]$ network with value v^* and

$$K_1 = \max_H \{f_1(x,y)\}=f_1(x_1,y_1).$$

Termination occurred because there is no unsaturated path from x_1 to y_1 according to the modified capacities. It means that we have a cut (S,S') such that

$$s, \quad x_1\in S \quad s', \quad y_1\in S'$$
$$f_1(x,y)\geq K_1-1 \quad \text{if} \quad (x,y)\in H \quad \text{and} \quad (x,y)\in(S,S')$$
$$f_1(x,y)=c(x,y) \quad \text{if} \quad (x,y)\notin H \quad \text{and} \quad (x,y)\in(S,S')$$
$$f_1(x,y)=0 \qquad \text{if} \quad (x,y)\in(S',S).$$

Denote by

$$H'=\{(x,y)\in H:(x,y)\in(S,S')\,f_1(x,y)=K_1\}$$
$$H''=\{(x,y)\in H:(x,y)\in(S,S')\,f_1(x,y)=K_1-1\}.$$

282

Clearly

(i)
$$v^* = f_1(S, S') - f_1(S', S) = |H'|K_1 + |H''|(K_1 - 1) + \sum_{\substack{(x,y) \in (S, S') \\ (x,y) \notin H}} c(x, y).$$

If there exists a flow g in $[N, A, c]$ network with value v^* such that

$$\max_{H} \{g(x, y)\} \leq K_1 - 1$$

means that

(ii)
$$v^* = g(S, S') - g(S', S) \leq |H'|(K_1 - 1) + |H''|(K_1 - 1) + \sum_{\substack{(x,y) \in (S, S') \\ (x,y) \notin H}} c(x, y).$$

From (i) and (ii) we get $|H'| = 0$ which contradicts to $(x_1, y_1) \in H'$. This completes the proof.

3. SUMMARY

Now go back to the original problem. Let us consider the ordering restrictions represented by a precedence network.

Define e_i, $i = 1, \ldots, n$ non-negative integers as the shortest execution times for the job J_i, that is $e_i := \left[\dfrac{d_i}{k_i} \right]$, where $[x]$ means the smallest integer which is greater than or equal to x.

Clearly, the longest path in the precedence network—according to the weight e_i—determines the earliest time in which all jobs can be completed. Denote by T the lenght of the longest path.

We need now an assuption according to the precedence network under which the intervals $[q_i, r_i]$ $i = 1, \ldots, n$ can be fixed.

Assumption: Suppose that every node is preceded and followed by nodes belonging to the longest path.

Under this assuption we can determine fixed $[q_i, r_i]$ $i = 1, \ldots, n$ time intervals by any critical path method. In this case we can formulate the appropriate (P2) type problem and it can be solved by the proposed algorithm.

REFERENCES

[1] Hu, T. C.: Parallel Sequencing and Assembly time Problems, *Operation Research* 9 (1961), 841–848.

[2] Ford, L. R., Fulkerson, D. R.: *Flows in Networks*, Princeton University Press, (1962).

[3] Moder, E. E. Phillips, C. R.: *Project Management with CPM and PERT*, Van Nostrand Reinhold, New York (1964).

A NETWORK SOLUTION TO A GENERAL VEHICLE SCHEDULING PROBLEM

J. MÁRQUEZ DIEZ CANEDO and O. MEDINA-MORA ESCALANTE

(Mexico City, Mexico)

I. STATEMENT OF THE PROBLEM

This problem arose in the Central Bank of Mexico, which requires a fleet of vehicles to mobilize currency throughout the country.

Thus, there is a central office which requires a fleet of vehicles. There are various types of vehicles amongst which an optimal fleet must be selected. All trips made by the vehicles are round trips. The objective is to find the composition of vehicles which will yield an optimal fleet: the number of vehicles of each type required to meet demand at minimum cost.

The following information is available:

t_{ik} = time at which a vehicle of type i must initiate trip k,
a_{ik} = time required by a vehicle of type i to make trip k,
p_i = price of a type i vehicle,
r_{ik} = cost of trip k if done by a type i vehicle.

Define

$T_{ik} = t_{ik} + a_{ik}$ = time at which a vehicle of type i would finish trip k and be available
for reassignment.

$V_k = \{i \mid$ a type i vehicle is capable of making trip $k\}$

$$i = 1, 2, \ldots, I; \quad k = 1, 2, \ldots, K.$$

There are considerations which may rule out a certain type of vehicle for a trip. For example, the payload, communication routes, etc. All these are accounted for in V_k.

II. THE ASSOCIATED TRANSPORTATION PROBLEM

Let N denote a set of nodes and A a set of ordered pairs of elements from N called arcs. A directed bipartite graph $G[N, A]$ with sets of nodes G, X, Y will be specified such that solving a transportation problem on the graph, with some aditional conditions will yield the composition of the optimal fleet along with schedules for each vehicle. These aditional conditions will be discussed in Section II.7.

II.1. The Nodes $N = \{G, X, Y\}$ of the Graph

Sources. The nodes $g_i \in G$ will represent vehicle generating nodes. These are the sources from which the vehicles originate, and there is one for each type of vehicle considered.

The nodes $\beta_{iT} \in X$ are defined by unique pairs of positive integers (i, T). Such a node will exist if and only if there exists at least one $T_{ik} = T$. This means that at least one vehicle of type i could be available for reassignment at time T.

Sinks. The nodes $\alpha_k \in Y$ represent the trips required, which are assumed to be a finite number.

II.2. The Arcs A in the Graph

Since a vehicle can only be used at time t if it was available at some time $T \leqq t$, then

$$(\beta_{iT}, \alpha_k) \in A \Leftrightarrow T \leqq t_{ik} \quad \text{and} \quad i \in V_k.$$

The capacity of such arcs is unlimited.

The vehicle generating nodes g_i can supply any node α_k, provided that such a vehicle can make the trip. Thus

$$(g_i, \alpha_k) \in A \Leftrightarrow i \in V_k$$

and such arcs have infinite capacity.

II.3. The Costs

Let $C(x, y)$ denote the cost of a unit of flow through arc $(x, y) \in A$. Then

(a) $$C(g_i, \alpha_k) = p_i + r_{ik}.$$

That is, whenever a vehicle is "taken" from a generating node its price must be paid, plus the cost of such a vehicle making the trip. This is equivalent to buying a type i vehicle.

(b) $$C(\beta_{iT}, \alpha_k) = r_{ik}.$$

286

II.4. Demands at the Sinks

The demand at each sink α_k is 1. That is, we require exactly one vehicle to make each trip.

II.5. The Decision Variables

Let $f(x, y)$ be the flow through arc $(x, y) \in A$. In terms of the vehicle scheduling problem it has the following interpretation

$$f(\beta_{iT}, \alpha_k) = \begin{cases} 1 & \text{if a vehicle of type } i \text{ available at time } T \text{ was used for trip } k, \\ 0 & \text{else,} \end{cases}$$

$$f(g_i, \alpha_k) = \begin{cases} 1 & \text{if a vehicle of type } i \text{ was bought to make trip } k, \\ 0 & \text{else.} \end{cases}$$

II.6. Supply at each Source

Each generating node g_i has unlimited supply, since it is always possible to buy a new vehicle for a particular trip, if one is willing to pay the price of the vehicle.

A source node β_{iT} will have a vehicle available for reassignment, if a vehicle of type i has made a trip k such that $T_{ik} = T$.

Let $\tau_{iT} = \{(x, \alpha_k) \mid x \in G \cup X; T_{ik} = T\}$. Then, the supply of vehicles at node β_{iT} is

$$q(\beta_{iT}) = \sum_{(x, y) \in \tau_{iT}} f(x, y).$$

II.7. The Model

Thus, what we have is the following problem

$$\min \sum_{(x, y) \in A} C(x, y) f(x, y)$$

$$\text{S.t. } \sum_{x \in G \cup X} f(x, y) = 1 \, \forall \, y \in Y$$

$$\sum_{y \in Y} f(x, y) \leq q(x) \, \forall \, x \in X$$

$$f(x, y) \in \{0, 1\} \, \forall \, (x, y) \in A.$$

This is a transportation problem except for the fact that the supply at the sources depends on there being flow in certain arcs, and the explicit restriction to integer flows.

It is easily seen that the model reduces to the one used by Dantzig and Fulkerson in the case, where only one type of vehicle is used to form the fleet. The solution when decoded in the same manner as the case for only one type of vehicle, will yield the number of vehicles of each type required, along with a trip schedule for each vehicle.

Example. Consider a problem with two types of vehicles and seven trips:

$$t=$$

i \ k	1	2	3	4	5	6	7
1	0	*	4	4	8	8	10
2	0	0	4	*	8	8	10

$$T=$$

i \ k	1	2	3	4	5	6	7
1	3	*	8	10	10	12	12
2	2	8	9	*	9	10	11

,

where $t_{ik}=T_{ik}=$ * implies $i \notin V_k$.

This problem yields the graph shown in Fig. 1. Note that $T_{16}T_{17}T_{27} \geqq t_{ik}$ for all i and all k. Therefore, source nodes β_{iT} with $T>10$ are not required.

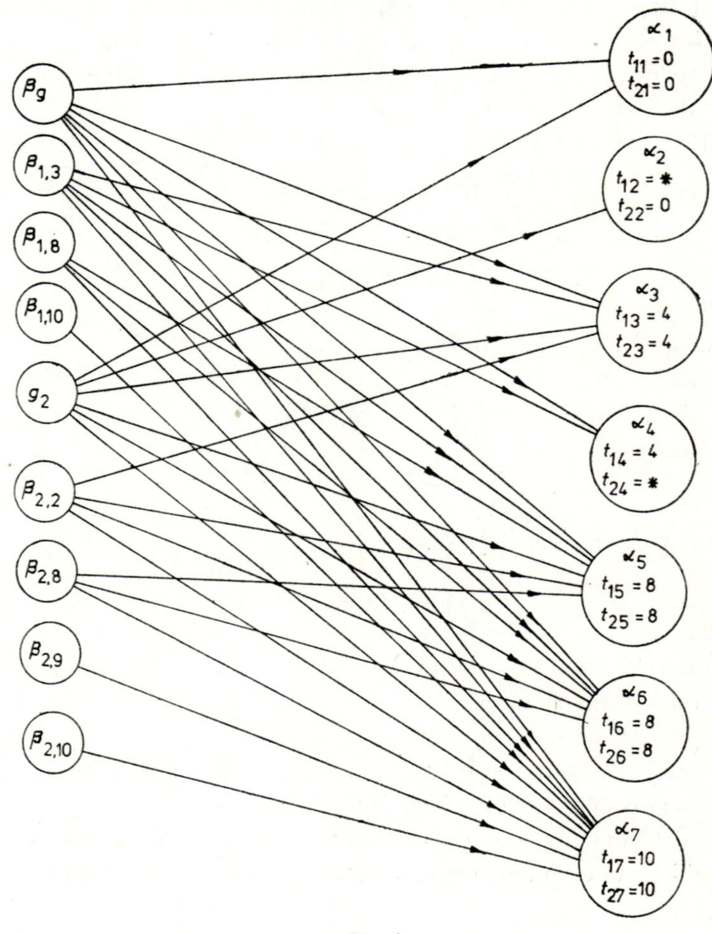

Fig. 1

Table 1. Matrix for the example

	g_1 α_1	g_1 α_3	g_1 α_4	g_1 α_5	g_1 α_6	g_1 α_7	g_2 α_1	g_2 α_2	g_2 α_3	g_2 α_5	g_2 α_6	g_2 α_7	β_1 3 α_3	β_1 3 α_4	β_1 3 α_5	β_1 3 α_6	β_1 3 α_7	β_1 8 α_5	β_1 8 α_6	β_1 8 α_7	β_1 10 α_7	β_2 2 α_3	β_2 2 α_5	β_2 2 α_6	β_2 2 α_7	β_2 8 α_5	β_2 8 α_6	β_2 8 α_7	β_2 9 α_7	β_2 10 α_7	$q\,\beta_1$ 3	$q\,\beta_1$ 8	$q\,\beta_1$ 10	$q\,\beta_2$ 2	$q\,\beta_2$ 8	$q\,\beta_2$ 9	$q\,\beta_2$ 10
Demands at the Sinks	+1						+1																														
								+1																													
		+1							+1				+1									+1															
			+1											+1																							
				+1						+1					+1			+1					+1			+1											
					+1						+1					+1			+1					+1			+1										
						+1						+1					+1			+1	+1				+1			+1	+1	+1							
Definition of $q(\beta_{iT})$													−1	−1	−1	−1	−1														+1						
																		−1	−1	−1												+1					
																					−1												+1				
																						−1	−1	−1	−1									+1			
																										−1	−1	−1							+1		
																													−1							+1	
																														−1							+1
Conditioned Supply													+1	+1	+1	+1	+1														−1						
																		+1	+1	+1												−1					
																					+1												−1				
																						+1	+1	+1	+1									−1			
																										+1	+1	+1							−1		
																													+1							−1	
																														+1							−1

The constraint matrix for this problem is shown in Table 1 and the solution of the problem for the following costs is given in Table 2.

$$r=$$

i \ k	1	2	3	4	5	6	7
1	2	∞	4	8	1	4	1
2	3	7	4	∞	3	4	3

$$P_1 = 100 \qquad\qquad P_2 = 300$$

Table 2. Results and decoding of the example

Variable	Value after L.P.
$(g_1\alpha_1)$	1.0
$(g_1\alpha_3)$	1.0
$(g_2\alpha_2)$	1.0
$(\beta_{1,3}\alpha_4)$	1.0
$(\beta_{1,10}\alpha_7)$	1.0
$(\beta_{1,8}\alpha_5)$	1.0
$(\beta_{2,8}\alpha_6)$	1.0
$q(\beta_{1,3})$	1.0
$q(\beta_{1,8})$	1.0
$q(\beta_{1,10})$	2.0
$q(\beta_{2,8})$	1.0
$q(\beta_{2,10})$	1.0

Vehicle and type	Trips done	Days of utilization
No. 1 Type 1	1, 4, 7	11
No. 2 Type 1	3, 5	6
No. 1 Type 2	2, 6	10

III. IMPLEMENTATION AND EMPIRICAL RESULTS

III.1. The Size of the Problem and Practical Difficulties that Arise

It can readily be seen that an upper bound on the number of variables is $B = I^{\cdot}K^{\cdot}(K+3)/2$; the actual number depends of course on the real data. For our particular problem it was estimated that every run would have about $0.8B$ variables.

Thus, in practice, there is a restriction imposed on the number of trips to consider in order to have a problem of manageable size. Therefore, the planning horizon could not be excessively long, which leads us to contemplate a simulation-like procedure.

Under such a scheme, the horizon is divided into smaller periods, and the results (fleet and scheduling) for a given period will be input to the next period together with its demand requirement.

Thus, in our simulation scheme, the model behaves as a central planning office which has a mechanism that permits an optimal assignment of vehicles to trips for any period where demand is known. In fact, it can be used in practice for precisely this purpose.

The following remarks are in order:

(1) In our particular application, and probably most others, the actual demand cannot be known long in advance, so that breaking up the horizon into small periods where the demand is assumed to be known, is not a serious limitation.

(2) On the other hand, shortening the planning horizon will reduce the scope of the objective function and the results obtained are sub-optimal. In order to diminish this unwanted but unavoidable effect, it is necessary to link the results from one period to the next within the planning horizon. We now discuss how this was done for our application.

III.2. *Modifying the Model in order to Link Solutions from One Period to the Next*

To achieve the desired interdependence between results of consecutive periods, some slight modifications to the original formulation are required. Namely, for a given intermediate period, vehicles which started trips the preceding period that end some time during the present period, must be accounted for in the model.

Thus, the model is modified as follows:

(1) A set of nodes $H = \{h_1, h_2, \ldots, h_I\}$ which shall be called "hangar nodes" is introduced.

A vehicle of type i will be available for reassignment at its hangar node h_i in the beginning of the present intermediate period, if its last trip in the previous period ended some time before, or on the first day of the present period.

In other words, these hangar nodes represent supply of vehicles "bought" during some previous period, which are available for reassignment at the beginning of the present period, at no extra cost.

Thus, these source nodes h_i can supply any sink node α_k at operating costs only, and (h_i, α_k) is an arc of the associated graph, if and only if $i \in V_k$.

The supply at a node h_i, is equal to the number of type i vehicles N_i which started trips on some previous period and are available for reassignment at the beginning of the present period.

(2) For vehicles of type i which initiated trips on some preceding period, that end at a time T in the present period, one of two things is done:

(a) If a node β_{iT} is required by the demand of the present period, the supply at this node will be increased by a constant amount N_{iT} equal to the number of vehicles of type i which started trips on some previous period that end at a time T in the present.

Thus, for such nodes:

$$q(\beta_{iT}) = \sum_{(x,y) \in \tau_{iT}} f(x, y) + N_{iT}.$$

(b) If such a node β_{iT} is not required by the demand of the present period, it is included in the model with

$$q(\beta_{iT}) = N_{iT}.$$

The arcs (β_{iT}, α_k) are generated according to the same rules as the original model.

III.3. The Problem of Costs for Short Periods

In models run for short periods of time, the purchase price of a vehicle outweighs its operating costs, so the model tends to select the cheapest types of vehicles whenever possible. This may not be desirable, since on the long run, it may be more costly to operate than its more expensive competitor.

In order to compensate for the above, an equal purchase price was assigned to all types, and the real price was accounted for in the operating cost through depreciation.

Finally, it is also possible to complement the above, by penalizing idle vehicles by their depreciation costs. Thus, if a vehicle available for reassignment at time T, is assigned to its following trip at time $t > T$, it is penalized with a cost proportional to $t - T$, based on its depreciation.

III.4. The Simulation Process

The models described were used in a simulation process. Demand for trips was estimated via census and probability distributions were inferred.

From the questionaires it appeared that at best, because of the random nature of the demand for trips, it would be possible to know demand one week in advance. Also, the data showed that demand was seasonal with peaks on certain months. Thus, the planning horizon was fixed as one year, and the length of the periods which divide the horizon was left as a parameter for the simulation process, not to exceed one week.

The number of simulation runs (times the planning horizon is simulated) was determined, based on how fast the average demands simulated converged to average demands estimated from the data. Thus, the simulation process used was the following, for each year simulated:

(1) Having selected the length of the period into which the horizon is divided, demand for each period is sampled from the corresponding probability distributions.

292

(2) The model for each period is set up and run in sequence, using the results of the model for the previous period, to determine the hangar nodes required and when vehicles which started trips the preceding period, will be available for reassignment during the period under consideration.

(3) Once the models for all periods in the horizon are run, the solutions are decoded; the number of vehicles of each type used during the horizon and their schedules are recorded, in order to obtain statistics relevant to the decision maker.

(4) The process is repeated for another horizon, until the number of simulation runs is met.

(5) From the results of the runs for each horizon, statistics are obtained on the characteristics of an optimal fleet.

IV. RESULTS

The first result we wish to discuss is about the values taken by the variables. A look at the matrix of the last example (Table 1) will clearly show that it is not unimodular, opening then the possibility of having non-integer solutions and the need to use time consuming branch and bound algorithms. However, each time the model was called, all variables were integer in the optimal solution to the associated L.P. Thus, in approximately 2000 runs of the model, it never failed to produce integer solutions, and from a practical point of view, this generalization of the tanker scheduling problem with conditional supply was quite effective. This can be observed in the results of our simple example in Table 2. Because of this, it is our feeling that the structure of these models deserves further research to see if it can be exploited in other situations.

The process was simulated under two situations: demand is assumed known three days in advance and one week in advance. This was done in the hope of gaining some insight about the additional cost due to less planning facilities implied by better knowledge of the demand.

All problems were solved on a UNIVAC 1106 using the UMPIRE code (4). The computer results are summarized in Table 3.

Table 3

Period	Average Size of Problems		Average CPU Running Time (seconds)
	Number of variables	Number of rows	
3 days	185	61	2.2
1 week	840	124	6.7

As concerns the simulation process, it has been indicated that, by the end of each year (horizon) simulated, a statistical record of every vehicle is available. Based on

technical data it is easy to evaluate whether a given vehicle was over—or under—
utilized, or if its performance was above the break-even point. Fleet estimates are
altogether computed.

Taking averages over the different years simulated, one is able to form a so-called
decision table, where each entry has information of fleet composition, operating costs
and fraction of satisfied demand. The decision process was greatly enhanced by this
approach. All alternatives were then weighted and the final decision was easy to make.
A skeleton of the actual decision table used is presented in Fig. 2.

Table entry for decision

N_1	N_2	N_3
	FU	
	SD	
	PC	
	CO	

N_i = Number of vehicles type i
FU = Estimate of fleet utilization
SD = Percentage of satisfied demand
PC = Purchase cost
CO = Cost of yearly operation

Fig. 2

REFERENCES

Bellmore, M., Bennington, G., Lubore, S.: A Maximum Utility Solution to a Vehicle Constrained
Tanker Scheduling Problem, *Nav. Res. Log. Quart.* 15, (1968), No. 3, Sept.

Dantzig, G. B., Fulkerson, D. R.: Minimizing the Number of Tankers to meet a fixed Schedule,
Nav. Res. Log. Quart. 1, (1954), 215–222.

Ford, L. R., Fulkerson, D. R.: *Flows in Networks*, Princeton University Press, 1962.

Forrest, J. J. H., Hirst, J. P. H., Tomlin, J. A.: Practical Solution to Large Mixed Integer Programming Problems with UMPIRE, *Man. Sci.* 20, (1974), No. 5, January.

Hitch, C. J.: Planning the Composition of an Air Transport Fleet, Case Study C in: F. Hanssmann;
Operations Research Techniques for Capital Investment, John Wiley, 1968.

PROBLEMS OF OPTIMIZATION IN A NETWORK MODEL

N. SIEBER

(Leipzig, GDR)

1. INTRODUCTION

This contribution deals with a new complex model of network analysis and presents the forthcoming problems of optimization.

We are of the opinion that the task of the network analysis is the terminal arrangement of activities to an optimal process considering their technological dependences and constrained resources, with regard to suitable objective functions.

In the practical applications, the concepts of activity, logical structure, technological dependences, optimal cost, or utilization of resources are interpreted in a very different way, according to the conditions of management or level of planning.

The model, which was developed in our department, is more general than comparable others because of its extended mathematical description and flexibly accommodated to the various demands of the management practice.

Moreover, an extensive system of computer programs with seven main parts tested and used for management projects in several firms is available.

2. TERMINAL PROBLEM

We consider a directed, circuitless graph, $N=(V, B)$, with the set of activities, $V=\{V_1, V_2, \ldots, V_n\}$ (set of nodes of N), and the set of dependences (set of arcs of N), $B=\{V_i, V_j\}$, $B=V \times V$, (Fig. 1) to be a network.

As is well known, the earliest times for beginning the activities and the critical path can be determined from the following problem:

Given are the durations, $D_i \geqq 0$, $i=1, 2, \ldots, n$, for each activity, $V_i \in V$, and the coefficients of distance $(a_{ij}^{*}, a_{ij}) \geqq 0$ for all $(V_i, V_j) \in B$.

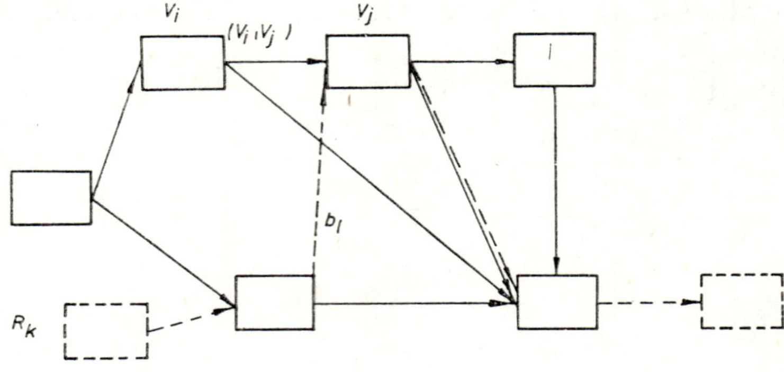

Fig. 1

Required are the times of beginning the activities, $t_i \geqq 0$, $i = 1, 2, \ldots, n$, so that the endtime $t = t_e$ will be minimal and the following restrictions are satisfied:

$$a_{ij} \leqq t_j - t_i \leqq a_{ij}^*, \quad \forall (V_i, V_j) \in B$$

$$t_i + D_i \leqq t_e, \quad i = 1, 2, \ldots, n.$$

This is the basic model for the METRA-Potential-Method; if $a_{ij} = D_i$ and $a_{ij}^* = \infty$, the basic model for CPM is obtained.

Both tasks can be formulated as problems of linear programming. Because of their special structure, numerous simple algorithms of solution are available, mostly on the basis of the theory of graphs.

The essential extension in the developed model is the introduction of variable durations for the activities. Then the terminal problem is formulated by the following problem of optimization:

Given are, for each activity $V_i \in V$, upper and lower bounds (MAX D_i, MIN $D_i) \geqq 0$ for the durations, D_i, $i = 1, 2, \ldots, n$, and the coefficients of distance $(v_{ij}^*, v_{ij}, W_{ij}^*, W_{ij}) \geqq 0$ for all arcs $(V_i, V_j) \in B$.

Required is a time-duration-system $(D, t) = \{D_1, D_2, \ldots, D_n, t_1, t_2, \ldots, t_n\}$, so that the restrictions

$$v_{ij} \cdot D_i + W_{ij} \leqq t_j - t_i \leqq v_{ij}^* \cdot D_i + W_{ij}^* \quad \forall (V_i, V_j) \in B$$

$$\text{MIN } D_i \leqq D_i \leqq \text{MAX } D_i$$

$$(i = 1, 2, \ldots, n)$$

$$t_i + D_i \leqq t_e$$

are satisfied and the endtime $t = t_e$ becomes minimal.

Because of the introduction of variable durations, the problem at first becomes more difficult. But some cases that cannot be covered by the METRA-Potential-Method are solvable. For example, it is possible to include the additional conditions,

$t_i + D_i \leqq GSET_i$ (latest endtime given). The model has favourable conditions for including the balancing of resources.

For this extended problem, Nägler [3] has developed an algorithm of solution. A generalization of Kelley's method is offered. Using the theory of graphs, the dual optimization problem leads to a maximal flow problem for which the algorithm has been developed.

First it is started with MIN D_i or MAX D_i for $i = 1, 2, \ldots, n$. The algorithm tries to find a time system with these input data. If it is not successful—for example, endtimes given are overstepped with these durations—then they will be changed in such a manner that a time system satisfying all restrictions is obtained and, furthermore, the sum of deviations, $\sum_{t=1}^{n} (\text{MAX } D_i - D_i)$, will be minimal. It is also possible to assume a convex (piecewise linear) duration-cost-curve as an objective function for each activity.

Of course, with too strong restrictions, the case may occur that, in spite of these above procedure, all requirements cannot be fulfilled. At first this fact is printed out and then, by increasing the times given, the algorithm computes a system with minimal exceeding. If this is impossible, the computation is stopped. The corresponding part of the computer program permits to find an optimal time-duration-system for networks with several hundred activities in 0.5 or 3 minutes. Hence this time algorithm is a good condition for the optimization of the balancing of resources in a complicated network structure.

3. BALANCING OF RESOURCES

Now we include a set of resources that cannot be stored (machines, workers), $R = \{R_1, R_2, \ldots, R_m\}$, in our considerations. Let $p_k \geqq 0, k = 1, 2, \ldots, m$, be the upper bounds of the number of resource-units of R_k being available per time-unit.

Further $R_k = \varphi(V_i)$ be a single-valued mapping over V in $R, (\varphi : V \rightarrow R)$, i.e. each activity $V_i \in V$ is uniquely assigned to one resource $R_k \in R$. This is no restriction of generality. If several types of resources are used for one activity, the model allows to suitably divide this activity and after this to parallelly reconnect the parts.

Each activity shall not be interrupted and each resource shall work during the whole activity duration with constant intensity.

Then the following terms are introduced:

u_i extent of work of R_k to V_i,

$r_i = \text{intround}\left(\dfrac{u_i}{D_i}\right)$ intensity of R_k to V_i

$$\text{with } r_i = \min \{r \mid r \cdot D_i \geqq u_i, \; r \text{ integer}\} \underset{\text{def}}{=} \text{intround}\left(\frac{u_i}{D_i}\right).$$

The quantities u_i are given; r_i, D_i are generally required. At last the set of arcs, $\bar{B} = (b_1, b_2, \ldots, b_q)$, for the flow of resources is introduced.

By means of the valuation by cost in the objective function [4], the balancing of resources can be formulated by the following general optimization problem:

Required is a flow of resources, $x = (x_1, x_2, \ldots, x_q)$, $x_i \geqq 0$ integer, and a terminal vector of tension, $y = (y_1, y_2, \ldots, y_s)$, $y_{i_1} = t_j - t_i$, $y_{i_2} = D_i$, so that the function of cost

$$z = ax^T + by^T + xCy^T$$

becomes minimal.

Moreover, the restrictions hold:

(a) $x \in X$, X is a linear polyhedron (of the type of transport problem),

(b) $y \in Y$, Y is a linear polyhedron (see terminal problem),

(c) $D_i \cdot \sum_{k \in I_i} x_k \geqq u_i$, I_i is the set of indices, $i = 1, 2, \ldots, n$ (conditions of connection which determine the use of reources for the activity).

Further:

    ax^T: cost for the units of resources required,
    by^T: cost for exceeding planned endtimes,
  xCy^T: cost for the standstill of resources.

This optimization problem is generally nonlinear and nonconvex (objective function). If $D_i = \text{const.}$, we speak of a problem of bilinear optimization.

For this case, $z^*(y) = \min_{x \in X} z(x, y)$ is concave over Y. Then the optimum lies on a boundary point of the polyhedron Y. For solution we apply a branch-and-bound-method which is controlled by a linear substitution function, $E(y) \leqq Z^*(y)$, $y \in Y$, so that in each node of the branched tree (vertex of Y) those neighbouring corners, for which the function, $E(y)$, increases (rule of branch) and does not exceed the actual optimum $z^*(y_0)$ are required.

4. COMPUTER REALIZATIONS AND PRACTICABLE APPLICATIONS

Today no practicable algorithms to solve the total model are available. However, there are four computer programs with heuristic algorithms for some slightly modified and practically important special cases.

(a) The endtime of the total plan shall be minimal by satisfying the bounds of resources [1]:
    $a = 0$, $b = (0, 0, \ldots, 1)$ or b arbitrary.

There exists a heuristic algorithm of solution as the computer program "balance optimization 1".

(b) The utilization of the capacities p_k shall be as uniform as possible by observing the endtimes [1]:

$$b=0.$$

For this a heuristic algorithm of solution as the computer program "balance optimization 2" is available.

(c) Sequencing problem

Case 1. There is only one executing unit in every type of resources. The following reductions are obtained:

$$p_k=1, \quad r_i=1, \quad D_i=\text{const.},$$

X is a linear polyhedron of the type of assignment problem.

An exact algorithm of solution as the computer program "sequence optimization 1" is available.

Case 2. The above restriction does not hold. There hold additional postulates of discreteness of the x_l:

$$x_l \in P = \left\{ p_{\gamma k} \;\middle|\; p_k = \sum_{\gamma=1}^{n_k} p_{\gamma k}, \, p_{\gamma k} \geqq 0 \text{ integer} \right\}.$$

For that exists a heuristic algorithm [6] with the computer program "sequence optimization 2".

In practical applications, the program "balance optimization 2" plays the most important role whereas the program "sequence optimization 1" finds almost no application, although for that the only exact algorithm is available at present. The first program is mainly used for balancing the capacity of production for periods of a quarter or one year.

In further investigations of the model, also resources which can be stored [8] were considered or random durations (stochastic terminal problem) assumed [7].

REFERENCES

[1] Suchowizki, S. I., Radtschik, I. A.: *Mathematische Methoden der Netzplantechnik*, Teubner-Verlag, Leipzig 1969.

[2] Sieber, N.: Mathematische Optimierung und Anwendungen im Bauwesen, *Berichte zum VII. IKM*, Verlag für Bauwesen, Berlin 1976.

[3] Nägler, G.: Ein allgemeines Grundmodell für Ablaufpläne, *Mathematische Operationsforschung und Statistik* 6, (1975), Heft 1, Berlin.

[4] Nägler, G.: Probleme der Modellierung der Ablaufplanung, *Wiss. Zeitschrift der HfB Leipzig* (1974), Heft 2.

[5] Schallehn, W., Sebastian, H.-J.: Ein Reihenfolgeproblem in der Netzplantechnik, *Elektron. Informationsverarbeitung und Kybernetik* 8, (1972), Heft 1, Berlin.

[6] Merkel, G.: Optimierung von Ressourceneinsatzplänen bei geographischen Nebenbedingungen, *Wiss. Zeitschrift der HfB Leipzig* (1974), Heft 2.

[7] Merkel, G.: Probleme der stochastischen Ressourceneinsatzplanung und deren Zusammenhang mit den entsprechenden deterministischen Problemen, *Materialien zur Arbeitstagung "Math. Optimierung"* Vitte/Hiddensee 1975, Akademie der Wiss, der DDR, Zentrum für Rechentechnik.

[8] Teubener, H.: Ein Verfahren zur komplexen Einbindung nichtspeicherfähiger und speicherfähiger Ressourcen sowie ökonomischer Kennzahlen in die Baubilanzierung der mittelfristen Planungsphase bei Anwendung der Netzplantechnik, Dissertation A, Hochschule für Bauwesen Leipzig 1974.

[9] Sieber, N.: Einige aktuelle Probleme der Netzplantechnik, *Wiss. Berichte der TH. Leipzig* (1977) Heft 5.

ON OPTIMAL PATTERN FLOWS

KAORU TONE

(Yokohama, Japan)

1. PRELIMINARIES. BASIC THEOREMS OF BENDERS DECOMPOSITION

Benders [1] has presented a partitioning procedure for solving mixed variables programming problems of the type

$$(1.1) \qquad \text{Max } \{cx+f(y) \mid Ax+F(y) \leqq b,\ x \in R^p,\ y \in S\},$$

where $x \in R^p$ (the p-dimensional Euclidean space), $y \in R^q$ and S is an arbitrary subset of R^q. Furthermore, A is an (m, p) matrix, $f(y)$ is a scalar function and $F(y)$ an m-component vector function both defined on S, and b and c are fixed vectors in R^m and R^p, respectively.

His basic idea is a partitioning of the given problem into two subproblems: a programming problem (which may be linear, nonlinear, discrete, etc.) defined on S, and a linear programming problem defined on R^p.

In this connection, he defines two sets C and G as follows:

(a) A polyhedral convex cone C in R^{m+1}

$$(1.2) \qquad C = \{(u_0, u) \mid A'u - cu_0 \geqq 0,\ u \geqq 0,\ u_0 \geqq 0\},$$

(b) A set G in R^{q+1}

$$(1.3) \qquad G = \bigcap_{(u_0, u) \in C} \{(x_0, y) \mid u_0 x_0 + uF(y) - u_0 f(y) \leqq ub,\ y \in S\}.$$

Then, he states the basic theorem for a partitioning procedure:

Theorem 1.1. *(1) Problem* (1.1) *is not feasible, if and only if the programming problem*

$$(1.4) \qquad \text{Max } \{x_0 \mid (x_0, y) \in G\}$$

is not feasible, i.e. if and only if the set G is empty.

(2) Problem (1.1) *is feasible without having an optimum solution, if and only if Problem* (1.4) *is feasible without having an optimum solution.*

301

(3) If $(\bar{x}, \bar{y})$ is an optimum solution of Problem (1.1) *and*

$$\bar{x}_0 = c\bar{x} + f(\bar{y}),$$

then (x_0, y) is an optimum solution of Problem (1.4) *and $\bar{x}$ is an optimum solution of the linear programming problem*

$$(1.5) \qquad \text{Max } \{cx \mid Ax \leqq b - F(\bar{y}), \ x \geqq 0\}.$$

(4) If $(\bar{x}_0, \bar{y})$ is an optimum solution of Problem (1.4), *then Problem* (1.5) *is feasible and the optimum value of the objective function in this problem is equal to $\bar{x}_0 - f(\bar{y})$. If $\bar{x}$ is an optimum solution of Problem* (1.5), *then $(\bar{x}, \bar{y})$ is an optimum solution of Problem* (1.1), *with optimum value $\bar{x}_0$ for the objective function.*

Based on this theorem, he has designed two multi-step procedures for solving Problem (1.1). The two procedures differ only in the way the linear programming problem is solved.

Both procedures start from a subset Q of C and solve a programming problem

$$(1.6) \qquad \text{Max } \{x_0 \mid (x_0, y) \in G(Q)\},$$

where $G(Q)$ is a set defined by

$$(1.7) \qquad G(Q) = \bigcap_{(u_0, u) \in Q} \{(x_0, y) \mid u_0 x_0 + uF(y) - u_0 f(y) \leqq ub, \ y \in S\}.$$

Let an optimum solution of (1.6) be $(\bar{x}_0, \bar{y})$. In the dual type procedure, first solve the problem

$$(1.8) \qquad \text{Min } \{(b - F(y))u \mid A'u \geqq c, \ u \geqq 0\}.$$

Let an optimum solution of (1.8) be $\bar{u}$. Then, we have:

Theorem 1.2. *If $(\bar{x}_0, \bar{y})$ is an optimum solution of Problem* (1.6), *it is also an optimum solution of Problem* (1.4) *if and only if*

$$(b - F(\bar{y}))\bar{u} = \bar{x}_0 - f_n(\bar{y}).$$

And, if equality holds, we can get an optimum solution $(\bar{x}, \bar{y})$ of Problem (1.1) *where $\bar{x}$ is an optimum solution of the linear programming problem* (1.5).

On the other hand, if

$$(b - F(\bar{y}))\bar{u} < \bar{x}_0 - f(\bar{y}),$$

then extend the set Q by adding a certain vertex of the feasible region of Problem (1.8) and/or a certain extremal ray of the convex cone C. And return to Problem (1.6). Repeat the above procedures until an optimum solution of Problem (1.1) is found or Problem (1.6) and hence (1.1) are decided to have no feasible solution.

The second procedure which is of primal type, solves Problem (1.5) instead of (1.8), because it is often more convenient to solve (1.5) rather than (1.8). And from an

302

optimum solution of Problem (1.5), we can get necessary informations on the feasible region of Problem (1.8) and the convex cone C. But, it may happen that (1.5) is not feasible. So, artificial variables are introduced in order to avoid infeasibility.

Among the three problems which we are going to solve, the first two—the maximal pattern flow problem and the minimal cost pattern flow problem—will be treated by the primal type procedure in which we shall take special considerations on the introduction of artificial variables.

And the last one—the extended Critical Path Method—will be solved by the dual type procedure in which we shall exhibit a parametric analysis of Benders Decomposition.

2. MAXIMAL PATTERN FLOWS

2.1 Notations

N: A network with node set V and directed arc set L. V contains n nodes, numbered 1 through n. We use the notation (ij) to denote the arc leading from node i to node j. And we assume that the correspondence between (ij) and arc is one to one within each subset E, P and A (see below). $L = E + P + A$.

E: Set of ordinary arcs in the problem i.e. arcs associated only with flow constraints (capacity and flow conservation constraints).

$x = (x_{ij})$: Set of flows on the arcs in the set E.
  P: Set of special arcs associated with pattern constraints.

$y = (y_{ij})$: Set of flows on the arcs in the set P.
  S: Set of y, satisfying the given pattern constraints.
  A: Set of artificial arcs which shall be introduced in the course of solution.

$z = (z_{ij})$: Set of flows on the arcs in set A.
  $A(i)$: $A(i) = \{j \mid (ij) \in L\}$. Set of nodes after the node i.
  $B(j)$: $B(j) = \{i \mid (ij) \in L\}$. Set of nodes before the node j.
  c_{ij}: The capacity of the arc (ij) in the set E or P.

2.2 Problems

The problem of which we are going to find the optimum solution is the following flow problem from source node 1 to sink node n.

[*Problem I*] (with variables v, x and y)

(2.1) $$\text{Maximize } v$$

subject to

(2.2)
$$\left(\sum_{\substack{j\in A(i)\\(ij)\in E}} x_{ij}+\sum_{\substack{j\in A(i)\\(ij)\in P}} y_{ij}\right)-\left(\sum_{\substack{j\in B(i)\\(ji)\in E}} x_{ji}+\sum_{\substack{j\in B(i)\\(ji)\in P}} y_{ji}\right)=\delta_i v. \quad \text{(all } i\in V),$$

(2.3)
$$0\leqq x_{ij}\leqq c_{ij} \quad \text{(all } (ij)\in E),$$

(2.4)
$$0\leqq y_{ij}\leqq c_{ij} \quad \text{(all } (ij)\in P),$$

(2.5)
$$y\in S: \textit{pattern constraints},$$

where $\delta_i=1$ (if $i=1$), -1 (if $i=n$) and 0 (otherwise).

By *pattern constraints* (2.5), y must be in a set S associated with flow patterns. More exactly, let the vector $y=(y_{ij})$ have k components. Then y must be in a certain subset S in the nonnegative orthant of the k-dimensional Euclidean space R^k. The constraints which decide S, may be linear, nonlinear or combinatorial. For example, $y_{ab}=y_{cd}$, $y_{ef}y_{gh}=0$, etc. where $(ab),(cd)$, (ef) and (gh) are arcs in P. We call a feasible solution of Problem I *a pattern flow*, because in many applications they are associated with certain patterns of the flows in special arcs in the network. *A maximal pattern flow* is a pattern flow with the maximum of the flow value v. We assume that the capacity c_{ij} is finite for every (ij). And then, the flow value v is finite for every feasible pattern flow.

For a given y satisfying (2.4) and (2.5), we consider the following:

[Problem II (y)] (with variables v and x)

(2.6)   Maximize v, *subject to* (2.2) *and* (2.3).

Also, we cosider the dual problem of Problem II (y).

[Problem III (y)] (with variables u and t)

(2.7)
$$\text{Minimize } \sum_{(ij)\in P} y_{ij}(u_j-u_i)+\sum_{(ij)\in E} c_{ij}t_{ij},$$

subject to
$$u_i-u_j+t_{ij}\geqq 0 \quad \text{(all } (ij)\in E), \quad -u_1+u_n\geqq 1, \quad t_{ij}\geqq 0 \quad \text{(all } (ij)\in E).$$

Next, for each $(ij)\in P$, add an arc $(1i)$ to the original network if $i\neq 1$, and add an arc (jn) if $j\neq n$. Let A be the set of the additional or artificial arcs and $z=(z_{ij})$ be the flows in arcs in the set A.

Thus, we define the following

[Problem IV (y)] (with variables v, x and z)

$$\text{Maximize } v-M\left(\sum_{(ij)\in A} z_{ij}\right),$$

subject to

$$0 \leq x_{ij} \leq c_{ij} \quad \text{(all } (ij) \in E), \quad 0 \leq z_{ij} \quad \text{(all } (ij) \in A),$$

$$\left(\sum_{j \in A(i)} x_{ij} + \sum_{j \in A(i)} z_{ij} \right) - \left(\sum_{j \in B(i)} x_{ji} + \sum_{j \in B(i)} z_{ji} \right) - \delta_i v =$$

$$= -\left(\sum_{j \in A(i)} y_{ij} - \sum_{j \in B(i)} y_{ji} \right) \quad \text{(all } i \in V),$$

where $\delta_i = 1$ (if $i=1$), -1 (if $i=n$), 0 (otherwise), *and M is a sufficiently large positive number.*

The dual problem of Problem IV (y) is

[Problem V (y)] (with variables u and t)

$$\text{Minimize} \sum_{(ij) \in P} y_{ij}(u_j - u_i) + \sum_{(ij) \in E} c_{ij} t_{ij},$$

subject to

$$u_i - u_j + t_{ij} \geq 0 \quad \text{(all } (ij) \in E), \quad u_i - u_j \geq -M \quad \text{(all } (ij) \in A),$$
$$-u_1 + u_n \geq 1, \quad t_{ij} \geq 0 \quad \text{(all } (ij) \in E).$$

Next, we define a polyhedral convex cone C and a set G respectively, as follows:

$$(2.8) \quad C = \{(u_0, u, t) \mid u_i - u_j + t_{ij} \geq 0 \text{ (all } (ij) \in E), \; -u_1 + u_n \geq u_0, \; t_{ij} \geq 0 \text{ (all } (ij) \in E),$$
$$u_0 \geq 0\},$$

and

$$(2.9) \quad G = \bigcap_{(u_0, u, t) \in C} \left\{ (w, y) \;\middle|\; u_0 w + \sum_i u_i \left(\sum_j y_{ij} - \sum_j y_{ji} \right) \leq \sum_{(ij)} c_{ij} t_{ij}, \; y \in S, \; 0 \leq y_{ij} \leq c_{ij} \right\}.$$

And we have a programming on the set G.

[Problem VI (G)] (with variables w and y)

$$\text{Max } \{w \mid (w, y) \in G\}.$$

Similarly, for a subset Q of C, we define a set $G(Q)$ as follows:

$$(2.10) \quad G(Q) = \bigcap_{(u_0, u, t) \in Q} \left\{ (w, y) \;\middle|\; u_0 w + \sum_i u_i \left(\sum_j y_{ij} - \sum_j y_{ji} \right) \leq \sum_{(ij)} c_{ij} t_{ij}, \; y \in S, \; 0 \leq y_{ij} \leq c_{ij} \right\}.$$

And finally, we have a programming on the set $G(Q)$.

[Problem VII $(G(Q))$] (with variables w and y)

$$\text{Max } \{w \mid (w, y) \in G(Q)\}.$$

We show the algorithm by Fig. 1 below.

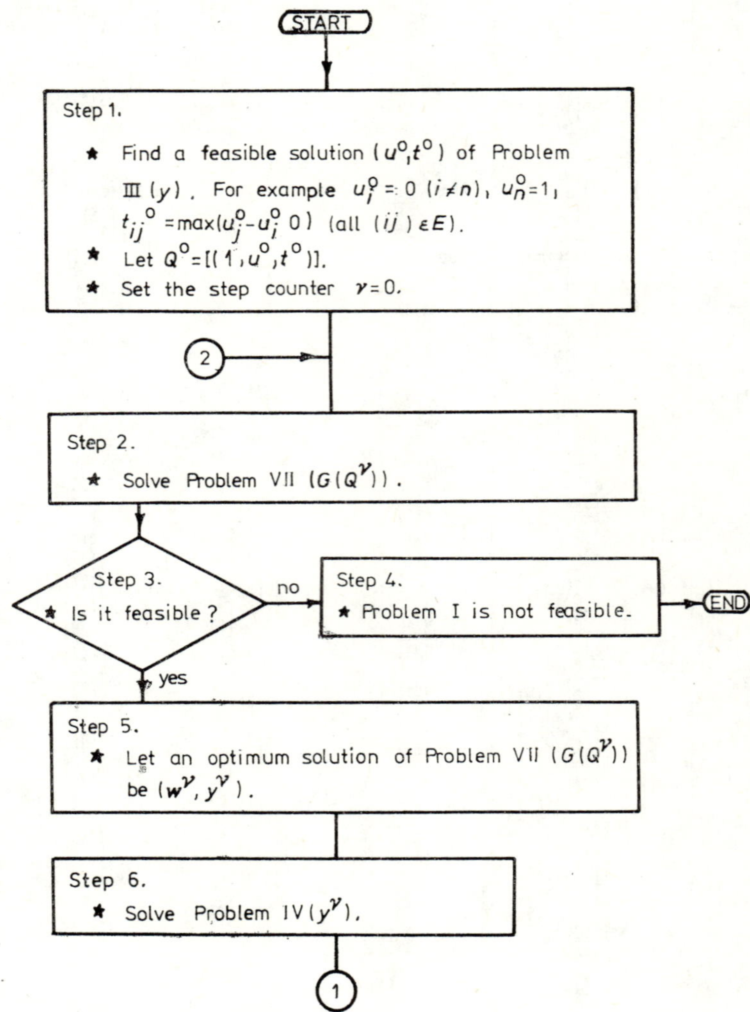

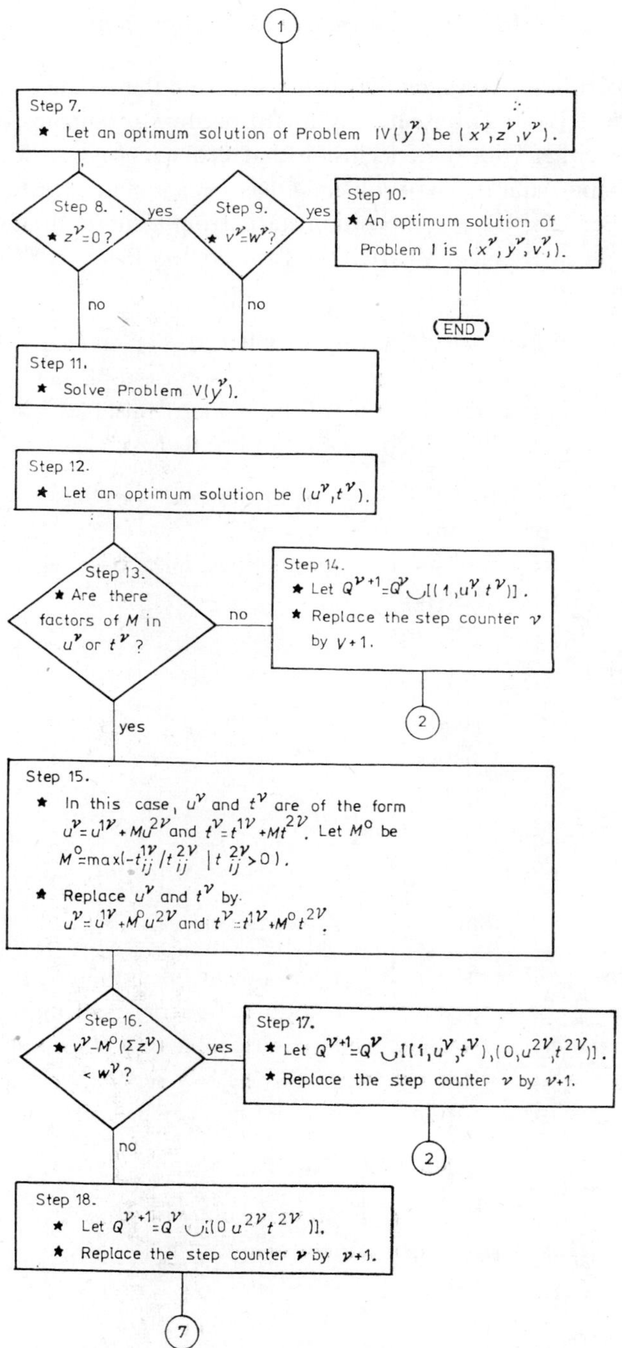

Fig. 1. Algorithm for finding Maximal Pattern Flow

2.4 Validity and Details of the Algorithm

The above algorithm follows, *mutatis mutandis*, from the Benders general Decomposition techniques [1] and consequently this procedure terminates, within a finite number of steps, either with the conclusion that Problem I is not feasible (Step 4), or that an optimum solution of Problem I has been obtained (Step 10). As was pointed out by Benders, his decomposition starts from a subset Q of C and extends Q at every time when Problem VII $(G(Q))$ is solved and finally we get an optimum solution of Problem I, if it exists.

In the course of solution, we must solve Problem III (y), given a y. But, in case of network flow problems, it is more convenient to solve Problem II (y) rather than Problem III (y). And the author introduced the artificial variables z in order to make Problem II (y) always feasible which resulted in Problem IV (y).

When an optimum solution (x^v, z^v, v^v) of Problem IV (y^v) has $z^v = 0$ and $v^v = w^v$, then we already have had an optimum solution (x^v, y^v, z^v) of Problem I by Theorem 1.2 and the duality theorem of linear programmings. Otherwise, we must extend the set Q in Step 14, Step 17, or Step 18.

Proposition 2.1. *Problem IV (y) is feasible for every nonnegative y and has a finite optimum.*

Proof. A feasible solution is given by $z_{1i} = y_{ij}$ (all $(ij) \in P$), $z_{jn} = y_{ij}$ (all $(ij) \in P$) and $x_{ij} = 0$ (all $(ij) \in E$). Finiteness follows from finiteness of the arc capacity.   Q.E.D.

Now, we shall describe some details of the procedures.

(1) Problem VII $(G(Q^v))$ in Step 2 may be a linear programming problem, a nonlinear programming, or a combinatorial optimization problem in accordance with the kind of given pattern constraints. And we must solve it by some known techniques respectively. But the size of the problem will be fairly reduced, compared with the original problem. Also the optimum solution of Problem VII $(G(Q^v))$ is always bounded, because we have, among its constraints, $w + \sum_i u_i^0 \left(\sum_j y_{ij} - \sum_j y_{ji} \right) \leqq \sum_{(ij)} c_{ij} t_{ij}^0$ and y_{ij} is bounded for all $(ij) \in P$. The concrete meaning of constraints of Problem VI (G) and Problem VII $G(Q)$ will be shown below.

(2) As was pointed out in Proposition 2.1, Problem IV (y^v) in Step 5 is always feasible and we can solve it by any minimal cost flow algorithms among which we will recommend a primal-dual type one, because we can get an optimum solution of the dual Problem V (y^v) at the same time when an optimum solution of Problem IV (y^v) is obtained.

308

In our pattern flow problem, so-called Max-Flow Min-Cut Theorem of network flows is not valid in general. But, a weak analogy exists. By Theorem 1.1, there is a correspondence between an optimum solution of Problem VI (G) and that of Problem I. The constraints of Problem VI (G) are as follows:

$$(2.11) \quad u_0 w + \sum_i u_i \left(\sum_j y_{ij} - \sum_j y_{ji} \right) \le \sum c_{ij} t_{ij}, \quad y \in S \text{ and } 0 \le y \le c, \text{ for every } (u_0, u, t) \in C.$$

Here, we are interested only in extremal rays of the polyhedral convex set C, because any other ray can be expressed by a convex combination of extremal rays and if (w, y) satisfies (2.11) for every extremal ray of C, it does also for every point of C. An extremal ray of C is, if $u_0 = 1$, of the form $u_1 = 0$, $u_n = 1$, $u_i = 0$ or 1 $(i=2, \ldots, n-1)$ and $t_{ij} = \max \{0, u_j - u_i\}$ (all $(ij) \in E$). This corresponds to a cut $(X, \bar{X})$ which separates source node 1 $(\in X)$ and sink node n $(\in \bar{X})$. That is, let $u_i = 0$ $(i \in X)$ and $u_i = 1$ $(i \in \bar{X})$. Then, inequality (2.11) results in

$$w \le \sum_{\substack{i \in X \\ j \in \bar{X}}} c_{ij} + \sum_{\substack{i \in X \\ j \in \bar{X}}} y_{ij} - \sum_{\substack{i \in \bar{X} \\ j \in X}} y_{ij}.$$

Thus, for every feasible pattern flow with flow value v and for every cut $(X, \bar{X})$ separating source node 1 and sink node n, we have

$$v \le \sum_{\substack{i \in X \\ j \in \bar{X}}} c_{ij} + \max_{\substack{y \in S \\ 0 \le y \le c}} \left\{ \sum_{\substack{i \in X \\ j \in \bar{X}}} y_{ij} - \sum_{\substack{i \in \bar{X} \\ j \in X}} y_{ij} \right\}.$$

This is a sort of flow-cut inequality, but equality does not hold in general. To make it equal, we must consider extremal rays of C with $u_0 = 0$. In this case, a meaningful extremal ray related with optimization is of the form $u_1 = 0$, $u_i = 0$ or 1 $(i=2, \ldots, n)$ and $t_{ij} = \max \{0, u_j - u_i\}$. And it corresponds to a cut $(X, \bar{X})$ separating node set X $(1 \in X)$ and $\bar{X}$. Then, inequality (2.11) results in

$$\sum_{\substack{i \in \bar{X} \\ j \in X}} y_{ij} \le \sum_{\substack{i \in X \\ j \in \bar{X}}} c_{ij} + \sum_{\substack{i \in X \\ j \in \bar{X}}} y_{ij}.$$

That is, for every cut $(X, \bar{X})$ with $1 \in X$ (and not always $n \in \bar{X}$), the sum of pattern flows from $\bar{X}$ to X is less than or equal to the sum of pattern flows from X to $\bar{X}$ and capacities of ordinary arcs from X to $\bar{X}$. The above-mentioned constraints define the set G. In our decomposition procedure, only effective constraints at each stage are chosen, instead of enumerating all.

3. MINIMAL COST PATTERN FLOWS

We consider the minimal cost pattern flow from source node 1 to sink node n, with flow value v. But, we transfer this problem into an equivalent minimal cost pattern circulation problem, by adding an arc $(n1)$ leading from n to 1. Let x_{n1} be the flow in arc $(n1)$. We add the constraint $x_{n1}=v$ to the problem. Thus, we have

[Problem I′] (with variables x and y)

(3.1)
$$\text{Minimize } \sum_E d_{ij}x_{ij}+f(y),$$

subject to

(3.2)
$$\left(\sum_j x_{ij}+\sum_j y_{ij}\right)-\left(\sum_j x_{ji}+\sum_j y_{ji}\right)=0 \quad (\text{all } i\in V,$$

(3.3)
$$0\leqq x_{ij}\leqq c_{ij} \quad (\text{all } (ij)\in E),$$

(3.4)
$$x_{n1}=v,$$

(3.5)
$$0\leqq y_{ij}\leqq c_{ij} \quad (\text{all } (ij)\in P),$$

(3.6)
$$y\in S,$$

where $d_{ij}(\geqq 0)$ is the unit cost of shipment from i to j and $f(y)$ is the cost of flows in arcs in P.

We can get an optimum solution of Problem I′ by an algorithm quite analogous to the preceding. Now, we define several problems and sets which correspond to those in Section 2.

[Problem III′ (y)] (with variables u and t)
$$\text{Minimize } v(u_1-u_n)+\sum_E c_{ij}t_{ij}+\sum_P y_{ij}(u_j-u_i),$$

subject to
$$u_i-u_j+t_{ij}\geqq -d_{ij} \quad (\text{all } (ij)\in E), \quad t_{ij}\geqq 0 \quad (\text{all } (ij)\in E).$$

Next, in order to make the dual problem of Problem III′ (y) always feasible for every v and y, add, for each arc $(ij)\in P$, an arc $(1i)$ to the original network if $i\neq 1$, and add an arc (jn) if $j\neq n$. Also, add arcs $(1n)$ and $(n1)$. Let A be the set of the additional or artificial arcs and $z=(z_{ij})$ be the flows in arcs in the set A.

[Problem IV′ (y)] (with variables x and z)
$$\text{Minimize } \sum_E d_{ij}x_{ij}+M\left(\sum_A z_{ij}\right),$$

subject to
$$0\leqq x_{ij}\leqq c_{ij} \quad (\text{all } (ij)\in E), \quad x_{n1}=v, \quad 0\leqq z_{ij} \quad (\text{all } (ij)\in A),$$
$$\left(\sum_j x_{ij}+\sum_j z_{ij}\right)-\left(\sum_j x_{ji}+\sum_j z_{ji}\right)=-\left(\sum_j y_{ij}-\sum_j y_{ji}\right) \quad (\text{all } i\in V).$$

[Problem V′ (y)] (with variables u and t)
$$\text{Minimize } v(u_1-u_n)+\sum_E c_{ij}t_{ij}+\sum_P y_{ij}(u_j-u_i),$$

310

subject to

$$u_i - u_j + t_{ij} \geq -d_{ij} \quad (\text{all } (ij) \in E), \quad u_i - u_j \geq -M \quad (\text{all } (ij) \in A), \quad t_{ij} \geq 0 \quad (\text{all } (ij \in E).$$

[Convex Cone C]

$$C = \{(u_0, u, t) \mid u_i - u_j + t_{ij} \geq -d_{ij}u_0 \ (\text{all } (ij) \in E), \ t_{ij} \geq 0 \ (\text{all } (ij) \in E), u_0 \geq 0\}.$$

[Set G]

$$G = \bigcap_{(u_0, u, t) \in C} \left\{ (w, y) \ \middle| \ u_0 w + \sum_i u_i \left(\sum_j y_{ij} - \sum_j y_{ji} \right) + u_0 f(y) \leq \sum_E c_{ij} t_{ij} + v(u_1 - u_n), \right.$$
$$\left. 0 \leq y \leq c, \ y \in S \right\}.$$

[Sets Q and G(Q)]
Q = A subset of convex cone C,

$$G(Q) = \bigcap_{(u_0, u, t) \in Q} \left\{ (w, y) \ \middle| \ u_0 w + \sum_i u_i \left(\sum_j y_{ij} - \sum_j y_{ji} \right) + u_0 f(y) \leq \sum_E c_{ij} t_{ij} + v(u_1 - u_n), \right.$$
$$\left. 0 \leq y \leq c, \ y \in S \right\}.$$

[Problem VII′ (G(Q))] (with variables w and y)

$$\text{Max } \{w \mid (w, y) \in G(Q)\}.$$

Now, we describe the algorithm for finding a minimal cost pattern flow. However, the flow chart in Fig. 1 is also applicable to this case. So, we will only point out the steps which differ from the corresponding steps in the preceding algorithm. Of course, each Problem in Fig. 1 must be put a dash after the number.

[Steps which differ from the corresponding steps in the preceding algorithm]

Step 1.   * *Find a feasible solution (u^0, t^0) of Problem III′ (y).*

       For example, $u_i^0 = 0$ (all i), $t_{ij}^0 = 0$ (all (ij)).

   * *Let $Q^0 = \{(l, u^0, t^0)\}$.*
   * *Let $v = 0$.*

Step 7.   * *Let an optimum solution of Problem IV′ (y^v) be (x^v, z^v).*

Step 9.   * $-\Sigma d_{ij} x_{ij}^v - f(y^v) = w^v$?

Step 10. * *An optimum solution of Problem I′ is (x^v, y^v).*

Step 16. * $-\Sigma d_{ij} x_{ij}^v - M^0(\Sigma z_{ij}^v) - f(y^v) < w^v$?

4. AN EXTENSION OF THE CRITICAL PATH METHOD

J. E. Kelley [2] has defined the Critical Path Method (CPM) as follows:

$$\text{(4.1)} \qquad \text{Maximize } \sum_P c_{ij} y_{ij},$$

subject to

$$\text{(4.2)} \qquad y_{ij} + t_i - t_j \leq 0 \quad (\text{all } (ij) \in P),$$

$$\text{(4.3)} \qquad d_{ij} \leq y_{ij} \leq D_{ij} \quad (\text{all } (ij) \in P),$$

$$\text{(4.4)} \qquad -t_1 + t_n \leq \lambda,$$

where P: set of activities in the project network, y_{ij}: the duration of activity (ij), D_{ij}: the normal duration of (ij), d_{ij}: the crash duration of (ij), λ: the project duration, and c_{ij}: the cost slope of activity (ij).

We consider an extension of CPM in the sense that some additional conditions are imposed on the durations of special activities. Examples of such conditions are as follows:

$$\text{(1)} \qquad y_{ab} + y_{cd} = k.$$

The sum of the duration of activity (ab) and that of (cd) must be equal to a constant k. To shorten y_{ab}, one must lengthen y_{cd} and vice versa, because both activities use a common resource.

$$\text{(2)} \qquad y_{ab} - y_{cd} = k.$$

The difference of y_{ab} and y_{cd} must be equal to k. That is, to shorten y_{ab}, one must shorten y_{cd} at the same time, because both activities have the same tendency as to activity duration.

$$\text{(3)} \qquad y_{ab} = d_{ab} \quad or \quad D_{ab}.$$

The duration of activity (ab) must be either D_{ab} or d_{ab}.
Cases *(1)* and *(2)* can be generalized as

$$\alpha y_{ab} + \beta y_{cd} \lesseqgtr \gamma.$$

This additional condition causes the dual problem of the extended CPM to be a pattern flow problem and we can solve it by a variant of decomposition techniques developed in Sections 2 and 3. Also, the parametric analysis regarding λ, can be handled by Kelley's Primal-Dual Method [3]. Therefore, we do not go into it further.

In the following, we only consider the discrete variable case *(3)* where durations of certain activities can take only discrete values.

312

$$4.1 \ Problem$$

[Problem I (λ)] (with variables y, z and t)

(4.5) $$\text{Maximize } u_0 = \Sigma c_{ij} y_{ij} + f(z),$$

subject to

(4.6) $$y_{ij} + t_i - t_j \leqq 0 \quad (\text{all } (ij) \in P),$$

(4.7) $$z_{ij} + t_i - t_j \leqq 0 \quad (\text{all } (ij) \in R),$$

(4.8) $$d_{ij} \leqq y_{ij} \leqq D_{ij} \quad (\text{all } (ij) \in P),$$

(4.9) $$-t_1 + t_n \leqq \lambda,$$

(4.10) $$z \in S,$$

where R is the set of discrete variable activities, $z = (z_{ij})$ is the vector of durations of activities in R, and S is the region of z which satisfies the given additional constraints. $f(z)$ is the utility of duration vector z.

In what follows, we will investigate a parametric analysis, of this problem with regard to the project duration λ.

For given λ and z, we consider two mutual dual problems:

[Problem II ($z|\lambda$)] (with variables y and t)

$$\text{Maximize } u_1 = \sum_P c_{ij} y_{ij},$$

subject to

$$y_{ij} + t_i - t_j \leqq 0 \quad (\text{all } (ij) \in P), \quad t_i - t_j \leqq z_{ij} \quad (\text{all } (ij) \in R),$$
$$d_{ij} \leqq y_{ij} \leqq D_{ij} \quad (\text{all } (ij) \in P), \quad -t_1 + t_n \leqq \lambda.$$

[Problem III ($z|\lambda$)] (with variables f, f', g, h and v)

$$\text{Minimize } u_2 = \lambda v + \sum_P D_{ij} g_{ij} - \sum_P d_{ij} h_{ij} - \sum_R f'_{ij} z_{ij},$$

subject to

$$f_{ij} + g_{ij} - h_{ij} = c_{ij} \quad (\text{all } (ij) \in P),$$

$$\sum_j (f_{ij} + f'_{ij}) - \sum_j (f_{ji} + f'_{ji}) = \delta_i v \quad (\text{all } i \in V),$$

$$f_{ij}, f'_{ij}, g_{ij}, h_{ij}, v \geqq 0,$$

where $\delta_i = 1$ (if $i = 1$), -1 (if $i = n$), 0 (otherwise).
We define several sets and a problem as follows.

[Polyhedral Convex Cone C]

$$(4.11) \qquad C=\left\{(f_0,f,f',g,h,v)\,\middle|\,f_{ij}+g_{ij}-h_{ij}=c_{ij}f_0 \quad (\text{all } (ij)\in P),\right.$$

$$\sum_j (f_{ij}+f'_{ij})-\sum_j (f_{ji}+f'_{ji})-\delta_i p=0 \quad (\text{all } i\in V),$$

$$\left. f_0,f,f',g,h,v\geqq 0\right\}.$$

[Set $G(\lambda)$]

$$(4.12) \qquad G(\lambda)=\bigcap_{(f_0,f,f',g,h,v)\in C}\left\{(z_0,z)\,\middle|\,f_0 z_0+\sum_P f'_{ij}z_{ij}-f_0 f(z)\leqq\sum_P d_{ij}g_{ij}-\right.$$

$$\left.-\sum_P d_{ij}h_{ij}+\lambda v,\,z\in S\right\}.$$

[Set Q]

Q is a subset of C.

[Set $G(Q\,|\,\lambda)$]

$$(4.13) \qquad G(Q\,|\,\lambda)=\bigcap_{(f_0,f,f',\kappa,h\,v)\in Q}\left\{(z_0,z)\,\middle|\,f_0 z_0+\sum_P f'_{ij}z^i{}_j-f_0 f(z)\leqq\sum_P d_{ij}g_{ij}-\right.$$

$$\left.-\sum_P d_{ij}h_{ij}+\lambda v,\,z\in S\right\}.$$

[Problem IV $\left(G(Q|\lambda)\right)$] (with variables z_0 and z)

$$\text{Max } \{z_0\,|\,(z_0,z)\in G(Q|\lambda)\}.$$

First, we assume that we can find an optimum solution (y^*,z^*,t^*) of Problem I (λ) for a sufficiently large λ. For example, let

$$y^*_{ij}=D_{ij} \quad (\text{all } (ij)\in P), \quad f(z^*)=\max_{z\in S}\{f(z)\}, \quad t^*_1=0,$$

and

$$t^*_j=\max\left\{\max_i (t^*_i+D_{ij}),\,\max_i (t^*_i+z^*_{ij})\right\} \quad (j=2,\ldots,n).$$

Then, (y^*,z^*,t^*) is an optimum solution for $\lambda=t^*_n$. Correspondingly, an optimum solution of Problem II $(z^*\,|\,t^*_n)$ is (y^*,t^*) and that of Problem III $(z^*\,|\,t^*_n)$ is $(f^*=0$ $f'^*=0,\,g^*=c,\,h^*=0,\,v^*=0)$. Take

$$(4.14) \qquad Q=\{(1,f^*,f'^*,g^*,h^*,v^*)\}.$$

Then, an optimum solution of Problem IV $\left(G(Q\,|\,t^*_n)\right)$ is

$$(z^*_0=\Sigma D_{ij}c_{ij}+f(z^*),\,z^*).$$

In the following algorithm, we shall show a method by which we can find an optimum solution of Problem I $(\bar{\lambda}-\Theta)$ $(\Theta\geqq 0)$, in the knowledge of that of Problem I $(\bar{\lambda})$.

314

We shall use the ordinary CPM algorithm [2], as a subroutine, in order to solve Problem III $(z|\lambda)$ and hence Problem II $(z|\lambda)$, at each stage of our algorithm. In particular, the max-flow and min-cut thus found, plays an important role.

4.2 Algorithm

Step 1. Initialization. Let an optimal solution of Problem I $(\bar{\lambda})$, Problem II$(\bar{z}|\bar{\lambda})$, Problem III $(\bar{z}|\bar{\lambda})$ and Problem IV $(G(Q|\bar{\lambda}))$ be $(\bar{y}, \bar{t}, \bar{z})$, $(\bar{y}, \bar{t})$, $(1, \bar{f}, \bar{f}', \bar{g}, \bar{h}, \bar{v})$ and $(\bar{z}_0, \bar{z})$, respectively. *(Go to Step 2.)*

Step 2. Finding the bound of decrease of $\bar{\lambda}$.

Step 2.1. Try the parametric analysis of Problem IV $(G(Q|\bar{\lambda}))$ with regard to $\bar{\lambda}$ and determine the range $[\bar{\lambda} - \Theta_1, \bar{\lambda}]$ $(\Theta_1 \geqq 0)$ where the solution $\bar{z}$ remains optimal. If there is no feasible solution of Problem IV $(G(Q|\lambda))$ for $\lambda < \bar{\lambda}$, then, Problem I (λ) has no feasible schedule for project duration less than $\bar{\lambda}$. *(The end.)* Otherwise, *go to Step 2.2.*

Step 2.2. By applying the parametric analysis to Problem III $(\bar{z}|\bar{\lambda})$ with regard to $\bar{\lambda}$, determine the range $[\bar{\lambda} - \Theta_2, \bar{\lambda}]$ $(\Theta_2 \geqq 0)$ where the solution $(1, \bar{f}, \bar{f}', \bar{g}, \bar{h}, \bar{v})$ remains optimal. *(Go to Step 2.3.)*

Step 2.3. Let $\Theta_0 = \min \{\Theta_1, \Theta_2\}$. *(Go to Step 3.)*

Step 3. Determining the optimal schedule for $\lambda = \bar{\lambda} - \Theta_0$. If $\Theta_0 = 0$, then *go to Step 4.* Otherwise, at the end of the parametric analysis in Step 2.2 by the primal-dual method, we can find a cut-set $(X, \bar{X})$ of the project network which has the minimum cut value at the time of the project duration $\bar{\lambda}$. Then, an optimal schedule $(y(\Theta), z(\Theta), t(\Theta))$ for $\lambda = \bar{\lambda} - \Theta$ $(0 \leqq \Theta \leqq \Theta_0)$ is as follows:

$$
\begin{aligned}
y_{ij}(\Theta) &= \bar{y}_{ij} - \Theta && (\text{if } i \in X, j \in \bar{X}, (ij) \in P), \\
&= \bar{y}_{ij} + \Theta && (\text{if } i \in \bar{X}, j \in X, (ij) \in P), \\
&= \bar{y}_{ij} && (\text{otherwise}), \\
z_{ij}(\Theta) &= \bar{z}_{ij} && ((ij) \in R), \\
t_i(\Theta) &= \bar{t}_i && (\text{if } i \in X), \\
&= \bar{t}_i - \Theta && (\text{if } i \in \bar{X}).
\end{aligned}
$$

(Go to Step 4.)

Step 4. Finding new solution of Problem I (λ) for $\lambda = \bar{\lambda} - \Theta_0^+$. In this step, we get an optimum solution of Problem I (λ) for $\lambda = \bar{\lambda} - \Theta_0 - \varepsilon$ where ε is a sufficiently small positive number and we use the notation Θ_0^+ instead of $\Theta_0 + \varepsilon$.

The method is mainly along Benders decomposition.

Step 4.1. Let $v = 0$ and $Q^v = Q$. *(Go to Step 4.2.)*

Step 4.2. Get an optimum solution (z_0^v, z^v) of Problem IV $(G(Q^v|\bar{\lambda} - \Theta_0^+)$. If $\Theta_0 = \Theta_1 < \Theta_2$ in Step 2, then we have not to solve the problem for $v = 0$, since $(z_0^v, z^v) =$

$=(\bar{z}_0 - \Theta_0^+ \bar{v}, \bar{z})$. If Problem IV $(G(Q^v) \mid \bar{\lambda} - \Theta_0^+)$ has no feasible solution, then Problem I (λ) has no feasible schedule, for project duration less than $\bar{\lambda} - \Theta_0$. *(The end.)* Otherwise, *go to Step 4.3.*

Step 4.3. Solve Problem III $(z^v \mid \bar{\lambda} - \Theta_0^+)$. (a) If it has an optimum solution $(f^v, f'^v, g^v, h^v, v^v)$ and the optimum value u_2^v of its objective function is equal to $z_0^v - f(z^v)$, then an optimum solution of Problem I $(\bar{\lambda} - \Theta_0^+)$ is (y^v, z^v, t^v) where (y^v, t^v) is an optimum solution of Problem II $(z^v \mid \bar{\lambda} - \Theta_0^+)$. Replace $\bar{\lambda}$ by $\bar{\lambda} - \Theta_0$, $(1, \bar{f}, \bar{f}', \bar{g}, \bar{h}, \bar{v})$ by $(1, f^v, f'^v, g^v, h^v, v^v)$ and $(\bar{z}_0, \bar{z})$ by (z_0^v, z^v), respectively. *(Go back to Step 2.)* Otherwise, if $u_2^v < z_0^v - f(z^v)$, then let

$$Q^{v+1} = Q^v \cup \{(1, f^v, f'^v, g^v, h^v, v^v)\}$$

and replace the step counter v by $v+1$. *(Go back to Step 4.2.)* (b) If an optimum solution of Problem III $(z^v \mid \bar{\lambda} - \Theta_0^+)$ is unbounded, then a critical path of the project network for $\lambda = \bar{\lambda} - \Theta_0$ with $z = z^v$, is composed of only activities with crash duration and/or z^v and infinite flow value is permitted on this path. Let

$$f_{ij}^v = 1 \quad \text{(on the critical path)} \quad = 0 \quad \text{(otherwise)},$$

$$f_{ij}'^v = 1 \quad \text{(on the critical path)}, \quad = 0 \quad \text{(otherwise)},$$

$$g_{ij}^v = 0 \quad \text{(all } (ij) \in P),$$

$$h_{ij}^v = 1 \quad \text{(on the critical path)}, \quad = 0 \quad \text{(otherwise)},$$

$$v^v = 1.$$

And let

$$Q^{v+1} = Q^v \cup \{(0, f^v, f'^v, g^v, h^v, v^v)\}.$$

Replace the step counter v by $v+1$.
(Go back to Step 4.2.)

4.3 Validity and Details of the Algorithm

Proposition 4.1. Problem I (λ) is not feasible if Problem IV $(G(Q) \mid \lambda)$ is not feasible.

Proof. By Theorem 1.1, Problem I (λ) is not feasible if and only if Problem IV $(G \mid \lambda)$ is not feasible. Since $Q \subset C$, we have $G(Q) \supset G$. Hence, if $G(Q)$ is empty, then also G.

$$\text{Q.E.D.}$$

This proposition works in Step 2.1 and Step 4.2 as the termination criterion.

Proposition 4.2. *Let the maximum value for the objective function z_0 of Problem IV $(G(Q) / \bar{\lambda} - \Theta)$ $(0 \leq \Theta \leq \Theta_0)$ be $z_0(\Theta)$. Then, $z_0(\Theta)$ satisfies*

$$z_0(\Theta) = \bar{z}_0 - \Theta \bar{v}.$$

316

Proof. By the definition of Θ_0, $\bar{z}$ is an optimum solution of Problem IV $(G(Q)$ $\bar{\lambda}-\Theta)$ $(0 \leq \Theta \leq \Theta_0)$. Therefore, for every $(f_0, f, f', g, h, v) \in Q$, we have

$$f_0 z_0(\Theta) + \Sigma f'_{ij} z_{ij} - f_0 f(\bar{z}) \leq \Sigma D_{ij} g_{ij} - \Sigma d_{ij} h_{ij} + (\bar{\lambda} - \Theta)v.$$

And since $(1, \bar{f}, \bar{f}', \bar{g}, \bar{h}, \bar{v}) \in Q$, we have

$$z_0(\Theta) + \Sigma \bar{f}'_{ij} \bar{z}_{ij} - f(\bar{z}) \leq \Sigma D_{ij} \bar{g}_{ij} - \Sigma d_{ij} \bar{h}_{ij} + (\bar{\lambda} - \Theta)\bar{v}.$$

But, for $\Theta = 0$, from the optimality of $(\bar{z}_0, \bar{z})$ and $(1, \bar{f}, \bar{f}', \bar{g}, \bar{h}, \bar{v})$, we have

$$\bar{z}_0 + \Sigma \bar{f}'_{ij} \bar{z}_{ij} - f(\bar{z}) = \Sigma D_{ij} \bar{g}_{ij} - \Sigma d_{ij} \bar{h}_{ij} + \bar{\lambda}\bar{v}.$$

Hence, $z_0(\Theta) \leq \bar{z}_0 - \Theta \bar{v}$. Now, we demonstrate the equality.

Suppose that the equality does not holds. Then, for some $\Theta (0 \leq \Theta \leq \Theta_0)$, there exists at least a point $(1, \bar{\bar{f}}, \bar{\bar{f}}', \bar{\bar{g}}, \bar{h}, \bar{\bar{v}}) \in Q$ such that

(4.15)
$$z_0(\Theta) = \Sigma D_{ij} \bar{\bar{g}}_{ij} - \Sigma d_{ij} \bar{h}_{ij} - \Sigma \bar{\bar{f}}'_{ij} \bar{z}_{ij} + f(\bar{z}) + (\bar{\lambda} - \Theta)\bar{\bar{v}} <$$
$$< \Sigma D_{ij} \bar{g}_{ij} - \Sigma d_{ij} \bar{h}_{ij} - \Sigma \bar{f}'_{ij} \bar{z}_{ij} + f(\bar{z}) + (\bar{\lambda} - \Theta)\bar{v}.$$

But, since $(\bar{\bar{f}}, \bar{\bar{f}}', \bar{\bar{g}}, \bar{h}, \bar{\bar{v}})$ is a feasible solution of Problem III $(\bar{z} \mid \bar{\lambda} - \Theta)$ and $(\bar{f}, \bar{f}', \bar{g}, \bar{h}, \bar{v})$ is an optimum solution, we must have

$$(\bar{\lambda} - \Theta)\bar{v} + \Sigma D_{ij} \bar{g}_{ij} - \Sigma d_{ij} \bar{h}_{ij} - \Sigma \bar{f}'_{ij} \bar{z}_{ij} \leq (\bar{\lambda} - \Theta)\bar{\bar{v}} + \Sigma D_{ij} \bar{\bar{g}}_{ij} - \Sigma d_{ij} \bar{h}_{ij} - \Sigma \bar{\bar{f}}'_{ij} \bar{z}_{ij}.$$

This contradicts (4.15). Q.E.D.

From the result of this proposition, an optimum solution $(z_0(\Theta), \bar{z})$ of Problem IV $(G(Q) \mid \bar{\lambda} - \Theta)$ and an optimum solution $(\bar{f}, \bar{f}', \bar{g}, \bar{h}, \bar{v})$ of Problem III $(\bar{z} \mid \bar{\lambda} - \Theta)$ satisfy

$$z_0(\Theta) - f(\bar{z}) = (\bar{\lambda} - \Theta)\bar{v} + \Sigma D_{ij} \bar{g}_{ij} - \Sigma d_{ij} \bar{h}_{ij} + \Sigma \bar{f}'_{ij} \bar{z}_{ij}.$$

Hence, by Theorem 1.2, we can get an optimum solution $(y(\Theta), \bar{z}, t(\Theta))$ of Problem I $(\bar{\lambda} - \Theta)$, where $(y(\Theta), t(\Theta))$ is an optimum solution of Problem II $(\bar{z} \mid \bar{\lambda} - \Theta)$. To solve Problem II $(\bar{z} \mid \bar{\lambda} - \Theta)$, we use the minimum cut $(X, \bar{X})$ of the network which can be found by the labelling method of the ordinary CPM.

Now, we have

Proposition 4.3. *An optimum schedule $(y(\Theta), z(\Theta), t(\Theta))$ for the project duration $\lambda = \bar{\lambda} - \Theta$ $(0 \leq \Theta \leq \Theta_0)$ is as follows:*

$$
\begin{aligned}
y_{ij}(\Theta) &= \bar{y}_{ij} - \Theta && (if\ i \in X, j \in \bar{X}, (ij) \in P), \\
&= \bar{y}_{ij} + \Theta && (if\ i \in \bar{X}, j \in X, (ij) \in P), \\
&= \bar{y}_{ij} && (otherwise), \\
z_{ij}(\Theta) &= \bar{z}_{ij} && ((ij) \in R), \\
t_i(\Theta) &= \bar{t}_i && (if\ i \in X), \\
&= \bar{t}_i - \Theta && (if\ i \in \bar{X}),
\end{aligned}
$$

where $(X, \bar{X})$ is a min-cut obtained by applying the ordinary CPM to Problem II $(\bar{z} \mid \bar{\lambda})$.

Next, in Step 4, we use Benders decomposition to find an optimum solution of Problem I $(\bar{\lambda} - \Theta_0^+)$.

From what we have mentioned above, we can see the validity of the algorithm.

Now, we will explain some details of the algorithm.

(1) The set Q may be extended at each time when we solve Step 4.3. But this will make the number of elements of Q very large and large memory size may be needed. However, we can set $Q = \{(1, f^*, f'^*, g^*, h^*, v^*)\}$ in Step 4.1 at each time when we solve Step 4, instead of extending Q, where $(1, f^*, f'^*, g^*, h^*, v^*)$ is given by (4.14). This will make the computing time longer.

(2) The parametric analysis of Problem IV $(G(Q)|\bar{\lambda})$ in Step 2.1, must be done by some techniques appropriate to the problem among which is included the enumeration of the discrete variable z. Also the discreteness of z guarantees the discrete changes of the interval of λ where z remains optimal and hence finiteness of the algorithm with respect to λ.

(3) The parametric analysis of Problem III $(\bar{z}|\bar{\lambda})$ in Step 2.2, can be done by the ordinary CPM, after fixing $z = \bar{z}$. And we have a minimum cut $(X, \bar{X})$ when we have found a maximum flow by the flow algorithm of CPM. We use this cut $(X, \bar{X})$ in Step 3.

(4) If the optimum solution of Problem III $(z^v \mid \bar{\lambda} - \Theta_0^+)$ is unbounded, then Problem II $(z^v \mid \bar{\lambda} - \Theta_0^+)$ is not feasible. And the maximum flow is unbounded on a critical path which consists of only activities with the crash duration and/or z^v. Let

$f_{ij}^v = 1$ (on the critical path), $= 0$ (otherwise),
$f_{ij}'^v = 1$ (on the critical path), $= 0$ (otherwise),
$g_{ij}^v = 0$ $\big($all $(ij) \in P\big)$,
$h_{ij}^v = 1$ (on the critical path), $= 0$ (otherwise),
$v^v = 1$.

Then, $(0, f^v, f'^v, g^v, h^v, v^v)$ is on an extremal ray of C which is not taken into Q^v, yet. The corresponding constraint of G is

$$\sum_{CP} z_{ij} \leqq (\bar{\lambda} - \Theta_0^+) - \sum_{CP} d_{ij},$$

where CP is the set of the activities on the critical path. And this constraint forces us to shorten some z_{ij} on the critical path.

REFERENCES

[1] Benders, J. F.: Partitioning Procedures for Solving Mixed-Variables Programming Problems, *Numerische Mathematik* 4, (1962), 238–252.

[2] Kelley, J. E. Jr.: Critical-Path Planning and Scheduling; Mathematical Basis, *Opns. Res.* 9, (1961), 296–320.

[3] Kelley, J. E. Jr.: Parametric Programming and the Primal-Dual Algorithm, *Opns. Res.* 7, (1959), 327–334.

318

PART 7

GRAPHS AND COMBINATORICS

POLYNOMIAL BOUNDING FOR NP-HARD PROBLEMS

P. M. CAMERINI and F. MAFFIOLI

(Milano, Italy)

1. INTRODUCTION

1.1 Let $\Gamma = (A, \mathcal{I})$ be an independence system (IS) defined over the set of arcs of a given digraph $D = (N, A)$. Let $n = |N|$, $m = |A|$ and c_{ij} be the cost of arc (i, j).

1.2 We will consider the following constrained assignment problem:

$$\min \sum_{ij} c_{ij} x_{ij}$$

$$\text{s. t.} \sum_{i} x_{ij} = 1 \quad j = 1, 2, \ldots, n$$

$$\sum_{j} x_{ij} = 1 \quad i = 1, 2, \ldots, n$$

$$\{(i, j) : x_{ij} = 1\} \in \mathcal{I}$$

$$x_{ij} \in \{0, 1\}.$$

1.3 We shall assume that a polynomial bounded algorithm exists for testing if any subset I of A belongs to $\mathcal{I}$ and that its complexity is $0(c(m))$.

1.4 A minimal dependent subset S os A is called a *circuit* of Γ. Because of (1.6) it is always possible to find a circuit, if any, in $I \subseteq A$ by a polynomial bounded computation.

1.5 This work presents two polynomial bounded methods, respectively in sections 2 and 3, for calculating probably tight lower bounds to the value of the solution of problem 1.2.

1.6 The interest of problem 1.2 arises from the fact that many of the NP-hard problems in the sense of Karp [6], such as for instance the traveling salesman, 3-dimensional assignment and several scheduling problems are particular cases of this problem.

1.7 It is very unlikely therefore that polynomial bounded methods for the *exact* solutions of problem 1.2 will ever be found [8]. As a consequence efficient and possibly tight bounding techniques may be very useful both for guiding implicit

319

search methods towards the solution of a given problem and/or to evaluate heuristically obtained suboptimum solutions [1, 2, 10].

2. A CONTRACTION ALGORITHM

2.1 *Step 0* (Start). Let Γ, D and C be respectively the given IS, digraph and cost matrix. $L \leftarrow 0$, $k \leftarrow 0$.

Step 1 (Assignment). Find a minimum cost assignment I of D by, say, the Hungarian method [7]. Let C' be the cost matrix at the end of this computation (i.e. each arc of D has a cost $c'_{ij} \geqq 0$ which would affect the solution if that arc is forced into it).

Step 2 (Bounding). $L \leftarrow L + \sum_I c_{ij}$, $k \leftarrow k + 1$.

Step 3 (Testing). Find a circuit S of Γ in I. If none exists stop: L is the best bound obtained.

Step 4 (Contraction). Let F be the set of nodes of D incident to arcs of S. Form a new digraph $D' = (N', A')$ as follows. Let $N' = N - F + \{k'\}$ and

$$A' = H + \{(i, k') : i \notin F \text{ and } \exists (i, j) \in A \text{ s.t. } j \in F\} +$$
$$+ \{(k', j) : j \notin F \text{ and } \exists (i, j) \in A \text{ s.t. } i \in F\}.$$

H being the set of arcs of $\langle N - F \rangle$, the subgraph induced by $N - F$ on D.

Step 5 (Cost updating). Adjoin to C' a new row and column indexed by k'. Let

$$\left. \begin{array}{l} c'_{jk'} = \min_{i \in F} c'_{ji} \\[2ex] c'_{k'j} = \min_{i \in F} c'_{ij} \end{array} \right\} \quad \forall j \in N - F.$$

Erase from C' the rows and columns corresponding to nodes of F.

$$C \leftarrow C', \quad D \leftarrow D'.$$

Step 6 (Bound improving). Solve the following problem:

$$\min \sum_{(i, j) \in A} x_{ij} c_{ij}$$

$$\text{s.t.} \sum_{i \in N} x_{ij} \begin{cases} = 1 & \forall \text{ nonprimed } j \in N \\ \geqq 1 & \forall \text{ primed } \quad j \in N \end{cases}$$

$$\sum_{j \in N} x_{ij} \begin{cases} = 1 & \forall \text{ nonprimed } i \in N \\ \geqq 1 & \forall \text{ primed } \quad i \in N \end{cases}$$

$$x_{ij} \in \{0, 1\}.$$

This can be done for instance by the out-of-kilter method*. Let c'_{ij} be equal to the value of the corresponding dual variable λ_{ij} at the end of the out-of-kilter method

---

* The authors have referred to [9] as a particularly clear outline of this method.

and therefore with the same property obtained at the end of step 1. Let I be the set of arcs of D corresponding to the just found solution. Go to step 2.

2.2 To prove that L at the end of algorithm 2.1 is a lower bound to the solution of problem 1.2, note that such an optimum solution corresponds to an assignment $\bar{J}$ of D and therefore step 1 is correct. Furthermore, $\bar{J}$ must contain for each circuit S of Γ at least one arc not belonging to $\langle F \rangle$ which is directed into a node of F and at least one arc not belonging to $\langle F \rangle$ which is directed out of a node of F. $\bar{J}$ must also contain exactly one arc directed out of and exactly one arc directed into each node not belonging to F. Therefore, solving the problem of step 6 with the new costs gives a lower bound to the difference between the cost of an optimum solution and the cost of I. Obviously, the procedure can be iterated until circuits are found in step 3 and the algorithm terminates. This must eventually occur since at each iteration D is smaller and Φ is a member if any IS.

2.3 The out-of-kilter method has been shown to be $O(m^2p)$ in complexity, where $p=[\log_2 \max_{(i,j)\in A} (c_{ij})]$ [5]. The most naif approach to step 3 yields a complexity of order $m^2c(m)$, while at most n iterations of algorithm 2.1 are required before termination. Hence L can be computed by an algorithm which is at worst $O[n(m^2p+ + m^2c(m))]$.

3. A BORDERING ALGORITHM

3.1 This method finds a lower bound by solving a transportation problem at each iteration. It is therefore convenient to consider problem (1.2) without IS constraints as a particular transportation, rather than assignment problem.

3.2 *Step 0* (Start). (As for the algorithm of section 2.)

Step 1 (Bounding). Solve the transportation problem without IS constraints by the ad hoc variation of the Hungarian method (see for instance [4]). Let $I=\{(i,j):x_{ij}=1\}$ and c'_{ij} be the costs at the end of the computation.

$$L \leftarrow L + \Sigma\, c_{ij} x_{ij}, \quad k \leftarrow k+1.$$

Step 2 (Testing.) Find a circuit S of Γ in $I \cap A$. If none exists, stop. L is the best bound obtained.

Step 3 (Updating.) Adjoin to C' a new row and column indexed by k'. Let

$$c'_{ij} = \infty \quad \text{for all} \quad (i,j) \in S,$$

$$c'_{k'j} = \begin{cases} 0 & \text{if } j \in F \\ \infty & \text{if } j \notin F, \end{cases}$$

$$c'_{ik'} = \begin{cases} 0 & \text{if } i \in F \\ \infty & \text{if } i \notin F, \end{cases}$$

$$c'_{k'k'} = 0,$$

where F is the set of nodes incident with arcs of S. $C \leftarrow C'$. Adjoin to the set of constraints the following:

$$\sum_i x_{ik'} = n - |F|,$$

$$\sum_j x_{k'j} = n - |F|,$$

$$0 \leq x_{k'k'} \leq n - |F| \quad \text{integer.}$$

Go to step 1.

3.3 The proof of correctness of this method arises from considerations which are very similar to those of section 2.2 and is omitted here for the sake of brevity.

Since the network flow problem is known to be $O(n^3)$, this is also the complexity for solving the transportation problem. At each step no less than two arcs become of ∞ cost, hence no more than $m/2$ steps are required before termination and the overall complexity of the method is $O\big(m(n^3 + c(m))\big)$.

4. CONCLUSIONS

4.1 If the given IS is the matroid having as bases the 1-trees of D, problem 1.2 becomes the well-known traveling salesman problem. The bounding methods presented here will yield lower bounds which are never worse (and possibly better) than those obtainable by the polynomial bounded method of [3], since they proceed considering a cycle at a time and they do not need to constrain the matrix C to satisfy the triangular inequality.

4.2 Note also that S is not uniquely defined, since in general many circuits may be found. Obviously, the final value of L depends on the choice of such a circuit. This fact would suggest to find, at each iteration the circuit which is likely to give the highest value of L. However, it does not seem to be a simple matter to establish a valid and computationally efficient policy for this choice, fundamentally because of the generality of the IS structure.

4.3 Research is in progress to evaluate the method proposed here by implementing it and testing its performances on a computer versus other bounding methods, such as e.g. subgradient methods [2, 10, 11].

Acknowledgment

Many insightful discussions with prof. E. L. Lawler are gratefully acknowledged, without them section 3 would probably not exists and section 2 would be incorrect.

322

REFERENCES

[1] Camerini, P. M. and Maffioli, F.: Bounds for 3-matroid intersection problems, *Inf. Proc. Letters* 3 (1975), 81–83.

[2] Camerini, P. M. and Maffioli, F.: Heuristically guided algorithm for k-parity matroid problems, *Discrete Mathematics* (to appear).

[3] Christofides, N.: Bounds for the traveling salesman problem, *Opns. Res.* 20 (1972), 1044–1056.

[4] Christofides, N.: *Graph theory: an algorithmic approach,* Academic Press (1975), 381.

[5] Edmonds, J. and Karp, R. M.: Theoretical improvements in algorithmic efficiency for network flow problems, *J. ACM* 19 (1972), 248–264.

[6] Karp, R. M.: On the computational complexity of combinatorial problems, *Networks* 5 (1975), 45–68.

[7] Kuhn, H. W.: The Hungarian Method for the Assignment Problem, *Nav. Res. Log. Quart.* 2 (1955), 83–97.

[8] Lawler, E. L.: Polynomial Bounded and (apparently) non-polynomial bounded matroid computations, in *Combinatorial Algorithms* (R. Rustin ed.) Algorithmics Press (1973), 49–57.

[9] Lawler, E. L.: *Combinatorial Optimization: Networks and Matroids,* to be published by Holt, Rinehart and Winston Inc.

[10] Maffioli, F.: Subgradient optimization, matroid problems and heuristic evaluation, in *Optimization Techniques,* Proc. 7th IFIP Conf. part 2, Springer Verlag (1976), 389–396.

[11] Wolfe, P., Held, M. and Crowder, H. P.: Validation of subgradient optimization, *Math. Programm.* 6 (1974), 62–88.

CLUSTER ANALYSIS AND GRAPH COLORING

M. DELATTRE and P. HANSEN

(Mons, Belgium and Lille, France)

1. INTRODUCTION

Cluster analysis [1], [4], [11], [17] is concerned with the problem of partitioning a given set $O = \{O_1, O_2, \ldots, O_N\}$ of N entities into well-separated homogeneous subsets $C_1, C_2, \ldots, C_M$ called clusters. Thus similar entities should belong to the same cluster and dissimilar entities to different clusters. The concept of similarity (or dissimilarity) is usually made precise by the introduction of a dissimilarity index d_{ij} defined on the set of all pairs of entities. This index verifies $d_{ij} \geqq 0$, $d_{ii} = 0$, $d_{ij} = d_{ji}$ and often some form of a triangular inequality. For a discussion of the best way to construct such an index in a given situation see e.g. Benzecri [1] Jardine and Sibson [11] or Sokal and Sneath [17].

The results given by several cluster analysis methods, including all those considered in this paper, the results are invariant under a monotone transformation of the dissimilarities, i.e. only the order of the dissimilarities between pairs of objects matters.

Many criteria of homogeneity and of separation criteria of the clusters have been proposed and the selection of such entities depends on the objectives of the clustering. In this paper we consider some treshold-type criteria which appear or are implicit in well-known methods.

Graph theoretic concepts and techniques (see Berge [2] for definitions) are increasingly being used in cluster analysis. Indeed it is convenient to associate to the given entities $O_1, O_2, \ldots, O_N$ the vertices $x_1, x_2, \ldots, x_N$ of a complete graph $G = (X, E)$ and to consider the dissimilarities d_{ij} as weights given to that graph edges. Then to a given treshold t on the dissimilarities corresponds a partial graph $G_t = (X, E_t)$ of G where E_t is the set of edges (x_i, x_j) of G such that $d_{ij} \geqq t$. An excellent survey of applications of graph theory to clustering has been given by Hubert [10].

Let $P_M = (C_1, C_2, \ldots, C_M)$ denote a partition of $O = \{O_1, O_2, \ldots, O_N\}$ into M clusters, i.e. $C_j \neq \emptyset$, $C_i \cap C_j = \emptyset$, $i, j = 1, 2, \ldots, M$, and $C_1 \cup C_2 \cup \ldots \cup C_M = 0$.

Let us call *split* of P_M and denote by $s(P_M)$ the minimum distance between entities in different clusters of P_M:

$$s(P_M) = \min_{i,\,j \neq i} \quad \min_{O_k \in C_i,\, O_l \in C_i} \quad d_{kl}.$$

Let us call *diameter* of P_M and denote by $d(P_M)$ the maximum dissimilarity between entities in the same cluster of P_M:

$$d(P_M) = \max_{i} \quad \max_{O_k,\, O_l \in C_i} \quad d_{kl}.$$

Let $\bar{P}$ denote the set of partitions of O and $\bar{P}_M$ the set of such partitions which contain exactly M clusters (or classes). Let us call a partition P_M of O into M clusters *efficient* iff no partition P'_M into M clusters has a strictly larger split and a not larger diameter or a strictly smaller diameter and a not smaller split:

$$P_M \text{ efficient} \Leftrightarrow \nexists\, P'_M \in \bar{P}_M \mid s(P'_M) > s(P_M) \text{ and } d(P'_M) \leqq d(P_M)$$

$$\text{or} \qquad s(P'_M) \geqq s(P_M) \text{ and } d(P'_M) < d(P_M).$$

Two efficient partitions P_M and P'_M will be called *equivalent* iff $s(P_M) = s(P'_M)$ and $d(P_M) = d(P'_M)$. A *complete set* of efficient partitions is a set such that any efficient partition not included in that set is equivalent to an efficient partition of that set.

The following problems will be considered in the sequel:

Problem 1. Determine a partition P_M^* of O into M clusters which maximizes the split:

$$s(P_M^*) = \max_{P_M \in \bar{P}_M} s(P_M).$$

Problem 2. Determine a partition P_M^* of O into M clusters which minimizes the diameter:

$$d(P_M^*) = \min_{P_M \in \bar{P}_M} d(P_M).$$

Problem 3. Determine a complete set of efficient partitions P_M^* of O into M clusters. The resolution of these problems allows to obtain optimal partitions and efficient partitions for all possible values of M; indications regarding the best, or most natural, value of M may also be obtained.

Hierarchical clustering methods yield chains of partitions $P_1, P_2, \ldots, P_N$ such that the partition P_{M+1} is a refinement of the partition P_M for $M = 1, 2, \ldots, N-1$. Such methods are considered in the next section, where it is shown they give an exact solution to Problem 1 and approximate solutions to Problem 2.

An exact solution to Problem 2 can only be obtained by a nonhierarchical clustering method as the one of section 3. A bicriterion approach described in section 4 allows to solve Problem 3. Finally, it is shown, following Hubert [9] and Peay [13], [14], that several other cluster analysis problems are reducible to Problems 1 and 2.

2. HIERARCHICAL CLUSTERING ALGORITHMS

As mentioned above, hierarchical clustering methods yield a chain of partitions $P_1, P_2, \ldots, P_N$ such that each partition is a refinement of the previous ones; in many hierarchical clustering methods P_{M+1} is an immediate refinement of P_M for $M = 1, 2, \ldots, N-1$, i.e. there exists no partition P' in $\bar{P}$ which is a refinement of P_M and of which P_{M+1} is a refinement. Hierarchical clustering algorithms may be agglomerative, i.e. proceed from individual entities towards larger and larger clusters by successive agglomerations of pairs of clusters, or divisive, i.e. proceed from the set O of all entities to smaller and smaller clusters by successive divisions of a cluster. The two most well-known hierarchical clustering methods are the single-link (or nearest neighbour) and complete link (or furthest neighbour) methods.

Agglomerative algorithms implementing these methods are described in a classical paper by Johnson [12]. They work as follows:

(a) Consider all individual entities as clusters.

(b) Consider all pairs of clusters; aggregate those two clusters such that the minimum (respectively maximum) dissimilarity between pairs of entities belonging one to each of these clusters is minimized. Iterate step (b) until a single cluster is obtained.

Clearly, at each iteration the single link algorithm aggregates those two clusters which contain the entities defining the split of the current partition, and the complete link algorithm aggregates those two clusters which contain the entities defining the diameter of the resulting partition so as to minimize that diameter.

These methods are thus locally optimal, i.e. they make an optimal choice at each iteration. It is not obvious a priori whether they are globally optimal or not, i.e., if the split or diameter obtained after several iterations is optimal or is not.

The following result, which generalizes a theorem of Zahn [20] is easily proven by contradiction:

Theorem 1. *For all $M = 1, 2, \ldots, N$, the partition P_M^* defined by the single link algorithm verifies*

$$s(P_M^*) \geqq s(P_M) \quad \forall \quad P_M \in \bar{P}_M.$$

Gower and Ross [5] have shown that the partition P_M^* given by the single link algorithm could be determined in a quicker way by first computing a shortest spanning tree of $G = (X, E)$ and then ranking its edges in order of increasing weights. If t denotes the weight of the $N-M+1$-th edge of the shortest spanning tree, P_M^* is given by the connected components of the complementary graph $\bar{G}_t = (X, E_t)$ of the partial graph $G_t = (X, E_t)$ of G (or, which is equivalent, by the connected components of the partial graph of G the edges of which are the $N \sim M$ first of the shortest spanning tree).

Using the algorithm of Prim [15] the shortest spanning tree of G may be computed in $O(N^2)$ operations; moreover the matrix of dissimilarities must be computed but only a column of that matrix need be stored at a time. Efficient algorithms have been proposed for the single link method with such a computational scheme by Sibson [16] and by Rohlf [19].

A result similar to theorem 1 does not hold for the complete link method, as shown by the following counter-example: consider four entities O_1, O_2, O_3, O_4 and dissimilarities $d_{12}=1$, $d_{13}=2$, $d_{24}=3$, $d_{23}=4$, $d_{14}=5$ and $d_{34}=6$. The partitions given by the hierarchical complete link algorithm are $P_4^*=\{O_1\}, \{O_2\}, \{O_3\}, \{O_4\}$ with $d(P_4^*)=0$, $P_3^*=\{O_1, O_2\}, \{O_3\}, \{O_4\}$ with $d(P_3^*)=1$, $P_2^*=\{O_1, O_2, O_3,\} \{O_4\}$ with $d(P_2^*)=4$ and $P_1^*=\{O_1, O_2, O_3, O_4\}$ with $d(P_1^*)=6$. However, the partition $P_2'=\{O_1, O_3\}, \{O_2, O_4\}$ has a diameter $d(P_2')=3$ which is less than that of P_2^*.

Therefore, if partitions of minimum diameter are sought for all M, a nonhierarchical algorithm is required; such an algorithm is briefly described in the next section.

If there are some strong reasons to believe that a hierarchical structure applies to the data, such as in the clustering of data on an evolutionary process, one may be content with the agglomerative complete link algorithm. An alternative is to consider a divisive hierarchical algorithm. Three such algorithms have been proposed by Hubert [8]. At each iteration the cluster C_i defining the diameter of the current partition is selected and divided into two clusters C_i' and C_i'' according to one of the following rules:

(a) Let O_k and O_1 be the entities defining $d(P_M)$. Assign O_k to C_i' and O_1 to C_i''. Choose recursively the most dissimilar pair of entities $\{O_p, O_q\}$ of C_i with exactly one entity assigned to C_i' or to C_i''. Assign the other entity to C_i'' or to C_i' respectively.

(b) Choose and assign O_k and O_1 as in (a). Choose recursively the most similar pair of entities $\{O_p, O_q\}$ of C_i with exactly one entity assigned to C_i' or to C_i''. Assign the other entity to C_i' or to C_i'' respectively.

(c) Choose and assign O_k and O_1 as in (a). Choose and assign recursively to C_i' or to C_i'' the entity O_p such that the maximum dissimilarity between pairs of entities in $C_i' \cup \{O_p\}$ or in $C_i'' \cup \{O_p\}$ is minimum.

These three methods are not locally optimal, i.e. none of them guarantees that the maximum dissimilarity between entities in C_i' or between entities in C_i'' will be minimum, or that the split between C_i' and C_i'' will be maximum. An algorithm for bipartition of a given set of entities which has the first of these properties has been proposed by Rao [18] and is described in the next section. This algorithm could be used instead of one of the rules of Hubert in a divisive hierarchical complete link algorithm. No systematic experimental comparison of divisive and agglomerative algorithms for the complete link method and its variants does yet seem to have been published, although Hubert [8] stressed the interest of such a study.

328

3. NONHIERARCHICAL CLUSTERING ALGORITHMS

In order to obtain a partition P_M^* with minimum diameter $d(P_M^*)$ in $\bar{P}_M$ a non-hierarchial algorithm must be constructed. Such an algorithm may be obtained by exploiting concepts and algorithms of graph coloring. Let us consider the partial graphs G_t of $G=(X, E)$ where t takes as values the weights of the edges of G in decreasing order. Without loss of generality these weights may be assumed to be different as a small perturbation will render them so, should it not be the case. Let these graphs be colored by an exact coloring algorithm.

Theorem 2. *The color classes of the optimal coloration of the last graph G_t which can be colored in M colors define a partition P_M^* of O such that $d(P_M^*) \leqq d(P_M) \forall P_M \in \bar{P}_M$.* An optimal algorithm for obtaining P_2^* has been proposed by Rao [18] and works as follows:

(a) Rank the edges of $G=(X, E)$ in order of decreasing values of their weights.

(b) Consider the edges $e_j=(x_k, x_l)$ of G in that order and color their vertices according to the following rules:

(b.0) For the first edge color x_k in color 1 and x_l in color 2.

(b.1) If x_k has color 1 (resp. 2) and x_l is uncolored color x_k in color 2 (resp. 1).

(b.2) If x_k has color 1 (resp. 2) and x_l has color α (resp. β) color x_l in 2 (resp. 1) as well as all the vertices colored α (resp. β) in the connected component of x_l in G_t; then color in color 1 (resp. 2) all the vertices colored β (resp. α) in that connected component.

(b.3) If x_k has color α (resp. β) and x_l is uncolored color x_k in color β (resp. α)

(b.4) If x_k and x_l have color α (resp. β) exchange the colors α and β in the connected component of x_l.

(b.5) If x_k and x_l are uncolored, color x_k in color α and x_l in color β.

(b.6) If x_k and x_l have color 1 (resp. 2) go to (c).

(c) Omit the last edge added. Color the remaining vertices in colors 1 or 2, one connected component of G_t at a time, setting color α equal to color 1 and color β to color 2. The ranking of the edges takes $O(N^2 \log_2 N^2)$ operations and, if the edges which involve the coloration of a new vertex in color 1, 2, α or β are noted the rest of the algorithm works in $O(N^2)$ operations. It is hence very efficient.

An algorithm for the general case is given in [6]. Its principle is the following:

(a) Rank the edges of $G=(X, E)$ in order of decreasing values of their weights. Consider a graph G_t with no edges and color all its vertices in color 1.

(b) Decrease t so as to add a new edge $l_j=(x_k, x_l)$ to G_t.

(c) If the colors assigned to x_k and x_l are different, go to (b).

(d) If the number of colors forbidden to x_k or to x_l (i.e. given to adjacent vertices of x_k or of x_l in G_t) is smaller than the current number of colors used minus one, recolor x_k or x_l in an allowed color and go to (b).

(e) Attempt to color G_t in the current number of colors used with a graph coloring routine. As soon as this has been done, go to (b). If this routine shows it cannot be done output the coloration of the previous graph. Then return to (b) noting the current number of colors used has increased by one.

The graph coloring routine used in [6] is based on a branch-and-bound algorithm similar in spirit to the quite successful algorithm of Brown [3]. Improvements are that the choice of the next vertex to be colored takes into account the effects of the previous vertex colorations and that tests using identification of vertices of the same color and temporary suppression of vertices are introduced.

Because such a branch-and-bound routine is used, the algorithm is not polynomially bounded. This seems hard to improve in view of the following result:

Theorem 3. *For* $M \geq 3$, *the problem of determining* P_M^* *such that*

$$d(P_M^*) \leq d(P_M) \forall P_M \in \bar{P}_M$$

is NP-complete.

A FORTRAN IV code called CLUSTERGRAPH 1 has been written for the previous algorithm and has allowed to solve cluster analysis problems with up to 270 entities. This was possible because the partial graphs G_t obtained for real-world applications are much easier to color than randomly generated graphs.

4. BICRITERION CLUSTER ANALYSIS

Instead of considering only the split or the diameter of a partition P_M, it may be of interest to consider both these values simultaneously. Indeed, they correspond to the requirements stated in the introduction that dissimilar entities should belong to different clusters and similar entities to the same cluster. This suggests that a trade-off partition with sufficiently large split and small diameter could be prefered to an optimal partition for the split (or diameter) with too large a diameter (or too small a split). Such partitions belong to, or are equivalent to an element of a complete set of efficient partitions of $O = \{O_1, O_2, \ldots, O_N\}$. The following result [7] shows that the number of efficient partitions in such a set is not too large for moderate N:

Theorem 4. *The number of distincts values of the splits* (P_M) *for all* $P_M \in \bar{P}$ *(i.e. for all* $P_M \in \bar{P}_M$ *for all* M) *is at most* $N-1$. *These values are equal to the weights of the edges of a shortest spanning tree of* $G = (X, E)$.

In view of theorem 4, an algorithm for determining a complete set of efficient partitions can be readily obtained. Its principle is:

(a) Determine the shortest spanning tree of $G = (X, E)$ and rank the edges of that tree in order of increasing values. Let $k = 0$.

(b) Consider the partial graph $G_k' = (X, E_k')$ of G the edges of which are the k smallest edges of the shortest spanning tree of G. Determine the connected components of

330

G'_k. Construct the complete reduced graph $G^*_k = (X^*_k, E^*_k)$ of G'_k, the vertices of which correspond to the connected components of G'_k. Assign to each edge $e^*_j = (x^*_k, x^*_l)$ of G^*_k a weight equal to the largest weight of an edge joining vertices of G from both the connected components corresponding to x^*_k and x^*_l.

(c) Determine by applying the algorithm of the previous section to G^*_k partitions P_M optimal for the diameter for $M = 1, 2, \ldots, N-k$. Set $k := k+1$ and go to (b).

This algorithm may be accelerated in various ways by taking into account the values of the split and the diameter of the efficient partitions already found in order to avoid some useless computations.

As noted by Zahn [20], the distribution of the weights of the edges of the shortest spanning tree gives useful indications as to the number of clusters in the most natural partition. Indeed large zones where this distribution has nul frequency correspond to large increases in the split of the successive partitions. The partitions defined by the sets of edges included just before such zones appear to be the most interesting ones. The distribution of the values of the diameter for successive partitions obtained by the algorithm of section 3 yields similar information. A two-ways table of the values of split and diameter for the efficient partitions of a complete set could be even more illuminating.

5. EQUIVALENT CLUSTERING PROBLEMS

The algorithms of the previous sections yield partitions optimal for the split or for the diameter by considering the partial graphs G_t of $G = (X, E)$. As has been noted, x_i and x_j are joined by an edge in G_t iff $d_{ij} \geqq t$. Quite a few problem in cluster analysis appear to be reducible to the problems considered above in the sense that applying the proposed algorithms to a graph G' yields partitions which are optimal in G according to their homogeneity criteria.

For these problems a graph $G' = (X, E')$ can be constructed such that x_i and x_j are joined by an edge in the partial graph G'_t of G' iff the couple of vertices (x_i, x_j) satisfies some specified property in G_t.

Hubert [9] has noted that in some situations the dissimilarity index d_{ij} cannot be assumed to be symmetric. If at a threshold level t, O_i and O_j are considered to be dissimilar if $d_{ij} \geqq t$ or $d_{ji} \geqq t$, G' is obtained by setting $d_{ij} = d'_{jt} = \max(d_{ij}, d_{ji})$; if both $d_{ij} \geqq t$ and $d_{ji} \geqq t$ must hold for O_i and O_j to be considered dissimilar setting $d'_{ij} = d'_{ji} = \min(d_{ij}, d_{ji})$ yields G'.

Some form of transitivity may be assumed of a similarity relation, as suggested by Peay [13]: O_i and O_j may be considered similar if there exists a chain of length at most r (a specified integer parameter) joining x_i and x_j in $\bar{G}_t$, the complementary graph of G_t. An efficient algorithm to obtain directly G' from G in that case has been proposed by Peay [13], who has also shown [14] his results may be extended to the case where d_{ij} is not symmetric.

REFERENCES

[1] Benzecri, J. P.: *L'analyse de données.* Tome 1. *La Taxinomie,* Paris: Dunod, 1973.

[2] Berge, C.: *Graphes et hypergraphes,* Paris: Dunod, 1970.

[3] Brown, J. R.: Chromatic Scheduling and the Chromatic Number Problem, *Management Science,* 19 (1972), 456–463.

[4] Duran, B. S. and Odell, P. L.: Cluster Analysis. A Survey, *Lecture Notes in Economics and Mathematical Systems* n° 100, Berlin–Heidelberg–New York: Springer, 1974.

[5] Gower, J. C. and Ross, G. T. S.: Minimum Spanning Trees and Single Linkage Cluster Analysis, *Applied Statistics,* 18 (1969), 54–64.

[6] Hansen, P. and Delattre, M.: Complete link cluster analysis by graph coloring, *Journal Amer. Stat. Ass.* 73 (1978), 397–403.

[7] Hansen, P. and Delattre, M.: Bicriterion Cluster Analysis, in preparation.

[8] Hubert, L.: Monotone Invariant Clustering Procedures, *Psychometrika,* 38 (1973), 47–62.

[9] Hubert, L.: Min and Max Hierarchical Clustering Using Asymmetric Similarity Measures, *Psychometrika,* 38 (1973), 63–72.

[10] Hubert, L.: Some Applications of Graph Theory to Clustering, *Phychometrika,* 39 (1974), 283–309.

[11] Jardine, N. and Sibson, R.: *Numerical Taxonomy* New York: Wiley, 1971.

[12] Johnson, S. C.: Hierarchical Clustering Schemes, *Psychometrika,* 32 (1967), 241–254.

[13] Peay, E. R.: Nonmetric Grouping: Clusters and Cliques, *Psychometrika,* 40 (1975), 297–313.

[14] Peay, E. R.: Grouping by Cliques for Directed Relationships, *Psychometrika,* 40 (1975), 573–574.

[15] Prim, R. C.: Shortest Connection Networks and Some Generalizations, *Bell Systems Technical Journal,* 36 (1957), 1389–1401.

[16] Sibson, R.: SLINK: An Optimally Efficient Algorithm for the Single-Link Cluster Method, *Computer Journal,* 16 (1973), 30–34.

[17] Sokal, R. R. and Sneath, P. H. A.: *Principles of Numerical Taxonomy,* San Francisco: W. H. Freeman and Co, 1973.

[18] Rao, M. R.: Cluster Analysis and Mathematical Programming, *Journal Amer. Stat. Ass.,* 66 (1971), 622–626.

[19] Rohlf, F. J.: Hierarchical Clustering Using the Minimum Spanning Tree, *Computer Journal,* 16 (1973), 93–95.

[20] Zahn, C. T.: Graph-Theoretical Methods for Detecting and Describing Gestalt Clusters, *I.E.E.E. Transactions on Computers,* C-20 (1971), 68–86.

EIGENVALUES AND EIGENVECTORS IN SEMIMODULES AND THEIR INTERPRETATION IN GRAPH THEORY

M. GONDRAN and M. MINOUX

(Paris, France)

1. INTRODUCTION

The study of many problems in graph theory, in generating languages or automates leads to the resolution of linear equations systems (see [1], [3], [6], [7], [8], [11], [13]) over an appropiate algebraic structure.

Although this algebraic structure, the semiring $(S, \oplus, *)$ has very poor properties ($\oplus$ is associative and commutative with neutral element ε, $*$ is associative with neutral element e, distributive from right and left with respect to $\oplus$ and admits ε to be an absorbing element $a * \varepsilon = \varepsilon \, \forall \, a \in S$) assuming the existence of a quasiinverse for certain elements most of the classical algorithms for solving system of linear equations (Jacobi, Gauss–Seidel, Gauss, Jordan algorithm, etc.).

In this paper it will be shown that many path problems are possible to be extended as existence problems of eigenvectors and eigenvalues in corresponding semirings.

These results are a first step toward of an extension of the Perron–Frobenius theorem to semirings. The interpretation of these eigenvalues and eigenvectors will be very rich. So in [10] for a special structure $(R^+, \min, \max)$ the main results of the hierarchic classification could be interpreted in the same manner as those of the factor analysis.

In Section 2 a number of exact definitions will be given, which will be necessary in the following sections.

In Section 3 some theorems will be presented in connection with the existence of eigenvalues (and eigenvectors belonging to them).

In Section 4 results about the set of eigenvectors (eigen-semimodul) belonging to one eigenvalue will be presented.

Finally, in Section 5 the given results will be illustrated and interpreted through some examples.

The relations between eigenvalues and determinants, as well as the concept of linear dependence or independence in semirings (and more generally in diodes) are studied in [15].

2. DEFINITIONS

A *semiring* $(S, \oplus, *)$ will be considered, e.g. a set S on which two rules $\oplus$ and $*$ are defined in the following manner

— the operation "addition" $\oplus$ generates in S a *commutative monoid* structure (completeness, commutativity, associativity) and admits ε as neutral element

— the operation "multiplication" $*$ generates in S a *monoid* structure (completeness, associativity), admits e as neutral element (unity), ε as absorbing element $(a * \varepsilon = \varepsilon \; \forall a \in S)$ and is distributive from right and left with respect to $\oplus$.

A *preorder* relation, indicated by the operation $\oplus$, is defined in the following way:

$$a \geqq b \leftrightarrow \exists \subset \in S: \quad a = b \oplus c.$$

Often it will be assumed, that this relation is also an order relation.

This will be the case, for instance, if $\oplus$ is idempotent (remember that the idempotentity of $\oplus$ implies that $\geqq$ is an order relation, since $a \geqq b$ and $b \geqq a$ can be written $a = b \oplus c$ and $b = a \oplus d$, from which follows: $a \oplus b = a \oplus (a \oplus d) = a \oplus d = b = (b \oplus c) \oplus b = b \oplus c = a$).

Remark 1. Note that if $\geqq$ is an ordering relation then the following holds $a \oplus b = \varepsilon \Rightarrow a = b = \varepsilon$. On the other hand, if it is assumed that $*$ is invertible, then $a * b = \varepsilon \Rightarrow \Rightarrow a = \varepsilon$ or $b = \varepsilon$. This remark will be used in Section 3.

Addition and multiplication of square matrices of order n with elements from S is then defined with the operations $\oplus$ and $*$.

The set $M(n, S)$ of such matrices admits the same structure as S with the matrix

$$\Sigma = \begin{bmatrix} \varepsilon & \ldots & \varepsilon \\ \vdots & & \vdots \\ \vdots & & \vdots \\ \varepsilon & \ldots & \varepsilon \end{bmatrix}$$

as neutral element, and

$$E = \begin{bmatrix} e & & \varepsilon \\ & \ddots & \\ \varepsilon & & e \end{bmatrix}$$

as unit matrix.

We note that the operations $\oplus$ and $*$ are induced in $M(n, S)$, by the structure of S. Let be S^n the set of n-dimensional column vectors.

A set $T_d \subset S^n$ is called a *right semimodule* of S^n if:

$$\forall \lambda \quad \text{and} \quad \mu \in S, \quad \forall V \quad \text{and} \quad W \in T_d:$$

(2.1) $$V\lambda \oplus W\mu \in T_d,$$

where $V\lambda$ is the vector with the components $v_i * \lambda$.

A set $T_g \subset S^n$ is called a *left semimodule* of S^n if:

$$\forall \lambda \quad \text{and} \quad \mu \in S, \quad \forall V \quad \text{and} \quad W \in T_g$$

$$(2.2) \qquad \lambda V \oplus \mu W \in T_g.$$

If $*$ is commutative, then these two notions are equivalent and then it will simply be called a *semimodule* A vector $V \in S^n \left[V \neq \begin{pmatrix} \varepsilon \\ \vdots \\ \varepsilon \end{pmatrix} \right]$ is an *eigenvector* of the matrix A, $A \in M(n, S)$, if there is a $\lambda \in S$, *eigenvalue*, such that

$$(2.3) \qquad AV = \lambda V.$$

If λ is an eigenvalue let $v(\lambda)$ be the set of eigenvectors, belonging to λ.

It will be shown that $v(\lambda)$ is a *right semimodule*. For a given matrix $A \in M(n, S)$ as *associated graph* the directed graph $G(A) = (X, U)$ will be considered, where $X = \{1, 2, \ldots, n\}$ is the set of vertices and where $(i, j) \in U$ (U is the set of edges) if and only if $a_{ij} \neq \varepsilon$. It can also be said that A is the generalized adjacency matrix of $G(A)$, a_{ij} is called the weight of the edge i, j and for every path

$$l = \{(i_1, i_2), (i_2, i_3), \ldots, (i_{p-1}, i_p)\}$$

the weight power $w(l)$ is defined as $w(l) = a_{i_1 i_2} * a_{i_2 i_3} * \ldots * a_{i_{p-1}, i_p}$.

Let A^k be the k-th power of $A \in M(n, S)$. A^+ denotes the limit (if it exists) of the matrix

$$A^{(k)} = A \oplus A^2 \oplus \ldots \oplus A^k$$

for $k \to \infty$ (see [5], [6], [9], [13] for a study of the convergence of A^k), and let further $A^* = E \oplus A^+$.

For the matrices A^* and A^+ the following fundamental relation holds:

$$(2.4) \qquad AA^* = A^+.$$

In [6], [7] it is shown that conditions for the existence of A^+ and A^* depend on the circuits of the graph $G(A)$.

If L_{ij} is the set of paths from i to j in $G(A)$, and A_i^{+j} is the element of A^+ in the i-th row and the j-th column, then the following relation holds:

$$(2.5) \qquad A_i^{+j} = \sum_{i \in L_{ij}} w(l).$$

3. EXISTENCE OF EIGENVALUES

Let L_{ii} denote the set of circuits with initial point i (the power of two circuits may not be the same unless their initial points are the same, because $*$ is not commutative: $w(a_{12}) * w(a_{21}) \neq w(a_{21}) * w(a_{12})$).

If A^* exists then A^+ also exists and:

$$(3.1) \qquad A_i^{+\,i} = \sum_{\gamma \in L_{ii}} w(\gamma).$$

Let $A^+{}_\mu^i$ resp. $A^*{}_\mu^i$ be the i-th column of the matrix $A^+ A^*$ resp. and $\mu \in S$.

Theorem 1. *If A^* exists, then A^{*i} and so $A^{+i}\mu$ also is an eigenvector of the matrix A for the eigenvalue e if the relation*

$$(3.2) \qquad \left(\sum_{\gamma \in L_{ii}} w(\gamma) \right) \mu \oplus \mu = \left(\sum_{\gamma \in L_{ii}} w(\gamma) \right) \mu \quad \text{holds.}$$

Proof. From (2.4) it follows

$$(3.3) \qquad A^{+i} = AA^{*i} = A(A^{+i} \oplus E^i)$$

and A^{+i} will be an eigenvector for e, because

$$(3.4) \qquad A_i^{+\,i}\mu \oplus \mu = A_i^{+\,i}\mu$$

in that case, and so $A^{*i}\mu = A_i^{+\,i}\mu$ will be fulfilled.

It will be seen, that this theorem has many consequences.

Corollary 1.1. *If A^* exists and if there is a circuit γ with initial point i and $\mu \in S$ exist that*

$$(3.5) \qquad w(\gamma)\mu \oplus \mu = w(\gamma)\mu$$

holds, then $A^{*i}\mu$ is an eigenvector of A for the eigenvalue e.

Theorem 1 and corollary 1.1 are often used for the special case $\mu = e$.

Corollary 1.2. *If A^* exists and if $E \oplus A = A$ holds, then every column of A^* is an eigenvector of A for the eigenvalue e.*

Proof. Immediate with $\mu = e$ and $w(\gamma) = a_{ii}$.

Corollary 1.3. (α) *if $*$ is invertible, if $(\lambda^{-1}A)^*$ exists and if:*

$$(3.6) \qquad (\lambda^{-1}A)_i^{*i} \oplus e = (\lambda^{-1}A)_i^{*i},$$

then $(\lambda^{-1}A)^{*i}$ is an eigenvector of the matrix A for the eigenvalue λ.

(β) *if $*$ is also commutative, then condition (3.6) can be replaced by*

$$(3.7) \qquad \sum_{\gamma \in L_{ii}} w(\gamma)\lambda^{-n(\gamma)} \oplus e = \sum_{\gamma \in L_{ii}} w(\gamma)^{-n(\gamma)}$$

where $n(\gamma)$ denotes the number of edges in the circuit γ.

336

Remark 2. *If $*$ is idempotent and commutative, then if V is an eigenvector of A for the eigenvalue e, then λV is an eigenvector of A for the eigenvalue λ.*

Proof. Immediate since from $AV=V$ follows

$$A(\lambda V)=\lambda AV=\lambda^2 V=\lambda V.$$

Important is the special case when $\oplus$ and $*$ are both idempotent.

Corollary 1.4. *If $\oplus$ and $*$ are idempotent and if A^* exists, then if*

$$(3.8) \qquad\qquad \mu=A_i^{+\,i}=\sum_{\gamma\in L_{ii}} w(\gamma),$$

$A^{*i}\mu$ is an eigenvector of A for the eigenvalue e.

Proof. Immediate, since $\mu^2\oplus\mu=\mu$ can be easily verified from (3.2).

Remark 3. In the case when $\oplus$ and $*$ are idempotent, $\forall a\in S$ the relation $a^{(2)}=$ $=e\oplus a\oplus a^2=e\oplus a=a^{(1)}$ holds, and so all the elements $a\in S$ are 1-regular (see [6], [7]) and hence there exists a quasi-inverse $a^*=e\oplus a$.

If now the rule $*$ is commutative, it is shown in [6] (proposition 4) that the matrix A has a quasi-inverse A^*.

If the rule $*$ is not commutative, an additional hypothesis is necessary for the existence of A^* (for instance that for every circuit: $\gamma\ e\oplus w(\gamma)=e$ holds [6] or the stronger supposition that $\forall a\in S\ e\oplus a=e$).

Corollary 1.5. *If $\oplus$ is idempotent, $*$ idempotent and commutative, then A^* exists and if*

$$\mu_i=A_i^{+\,i}=\sum_{\gamma\in L_{ii}} w(\gamma),$$

then $\lambda\mu_i A^{*i}$ is an eigenvector of A for the eigenvalue λ, $\forall\lambda\in S$.

Proof. The existence of A^* follows from remark 3 and the corollary can be deduced from corollary 1.4 applied to remark 2.

Lemma 1. *If A is irreducible $(G(A)$ is strongly connected) $*$ invertible, the pre-order relation $(\geqq)$ is an order relation and if $V=(v_i)$ is an eigenvector of A for the eigenvalue λ, then*

$$(3.9) \qquad\qquad \lambda>\varepsilon$$

$$(3.10) \qquad\qquad v_i>\varepsilon,\quad \forall_i.$$

Proof. First we prove (3.9).

If $\lambda=\varepsilon$, $AV=\varepsilon$. Then from remark 1 follows that $a_{ij}*v_j=\varepsilon\ \forall_i,\ \forall_j$. This implies $a_{ij}=\varepsilon$ or $v_j=\varepsilon$. But $v_j=\varepsilon\ \forall v_j$ is impossible (by the definition of eigenvectors).

Let j_0 be such an index that $v_{j_0} \neq \varepsilon$. Then it follows from $a_{ij_0} v_{j_0} = \varepsilon \ \forall_i$ that $a_{ij_0} = \varepsilon \ \forall_i$ in contradiction to the strong connectedness of $G(A)$.

Now the proof of (3.10) again is in an indirect manner. Suppose there is a v_j with $v_j = \varepsilon$, then let $X_1 = \{j \mid v_j = \varepsilon\}$ and $X_2 = X \backslash X_1$ (note that $X_1 \neq \emptyset$ and $X_2 \neq \emptyset$). Then A and V can be partitioned into

$$
A = \begin{array}{|c|c|} \hline A_{11} & A_{12} \\ \hline A_{21} & A_{22} \\ \hline \end{array}
\quad \text{and} \quad
V = \begin{array}{|c|} \hline V_1 \\ \hline V_2 \\ \hline \end{array}
\begin{array}{l} \} X_1 \\ \} X_2 \end{array}
$$

with column groups X_1, X_2.

$AV = \lambda V$ implies

$$A_{12} V_2 = \varepsilon$$

from which

$$
A_{12} = \begin{array}{|ccc|} \hline \varepsilon & \cdots & \varepsilon \\ \vdots & & \vdots \\ \vdots & & \vdots \\ \varepsilon & \cdots & \varepsilon \\ \hline \end{array}
$$

a contradiction with the irreducibility of A. ∎

Remark 4. *If the matrix A is not irreducible, then it can be partitioned into*

$$
A = \begin{array}{|c|c|} \hline A_{11} & \varepsilon \\ \hline A_{21} & A_{22} \\ \hline \end{array} .
$$

Then if V_2 is an eigenvector of A_{22} for λ, then $V = \begin{pmatrix} V_1 \\ V_2 \end{pmatrix}$, with $V_1 = \begin{pmatrix} \varepsilon \\ \vdots \\ \varepsilon \end{pmatrix}$ — is an eigenvector of A for λ.

In the following it will always be assumed that the matrix A is irreducible.

Because of lemma 1 for every eigenvector $V = (v_j)$ $v_j > \varepsilon$ must hold, if $\geqq$ is an ordering relation and if $*$ is invertible.

Theorem 1 shows that the existence of A^* implies (with certain hypotheses) the existence of eigenvectors for the eigenvalue e.

The following theorem constitutes the reverse of this result and shows clearly that the existence of A^* is closely connected to the existence of eigenvectors corresponding to the eigenvalue e.

Theorem 2. *If $\oplus$ is idempotent and $*$ invertible and if e is an eigenvalue of A, then the matrix A^* exists.*

338

Proof. Let us consider some circuit $\gamma = (i_1, \ldots, i_2, \ldots, i_k, i_1)$. If V is an eigenvector, then:

$$\sum_j a_{i_1 j} v_j = v_{i_1}$$

(3.11)
$$\sum_j a_{i_2 j} v_j = v_{i_2}$$

$$\cdots \cdots \cdots$$

$$\sum_j a_{i_K j} v_j = v_{i_K},$$

so that in particular:

(3.12)
$$a_{i_1 i_2} v_{i_2} \leqq v_{i_1}$$
$$a_{i_2 i_3} v_{i_3} \leqq v_{i_2}$$
$$\cdots \cdots \cdots$$
$$a_{i_K i_1} v_{i_1} \leqq v_{i_K}.$$

The first two inequalities imply

$$a_{i_1 i_2}(a_{i_2 i_3} v_{i_3}) \leqq a_{i_1 i_2}(v_{i_2}) \leqq v_{i_1}$$

and after k steps one gets

(3.13)
$$a_{i_1 i_2} a_{i_2 i_3} \ldots a_{i_k i_1} \leqq v_{i_1}$$

that means (from the definition of the preorder relation $\geqq$):

$$v_{i_1} = w(\gamma) v_{i_1} \oplus t.$$

Since $\oplus$ is idempotent, $\geqq$ is an order relation (remark 1) and since $*$ is invertible $v_{i_1} \neq \varepsilon$ (lemma 1). So v_{i_1} must be invertible from which follows

(3.14)
$$e = w(\gamma) \oplus t * v_{i_1}^{-1}.$$

Adding $w(\gamma)$ to both sides of (3.14) and using the fact that $\oplus$ is idempotent one gets:

(3.15)
$$w(\gamma) \oplus e = w(\gamma) \oplus w(\gamma) \oplus t * v_{i_1}^{-1} = w(\gamma) \oplus t * v_{i_1}^{-1} = e.$$

Equation (3.15) holding for every circuit γ, proposition 3 of [6] guarantees the existence of A^*. (Moreover in that case: $A^* = E \oplus A \oplus A^2 \oplus \ldots A^{n-1}$). ■

It follows:

Corollary 2.1. *If $\oplus$ is idempotent and $*$ is invertible, then if λ is an eigenvalue, $\lambda^{-1} A)^*$ exists.*

Remark 5. *If $\oplus$ is such that $a \oplus b = a$ or b if $*$ is invertible and e is an eigenvalue of A, then there exists a circuit γ with $w(\gamma) = e$.*

If $V = (v_j)$ is an eigenvector of A for e one gets from the definition

$$\sum_j a_{ij} * v_j = v_i \quad \forall_i,$$

and because $a\oplus b=a$ or b, to every i an index $j(i)$ can be associated such that

$$a_{ij(i)}\!*\!v_{j(i)}=v_i.$$

(If there are more such indeces j, then one can choose arbitrarily any of these.)

The directed graph $H=(X, Y)$, where $Y\subset U$ is the set of edges $(i, j(i))$ which contains n vertices and n edges. Its cyclomatic number is equal to its connection number (see [2]) and that must be at least equal to 1. There is at least one cycle γ which is also a circuit of $(G(A))$. On this circuit the relations (3.12) must hold with equalities and the relation (3.13) is then $w(\gamma)\!*\!v_{i_1}=v_{i_1}$, from wich $w(\gamma)=e$ follows (v_{i_1} must be invertible by lemma 1 and theorem 2). ∎

So it follows that *if there is no circuit γ with $w(\gamma)=e$, then e cannot be an eigenvalue of A.*

Remark 6. *If $*$ is invertible and if λ is an eigenvalue of A then for every circuit*

$$w(\gamma)\leqq\lambda^{n(\gamma)}$$

$\big(n(\gamma)$ *the number of vertices in the circuit*$\big)$.

Proof. Let V be an eigenvector of A for the eigenvalue λ ($AV=\lambda V$). Then equation (3.13) becomes: $w(\gamma)\!*\!v_{i_1}\leqq\lambda^{n(\lambda)}\!*\!v_{i_1}$, from which the statement follows (v_{i_1} is invertible by lemma 1).

Remark 7. *If for $\oplus$ $a\oplus b=a$ or b, and if $*$ is invertible, and if λ is an eigenvalue of A, then there exists a circuit γ such that*

$$w(\gamma)=\lambda^{n(\lambda)}$$

(the proof is the same as for remark 5).

4. EIGEN SEMI-MODULES

In Paragraph 3 sufficient conditions for the existence of eigenvectors and eigenvalues were given.

Now a characterisation of the set $v(\lambda)$ of eigenvectors (i.e. of the eigen semi-module), corresponding to an eigenvalue λ will be given. At first results for $v(e)$ will be deduced. After this it will be shown how these results can be obtained for general eigenvalues λ.

Lemma 2. *If $\oplus$ is idempotent and if A^* exists then:*

$$V\in v(e)\Rightarrow V=A^*V.$$

Proof.

$$V = V$$
$$AV = V$$
$$A^2V = V$$
$$\vdots$$
$$A^nV \doteq V,$$

from which

$$A^*V = V.$$

By this every vector $V \in v(e)$ is a linear combination of the columns of A^*. But the columns of A^* need not necessarily be elements of $v(e)$ (see theorem 1).

Under special conditions it will be shown that the set of vectors of the form $A^{*i}\mu_i$, which belong to $v(e)$ is a generating set of $v(e)$.

Theorem 3. *If $a \oplus b = a$ or b and A^* exists, then the following holds:*

$$V \in v(e) \Rightarrow V = \sum_i A^{*i}\mu_i \quad \text{with} \quad A^*\mu_i \in v(e).$$

Proof. $\oplus$ is idempotent and from lemma 2 it follows that $V = A^*V$ and so

(4.1)
$$V = \sum_i A^{*i}v_i.$$

On the other hand, the following holds:

$$V = A * V \Rightarrow \sum_j a_{ij} * v_j = v_i \quad (i = 1, \ldots, N).$$

Since $a \oplus b = a$ or b, we construct (as in remark 5) the directed partial graph $H = [X, Y]$, where

$$(i, j) \in Y \leftrightarrow a_{ij} * v_j = v_i.$$

Then the cyclomatic number of H is equal to its connection number. Since every vertex of H has an exterior semidegree equal to 1, more than one circuit cannot be in a connected component.

So in general H is composed of K connected components $(K \geq 1)$ H^k $(k = 1, \ldots, K)$, each of these containing exactly one circuit γ^k. Let $i \in \gamma^k$. On the circuit γ^k the relations (3.12) are satisfied with equalities and so the following holds:

$$w(\gamma^k) * v_i = v_i.$$

Then $w(\gamma^k) * v_i \oplus v_i = v_i = w(\gamma^k) * v_i$ and from theorem 1 $A^{*i}v_i \in v(e)$.

Let $j \in H^k$ $(j \neq i)$. Then there is one and only one path $l_{ji} \in L_{ji}$ between j and i in H^k. On this path the relations (3.12) are satisfied with equalities and this gives

$$w(l_{ji}) * v_i = v_j.$$

The sum (4.1) contains the two terms:

$$A^{*j}v_j \quad \text{and} \quad A^{*i}v_i$$

and so

$$A^{*j}v_j \oplus A^{*i}v_i = [A^{*j}w(l_{ji}) \oplus A^{*i}] \divideontimes v_i = A^{*i}v_i$$

(from (2.5) and the idempotentity of $\oplus$).

In every connected component H^k a vertex $i_k \in \gamma^k$ can be chosen so that

$$V = \sum_{\kappa=1}^{K} A^{*i_k} \divideontimes v_{i_k} \quad \text{and} \quad A^{*ik} \divideontimes v_{ik} \in v(e) \, \forall k.$$

from which the theorem follows. ∎

Corollary 3.1. *If $a \oplus b = a$ or b, if A^* exists and if $\divideontimes$ is invertible, then $v(e)$ is a right semi-module generated by the columns of A^*, which are eigenvectors of A for $e \cdot \{A^{*i} | A^{*i} \in v(e)\}$ is a minimal generating set (a basis) of $v(e)$.*

(If two columns $\{A^{*i}$ and $A^{*j} \in v(e)\}$ are equal up to a constant, then one of them is omitted.)

Corollary 3.2. *If $a \oplus b = a$ or b, $e \oplus a = e \, \forall a \in S$ and $\divideontimes$ is idempotent, then A^* exists and $v(e)$ is a (right) semi-module generated by the vectors $A^{*i}\mu_i$, with*

$$\mu_i = \sum_{\gamma \in L_{ii}} w(\gamma) = A_i^{+i}$$

which constitute a minimal generating set (basis) of $v(e)$.

Proof. A^* exists by remark 3. From corollary 1.4 the $A^{*i}\mu_i$ are elements of $v(e)$. On the other hand, $\mu_i = \sum_{\gamma \in L_{ii}} w(\gamma)$ is the greatest element $\mu \in S$ for which the relation (3.2) holds.

In fact (3.2) is $\mu_i\mu \oplus \mu = \mu_i\mu$, from which $\mu \leq \mu_i\mu$, since from $e \oplus \mu = \mu$ follows $\mu_i\mu \oplus \mu_i = \mu_i$ from which $\mu_i\mu \leq \mu_i$, i.e. $\mu \leq \mu_i$.

According to theorem 3 every vector $V \in v(e)$ can be written as $V = \Sigma A^{*i}v_i$ with $A^{*i}v_i \in v(e)$. From the preceding remark it is known that $v_i \leq \mu_i$ and consequently

$$V = \sum_i (A^{+i}\mu_i)v_i,$$

since $\mu_i v_i = o_i v_i$ (these equalities hold, because v_i satisfies (3.2) $\mu_i v_i \oplus v_i = \mu_i v_i$ and further $e \oplus \mu_i = e$ implies $\mu_i v_i \oplus v_i = v_i$). ∎

Now these results make the study of $v(\lambda)$ possible.

Lemma 3. *If $\divideontimes$ is commutative and if $\forall a \in S$, $e \oplus a = e$, then A^* exists and*

$$V \in v(\lambda) \Rightarrow V = A^*V.$$

Proof. If $AV=\lambda V$ then holds for $\forall k\in N$: $A^kV=\lambda^kV$ (x is commutative). Now $e\oplus\lambda=e$ implies $e\oplus\lambda\oplus\lambda^2=e\oplus\lambda(e\oplus\lambda)=e\oplus\lambda$ from which recursively

$$e\oplus\lambda\oplus\lambda^2\oplus\ldots\oplus\lambda^k=e$$

and so

$$(E\oplus A\oplus A^2\oplus\ldots\oplus A^k)V=V$$

and because $E\oplus A^{(k)}$ converges to A^* when $k\geq n-1$ (see [6]) the lemma is proved. ∎

Corollary 3.3. *If $a\oplus b=a$ or b and $e\oplus a=e$ $\forall a\in S$, if $*$ is idempotent and commutative then A^* exists and*

(α) every $\lambda\in S$ is an eigenvalue of A,

(β) $V\in v(\lambda)$ can be written as $V=\Sigma A^{*i}\lambda\mu_i$ with $A^{*i}\lambda\mu_i\in v(\lambda)$,

(γ) $v(\lambda)$ is a sub-module generated by the vectors $A^{*i}\lambda\mu_i$ with $\mu_i=A_i^{+i}$. These vectors form a basis of $v(\lambda)$.

Proof. (α) is proved by Corollary 1.5. The proof of (β) is more complex and will be given in a similar way to that of Theorem 3. Remark that $*$ is commutative and $e\oplus a=e$ and so from Lemma 3 it follows $V=A^*V$, i.e.

$$V=\sum_i A^{*i}v.$$

On the other hand,

$$AV=\lambda V\Rightarrow\sum_j a_{ij}v_j=\lambda v_i\quad(i=1,\ldots,N).$$

As in the proof of theorem 3 the partial graph $H=(X,Y)$ will be constructed, where $(i,j)\in Y\Leftrightarrow$

(4.2) $$a_{ij}v_j=\lambda v_i,$$

and we consider a connected component H^k ($k=1,\ldots,K$). Let $i\in\gamma^k$. On the circuit γ^k the relation (4.2) implies

$$w(\gamma^k)v_i=\lambda_i$$

because $*$ is idempotent.

And so one gets: $w(\gamma^k)\lambda v_i\oplus\lambda v_i=w(\gamma^k)\lambda v_i$ and from corollary 1.1 $A^{*i}\lambda v_i\in v(e)$.

From this follows (remark 2) that $A^{*i}\lambda v_i\in v(\lambda)$ also holds.

As in the proof of theorem 3, one shows that any other term of the form $A^{*j}\lambda v_j$, $j\in H^k$ ($j\neq i$) is absorbed by $A^{*i}\lambda v_i$.

Choosing in every connected component H^k one vertex $i_k\in\gamma^k$, it can be written

$$V=\sum_{k=1}^K A^{*i_k}\lambda v_{i_k},\quad\text{with}\quad A^{*i_k}\lambda v_{i_k}\in v(\lambda).$$

So (β) is proved.

The proof of (γ) is the same as that of corollary 3.2.

Remark 8. *If $*$ is invertible and if $\lambda \neq e$ is an eigenvalue of A, then every eigenvector of A relative to λ is an eigenvector of $\lambda^{-1} * A$ for the eigenvalue e.*

$$AV = \lambda V \Rightarrow (\lambda^{-1}A)V = V.$$

Remark 9. *If $\oplus$ is idempotent, if $e \oplus a = e \; \forall a \in S$, and if $a_{ii} = e \; \forall \, i$ then $v(\lambda) \subset v(e)$.*

Proof. Let $V \in v(\lambda)$, $AV = \lambda V$, from which $(A \oplus E)V = \lambda V \oplus V = AV = \lambda V$ and so for $\forall \; \lambda v_i = v_i \oplus \lambda v_i$.

On the other hand, $e \oplus \lambda = e$ implies $v_i = v_i \oplus \lambda v_i$ and finally: $V = \lambda V$ so that $AV = \lambda V = V$ and $V \in v(\lambda) \Rightarrow V \in v(e)$.

5. EXAMPLES

(E1): *Shortest path in a graph*

$$\begin{cases} S = \mathbf{R} \cup \{+\infty\} = \bar{\mathbf{R}} \\ \oplus = \text{Min (idempotent)}; \; \varepsilon = +\infty \\ * = + \; (\text{invertible}); \; e = 0. \end{cases}$$

If λ is an eigenvalue of A, then $(\lambda^{-1}A)^*$ exists (corollary 2.1). On the other hand, according to the remarks 6 and 7 the following holds:

$$\lambda = \underset{\gamma}{\text{Min}} \left\{ \frac{w(\gamma)}{n(\gamma)} \right\}$$

(considered in the sense of the division of real numbers and usual order relation), and there exists a γ_0 such that

$$\lambda = \frac{w(\gamma)_0}{n(\gamma_0)}.$$

So that

$$\lambda = \frac{w(\gamma_0)}{n(\gamma_0)} = \underset{\gamma}{\text{Min}} \left\{ \frac{w(\gamma)}{n(\gamma)} \right\}.$$

It follows that the eigenvalue λ is *unique and is equal to the minimum of the average weight of the circuit* of minimum average weight.

According to corollary 3.1 the columns $(\lambda^{-1}A)^{*i}$ (such that i belongs to a circuit with average weight λ) generate a basis of the eigen semi-module $v(\lambda)$.

If G has circuits with strictly negative weights, A^* does not exist. In this case, however an eigenvalue $\lambda < 0$ exists with eigenvectors belonging to it $\mu = \lambda^{-1}$ is so the smallest real number > 0 for which the matrix $(\mu A)^*$ exists.

On the other hand, if A^* does not exist, $(\lambda^{-1}A)^*$ can be interpreted as the "best of the feasible approximations for A^*" (in a certain sense).

344

The problem of the determination of the circuit with minimal $\dfrac{w(\gamma)}{n(\gamma)}$ was considered by Dantzig, Blattner, Rao [4] and Lawler [12]. In [12] an algorithm is given
(its complexity is of the order n^4), which simply calculates

$$A' = E \oplus A \oplus \frac{A^2}{2} \oplus \ldots \oplus \frac{A^n}{n} \qquad \text{(in the sense of division of real numbers)}$$

(since the circuit with minimal average weight is necessarily *elementary*, it is sufficient
to calculate up to the order n).

Interesting applications of this problem are:

—the transportation of goods [4]

—the assignment problems [12].

(E2): *Maximal capacity path*

$$\begin{cases} S = \mathbf{R}^+ \cup \{+\infty\} = \overline{\mathbf{R}}^+ \\ \oplus = \text{Max (idempotent and } a \oplus b = a \text{ or } b); \ \varepsilon = 0 \\ * = \text{Min (idempotent and commutative)}; \ e = +\infty (e \oplus a = e). \end{cases}$$

A^* exists according to remark 3.

Corollary 3.3 will be applied. Every $\lambda \in S$ is then an eigenvalue. If

$$\mu_i = \sum_{\gamma \in L_{ii}} w(\gamma)$$

is the capacity of the circuit with maximal capacity passing through i, then $A^{*i}\lambda\mu_i$ is
an eigenvector of A for the eigenvalue λ (corollary 1.5) and the different vectors of
the form $A^{*i}\lambda\mu_i$ form a basis of $v(\lambda)$.

So one can deduce here certain results of [10].

(E3): *Existence of paths and connectedness*

This a special case of (E2).

$$\begin{cases} S = \{0, 1\} \\ \oplus = \text{Max}; \quad \varepsilon = 0 \\ * = \text{Min}; \quad e = 1. \end{cases}$$

Only e can be eigenvalue.

If the graph $G(A)$ is completely connected (A is irreducible), then

$$A^* = \begin{pmatrix} e \ldots \ldots e \\ e \ldots \ldots e \end{pmatrix}$$

and $v(e)$ contains only the eigenvector

$$V = \begin{pmatrix} e \\ \vdots \\ e \end{pmatrix}$$

(lemma 1).

Generally, if $G(A)$ is not completely connected, it can be decomposed into k connected components:

$$G^1 = [X^1, U^1] \quad G^2 = [X^2, U^2] \ldots G^k = [X^k, U^k].$$

Let

$$V = \begin{pmatrix} V^{(1)} \\ V^{(2)} \\ V^{(k)} \end{pmatrix}$$

be the corresponding partition of a vector $V \in S^n$. Then every vector of the form $V^{(i)} = \begin{pmatrix} e \\ \vdots \\ e \end{pmatrix}$ and $V^{(j)} = \begin{pmatrix} \varepsilon \\ \vdots \\ \varepsilon \end{pmatrix}$ $(\forall_j \neq i)$ is an eigenvector of A for the eigenvalue e (see Remark 4).

(E4): *Maximal probability path*:

$$\begin{cases} S = \{a \in \mathbf{R} \mid 0 \le a \le 1\} \\ \oplus = \text{Max}; \ \varepsilon = 0 \\ * = \times \ \text{invertible}; \ e = 1. \end{cases}$$

This can be reduced to the case of shortest path by choosing the logarithm of the weight a_{ij} for every edge $u = (i, j)$. The results are analogous to those obtained for the shortest path problem. Specially, there exists one eigenvalue λ corresponding to a circuit γ_0 such that

$$\lambda = w(\gamma_0)^{\frac{1}{n(\gamma_0)}} = \text{Min}_{\gamma} \ \{w(\gamma)^{\frac{1}{n(\gamma)}}\}.$$

(E5): *Markov chains*

$$\begin{cases} S = \{a \in \mathbf{R} / 0 \le a \le 1\}. \\ \oplus = + \ (\text{non idempotent}); \ \varepsilon = 0 \\ * = \times \ (\text{invertible}); \ e = 1. \end{cases}$$

A is a stochastic matrix $\left(\sum_j a_{ij} \le 1 \right)$. Then if A admits an eigenvector $V \ge 0$ for the *real* eigenvalue $\lambda \ge 0$ and if A is *irreducible* ($G(A)$ is completely connected), then from lemma 1 follows

$$\lambda > 0$$

and

$$V_i > 0 \qquad (\forall_i = 1, \ldots, n),$$

a classical result about the stationary nature of a Markov chain (see also the theorem of Perron–Frobenius [14] about eigenvalues and eigenvectors of matrices with positive elements).

346

(E6): *Regular language*

$$\begin{cases} S = \text{set of words} \\ \oplus = \text{union of set elements}; \ \varepsilon = \emptyset \\ * = \text{concatenation}, \ e = \text{void word}; \end{cases}$$

There will be no eigenvalue and eigenvector, if the language is not degenerated. On the other hand, if there is a void circuit, Corollary 1.1 is applicable with $\mu = e$.

(E7): *The k-th shortest path* (see [6], pp. 91 and 92)

$$\begin{cases} S = \overline{\mathbf{R}}^k \\ \oplus = \text{the } k \text{ least terms of two vectors}, \ \varepsilon = (+\infty)^k \\ * = \text{the } k \text{ least terms of the couplés sum } \varepsilon = \{0, +\infty, +\infty\}. \end{cases}$$

$\oplus$ is not idempotent and only Theorem 1 can be applied. If the graph contains no circuit with negative length, A^* exists.

So if k circuits with length zero pass through the vertex i, then

$$\sum_{\gamma \in L_{ii}} w(\gamma) = (O)^k \quad \text{and} \quad \sum_{\gamma \in L_{ii}} w(\gamma) \oplus e = (O)^k = \sum_{\gamma \in L_{ii}} w(\gamma)$$

and A^{*i} is an eigenvector of A for the eigenvalue e. One would define in the same way the eigenvectors for the k-th maximal capacity path and k-th maximal feasible path problems.

(E8): *Multiobjective problems* (see [6], pp. 89 and 90)

For simplicity (in the notions) here two objectives will be considered: shortest length and maximal capacity. Then, if $S' = \{\overline{\mathbf{R}}, \overline{\mathbf{R}}^+\}$ where the first component corresponds to the length and the second to the capacity, then

$$\begin{cases} S = P(S') \\ \oplus = \text{the set of efficient vectors for the union } \in = (+\infty, 0) \\ * = \text{is defined in the following way}: \end{cases}$$

$$\begin{aligned} \text{if} \quad & u = (u_\alpha) \quad \text{with} \quad u_\alpha S \\ & v = (v_\beta) \quad \text{with} \quad v_\beta \in S \end{aligned}$$

then $u * v$ is the set of those efficient vectors $w = (w_\gamma)$ for which $w_{\gamma_1} = u_{\alpha_1} + u_{\beta_1}$ and

$$w_{\gamma_2} = \min(u_{\alpha_2}, u_{\beta_2}); \quad e = (0, +\infty).$$

If $G(A)$ has no circuit with negative length, then A^* exists.

Then if through the vertex i a circuit γ with length zero passes, then the following holds:

$$(O, c(\gamma)) * (O, c(\gamma)) \oplus (O, c(\gamma)) = (O, c(\gamma)) * (O, c(\gamma)),$$

where $c(\gamma)$ is the capacity of the circuit γ. This shows that $A^{*i} * (O, c(\gamma))$ is an eigenvector of A for the eigenvalue e (according to corollary 1.1).

REFERENCES

[1] Backhouse, R. C. and Carre, B. A.: Regular Algebra Applied to Path-finding Problems, *J. Inst. Maths. Applics* (1975), n° 15, 161–186.

[2] Berge, C.: *Graphes et hypergraphes,* Dunod 1970.

[3] Carre, B. A.: An Algebra for network routing problems, *J. Inst. Maths. Applics* (1971), n° 7, 273–294.

[4] Dantzig, G. B., Blattner, W. O. and Rao, M. R.: Finding a cycle in a graph with minimum cost to time ratio, in *Théorie des graphes,* Rome 1966, ed. Dunod 1967, 77–83.

[5] Faucitano, G. et Nicaud, J. F.: Algorithme de Gauss pour la quasi-inversion de matrices, rapport de DEA sur la complexité des Algorithmes (professeur J. Berstel), Université de Paris VI, juin 1975.

[6] Gondran, M.: Algèbre linéaire et cheminement dans un graphe, note EDF HI 1137/02 du 9/7/73 paru dans *R.A.I.R.O.,* 9ième année, V-1, janvier 1975, 77–99.

[7] Gondran, M.: Path Algebra and Algorithms, in *Combinatorial Programming: Methods and Applications,* B. Roy (ed), 137–148, Copyright 1975 by D. Reidel Publishing Company, Dordrecht — Holland.

[8] Gondran, M.: L'algorithme glouton dans les algebres de chemins, *Bulletin des Etudes et Recherches EDF,* série C, n° 1, 1975, 25–32.

[9] Gondran, M.: Les éléments quasi-inversibles dans un S.A.B.O.T., note EDF (à paraître).

[10] Gondran, M.: Valeurs propres et vecteurs propres en classification hiérarchique, note EDF HI 1813/02 du 11/4/75, (à paraître), *R.A.I.R.O.*

[11] Hopcroft, J. E. and Ullman, J. D.: *Formal Languages and Their Relation to Automata,* Addison-Wesley Publishing Company, 1972.

[12] Lawler, E. L.: Optimal cycles in doubly weighted directed linear graphs, in *Théorie des graphes,* Rome 1966, éditeur Dunod 1967, pp. 209–213.

[13] Minoux, M.: Structure algébrique généralisée des problemes de cheminement dans les graphes-théorèmes, algorithmes et applications, (à paraître), *R.A.I.R.O. Rech. Op.* 10, n° 6, 33–62.

[14] Gantmacher, F. R.: *Théorie des matrices,* tome 2, Dunod 1966.

[15] Gondran, M. and Minoux, M.: L'indépendence linéaire dans le dioïds, *Bull. des Etudes et Recherches EDF Série C,* n° 1 (1977), 67–90.

AN ALGORITHM FOR TRANSFORMING
A SPANNING TREE INTO A STEINER TREE

P. KORHONEN

(Helsinki, Finland)

1. INTRODUCTION

A minimal (length) Steiner tree for a given set of points (A) is a solution of the following problem:

Construct the minimum-length tree containing a given set of points $A = \{a_1, a_2, \ldots, a_n\}$ in E^r within its set of vertices.

The solution tree may contain other points $S = \{s_1, s_2, \ldots, s_k\}$, $0 \leq k \leq n-2$ as vertices, too.

The points in the set S which are added to a tree to reduce its length are called Steiner points after J. Steiner, who first introduced the problem for the case $n = 3$.

An algorithm for finding the minimal Steiner tree in a finite number of steps has been given by Z. A. Melzak (1961) and further developed by E. J. Cockayne (1970). Melzak's algorithm involves the searching of such a large number of possibilities that the computation is only feasible for very small values of n. Although a variety of geometric criteria have been devised which are useful in reducing the number of possible alternatives the algorithm is still impractical for problems with n greater than 30. In practice even 8 points often seem to take quite much computing time.

This paper describes an algorithm (and its background) which finds a local solution for the Steiner problem in E^2. This fast algorithm transforms a given spanning tree into a Steiner tree shorter than the original tree. The algorithm converges in a finite number of steps and, if the initial tree for A is minimal, produces a sufficiently close approximation for the minimal Steiner tree. In the last section some experimental results are presented.

2. BASIC DEFINITIONS

It is easy to show that, for any Steiner point, the tree contains three edges leading to it, and the edges meet pairwise at 120°. From this fact we can deduce (as Gilbert and Pollak) five properties of the Steiner minimal tree, which define a set, that we call a *Steiner tree*.

Definition. A tree U with vertices $a_1, a_2, \ldots, a_n, s_1, s_2, \ldots, s_k$ is a Steiner tree on $a_1, a_2, \ldots, a_n$ if and only if it has the following properties:

P1: no pair of edges meet at less than 120^0 at each a_i

P2: $w(s_i)=3$, $1\leq i\leq k$, $(w(x)$ is the degree of a vertex $x)$

P3: $w(a_i)\leq 3$, $1\leq i\leq n$

P4: each s_i is the Steiner point of the triangle formed by the points in U with which it is directly connected

P5: $0\leq k\leq n-2$.

A *full* Steiner tree on A is a tree satisfying P1–P4 and also has $k=n-2$.

A subtree of a Steiner tree, which by itself is a full Steiner tree is called by Gilbert and Pollak a *full component*. A Steiner tree is a union of full components.

By the *topology* of a tree we (as Gilbert and Pollak) shall mean the connection matrix of the tree. (Two trees are said to have the same topology if their connection matrices are identical.) Thus the topology specifies connections, but not the positions of the points.

The topology of a tree satisfying P2–P3 will be called a *Steiner topology*.

If it is possible to construct a Steiner tree with a given Steiner topology, we call this topology a *feasible* Steiner topology.

3. GEOMETRY OF STEINER POINTS

The Steiner point for the triangle ABC (all angles are less than $120°$) is found by constructing three equilateral triangles built outward on the sides of ΔABC. Let the three corner points be K, L and M, as shown in Fig. 1. Connect KA, LB and MC. The intersection point S is the Steiner point. (It is a known result that the length of the lines KA, LB and MC are the same.) We call this expression the equilateral construction by E. J. Cockayne (2).

If we replace the tree with edges AS, BS and CS, for example, by the single line AK, so we have a full Steiner tree (the length is equal to the length of the original tree) with no Steiner point. We say that the branches SB and SC are "paired" to a single point K. The points K, L and M are called the *imaginary positions* of S (with respect) to A, B and C respectively.

At the same way in any n point full Steiner component (Fig. 2), any two branches, that leave some Steiner point (S), can be thought to be "paired" to a single point (K). The length of the line SK is equal to the total length of the "paired" branches. If further the "paired" branches are reduced to two lines that end to the points (M) and (L), the triangle KML is equilateral.

These imaginary positions of Steiner points are very useful for a computational process as will be explained later.

350

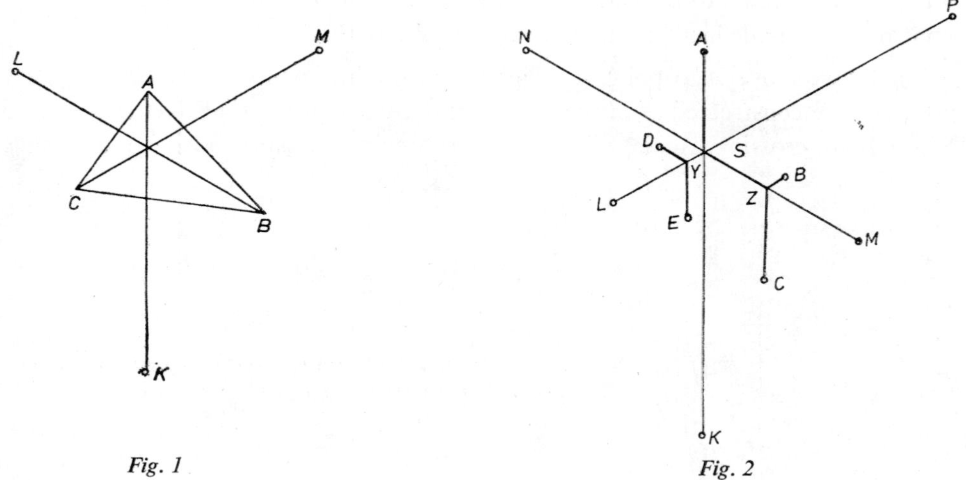

Let us examine closer the full component which is presented in Fig. 2. (The imaginary positions of $M(Z)$ and $L(Y)$ are drawn only to S.) We show now how we can reconstruct a Steiner tree when we disturb this stabile state by changing the position of A.

First, we move A to point A_1. We construct the new real and imaginary positions of S by triangle $\triangle A_1ML$, of Z by triangle $\triangle N_1BC$ and of Y by triangle $\triangle P_1DE$ as shown in Fig. 3. The tree is a Steiner tree. Next we move A to point A_2 and construct S, P_2

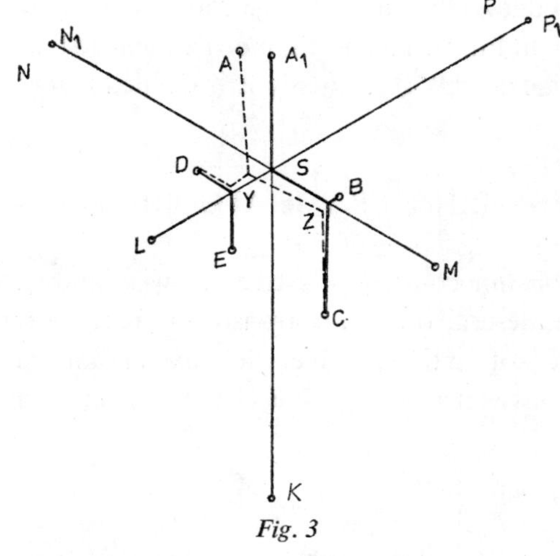

and N_2 as before. When we try to define Z, so we see that the triangle $\triangle N_2BC$ is no more a Steiner triangle, and the Steiner tree cannot be built by this topology assumption. We change it by destroying Z and removing edges SZ, CZ and BZ and connecting C to B and S to B. Now we can define S, P_2' and K_2' from the triangle ΔA_2BL and finally Y is defined from the triangle $\Delta P_2'ED$ (Fig. 4).

Finally we move A to point A_3 (Fig. 5). The Steiner points and their imaginary positions are constructed as in the first case, but we find that the edges that lead to S and Y are crossing. Again we see that our topology assumption is not a feasible

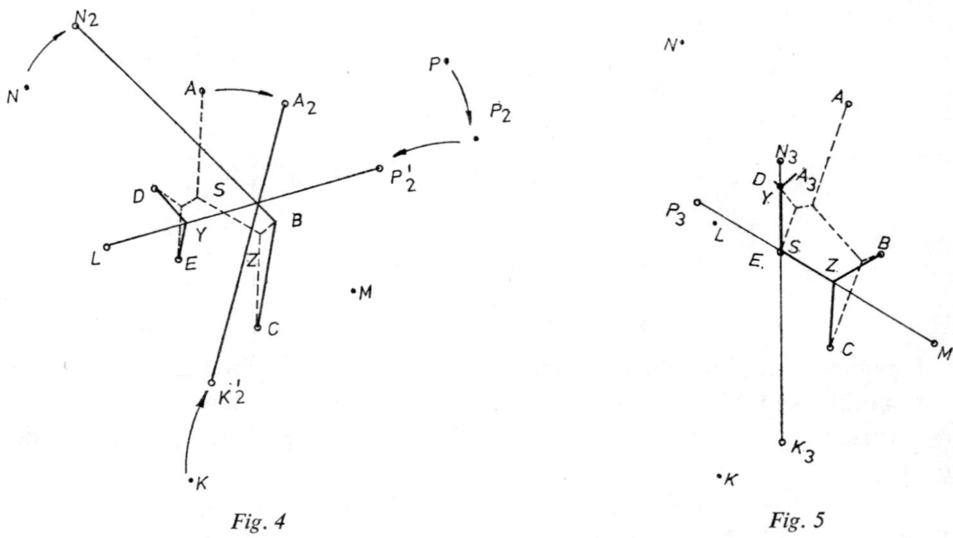

Fig. 4   Fig. 5

Steiner topology and, therefore, we change the assumption by replacing the edge AS by edge A_3Y and the edge EY by edge ES. We construct equilateral triangles ΔA_3DN_3 and ΔEMK_3 as shown in Fig. 5. The position of Y is defined from the triangle ΔK_3A_3D and of S from the triangle ΔN_3ME. Finally Z is defined from the triangle ΔP_3BC.

4. THE FRAME ALGORITHM

The FRAME algorithm constructs a tree stepwise and at each step calls the STEINER algorithm, described later, to transform the tree into a Steiner tree.

Let B_1 and B_2 be sets of vertices, and let the n fixed points be $a_1, a_2, \ldots, a_n$. Let F_1 and F_2 be two functions which specify at each step one element in the set B_1 and B_2 respectively.

The FRAME *algorithm*

step 1. $B_2 \leftarrow \{a_1, a_2, \ldots, a_n\}$, if $n=0$ then stop;
step 2. $k \leftarrow F_2$, $B_1\{a_k\}$, $B_2 \leftarrow B_2 - \{a_k\}$;
step 3. if B_2 is empty then stop;
step 4. $m \leftarrow F_1$, $k \leftarrow F_2$;

352

step 5. connect vertex m in B_1 and vertex k in B_2 with an edge

$$B_1 \leftarrow B_1 \cup \{a_k\}, \quad B_2 \leftarrow B_2 - \{a_k\};$$

step 6. call STEINER to transform the tree into a Steiner tree;
step 7. go to 3;
The solution tree for a given set of points depends only on functions F_1 and F_2.
By varying functions F_1 and F_2 suitably we can try to find even the global optimum. A good solution is found, if we define F_1 and F_2 so that they characterize a minimal spanning tree (e.g. Γ) by Prim's algorithm.

5. THE STEINER ALGORITHM

The STEINER algorithm checks whether a given directed connection may belong to a Steiner tree with an assumed topology, and, in the negative case, corrects the tree with elementary topology changes into a Steiner tree.

The algorithm also computes new real and imaginary positions of Steiner points. After each wrong topology assumption the algorithm stops, makes the topology change, and continues to check by this new topology assumption.

We denote by a capital letter (V) the name of a vertex, by V_t its real position and by V_U the imaginary position of V to U.

We define that the imaginary positions of any fixed point V are in a place V_t. The Boolean variables TOPF and TOPFF are used to indicate the feasibility of the assumed topology.

```
procedure STEINER (A, B, TOPF);
comment  A:        a source vertex
         B:        a destination vertex;
begin    procedure ADDS (A, B, C);
         comment  The new point is added and STEINER is called to check the tree
                  by this new topology;
         begin    remove edges AB and BC;
                  generate a new point S; (The position is not yet defined)
                  connect A, B and C to S;
                  TOPFF := true;
                  STEINER (B, S, TOPFF);
                  if TOPFF, then
                  STEINER (S, B, TOPFF);
         end      of ADDS;
         procedure REMS (A, B, C, S);
         comment  The Steiner point S is removed and STEINER is called to check
                  the tree by this new topology;
         begin    remove edges AS, BS and CS;
                  connect A and C to B;
```

                TOPFF := *true*;
                STEINER (*B*, *C*, TOPFF);
                if TOPFF *then*
                STEINER (*C*, *B*, TOPFF);
                if the edge *BC* exists *then*
                begin TOPFF := *true*;
                        STEINER (*B*, *A*, TOPFF);
                        if TOPFF *then*
                        STEINER (*A*, *B*, TOPFF);
                end
end   of REMS;
procedure ROTS (*S*, *Z*);
comment   Two Steiner points are rotated and STEINER is called to check
                the tree by this new topology;
begin   find $A \neq Z$ and $B \neq Z$ connected to S and $C \neq S$ and $D \neq S$ con-
                nected to Z;
                if the edges *AC* and *BD* are crossing *then* change *A* and *B*;
                compute the imaginary position of S to Z by drawing the equila-
                teral triangle outward on the side $B_S D_Z$ of the triangle $\Delta A_S B_S D_Z$;
                connect *A* to *Z* and *D* to *S* and remove *AS* and *DZ*;
                TOPFF := *true*;
                STEINER (*S*, *Z*, TOPFF);
                if TOPFF *then*
                STEINER (*Z*, *S*, TOPFF);
end   of ROTS;
comment   the main block is started;
                if B is a fixed point *then*
begin   find $P \neq A$ connected to *B* so that the improvement of the
                STEINER tree over the tree $A_B B_t$ and $B_t P_B$ is maximal.
                if $A_B B_t P_B \leqq 120^0$ *then*
                begin TOPF := *false*; ADDS (*A*, *B*, *P*)
                end
end   *else*
begin   find $P \neq A$ and $R \neq A$ connected to *B*;
                if $\Delta A_B P_B R_B$ is a Steiner triangle *then*
                begin define new B_t, B_P and B_R by drawing a Steiner construction
                        for the triangle $\Delta A_B P_B R_B$;
                        STEINER (*B*, *P*, TOPF);
                        if TOPF *then*
                        STEINER (*B*, *R*, TOPF);
                        if TOPF and $\sphericalangle A_t B_t B_A = 0°$ *then* ROTS (*A*, *B*);

$$end \quad else$$
$$begin \ if \ \sphericalangle A_B P_B R_B > 120^0 \ then \ change \ A \ and \ P$$
$$else \ if \ \sphericalangle A_B R_B P_B > 120^0 \ then \ change \ A \ and \ R;$$
$$if \ A \ is \ a \ fixed \ point \ then$$
$$REMS \ (P, \ A, \ R, \ B)$$
$$else$$
$$ROTS \ (A, \ B)$$
$$end$$
$$end$$
$$end \qquad of \ STEINER;$$

6. PROPERTIES OF STEINER

In this section we examine closer the behaviour and properties of the STEINER algorithm.

Firstly STEINER is called to check if two components can be connected, by examining the property P1 at the fixed point. If this property fails at the (fixed) common point (B) of these full components, the procedure ADDS is called to add a new Steiner point and STEINER is again called to check the feasibility of this new topology assumption. If no further topology changes are needed, the real and imaginary positions for all Steiner points of this extended full component are determined as shown in section 3.

When checking some full component, there are three possibilities that may force us to change the topology assumption:

1. At some fixed (end) point of this full component the property P1 fails.
2. A full component must be split into two components.
3. The property P4 fails.

In case 2 we call procedure REMS to remove the Steiner point that is impossible to locate and call STEINER to check the topology assumption for each component.

In case 3 we change two edges as explained in the procedure ROTS, and compute new real and imaginary positions for these two Steiner points and call STEINER to check this assumption. When no topology changes are needed the algorithm stops.

We cannot guarantee that STEINER ever finds the global optimum, because it does not try to seek "the best topology change" but assumes in every step that this topology is valid everywhere else and changes topology immediately, when the assumption fails. The quality of the solution depends on both the topology of the initial tree and its generating sequence.

We can afterwards try to improve a solution by checking some additionally required properties of the Steiner minimal tree, not defined for a Steiner tree, for example:

1. The solution tree must be a minimal tree.
2. The length of a line that connects two Steiner points S_1 and S_2 must be at least $(\sqrt{3} - 1) L_0$, where L_0 is the length of the shortest line leading to S_1 and S_2 (not $S_1 S_2$).

If the solution tree does not have properties 1 and 2, we make the necessary topology changes and use STEINER to find a feasible Steiner topology.

7. EXPERIMENTAL RESULTS

A program in Algol was written implementing our algorithm on the Burroughs B6700.

In order to test the algorithm, random points were generated uniformly distributed inside the square with corners at (0, 0), (0, 10), (10, 0) and (10, 10). Five replications were computed for each $n = 5, 10, \ldots, 50$. As the initial tree the minimal spanning tree generated by Prim's algorithm was used.

The results are summarized in the following table.

Summary of experimental results

n	t_{avg} (sec)	s_{avg} (%)	s_{max} (%)
5	0.112	1.45	3.7
10	0.327	1.50	2.9
15	0.628	3.76	6.2
20	0.945	2.25	2.9
25	1.404	2.89	4.4
30	1.706	2.51	3.3
35	1.832	2.39	3.0
40	2.280	2.54	3.1
45	2.381	2.89	3.4
50	2.757	2.39	2.8

t_{avg} = average computation time
s_{avg} = average percentage improvement
s_{max} = maximum percentage improvement
    (The averages are based on 5 samples)

We see that the computation time for the minimal tree is about $0.03 \times n$ sec and it makes possible to find the local Steiner solution even for 1000 points. The average improvement over the minimal tree is approximately 2.5 percent.

The algorithm was tested on the same data also with initial trees constructed by modifying Prim's algorithm so that in each step we choose the farthest point and connect it to the nearest point of the tree. Only with small n (n 10) better results are found. When n was great, the solution was even worse than the minimal tree.

In conclusion, the STEINER algorithm speedily transforms any initial spanning tree into a Steiner tree that is shorter than the original tree. The algorithm converges in a finite number of steps. The quality of the solution depends on the topology and the generating sequence of the initial tree.

REFERENCES

[1] Gilbert, E. N. and Pollak, H. O.: Steiner minimal trees, *SIAM J. Appl. Math.* 16, 1 (1968), 1–29.
[2] Cockayne, E. J.: Computation of minimal length full Steiner trees on the vertices of a convex polygon, *Math. Comp.* 23, 107 (1969), 521–531.
[3] Cockayne, E. J.: On the efficiency of algorithm for Steiner minimal trees, *SIAM J. Appl. Math.* 18, 1 (1970), 150–159.
[4] Chang Shi-Kuo: The generation of minimal trees with a Steiner topology, *J. ACM* 19, 4 (1972), 699–711.
[5] Boyce, W. M. and Seery, J. B.: STEINER 72, An improved version of Cockayne and Schiller's program STEINER for the minimal network problem, Bell laboratories, Computing Science Technical Report $\neq$ 35 (1975).

GENERALIZED PATH ALGEBRAS

M. MINOUX

(Issy, France)

I. INTRODUCTION

The interpretation of path-finding problems in (finite) graphs in terms of algebraic structures has been extensively studied by many authors [2], [6], [7], [11], [12], [13].

Recently, it has been shown (Carre [2], and Gondran [6], [7]) that many problems of this type are equivalent to solving systems of linear equations in some particular algebraic structures called *semi-rings*. It follows that path-finding algorithms may be viewed as specializations of other well-known algorithms in classical algebra: Jacobi, Gauss–Seidel, Gauss–Jordan, etc.

It turns out, however, that a number of path-finding problems cannot be interpreted within the framework of these "classical" path algebras, e.g.: the shortest path problem with time-varying lengths on the arcs ([3], [5]), or the shortest path problem with time constraints on movement and parking ([8]).

The purpose of this paper is to show that these problems (let us call them "non-classical") must be interpreted and solved in more general algebraic structures.

II. GENERALIZED PATH ALGEBRAS: DEFINITIONS AND PROPERTIES

Let $G=[X, U]$ be a given digraph with node set $X(|X|=N)$, and with arc set U.

Using the Markov chains terminology, we shall associate with each node $x_i \in X$ an element $E_i \in S$ called the *state* of x_i (S is the set of states), and with each arc $(x_i, x_j) \in U$ a mapping h_{ij} ($S \xrightarrow{h_{ij}} S$) called a *transition* from x_i to x_j.

II.1. Notations and definitions

Let S be a set (state-set) together with a given internal operation $\oplus$, associative and commutative (S is a monoid). The zero element is denoted ε.

Let H be the set of all endomorphisms of S, with unit element $\mathbf{e}$.

$$(h(a\oplus b)=h(a)\oplus h(b) \ \forall h\in H, \ a\in S, \ b\in S).$$

$\oplus$ induces on H an internal operation still denoted $\oplus$.

The zero element of $\oplus$ in H is: $h^\varepsilon(h^\varepsilon(a)=\varepsilon, \ \forall a\in S)$.

Consider a second law $\otimes$ in H defined by:

$$h\otimes g = g\circ h \ (\forall h, \ \forall g\in H),$$

where $\circ$ is the product of mappings.

$\otimes$ is associative, but generally not commutative. The unit element of H is $\mathbf{e}$. We note that H is closed for $\otimes$.

Notice that $\otimes$ is right and left distributive with respect to $\oplus$.

The structure: $(S, H, \oplus, \otimes)$ will be called *a generalized path algebra*.

(Observe that the triple $(H, \oplus, \otimes)$ is a "classical" path algebra in the sense of [2], [7].)

II.2. Generalized adjacency matrix of a graph

With each arc $(x_i, x_j)\in U$ of graph G, let us associate an endomorphism $h_{ij}\in H$ which we shall call: *transition of length 1 from x_i to x_j*.

Throughout this paper, we assume that: $h_{ij}(\varepsilon)=\varepsilon$.

(If there exist several (x_i, x_j) arcs, h_{ij} is defined for the ordered pair (x_i, x_j) as the sum $(\oplus)$ of all transitions associated with arcs (x_i, x_j).)

The *generalized adjacency matrix $A=(a_{ij})$* of G is then defined by:

$$\begin{cases} a_{ij}=h_{ij} & \text{if} \ (x_i, x_j)\in U, \\ a_{ij}=h^\varepsilon & \text{if} \ (x_i, x_j)\notin U. \end{cases}$$

The operations $\oplus$ and $\otimes$ on H induce on H^{N^2} operations still denoted $\oplus$ and $\otimes$.

II.3. Some properties of generalized adjacency matrices

Let $\pi_{ij}=\{(i_1, i_2), (i_2, i_3), \ldots, (i_{p-1}, i_p)\}$ be a path of length p (p arcs) between $i=i_1$ and $j=i_p$ in G.

The endomorphism:

$$t(\pi_{ij})=h_{i_1 i_2}\otimes h_{i_2 i_3}\otimes\ldots\otimes h_{i_{p-1} i_p}$$

will be called the *transition* associated with path π_{ij}.

The length of the transition $t(\pi_{ij})$ is the number of arcs in π_{ij}.

The following properties show that many path-finding problems on G reduce to computing the matrices A^k (k-th power of A) or: $A^{(k)}=A\oplus A^2\oplus A^3\oplus\ldots\oplus A^k$.

360

Property 1:

$$(a^k)_{ij} = \sum_{\pi} t(\pi)$$

where summation is taken on all $x_i - x_j$ paths π *with exactly k arcs.*

Property 2:

$$(a^{(k)})_{ij} = \sum_{\pi} t(\pi)$$

where summation is taken on all $x_i - x_j$ paths π *with no more than k arcs.*

Demonstration is made by recurrence on k, using the property that h^ε is right and left absorbing for $\otimes$ (see [7], [9]).

III. SOLUTION OF NON-CLASSICAL PATH-FINDING PROBLEMS

III.1. Existence of A^* quasi-inverse of A

For $h \in H$, let h^k be the k-th power of h, i.e.:

$$h^k = h \otimes h \otimes h \otimes \ldots \otimes h \quad (k \text{ times})$$

and let:

$$h^{(k)} = e \oplus h \oplus h^2 \ldots \oplus h^k.$$

Definition 1. An endomorphism $h \in H$ is said to be *p-regular* if: $h^{(p)} = h^{(p+1)}$.

Definition 2. We say that a graph G is without *p-absorbing circuit* if the transition $t(\mu)$ associated with each circuit μ of G is *p-regular*. A graph is without *O-absorbing circuit* if and only if: $t(\mu) \oplus e = e$ ($\forall \mu$ circuit of G).

Definition 3. A subset H' of H is said to be *K-stable* (where $K < +\infty$ is an integer) if

$$h_1 \otimes h_2 \otimes \ldots \otimes h_K = h^\varepsilon \quad (\forall h_1, h_2, \ldots, h_K \in H').$$

Notice that all elements of a *K*-stable subset H' of H are *K-regular*.

The existence of the matrix $A^* = \lim_{k \to +\infty} A^{(k)}$, called the *quasi-inverse of A*, can then be proved in the following cases:

(1) G is without *O*-absorbing circuit (see [7]);
(2) G is without *p*-absorbing circuit, and the h_{ij} *commute* ([7]);
(3) The subset $H' = \{a_{ij} / (x_i, x_j) \in U\}$ is *K-stable* ([9]).

Classical path-finding problems correspond to the special case where:
— S is given a second (internal) law $*$ (associative, unit element, left and right distributive with respect to $\oplus$);
— all the endomorphisms a_{ij} belong to a certain subset $H' \subset H$ (closed for $\oplus$ and $*$) such that $(H', \circ)$ and $(S, *)$ are isomorphic.

This allows explicit determination of A^* directly, without considering the states of the nodes in G.

III.2. Generalized algorithms

Here we are interested in ("non-classical") path-finding problems for which A^* exists, but *cannot be computed explicitly*, because the endomorphisms a_{ij}, or their products $a_{ij} \otimes a_{jk}$ are known only by their effect on elements of S.

We then must content ourselves with determining the images

$$(a^*)_{1,j}(E_1^0) \quad (j=1, \ldots, N)$$

of an element $E_1^0 \in S$ (initial state) associated with some node x_1 taken as origin of the paths.

When $\oplus$ is *not idempotent*, we generalize the Bellman's algorithm [1] in the following way [9]:

If E_j^t are the states of the nodes of G ($j=1, \ldots, N$) at iteration t, the E_j^{t+1} are computed by:

$$\begin{cases} E_1^{t+1} = \sum_i a_{i,1}(E_i^t) \oplus E_1^0 \\ E_j^{t+1} = \sum_i a_{i,j}(E_i^t) \qquad (j \neq 1). \end{cases}$$

If $A^* = A^{(k)}$, then convergence occurs in less than k iterations.

When $\oplus$ is *idempotent*, the generalized Bellman's algorithm can be improved, thus leading to the generalized Ford's and Moore's algorithms [9].

Moreover, if we assume that S is given a preorder relation $\leq$ with the following hypothesis:

H1:
$$\forall \alpha \in S, \quad \beta \in S, \quad \gamma \in S:$$
$$\alpha \geq \gamma \quad \text{and} \quad \beta \geq \gamma \Rightarrow \alpha \oplus \beta \geq \gamma$$

H2:
$$\forall a_{ij}: \quad a_{ij}(\alpha) \geq \alpha \quad (\forall \alpha \in S)'$$

then, the following algorithm gives, for each $x_j \in X$ an element $E_j \in S$ (state) which is *minimal* with respect to $\leq$.

(Note that, since $\leq$ is a preorder relation, the minimum element is not necessarily unique.)

362

Algorithm (A1)

(a) iteration $k=0$; $E_1=E_1^0$ (initial state of node x_1 taken as origin); $E_i=\varepsilon$ ($\forall i\neq 1$); $X1=\emptyset$, $X2=\{x_1\}$.

(b) iteration $k=k+1$. Select $x_m\in X2$ such that $E_m\leqq E_i$ ($\forall x_i\in X2$); set: $X1=X1\cup\{x_m\}$ $X2=X2-\{x_m\}$

(c) if $X1=X$, terminated: the E_j are minimal $\forall x_j\in X$. Otherwise, go to (d).

(d) for all the arcs $(x_m, x_j)\in U$:

    — compute $E_j'=E_j\oplus a_{m,j}(E_m)$

    — if $E_j'\neq E_j$, set: $X2=X2_\cup\{x_j\}$

    — set: $E_j=E_j'$

(e) if $X2=\emptyset$, terminated. Otherwise, go to (b).

In the special case, where $\leqq$ is the order relation defined by:

$$\alpha\leqq\beta\Rightarrow\exists\gamma\in S;\quad \alpha\oplus\gamma=\beta$$

at each step of the algorithm: $X1\cap X2=\emptyset$, and the algorithm converges in at most N iterations. ($A1$) then appears as a generalization of Dijkstra's algorithm [4].

IV. EXAMPLES

An important class of path-finding problems can be formulated and solved in the generalized path algebra $(S, H, \oplus, \otimes)$.

Example 1 (shortest path with time-varying lengths on the arcs)

In this case, we take: $S=\mathbf{R}\cup\{+\infty\}$, $\oplus=\mathrm{Min}$, $\varepsilon=+\infty$. Each arc (x_i, x_j) is associated with a real function of time $t_j=h_{ij}(t_i)$ giving the arrival time t_j in x_j when starting from x_i at time t_i. h_{ij} is an endomorphism of S iff it is nondecreasing, and this assumption is not restrictive, as shown in [5]. It is easy to see [9] that the generalized Dijkstra's algorithm can be applied to this problem, thus leading to an $O(N^2)$ procedure for the computation of all shortest paths originated at a given vertex x_0.

The k-th shortest path problem with time-varying lengths on the arcs can be in the same way [9].

Example 2 (circuitless graphs and generalized dynamic programming)

Another example, strongly related to the previous one, is Pert scheduling with time-varying lengths of the tasks (the time needed to complete task i depends on the instant t_i at which it begins). In this case: $S=\mathbf{R}\cup\{-\infty\}$, $\oplus=\mathrm{Max}$, $\varepsilon=-\infty$. The graph G associated with the Pert network is circuit-free. If we want to determine a critical length path, the generalized Bellman's algorithm can be applied to this problem (generalized dynamic programming).

363

Example 3 (shortest path with time-constraints)

In a railway network, for example, the tracks are busy at certain known time intervals, due to the existing traffic on the network. It is also possible to park at the nodes (the stations) at certain known time-intervals. Given an additional train, starting from x_i at time t_i towards x_j, the problem is to determine a shortest time path between x_i and x_j on the network (see [8]).

In this case, S is the set of all finite families of (closed) real intervals, $\oplus$ is the union of two families.

It can be shown [9] that the generalized Dijkstra's algorithm applies to this problem, leading to an efficient solution method (for computational results, refer to [8]). Unlike the classical Dijkstra's algorithm, more than N iterations may be necessary for convergence, and the optimal path may not be elementary.

Other examples

A number of additional examples will be found in [9], including:

— shortest path with constraints;
— shortest path with actualization and extensions;
— maximum probability path with time-depending probabilities on the arcs;
— automata and formal languages.

REFERENCES

[1] Bellman, R.: On a routing problem, *Quart. Appl. Math.* 16, (1958).

[2] Carre, B. A.: An algebra for network routing problems, *Journ. Inst. Maths. Appl.* 7, (1971), 273–294.

[3] Cooke, K. L., Halsey, E.: The shortest route through a network with time-dependent internodal transit times, *Journ. Math. Anal. Appl.* 14, (1966), 493–498.

[4] Dijkstra, E. W.: A note on two problems in connexion with graphs, *Numerische Mathematik* I, (1959), 269–271.

[5] Dreyfus, S. E.: An appraisal of some shortest path algorithms, *Ops. Res.* 17, N° 3, 395–412.

[6] Gondran, M.: Algèbre des chemins et algorithmes, in: *Programmation combinatoire*, B. ROY ed. (Reidel), 1975.

[7] Gondran, M.: Algèbre linéaire et cheminement dans un graphe, *Rev. Fr. Autom. Inform. Rech. Operat.* V–1, (1975).

[8] Halpern, J., Priess, I.: Shortest path with time constraints on movement and parking, *Networks* 4 (1974), 241–253.

[9] Minoux, M.: Structures algébriques généralisées des problèmes de cheminement dans les graphes: Théorèmes, algorithmes et applications, *Rev. Fr. Autom. Inf. Rech. Operat.* 10 (1976), 33–62.

[10] Moore, E. F.: The shortest path through a maze, *Proc. Int. Symp. Theory of Switching,* part II, 1957, 285–292.

[11] Peteanu, V.: An algebra of the optimal path in networks, *Mathematica* 9, (1967), 335–342.

[12] Roy, B.: Chemins et circuits: énumération et optimisation, in: *Programmation combinatoire*, B. ROY Ed. (Reidel), 1975.

[13] Tomescu, I.: Sur les méthodes matricielles dans la théorie des réseaux, *C. R. Acad. Sc. PARIS*, 263, Série A, 826–829.

364

TESTING OF A NETWORK
RELAXATION METHOD FOR 0–1 SET PARTITIONING
AND SET COVERING

J. M. MULVEY

(Boston, USA)

The pure network model can be formulated as follows:*

$$[N] \quad (1) \qquad \underset{x \geq 0}{\text{minimize}} \; \sum_{j=1}^{n} c_j x_j$$

subject to:

$$(2) \qquad \sum_{j \in A(i)} -1 x_j + \sum_{k \in B(i)} +1 x_k = b_i$$

$$\text{for } i = 1, \ldots, m$$

$$(3) \qquad \text{and } 0 \leq x_j \leq u_j \qquad \text{for } j = 1, \ldots, n$$

where m: number of nodes

n: number of arcs

$B(i)$: set of nodes with feasible arc $(., i)$

$A(i)$: set of nodes with feasible arc $(i, .)$

u_j: upper bounds on arc j

b_i: net supply or demand on node i

c_j: cost coefficient on arc j.

The classical assignment, transportation and transshipment models possess this structure.

Since the early 1950's specialized algorithms has been designed for these models. There have been marked reduction in processing time and a large increase in solvable problem size during the past three years (see Charnes, et al. [1973] and Mulvey [1975]). Problems with thousands of nodes and tens-of-thousands of arcs are routinely handled.

---

* Standard notation is employed: vectors are identified by an over-bar and transposition of vectors is implied for multiplicative conformity.

These improvements in network technology provide a strong motivation for connecting "non-network" models with the network framework. Thus, in this paper, the equivalence between 0-1 set partitioning/covering and a transportation model with multiple side restrictions is shown. Instead of using the conventional branch and bound approach (i.e., relaxing the integer requirements and employing linear programming), we transform the problem into a side constrained network model and initially relax the side restrictions as well as the integer requirements. We then solve the resulting *pure* network. The side constraints are subsequently satisfied in the course of performing an implicit enumeration to resolve the integer requirements. This methodology has been extended to general 0-1 integer programs (see Glover and Mulvey [1975]).

The network relaxation method has considerably more variables than the original formulation. Nonetheless this problem manipulation has several advantages over the usual LP relaxation. First of all, calculations are carried out in integer arithmetic and reinversions of large matrices are unnecessary. Second, the extreme sparsity which is a common trait in large scale examples is maintained throughout the algorithm. Additionally, it should be remembered that large networks can be efficiently solved.

The set covering problem is defined as follows:

$$[\text{SC}] \quad (4) \qquad \underset{y \geq 0}{\text{minimize}} \ \sum_{j=1}^{n} c_j y_j$$

$$(5) \qquad \text{subject to} \ \sum_{j=1}^{n} a_{kj} y_j \geq d_k \quad \text{for} \ k = 1, \ldots, m$$

$$(6) \qquad y_j = \{0 \text{ or } 1\} \quad \text{for all } j,$$

where a_{kj}: 0 or 1 all k, j
d_k: positive integer all k.

If constraints (5) are equalities, the problem is the special case of set partitioning [SP]. It should be noted that d_k is often unity for all k.

This model and its extensions have numerous real-life applications, such as maintenance routing. Various solution strategies have been developed for these integer programs, for example see Bellmore and Ratliff [1971], Lemke, et al [1971], and Marsten [1974]. Marsten's [1974] enumeration procedure can handle partitioning examples with one hundred constraints and several thousand variables. According to Marsten [1974], the principle bottleneck in progressing to larger cases is solving the underlying linear program. This is primarily due to its high degeneracy. This difficulty does not arise in network relaxation; in fact, as shown in Mulvey [1975], the specialized network algorithms can capitalize on degeneracy.

366

1. PROBLEM MANIPULATION

In order to show the equivalence between the network model and set covering/partitioning, we provide the following definitions:

$$(7) \qquad y_j \triangleq \sum_{i \in I_j} x_{ij}/e_{ij} \quad \text{for } j = 1, \ldots, m$$

where

$$(8) \qquad x_{ij} = 0 \text{ or } 1 \quad \text{for all } i, j$$

$$(9) \qquad x_{ij} = x_{i'j} \quad \text{for all } i, i' \in I_j$$

in which

$$I_j = \{i \mid a_{ij} \neq 0\}.$$

The index set I_j identifies, for column j, those rows which have non-zero coefficient elements, a_{ij}'s in constraints (5).

The e_{ij} coefficients are suitably defined so that whenever $x_{ij} = 1$ for all $i \in I_j$, then $y_j = 1$, and whenever $x_{ij} = 0$ for all $i \in I_j$, then $y_j = 0$. Several alternative methods exist for determining the values of the e_{ij} coefficients. We can let

$$(10) \qquad e_{ij} = |I_j| \quad \text{for all } i \in I_j,$$

for $j = 1, \ldots, n$. This formula is the simplest since the costs are equally distributed for all x variables associated with a particular y variable.

A second method would be to weight the e_{ij} coefficients by the number of non-zero elements in a row; thereby, coefficients which occur in rows with a small number of elements would be given a relatively high cost. Thus we allow

$$(11) \qquad e_{ij} = \frac{\sum_{k \in I_j} |J_k| - |J_i|}{\left((I_j| - 1) \sum_{k \in I_j} |J_k| \right)} \quad \text{for all } i, j$$

where

$$J_i = \{j \mid a_{ij} \neq 0\}.$$

This scheme is more effective in locating an initial integer feasible solution within the branch and bound procedure than the equal weighting scheme. See Section 4.2 for details.

It is important to recognize that there are always some e_{ij} coefficients which equate the values of the objective functions for the LP and the network relaxation models. This is easily shown by duality theory.

Constraints (7), (8) and (9) specify a conduit between the original system [SP] and the network model [N]. By substituting the right hand side of (7) for the y-variables in (4)–(6), the set partitioning/covering model becomes:

$$(12) \qquad \underset{x \geq 0}{\text{minimize}} \sum_{j=1}^{n} c_j \sum_{i \in I_j} x_{ij}/e_{ij}$$

subject to:

(13)
$$\sum_{j=1}^{n} a_{km} \sum_{i \in I_j} x_{ij}/e_{ij} \geq d_k \quad \text{for} \quad k=1, \ldots, m.$$

and

(14)
$$\sum_{i \in I_j} x_{ij}/e_{ij} = 0 \quad \text{or} \quad 1.$$

The terms in the objective function (12) can be rearranged as

$$\sum_{j=1}^{n} \cdot \sum_{i \in I_j} f_{ij} x_{ij},$$

where $f_{ij} = c_j/e_{ij}$ for all i, j.

Since the x_{ij}'s for any $i \in I_j$ can be used interchangeable (see constraint (9)), we can eliminate the i indices in (13), so that

$$\sum_{j=1}^{n} a_{kj} \sum_{i \in I_j} x_{kj}/e_{ij} \geq d_k \quad \text{for all } k.$$

But due to restrictions on e_{ij} (see formula (10)), this inequality can be further simplified to

(13i)
$$\sum_{j=1}^{n} a_{kj} x_{kj} \geq d_k \quad \text{for all } k.$$

These are demand restrictions in the network relaxation model.

This transformation can be performed on any 0–1 integer program, but a generalized network code (with gains) is required to solve it, in which the gain on arc (kj) is represented by the a_{kj} coefficient. For the set partitioning/covering problems, however, constraint (13i) can be redefined because of restricted values of $a_{kj}(+1$ or $0)$. We obtain

$$\sum_{j \in J_k} x_{kj} \geq d_k, \quad \text{for all } k$$

resulting in the usual transportation demand restrictions.

To derive the supply side of the transportation model, we relax integer requirements on the x-variables to obtain

$$0 \leq x_{ij} \leq 1.$$

Then summing the non-zero coefficients in the j-th column,

$$0 \leq \sum_{i \in I_j} x_{ij} \leq g_j \quad \text{for all } j,$$

where $g_j = |I_j|$, i.e., the cardinality of I_j.

368

In summary, the transportation model with multiple side conditions can now be presented:

$[T_S]$
$$\underset{x_{ij}=0,\,1}{\text{minimize}}\ \sum_{j=1}^{n} \sum_{i\in I_j} f_{ij}x_{ij}$$

subject to

$$\sum_{j\in Jk} x_{kj}\geqq d_k \quad \text{for } k=1,\ldots,m$$

$$\sum_{i\in I_j} x_{ij}\leqq g_j \quad \text{for } j=1,\ldots,n$$

with the side restrictions $x_{ij}=x_{i'j}$ all i, $i'\in I_j$ for $j=1,\ldots,n$.

We now present a small numerical example. Letting the A coefficient matrix, cost vector, and right hand side vector d be defined as

$$A=\begin{bmatrix} 1 & & 11 \\ 1 & & 1\ 1 \\ & & 111 \\ & & 1\ \ \ 1 \end{bmatrix} \qquad d=\begin{bmatrix} 1 \\ 1 \\ 1 \\ 1 \end{bmatrix} \qquad c=[10\ \ 10\ \ 10\ \ 400\ \ 14].$$

The underlying transportation model for this example is depicted in Fig. 1, including the transformed cost coefficients f_{ij}. Since equality of supplies and demands is required for feasibility, a slack node must be added to the network—node S in Fig. 1—as well

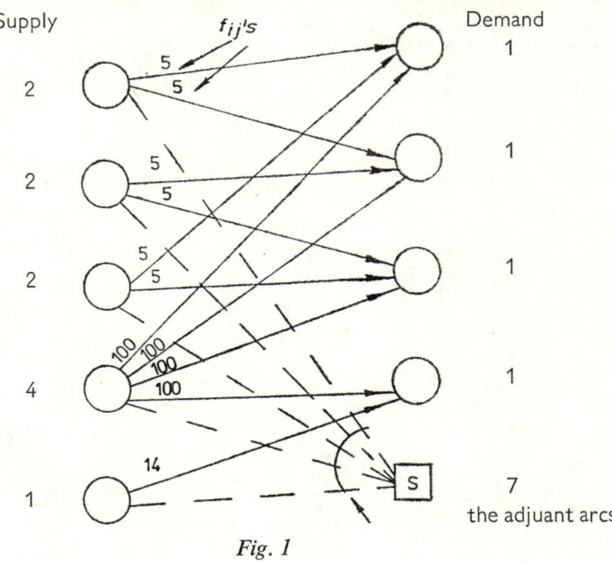

Fig. 1

as arcs between the supply nodes and slack node. It should be obvious that the side constraint in $[T_S]$ requires that flow on the arcs connected to this slack node, the adjuvant arcs, be either at their lower or upper bound. We call these $O-U$ restrictions.

Notice that the number of variables in the network relaxation model is equal to the number of non-zero elements in the original coefficient matrix. Moreover, the number of supply nodes equals the number of constraints, while the number of demand nodes is the number of original variables. For a large unsolved* partitioning example in Marsten [1974], the problem possesses 419 rows, 21,585 columns and approximately 80,000 non-zero coefficient elements. Although, this would result in a large-scale network, it is of manageable proportions for an advanced network code (see Mulvey [1975]).

2. GENERALIZED SETS

We illustrate the concept of generalized sets via the previous example. Notice that variables one through three define all pairs of items which "cover" the first three rows of this problem and that the remaining coefficient entries for these variables are zero. Whenever such a situation occurs, we can combine these variables (sets) into a generalized set with the following structure:**

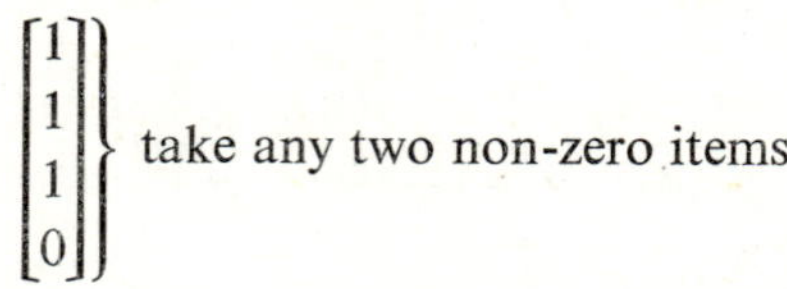

$$\left.\begin{bmatrix}1\\1\\1\\0\end{bmatrix}\right\}\text{ take any two non-zero items}$$

The condensed transportation network for this formulation is depicted in Fig. 2.

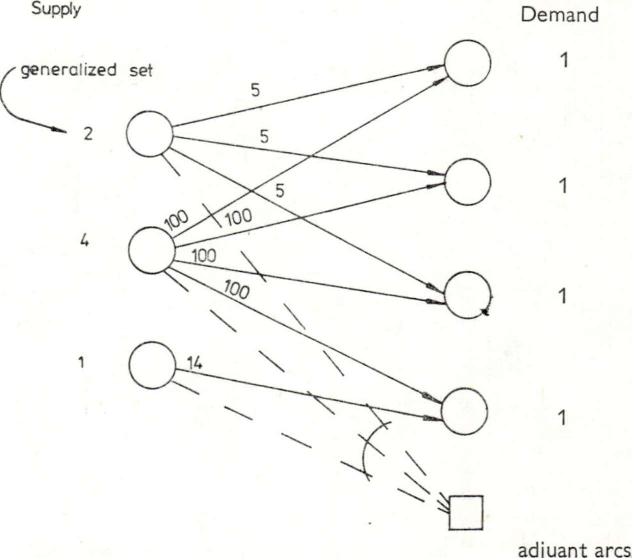

Fig. 2. An equivalent transportation model

* Over 3 hours of 360–91 CPU were expended in an attempt to solve the underlying linear program.
** The cost coefficients for the arcs representing generalized sets are not always immediately available.

As shown, any mix of generalized and regular sets can be employed. The primary advantage of generalized sets is the large number of regular sets which are implicitly treated, for instance, a situation with twenty non-zero elements — take any three represents $\binom{20}{3} = 1140$ regular sets! Yet these 1140 variables are handled in the network relaxation by simply adding one node with a supply of three and twenty arcs emitting out of this node.

Clearly, the generalized set idea is a powerful device which should be exploited by the management scientist. Since many real-life applications fits this concept, it is often convenient to employ generalized sets within the modeling framework, resulting in a sizeable reduction in execution times.

3. SOLUTION STRATEGIES

A strategy for solving $[T_S]$ is portrayed in Fig. 3. This is a variant of Geoffrion & Marsten's framework [1972], and its principle advantage over standard enumeration is encompassed in the usage of a network code for solving candidate subproblems. Since our network code is primal-based, it is important to utilize previous information for starting each subsequent candidate subproblem. We have taken this notion into account in designing the restarting procedure.

The standard fathoming criteria are shown in steps 8, 9, and 10 of Fig. 2. Fathoming criterion 1, infeasibility, arises whenever an artificial network variable remains in the basis at a positive flow or whenever a side restriction is violated for a fixed variable. If the network is feasible for all variables, i.e., *all* side restrictions are satisfied, then separation (Geoffrion and Marsten [1972]) cannot be performed and the subproblem has been subsequently fathomed by criterion 2. A standard bounding procedure is envisioned for fathoming criterion 3, that is, the subproblem is fathomed if the value of a relaxed version of the subproblem (lower bound) is greater than the value of the incumbent.

A disadvantage of the network relaxation approach is a loosening of the LP bounding mechanism. To minimize this difficulty, we have designed alternative "strengthened" network relaxations in which the structure of the problem is incorporated into the network. For example, side constrained set covering/partitioning problems can be handled in this manner. Instead of a transportation model, however, we employ a transshipment model for these strengthened representations, and these will be discussed in the future.

As an alternate strategy to that proposed in Fig. 3, Lagrangian multipliers λ for each side constraint can be employed. Let λ_j be the dual multiplier corresponding to the constraint:

$$x_{ij} = x_{i'j} \quad \text{for all} \quad i \in I_j \quad \text{and for} \quad j = 1, \ldots, m.$$

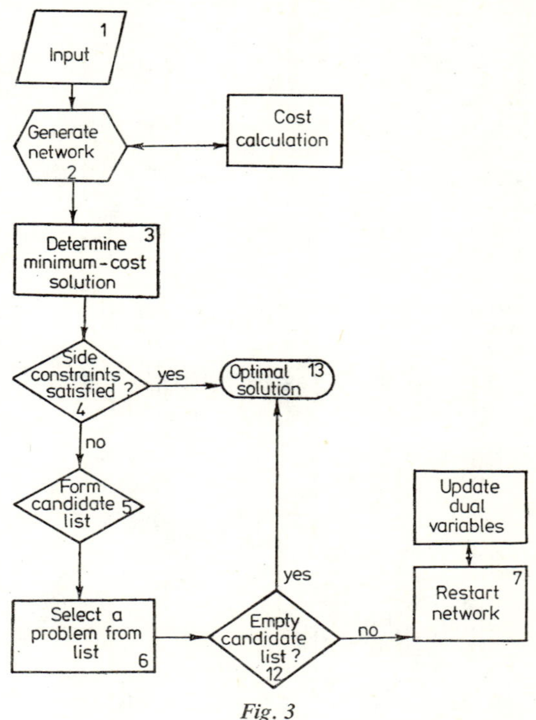

Fig. 3

Then dualizing with respect to these constraints brings them into the objective function, and this results in a valid bound for the enumeration process (see Geoffrion [1974] for further details). These multipliers are treated as adjusted costs on the adjuvant arcs in the network (see Fig. 1). This Lagrangian method may be more effective in locating a feasible integer solution than the usual branch and bound approach.

4. PRELIMINARY COMPUTATIONAL EXPERIENCES

Since the network relaxation method is aimed at quickly finding a good feasible point for large, sparse scheduling problems whereas the LP relaxation method is geared toward locating the optimal solution of smaller integer problems and since LP relaxation cannot handle generalized sets, it is difficult to empirically compare these methods. See Dembo and Mulvey [1975] for a discussion on comparing mathematical programming techniques. Nonetheless, the section describes experiences with two types of set partitionin/covering examples.

4.1. Three Steiner Triple Problems

In this subsection, we present computational results for solving three small, but difficult, set covering problems. The first is a set covering a problem which arises in computing the 1-width of incidence matrices for Steiner triple systems. This integer

372

problem is one member in a class of difficult problems; see Fulkerson, Nemhauser, and Trotter [1974]. A numerical example is shown in Fig. 4.

$$A_g = \begin{bmatrix} Z & I & 0 \\ 0 & Z & I \\ I & 0 & Z \\ I & I & I \end{bmatrix} 12 \times 9 \quad c_j = 3 \quad \text{for } j = 1, \ldots, n.$$

where

$$Z = \begin{bmatrix} 0 & 1 & 1 \\ 1 & 0 & 1 \\ 1 & 1 & 0 \end{bmatrix}.$$

Fig. 4. Set covering example

The corresponding network relaxation is depicted in Fig. 5. Notice the overall symmetry of this network and its small size: 21 nodes, 36 arcs. We concur with Fulkerson, Nemhauser, and Trotter [1974] that symmetry is the principle cause for excess calculations in the branch and bound procedure. But because of this symmetry the heuristic for setting cost coefficients on the arcs, i.e., the f's in $[T_S]$, is trivial. As seen in Fig. 5,

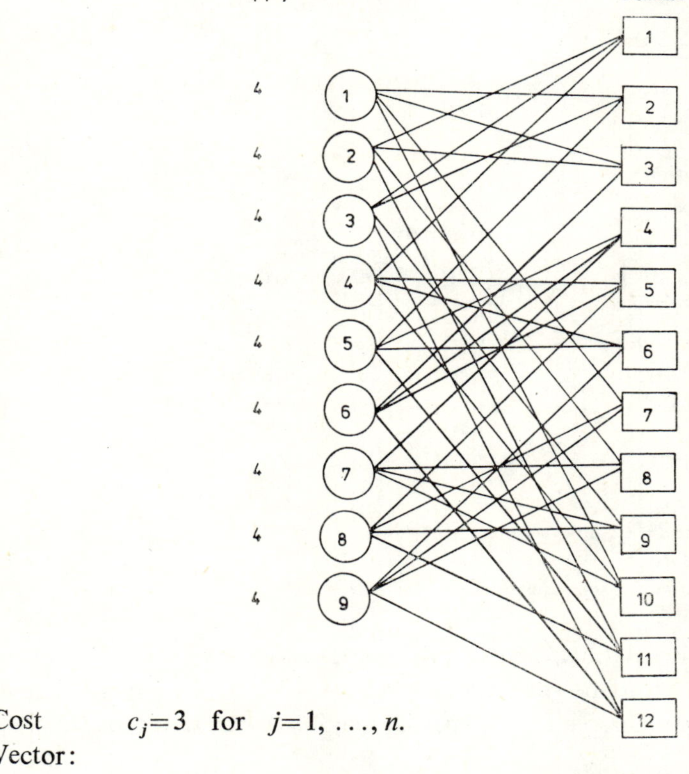

Cost Vector: $c_j = 3$ for $j = 1, \ldots, n.$

Fig. 5. Transportation relaxation model for Steiner triple example

each of the source nodes has four arcs leading out; hence, by equation (10), the cost coefficients for all real arcs are equal to $\frac{3}{4}$, since the original costs c_j equal three.

We have solved this problem with an elementary, static implicit enumeration code using a last-in, first-out (LIFO) strategy for traversing the branch and bound tree (selecting subproblems in the candidate list). Penalties are not used for selecting branches. Rather, the variable that is closest to its integer value is chosen as a separation variable. The bounding mechanism is the value of the objective function in the network relaxation model. A portion of the resulting branch and bound tree is shown in Fig. 6. The value on the nodes indicate the separation variable at each partial solution.

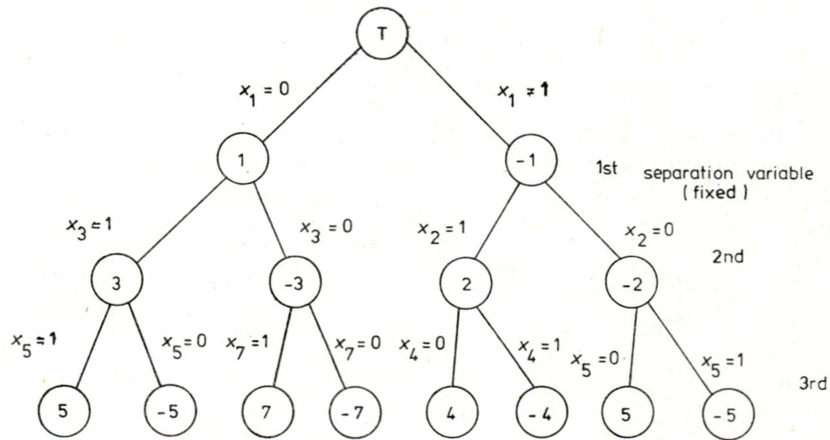

Fig. 6. Portion of branch and bound tree for set covering example

Here are the summary statistics for this problem:

	# of Times Called	Total Processing Times*	Average Processing Times	Worst (maximum) Processing Times
LP+0	313	8.074	.026	.173
NET	312	5.615	018	.098
START1	1	.005	.005	.005
RDC	1531	3.860	.003	.007

where   LP+0: Total program including $I/0$
        NET: Network processing routine
    START1: Start subroutine
     RDC: Subroutine for selecting entering basic-variable

It can be seen from this data that each subproblem takes approximately 5 pivots on the average to locate a solution. The total problem takes 8.074* seconds to discover and verify the integer optimal solution. 313 partial solutions were examined (313 nodes

* CDC 6500-RUN computer.

374

in the tree), resulting in an average processing time of .026 seconds per subproblem. According to Fulkerson, Nemhauser, and Trotter [1974], a cutting plane approach took approximately 20 seconds* to solve this problem after adding 44 cuts while an implicit enumeration took 5 seconds.*

To measure the effects of symmetry on this problem, we generated an identical problem with the following altered cost vector:

$$87 \quad 24 \quad 63 \quad 71 \quad 90 \quad 11 \quad 8 \quad 76 \quad 99.$$

The summary statistics for this branch and bound are

	# of Times Called	Total Processing Times*	Average Processing Times	Worst (maximum) Processing Times
LP+0	89	2.941	.033	.171
NET	88	2.155	.024	.138
START1	1	.005	.005	.005
RDC	644	1.375	.002	.007

where   LP+0: Total program including $I/0$
        NET: Network processing routine
    START1: Start subroutine
      RDC: Subroutine for selecting entering basic-variable.

which reveals that an asymmetric set covering problem is considerably easier to solve with the network relaxation approach.

The third example problem is slightly larger [9 sources, 36 sinks, and 147 arcs], but its size is partially offset by the asymmetry which is introduced in the cost coefficients. Table 1 provides the data for this example. Again a crude implicit enumeration code solved this problem and the summary statistics are tallied below:

Table 1. Third set covering example

Sinks	(Source) Corresponding from nodes					Sinks	(Source) Corresponding from nodes					Sinks	(Source) Corresponding from nodes					Sinks	(Source) Corresponding from nodes				
10	2	3	4			19	1	3	5			28	1	2	6			37	5	6	7		
11	4	6	8			20	4	5	9			29	1	8	9			38	2	7	9		
12	3	7	8			21	1	4	7			30	2	5	8			39	3	6	9		
13	2	3	4	6		22	1	3	5	7		31	1	2	6	3	8	40	5	6	7	9	
14	4	6	8	7	1	23	4	5	9	6	1	32	1	8	9	3	4	41	2	7	9	4	3
15	3	7	8	5	2	24	1	4	7	3	9	33	2	5	8	4	9	42	3	6	9	5	2
16	2	3	4	1	6	25	1	3	5	2	7	34	1	2	6	9	8	43	5	6	7	4	9
17	4	6	8	5		26	4	5	9	2		35	1	8	9	7		44	2	7	9	6	
18	3	7	8	6		27	1	4	7	4	9	36	2	5	8	4		45	3	6	9	5	2

Cost Vector   [26   42   70   63   29   84   11   100   32]

* Univac 1108 computer.

	# of Times Called	Toral Processing Times*	Average Processing Times	Worst (maximum) Processing Times
LP+0	135	22.055	.163	.585
NET	134	18.846	141	.564
START1	1	.009	.009	.009
RDC	3007	15.145	.005	.015

where   LP+0: Total program including $I/0$
          NET: Network processing routine
     START1: Start subroutine
        RDC: Subroutine for selecting entering basic-variable.

* Univac 1108 computer.

Although this network has four times as many arcs as the first example, its total processing time is less than twice as much as the first network's processing time. Here again the symmetry in the first example caused its excess run time. As a matter of speculation, we do not expect this type of symmetry to arise in large set partitioning problems.

4.2. Air Line Crew Scheduling

An important application of set partitioning is airline crew scheduling and routing. To test the power of the network relaxation method, we employed a real-life crew scheduling example from Marsten [1974]. This problem has 71 constraints, 545 variables and approximately 2270 non-zero entries in the A-coefficient matrix. This problem is translated into a transportation network by the process outlined in Section 1 and solved by the LPNET program (see Mulvey [1975] for a description of this code).

Computational results are shown in Table 2. The central processing units (CPU) for solving these networks are listed in columns 1 and 2. It is important to note that

Table 2. Computational Results for Airline Crew Scheduling Problem

Strategy	Total CPU*	Reduced cost CPU*	# pivots	cost coefficient heuristic	value of the objective function	# iolated side constraints
1	1.64	1.18	373	equal	7,181	43
2	1.54	1.02	327	weighted	10,494	37
3	1.26	0.84	277	(weighted)^2	10,509	43

* DEC 1070 Computer Seconds.

the adjuvant arcs become artificial variables during the starting subroutine of the simplex method. Otherwise, a much higher CPU figure will occur.

The only difference between these three problems is the heuristic for setting costs coefficients. On run 1, the equal cost heuristic which is defined in constraints (10) is employed. Run 2 employs the weighted cost heuristic, constraints (11), while run 3 utilizes the weighted-squared cost heuristic. The second strategy results in the fewest number of violated side constraints, yet the value of the objective function remains high. Since the network is the primary fathoming device in the branch and bound, it is important to maintain a strong bounding mechanism — i.e., a relatively high objective function value. For these reasons, strategy 2 is selected as better than either the first or third strategy.

5. CONCLUSIONS

Although, detailed experimentation with the network relaxation concept has not been completed, preliminary testing indicates that this approach will be competitive with the more traditional integer programming methods for certain problem classes. In particular, besides the experiences described herein, several large-scale scheduling and allocation problems have been formulated with a varient of network relaxation, and implementation of these models has demonstrated that good sub-optimal solutions can be produced in a reasonable amount of computer processing time. These results will be published in future reports. However, since it is difficult to properly evaluate optimization algorithms without a comprehensive test program (for some problems with developing such a program see Dembo and Mulvey [1975]), conclusive statements about network relaxation must be withheld until a thorough analysis is performed. A more exhaustive empirical analysis is clearly needed.

We should make one final point regarding the branch and bound enumeration procedure. This program can be easily converted into a fixed-charge network code, such as that described by Barr, Glover, and Klingman [1975]; therefore, any conclusions about network relaxation have implications for the fixed-charge network problem, and vice versa.

REFERENCES

Balas, E. and Padberg, M. W.: Set Partitioning, paper presented at the NATO Advanced Study Institute on Combinatorial Programming, Versailles, 1974.

Barr, R. S., Glover, F. and Klingman, D.: Optimization Methods for Large-Scale Fixed Charge Transportation Problems, presented at the ORSA/TIMS meeting, Chicago, 1975.

Bellmore, M. and Ratliff, H. D.: Set Covering and Involuntory Bases, *Management Science 18* (1971), 194–206.

Bowman, J. and Starr, J.: Set Covering by Ordinal Cuts I: Linear Objective Functions, Working Paper 97–72–3, Graduate School of Industrial Administration, Carnegie-Mellon University, June 1973.

Charnes, A., Glover, F., Karney, D., Klingman, D., Stutz, J.: Past, Present, and Future of Large-Scale Transshipment Computer Codes and Applications, Research Report CS 131, University of Texas at Austin, 1973.

Dembo, R. and Mulvey, M.: On the Analysis and Comparison of Mathematical Programming Algorithms and Software, Harvard University, HBS 76–19, 1976.

Fulkerson, D. R., Nemhauser, G. L. and Trotter, L. E. Jr.: Two Computationally Difficult Set Covering Problems that Arise in Computing the 1-Width of Incidence Matrices of Steiner Triple Systems, *Mathematical Programming Studies, 2* (1974), 72–81.

Geoffrion, A. M.: Lagrangean Relaxation for Integer Programming, *Mathematical Programming Studies, 2* (1974), 82–114.

Geoffrion, A. M. and Marsten, R. E.: Integer Programming Algorithms: A Framework and State-of-the-Art Survey, *Management Science, 18* (1972), 9, 465–491.

Glover, F., Karney, D. and Klingman, D.: Implementation and Computational Comparisons of Primal, Dual, and Primal–Dual Computer Codes for Minimum Cost Network Flow Problems, *Networks, 4* (1974), 191–212.

Glover, F., Klingman, D. and Stutz, J.: Implementation and Computational Study of a Generalized Network Code, presented at the 44th Nat. Meeting of ORSA, San Diego, Calif., November 1973.

Glover, F. and Mulvey, J. M.: Equivalence of the 0–1 Integer Programming Problem to Discrete Generalized and Pure Networks, Report No. 75–19, University of Colorado at Boulder, 1975, to appear in *Operations Research*.

Lemke, C. E., Salkin, H. M. and Spielberg, K.: Set Covering by Single Branch Enumeration with Linear Programming Subproblems, *Operations Research, 19* (1971), 998–1022.

Marsten, R.: An Algorithm for Large Set Partitioning Problems, *Management Science, 20* (1974), 5, 774–787.

Mulvey, J.: Special Structures in Network Models and Associated Applications, Ph. D. dissertation and Western Management Science Institute Working Paper, U.C.L.A., 1975.

OPTIMIZATION IN EDGE CHROMATIC SCHEDULING

D. DE WERRA

(Lausanne, Switzerland)

INTRODUCTION

Let $G=(X,E)$ be a multigraph with $|E|=m$ edges and without loops. A *k-coloring* $H=(H_1, \ldots, H_k)$ of the edges of G is a partition of E into k (non-empty) subsets H_i. H is *good for node x* if the number $k(x)$ of different colors appearing on edges adjacent to node x satisfies

$$k(x)=\begin{cases}d(x) & \text{if} \quad d(x)\leq k \\ k & \text{if} \quad d(x)\geq k\end{cases}$$

(here $d(x)$ is the degree of node x).

H is *good* if it is good for all nodes. With each k-coloring $H=(H_1, \ldots, H_k)$ we associate a sequence $h=(h_1, \ldots, h_k)$ such that a) $h_1\geq \ldots \geq h_k>0$ and b) $h_i=|H_i|$ $(i=1, \ldots, k)$. Clearly if $k\geq \max_x d(x)$ a good k-coloring is also a *usual* k-coloring (i.e. a coloring where any two adjacent edges have different colors).

An *edge chromatic scheduling problem* is by definition a problem which can be formulated in terms of edge coloring in a multigraph $G=(X, E)$. Some school timetable problems are of this type ([1], chap. 6). In some instances it is desired to determine a schedule (i.e. a usual edge k-coloring $H=(H_1, \ldots, H_k)$) which minimizes

a) $f_1(H)=k$ (i.e. the number of periods used)

b) $f_2(H)=\Sigma c_i h_i$ (where c_i is the unit cost of color i, i.e. the cost of assigning a job to period i).

In general f_1 and f_2 cannot be minimized simultaneously and so we have a multi-objective optimization problem. Our purpose is to try to characterize the non-dominated solutions of the problem.

379

Sometimes it will be convenient to introduce some zeroes at the end of a sequence h so that all sequences associated to the colorings of a given multigraph G have the same number of terms (for instance $m=|E|$ terms).

Given 2 sequences $h=(h_1, \ldots, h_m)$ and $h'=(h'_1, \ldots, h'_m)$ associated to colorings of G, we write $h>h'$ if

$$\sum_{i=1}^{p} h'_i \geq \sum_{i=1}^{p} h'_i \quad (p=1, \ldots, m).$$

This defines a partial ordering of the sequences associated to colorings of G.

A sequence h associated to a usual coloring (resp. good coloring) is a *usual sequence* (resp. *good sequence*); h is *maximum* if there is no usual sequence h' ($h' \neq h$) with $h'>h$.

For the above-mentioned scheduling problem one sees easily that all non-dominated solutions correspond to maximum usual sequences (provided the costs c_i are in non-decreasing order).

If G is a bipartite multigraph, then it has been shown [4] that given any good sequence $h=(h_1, \ldots, h_k)$ there exists a good sequence $h'=(h'_1, \ldots, h'_{k-1})$ with

$$(h'_1, \ldots, h'_{k-1}, 0)>(h_1, \ldots, h_k).$$

This implies that G has a good k-coloring for any k (since G obviously has a good m-coloring) and also that the chromatic index is equal to the maximum degree d. Furthermore all maximum usual sequences have at most d positive terms; this shows that if G is bipartite there is a solution of the scheduling problem which minimizes both f_1 and f_2.

Now if G is any multigraph, there may not exist a good k-coloring of G for each k. However J. C. Fournier has obtained recently a result which in the case of simple graphs (i.e. without multiple edges nor loops) may be formulated as follows [3]:

Let G be a simple graph; then for any k there exists an *almost good* (a.g.) k-coloring, i.e. a k-coloring which is good for all nodes x with degree $d(x) \neq k$ and which satisfies

$$k-1 \leq k(x) \leq k \quad \text{if} \quad d(x)=k.$$

Unfortunately, it is not true that for any almost good $h=(h_1, \ldots, h_k)$ (i.e. a sequence h associated with an e.g. k-coloring) there exists an almost good $h'=(h'_1, \ldots, h'_{k-1})$ with $(h'_1, \ldots, h'_{k-1}, 0)>(h_1, \ldots, h_k)$. An example is given in [5].

However, we have the following result whose proof is essentially the same as the proof of theorem 2.2 in [5].

Theorem 1. *Let G be a simple graph and $h=(h_1, \ldots, h_k)$ an almost good sequence; then there exists a $h'=(h'_1, \ldots, h'_{k-1})$ such that a) h' is associated with a $(k-1)$-coloring*

which is good for all nodes x with $d(x) \leq k-2$ and satisfies $k(x) \geq k-2$ for nodes x with
$d(x) \geq k-1$.

b) $$(h'_1, \ldots, h'_{k-1}, 0) > (h_1, \ldots, h_k).$$

Since for any $k \geq d+1$ (where d is the maximum degree), an a.g. k-coloring is a usual k-coloring, we obtain:

Corollary 1.1. *All maximum usual sequences h have d or $d+1$ positive terms.*

This implies also that the chromatic index is at most $d+1$. For multigraphs one could show similarly that $d+1$ must be replaced by $\max_x [d(x) + \max_y m(x, y)]$, where $m(x, y)$ is the number of edges between nodes x and y. So we have obtained a characterization of the non-dominated solutions of the scheduling problem which was mentioned previously. But the problem of generating all non-dominated solutions has not been solved yet even in the case of bipartite multigraphs.

FINAL COMMENTS

In this section, we mention a result on edge colorings and we derive a class of multigraphs for which there is a simple characterization of the maximum usual sequences.

An elementary odd cycle O of a multigraph $G=(X, E)$ is *quasiweak* if there exist 2 subsets C_1 and C_2 of consecutive nodes of O satisfying

   i) $|C_1|$ and $|C_2|$ are even
   ii) $|C_1 \cap C_2| = 1$, $|C_1 \cup C_2| = |O|$
   iii) in $G_0 = (X, E-O)$ each elementary chain joining C_1 to C_2 (without intermediate node on O) is of even length.

By using a proof technique which is essentially based on the method given in [6], we may state:

Theorem 2. *Let $G=(X, E)$ be a multigraph and S a subset of nodes which meet all elementary odd cycles which are not quasiweak.*

Then for any k there exists a k-coloring which is good for all nodes in $X-S$ and satisfies $k(x) \geq \min\left(k, [(1+d(x))/2]\right)$ for each node x in S.

(The fact that $k(x) \geq [(1+d(x))/2]$ is shown as follows: let $d(x)+1 = 2p+q$ (with $0 \leq q \leq 1$); then $[(1+d(x))/2] = p$; if $k(x) < p$, there is one color b which is missing at node x and one color which appears on at least 3 edges adjacent to x; in this case one can combine colors a and b and make both of them appear at x [2]).

As a consequence one sees that a multigraph G where all elementary odd cycles are quasiweak has a good k-coloring for each $k \geq 3$ (these graphs are thus similar to bipartite multigraphs).

However there is a difference which appears in the following statement:

Theorem 3. *Let G be a multigraph where all elementary odd cycles are quasiweak; then for any good sequence $h=(h_1, \ldots, h_k)$ there exists an "almost good" sequence*

$$h'=(h'_1, \ldots, h'_{k-1}) \quad \text{with} \quad (h'_1, \ldots, h'_{k-1}, 0) > (h_1, \ldots, h_k).$$

So although there is a good k-coloring for each k, one may not always associate a good sequence h' to a good sequence h but only on almost good sequence.

From Theorem 3 we deduce immediately the following:

Corollary 3.1. *All maximum usual sequences have d or $d+1$ positive terms if all elementary odd cycles are quasiweak.*

REFERENCES

[1] Bondy, J. A., Murty, U. S. R.: *Graph Theory with Applications,* Macmillan, 1976.

[2] Fournier, J. C.: Colorations des arêtes d'un graphe, *Cahiers du C.E.R.O.,* 15 (1973), 311–314.

[3] Fournier, J. C.: Méthode et théorème général de coloration des arêtes d'un multigraphe, *Journal de Mathématiques Pures et Appliquées,* 56 (1977), 437–453.

[4] De Werra, D.: Some remarks on good colorations, *J. of Comb. Theory* B, 21 (1976), 57–64.

[5] De Werra, D.: Remarks on the color-feasible sequences of a multigraph, *Networks* 8 (1978), 7–16.

[6] De Werra, D.: Multigraphs with quasiweak odd cycles, *J. of Comb. Theory,* 23 (1977), 75–82.

MATROID INTERSECTION PROBLEMS WITH GENERALIZED OBJECTIVES

U. ZIMMERMANN

(Cologne, FRG)

1. INTRODUCTION

For the theory of combinatorial optimization the introduction of matroids has turned out to be a powerful concept. Let be $N := \{1, 2, \ldots, n\}$ and S a nonempty set of subsets of N.

(1.1) Definition. $M := M(N, S)$ is called a *matroid over N* if the following two properties hold

$$(1.1.1) \qquad I \in S, \quad J \subseteq I \Rightarrow J \in S$$

$$(1.1.2) \qquad I, \quad J \in S, \quad |I| < |J| \Rightarrow \exists j \in J \setminus I : I \cup \{j\} \in S.$$

The elements of S are called *independent sets*.

If $M_B = M(N, S_B)$ and $M_R = M(N, S_R)$ denote two matroids over the same set N the elements of $S := S_B \cap S_R$ are called the *intersections* of M_B and M_R. The *weighted matroid intersection problem*

$$\text{(SP)} \qquad \max_{I \in S} \sum_{i \in I} c_i$$

with coefficients $c_i \in \mathbf{R}$, $i \in N$ has been considered by several authors (Edmonds [7], Lawler [11], Krogdahl [10]). An efficient algorithm (in the sense of Edmonds) for solving the problem (SP) is the primal weighted intersection algorithm published in [11]. A pure combinatorial proof of its validity is given in [10].

A generalization of the problem (SP) can be derived by changing the objective. The sum-objective of the usual weighted matroid intersection problem *(sum-intersection problem)* can be replaced by others which are of practical or theoretical interest. A problem with a trivial solution is the *bottleneck-intersection problem*

$$\text{(BP)} \qquad \max_{I \in S} \min_{i \in I} c_i$$

383

with coefficients $c_i \in \mathbf{R}$, $i \in N$. The solution of the problem (BP) is $I := \{\hat{\imath}\}$ with

$$c_{\hat{\imath}} := \max \{c_i \mid \{i\} \in S.$$

We define

$$r := \max \{|I| \mid I \in S\}$$

and

$$S_k := \{I \in S \mid |I| = k\}$$

for $1 \leq k \leq r$. Now we consider the *bottleneck-k-intersection problem*

$$(\text{BP})_k \qquad\qquad \max_{I \in S_k} \min_{i \in I} c_i.$$

The problem $(\text{BP})_k$ is a non-trivial problem. It can be solved by the threshold-method of Edmonds and Fulkerson [8]. This method works provided an element of S_k in an arbitrary $N' \subseteq N$ can be determined whenever it exists. Now it can easily be shown that the elements of S_k for $1 \leq k \leq r$ are intersections of maximal cardinality of two matroids. Thus an element of S_k in $N' \subseteq N$ can be determined by means of the cardinality intersection algorithm of Lawler [11] whenever there exists such an element in $N' \subseteq N$. Therefore, the problem $(\text{BP})_k$ can be solved with the aid of an efficient algorithm, too.

For the construction of efficient algorithms it suffices to discuss k-intersection problems. Every intersection problem can be solved by at first solving $r(\leq n)$ k-intersection problems and then comparing their solutions.

Besides the above-mentioned problems there are further examples of practical interest. For instance the *p-norm-k-intersection problem*

$$(\text{PP})_k \qquad\qquad \max_{I \in S_k} \sqrt[p]{\sum_{i \in I} (c_i)^p}$$

for $0 < p \leq \infty$ and the *lexicographic* or *multicriteria-k-intersection problem*

$$(\text{LexP})_k \qquad\qquad \operatorname*{lex\,max}_{I \in S_k} \begin{bmatrix} \sum\limits_{i \in I} c_i^1 \\ \dots \\ \sum\limits_{i \in I} c_i^m \end{bmatrix}$$

with $m \in \mathbf{N}$. A special example with a two-component objective is the *time-cost-k-intersection problem*

$$(\text{TP})_k \qquad\qquad \operatorname*{lex\,max}_{I \in S_k} \begin{bmatrix} \min\limits_{i \in I} c_i^1 \\ \sum\limits_{i \in I_{\min}} c_i^2 \end{bmatrix}$$

with

$$I_{\min} := \{i \in I \mid c_i^1 = \min \{c_j^1 \mid j \in I\}\}.$$

All these examples can be viewed as special cases of the *algebraic k-intersection problem*

$$(\text{AP})_k \qquad \max_{I \in S_k} \ast_{i \in I} c_i$$

with coefficients $c_i \in H$, $i \in N$, chosen from an ordered commutative semigroup $(H, \ast, \leq)$.

An algebraic approach of this kind has been applied to assignment problems by Burkard, Hahn and Zimmermann [2] in 1977. Subsequently matroidal restrictions (Zimmermann [12]), transportation problems (Burkard [1]) and network flow problems (Burkard, Hamacher and Zimmermann [3]) have been investigated with the aid of the same method.

2. ALGEBRAIC STRUCTURE

The coefficients of the algebraic k-intersection problem $(\text{AP})_k$ are chosen from an ordered commutative semigroup $(H, \ast, \leq)$ with internal composition "$\ast$" and order relation "$\leq$".

(2.1) Definition. $(H, \ast, \leq)$ is called *ordered commutative semigroup (group)* if the following three axioms hold

(2.1.1) $(H, \leq)$ is a nonempty ordered set

(2.1.2) $(H, \ast)$ is a commutative semigroup (group)

$$(2.1.3) \qquad a \leq b \Rightarrow a \ast c \leq b \ast c \quad \forall a, b, c \in H.$$

In the following $(H, \ast, \leq)$ is always an ordered commutative semigroup. Let $c_i \in H$ for $i \in I := \{1, 2, \ldots, t\}$. Then we define

$$(2.2) \qquad \ast_{i \in I} c_i := c_1 \ast c_2 \ast \ldots \ast c_t.$$

Especially, if $c_i = a$ for all $i \in I$ holds, we define

$$(2.3) \qquad a^{|I|} := \ast_{i \in I} c_i.$$

The following table describes the relationship between the semigroup and the examples of Section 1.

type	objective	$(H, *, \leq)$	internal composition	order relation
SP_k	$\displaystyle\sum_{i \in I} c_i$	$(\mathbf{R}, +, \leq)$	$a*b := a+b$	usual
BP_k	$\displaystyle\min_{i \in I} c_i$	$(\mathbf{R}, \min, \leq)$	$a*b := \min(a, b)$	usual
PP_k	$\displaystyle\sqrt[P]{\sum_{i \in I}(c_i)^P}$	$(\mathbf{R}\cdot\omega \cup \mathbf{R}_+, *, \leq)$ with $\omega := \sqrt[P]{-1}$	$a*b := \sqrt[P]{a^P + b^P}$	if $\alpha \leq \beta$ in $\mathbf{R}_+$ w.r.t. the usual order, then $\beta \cdot \omega \leq \alpha \cdot \omega \leq \alpha \leq \beta$
$LexP_k$	$\begin{bmatrix} \sum_{i \in I} c_i^1 \\ \cdots \\ \sum_{i \in I} c_i^m \end{bmatrix}$	$(R^m, +, \leq)$	$\begin{bmatrix} a^1 \\ \cdots \\ a^m \end{bmatrix} * \begin{bmatrix} b^1 \\ \cdots \\ b^m \end{bmatrix} := \begin{bmatrix} a^1+b^1 \\ \cdots \\ a^m+b^m \end{bmatrix}$	lexicographic
TP_k	$\begin{bmatrix} \min_{i \in I} c_i^1 \\ \sum_{i \in I\min} c_i^2 \end{bmatrix}$	$(\mathbf{R}\times\mathbf{R}_-, *, \leq)$	$\begin{bmatrix} a^1 \\ a^2 \end{bmatrix} * \begin{bmatrix} b^1 \\ b^2 \end{bmatrix} := \begin{cases} a & a^1 < b^1 \\ b & b^1 < a^1 \\ \begin{bmatrix} a^1 \\ a^2+b^2 \end{bmatrix} & a^1 = b^1 \end{cases}$	lexicographic

All examples fulfill two more axioms. The first one

$$(2.4) \qquad a \leq b \Rightarrow \exists c \in H : a = b*c \quad \forall a, \ b, \ c \in H$$

enables us to solve certain equations and the second one

$$(2.5) \qquad a*b = a*c \Rightarrow b = c \vee a*b = a \quad \forall a, \ b, \ c \in H$$

is a weak cancellation law. Finally, we consider the axiom

$$(2.6) \qquad a*b \leq a \quad \forall a, \ b \in H$$

which is not fulfilled in every case. For instance for positive $b \in \mathbf{R}_+ \backslash \{0\}$ and $a \in \mathbf{R}$ just the opposite inequality $a+b > a$ holds. In general axiom (2.6) holds in the *negative cone of H*

$$(2.7) \qquad \operatorname{neg} H := \{a \in H \mid a \leq e\}$$

with the neutral element e. A neutral element can always be adjoined to H, if (2.4) and (2.5) hold in $(H, *, \leq)$. For example adjoin ∞ to $(\mathbf{R}, \min, \leq)$ and $\binom{-\infty}{0}$ to $(\mathbf{R}\times\mathbf{R}_-, *, \leq)$.

The coefficients of an arbitrary algebraic k-intersection problem may be transformed in the following way:

386

$$\text{①} \qquad\qquad \alpha := \max\{c_i \mid i \in N\}$$

$$\text{②} \qquad\qquad \alpha * c_i' := c_i \quad \text{if} \quad \alpha > c_i,$$

$$c_i' := e \quad \text{if} \quad \alpha = c_i,$$

Denote by $(AP)_k'$ the transformed problem with coefficients c_i' for all $i \in N$.

(2.8) Lemma. *Every solution $I \in S_k$ of $(AP)_k$ is a solution of the transformed problem $(AP)_k'$ and vice versa. The optimal values z resp. z' fulfill the equation*

$$(2.8.1) \qquad\qquad z = \alpha^k * z'.$$

Proof. From ② in transformation T1 follows

$$\alpha * c_i' := c_i$$

for all $i \in N$. Let $I \in S_k$. Then $|I| = k$ and therefore

$$\underset{i \in I}{*}\ c_i = \alpha^{k} * \underset{i \in I}{*}\ c_i'.$$

This yields immediately (2.8).

The coefficients c_i' are elements of the negative cone for step ② of (T1) yields in the nontrivial case

$$\alpha * c_i' = c_i < \alpha$$

which implies by means of (2.1.3) $c_i' < e$.

In [2] the following result is proved (only interchanging $\leqq$ and $\geqq$):

(2.9) Lemma. *If $(H, *, \leqq)$ is an ordered commutative semigroup which fulfills (2.4) then $(\text{neg } H, *, \leqq)$ is an ordered commutative semigroup which fulfills (2.4).*

(2.9) immediately implies

(2.10) Corollary. *If $(H, *, \leqq)$ is an ordered commutative semigroup which fulfills (2.4) and (2.5) then $(\text{neg } H, *, \leqq)$ is an ordered commutative semigroup which fulfills (2.4), (2.5) and (2.6).*

In the following we consider only $(AP)_k$ with coefficients in an ordered commutative semigroup $(H, *, \leqq)$ which fulfills (2.4), (2.5) and (2.6); with regard to transformation (T1) and to (2.10) this can be done without loss of generality. Such semigroups have a unique canonical representation which can be derived from the results of Clifford [4], [5] and Conrad [6]. A more detailed discussion can be found in Zimmermann [13].

The semigroup $(H, *, \leqq)$ can be partitioned in a family $(H_j | j \in J)$ of ordered commutative semigroups which fulfill (2.6), a weaker form of (2.4)

$$(2.4)' \qquad a \leqq b \Rightarrow \exists c \in H_j : a = b * c \quad \forall a, \; b \in H_j$$

and the cancellation law

$$(2.5)' \qquad a * b = a * c \Rightarrow b = c \quad \forall a, \, b, \, c \in H_j.$$

Let $j(a)$ be the index of that semigroup of the family which contains $a \in H$. The internal compositions and order relations are connected by

$$(2.11) \qquad a * b = \begin{cases} \min (a, b) & \text{if} \quad j(a) \neq j(b) \\ a *_{j(a)} b & \text{if} \quad j(a) = j(b) \end{cases}$$

and

$$(2.12) \qquad a \leqq b \leftrightarrow j(a) < j(b) \vee \left(j(a) = j(b) \wedge a \leqq_{j(a)} b \right).$$

For the above-mentioned examples the following table specifies the partitions of the negative cone of the semigroups. Obviously, each element of the partitions in the family is part of or equal to the negative cone of a certain ordered commutative group. In general, each $\tilde{H}$ which fulfills (2.5)' can be embedded in an ordered commutative group (see Fuchs [9], Corollary 6). Therefore each $(H_j, *_j, \leqq_j)$ for $j \in J$ can be embedded in the negative cone of an ordered commutative group $(G_j, \oplus_j, \leqq_j)$ for $j \in J$. For example, in the case of the $(BP)_k$ each element $\{a\}$ coincides with the trivial group of one element which is neutral and in the case of the $(TP)_k$ the elements $\{a\} \times \mathbf{R}_-$ coincide with the negative cone of $(\{a\} \times \mathbf{R}, *, \leqq)$. Let e_j denote the neutral element of G' for $j \in J$.

type	negative cone	partition
SP_k	$(\mathbf{R}_-, +, \leqq)$	$(\mathbf{R}_- \backslash \{0\}, \{0\})$
BP_k	$(\mathbf{R} \cup \{\infty\}, \min, \leqq)$	$(\{a\} \mid a \in \mathbf{R} \cup \{\infty\})$
PP_k	$(\mathbf{R}_+ \cdot \omega, *, \leqq)$	$((\mathbf{R}_+ \backslash \{0\}) \cdot \omega, \{0\})$
$lexP_k$	$(\hat{\mathbf{R}}^m, +, \leqq)$	$(\hat{\mathbf{R}}^m \backslash \{0\}, \{0\})$ mit $\hat{\mathbf{R}}^m := \{a \in \mathbf{R}^m \mid a \leqq 0\}$
TP_k	$((\mathbf{R} \times \mathbf{R}_-) \cup \{(\begin{smallmatrix}\infty\\0\end{smallmatrix})\}, *, \leqq)$	$(\{a\} \times \mathbf{R}_- \mid a \in \mathbf{R}), \{(\begin{smallmatrix}\infty\\0\end{smallmatrix})\}$

Now we consider the following second transformation T2 of the coefficients of an $(AP)_k$:

Transformation T2: $\qquad\qquad c_i \to T2(c_i) := c_i''$

---

①
$$\beta := \max_{I\in S_k} \min_{i\in I} c_i$$

②
$$c_i'' := \begin{cases} e_{j(\beta)} & j(\beta) < j(c_i) \\ -\infty & j(\beta) > j(c_i) \\ c_i & j(\beta) = j(c_i) \end{cases}$$

All transformed coefficients $c_j'' \neq -\infty$ are elements of the ordered commutative group $G_{j(\beta)}$. If we define

$$N'' := \{ j\in N \mid c_j'' \neq -\infty \},$$

then we know from matroid theory that

$$S_k'' := \{ I\in S_k \mid I \subseteq N'' \}.$$

consists of the maximal intersections of two matroids over the same set N''. Now we consider the reduced problem

$(AP)_k''$
$$\max_{I\in S_k''} \; \underset{i\in I}{\text{\Large$*$}} \; c_i''.$$

Define for convenience for $a\in H$

(2.11)
$$a * -\infty := -\infty * a := -\infty \quad \text{and} \quad -\infty < a$$

(2.12) Theorem. *Every solution $I\in S_k$ of the $(AP)_k$ is a solution of the transformed problem $(AP)_k''$ and vice versa. The optimal values of both problems are equal.*

Proof. Define $j[I] := j(\underset{i\in I}{*}\, c_i)$. Let be $I, \tilde{I}$ solutions of the $(AP)_k$ resp. the corresponding $(BP)_k$ in step 1 of (T2). Assume $j[\tilde{I}] < j[I]$. Then from (2.11) and (2.12) we get

$$\beta := \underset{i\in\tilde{I}}{*}\, c_i < \underset{i\in I}{*}\, c_i$$

which yields a contradiction to the maximality of β; on the other hand,

$$\underset{i\in\tilde{I}}{*}\, c_i \leqq \underset{i\in I}{*}\, c_i$$

must hold which implies $j[\tilde{I}] \leqq j[I]$. Therefore, we find

$$j(\alpha) = j[\tilde{I}] = j[I].$$

Now assume that $I \notin S_k''$.

Then there exists an index $i\in I\backslash N'$ with $j(c_i) < j(\beta)$. This implies the contradiction $j[I] < j(\beta)$. Furthermore, if $\hat{I}\in S_k''$ then there exists an element $i\in\hat{I}$ such that $j(c_i) = j(\beta)$.

Otherwise $j(\beta) < j[\hat{I}]$ holds which contradicts the maximality of β. Therefore we find

$$\bigoplus_{i \in \hat{I}} c_i'' = \mathop{*}_{i \in \hat{I}} c_i \leqq \mathop{*}_{i \in I} c_i$$

for all $\hat{I} \in S_k''$. Thus I is a solution of the $(AP)_k''$. As $I \in S_k''$ holds we find

$$\mathop{*}_{i \in I} c_i \leqq \bigoplus_{i \in \hat{\hat{I}}} c_i$$

for each solution $\hat{\hat{I}}$ of the $(AP)_k''$. Thus $\hat{\hat{I}}$ is a solution of the $(AP)_k$. ∎

By means of transformation (T2) the problem $(AP)_k$ has been reduced to an equivalent one $(AP)_k''$, which has coefficients in an ordered communative group.

The combinatorial proof of Krogdahl [10] of Lawler's primal intersection method avoids the use of the duality theorem of linear programming and employs only the group structure of $(\mathbf{R}, +, \leqq)$. Therefore all theorems in [10] hold when we replace $(\mathbf{R}, +, \leqq)$ by an arbitrary ordered commutative group $(G, \oplus, \leqq)$. Thus the problem $(AP)_k''$ can be solved with the aid of the primal intersection algorithm formulated for such a group. A more detailed description can be found in [13].

3. THE GENERAL ALGORITHM

The general algorithm can roughly be stated in the following way.

Algorithm for solving the algebraic k-intersection problem

$(AP)_k$ $\qquad\qquad\qquad\qquad\qquad \max_{I \in S_k} \mathop{*}_{i \in I} c_i$

with coefficients in an ordered commutative semigroup $(H, *, \leqq)$ which fulfill (2.4) and (2.5)

---

① Transform the coefficients by (T1):

$\qquad c_i := (T1)(c_i)$ for $i \in N$

② Transform the coefficients by (T2):

$\qquad c_i := (T2)(c_i)$ for $i \in N$

③ Solve the algebraic-k-intersection problem in the group $(G_{j(\beta)}, \oplus_{j(\beta)}, \leqq_{j(\beta)})$.

$\qquad$ Let the solution be I.

④ $\quad z := \alpha * \bigoplus_{i \in I} {}_{j(\beta)} c_i$

The set $I \subseteq N$ determined in step ③ is a solution of the problem $(AP)_k$ with the optimal value z. The algorithm is efficient as each step can be performed with the aid of an efficient algorithm.

Example: A Time-Cost-4-Intersection Problem

For convenience we have ordered the coefficients of the $(TP)_4$ with respect to $\geqq$.

390

i	1	2	3	4	5	6	7	8	9	10
c_i	$\begin{pmatrix} 8 \\ -7 \end{pmatrix}$	$\begin{pmatrix} 3 \\ 0 \end{pmatrix}$	$\begin{pmatrix} 2 \\ 4 \end{pmatrix}$	$\begin{pmatrix} 1 \\ -2 \end{pmatrix}$	$\begin{pmatrix} 1 \\ -4 \end{pmatrix}$	$\begin{pmatrix} 1 \\ -4 \end{pmatrix}$	$\begin{pmatrix} 1 \\ -5 \end{pmatrix}$	$\begin{pmatrix} -5 \\ -2 \end{pmatrix}$	$\begin{pmatrix} -5 \\ -3 \end{pmatrix}$	$\begin{pmatrix} -7 \\ -3 \end{pmatrix}$

The coefficients are elements of the ordered commutative semigroup $(\mathbf{Z} \times \mathbf{Z}_-, *, \leqq)$. S_4 is chosen to consist of the *common transversals* of the two partitions

P_B	1, 8	2, 3, 6	4, 10	5, 7, 8

P_R	1, 9, 10	2, 3, 5	4, 8	6, 7

of $N = \{1, 2, \ldots, 10\}$. For example $\{1, 2, 4, 7\} \in S_4$. From matroid theory we know that common transversals are maximal intersections of two transversal matroids.

Step ① of the algorithm only changes the value of c_1:

$$c_1 := \begin{pmatrix} \infty \\ 0 \end{pmatrix}, \quad \alpha := \begin{pmatrix} 8 \\ -7 \end{pmatrix}.$$

In step ② of the algorithm the threshold method [8] stops with $\beta = c_6$, because the set $\{1, 2, \ldots, 5\}$ does not but the set $\{1, 2, \ldots, 6\}$ does contain a common transversal, namely $I = \{1, 4, 5, 6\}$. Therefore

$$\beta = \begin{pmatrix} 1 \\ -4 \end{pmatrix}, \quad j(\beta) = 1$$

and the transformed coefficients are given by

i	1	2	3	4	5	6	7	8	9	10
ci	$\begin{pmatrix} 1 \\ 0 \end{pmatrix}$	$\begin{pmatrix} 1 \\ 0 \end{pmatrix}$	$\begin{pmatrix} 1 \\ 0 \end{pmatrix}$	$\begin{pmatrix} 1 \\ -2 \end{pmatrix}$	$\begin{pmatrix} 1 \\ -4 \end{pmatrix}$	$\begin{pmatrix} 1 \\ -4 \end{pmatrix}$	$\begin{pmatrix} 1 \\ -5 \end{pmatrix}$	$-\infty$	$-\infty$	$-\infty$

The group $(G_1, \oplus_1, \leqq_1)$ is isomorphic with $(\mathbf{R}, +, \leqq)$, Therefore we replace

$$c_i = \begin{pmatrix} 1 \\ c_i' \end{pmatrix} \quad \text{by} \quad c_i' \in \mathbf{R}$$

and solve the remaining usual sum-k-intersection problem

$$\max_{I \in S_k''} \sum_{i \in I} c_i'$$

with $S_k'' := \{I \in S_k'' \mid I \subseteq \{1, 2, \ldots, 7\}\}$ in step ③.

A solution is $I = \{1, 2, 4, 7\}$. Therefore I is the solution of the $(TP)_4$ with optimal value (step ④)

$$z = \alpha * \bigotimes_{i \in I} c_i = \begin{pmatrix} 8 \\ -7 \end{pmatrix} * \begin{pmatrix} 1 \\ 0 \end{pmatrix} * \begin{pmatrix} 1 \\ 0 \end{pmatrix} * \begin{pmatrix} 1 \\ -2 \end{pmatrix} * \begin{pmatrix} 1 \\ -5 \end{pmatrix} = \begin{pmatrix} 1 \\ -7 \end{pmatrix}$$

REFERENCES

[1] Burkard, R. E.: A General Hungarian Method for the Algebraic Transportation Problem, Report 1976–6, *Discrete Maths* 22 (1978), 219–232.

[2] Burkard, R. E., Hahn, W., Zimmermann, U.: An Algebraic Approach to Assignment Problems, *Math. Progr.* 12 (1977), 318–327.

[3] Burkard, R. E., Hamacher, H., Zimmermann, U.: The Algebraic Network Flow Problem, Report 1976–7, Mathematisches Institut der Universität zu Köln (1976).

[4] Clifford, A. H.: Naturally Totally Ordered Commutative Semigroups, *Amer. J. Math.* 76 (1954), 631–646.

[5] Clifford, A. H.: Completion of Semi-Continuus Ordered Commutative Semigroups, *Duke Math. J.* 26 (1959), 41–59.

[6] Conrad, P.: Ordered Semigroups, *Nagoya Math. J.* 16 (1960), 51–64.

[7] Edmonds, J.: Matroids and the Greedy Algorithm, *Math. Prog.* 1 (1971), 127–136.

[8] Edmonds, J., Fulkerson, D. R.: Bottleneck Extrema, *J. Comb. Theory* 8 (1970), 299–306.

[9] Fuchs, L.: *Teilweise geordnete algebraische Strukturen,* Studia Mathematica, Mathematische Lehrbücher 19, Vandenhoek und Ruprecht, Göttingen (1966).

[10] Krogdahl, St.: A Combinatorial Proof for Lawler's Matroid Intersection Algorithm. Private Communication (to appear).

[11] Lawler, E.: Matroid Intersection Algorithms, *Math. Prog.* 9 (1975), 31–56.

[12] Zimmermann, U.: Some Partial Orders Related to Boolean Optimization and the Greedy Algorithm, *Annals of Discr. Math.* 1 (1977), 539–550.

[13] Zimmermann, U.: Boole'sche Optimierungsprobleme mit separabler Zielfunktion und matroidalen Restriktionen, Thesis, Universität zu Köln (1976).

PART 8

DISCRETE PROGRAMMING

SET COVERING WITH CUTTING PLANES FROM CONDITIONAL BOUNDS*

E. BALAS

(Pittsburgh, USA)

1. INTRODUCTION

We consider the set covering problem

$$\text{(SC)} \qquad \min \{cx \mid Ax \geqq e, \; x_j = 0 \text{ or } 1, j \in N\},$$

where $A = (a_{ij})$ is $m \times n$, $e \in R^m$, $e = (1, \ldots, 1)$, $c \in R^n$, and $a_{ij} \in \{0, 1\}$, $i \in M = \{1, \ldots, m\}$, $j \in N = \{1, \ldots, n\}$. We will denote by a^i and a_j the i-th row and j-th column of A, respectively. Without loss of generality, we assume that $c_j > 0$, $\forall j \in N$. Using established terminology, we call a vector x satisfying the constraints of (SC) a *cover*, and the set of indices j such that $x_j = 1$, the *support* of the cover. A cover is called *prime* if no proper subset of its support defines a cover.

This problem, and its equality-constrained counterpart, the set partitioning problem, are useful mathematical models for a great variety of scheduling and other important real world problems, like crew scheduling, truck delivery, tanker routing, information retrieval, fault detection, stock cutting offshore drilling platform location, etc., and a literature of considerable size exists on solution methods for these models (see [6] for a survey of set covering and set partitioning; [5] for a computational study and comparison of several solution techniques; [3], [8] and [9] for some of these methods; and [2] for a more recent survey of set partitioning, which also contains a bibliography of applications of both models).

In this paper we propose a new approach to set covering, based on the idea of conditional bounds. In Section 2 we introduce this concept for arbitrary mixed integer programs, and show how it can be used to derive valid disjunctions. The latter in turn can be used either to partition the feasible set in the framework of a branch and

* Research partially supported by the National Science Foundation and the U.S. Office of Naval Research.

bound approach, or to derive a family of valid cutting planes. In the case of a set covering problem, the cutting planes derived from conditional bounds are themselves of the set covering type. These cuts are discussed in Section 3, where the Bellmore–Ratliff inequalities [3] generated via involutory bases are shown to be a special case of the larger family of cutting planes defined in this paper. In Section 4 we examine the conditions under which a cutting plane derived from a conditional bound cuts off a specified prime cover. Since the family of cuts from conditional bounds is too large to be generated in its entirety, in Section 5 we discuss a procedure for generating "strong" members of the family. Sections 6 and 7 discuss heuristics for generating "good" prime covers and feasible solutions to the dual of the linear program associated with (SC), which are needed to generate convenient cuts. Next we state two algorithms based on cutting planes from conditional bounds (Section 8). Section 9 contains a numerical example, and in Section 10 we discuss some early computational experience.

2. CONDITIONAL BOUNDS

The central idea of our approach is to derive valid inequalities for the set covering problem from conditional bounds. Since this concept is meaningful for arbitrary mixed integer programs, we will introduce it in this more general context.

A *conditional lower bound* on the (objective function) value of a mixed integer program (P) is a number which is a valid lower bound *if* (P) is amended by some inequalities. The inequalities used to derive the conditional bound are called *conditional*. The purpose of these inequalities is to produce a conditional lower bound at least equal to a known upper bound. If this is achieved, then at least one of the conditional inequalities is violated by any solution better than the one associated with the upper bound, and this yields a valid constraint.

To be more specific, consider the integer program

$$(\text{P}) \qquad \min \{cx \mid Ax \geqq b, \, x \geqq 0, \, x_j \text{ integer}, \, j \in N\},$$

where A is an arbitrary $m \times n$ matrix ($n = |N|$) and b is an arbitrary m-vector. The pair of dual linear programs associated with (P) is

$$(\text{L}) \qquad \min \{cx \mid Ax \geqq b, \, x \geqq 0\}$$

and

$$(\text{D}) \qquad \max \{ub \mid uA \leqq c, \, u \geqq 0\}.$$

For any feasible optimization problem (S), let $z(\text{S})$ be the value of (an optimal solution to) (S).

Any feasible solution x to (P) provides an upper bound cx on $z(\text{P})$, and any feasible solution u to (D) provides a lower bound ub on $z(\text{D}) = z(\text{L})$, hence also on $z(\text{P})$.

Let u be a feasible solution to (D), i.e., such that

$$(1) \qquad uA \leqq c, \quad u \geqq 0,$$

and suppose the constraints of (L) and (P) are amended with the set of (conditional) inequalities

$$(2) \qquad Cx \geqq e_p,$$

where $C = (c_{ij})$ is a $p \times n$ matrix of 0's and 1's, and e_p is the p-vector of 1's, with $1 \leqq p \leqq n$. Suppose also that C has no zero rows, i.e.,

$$(3) \qquad \sum_{j=1}^{n} c_{ij} \geqq 1, \quad 1 = 1, \ldots, p.$$

Further, denote by (P_C) and (L_C) the problems obtained by amending the constraint sets of (P) and (L), respectively, with the set (2) of conditional inequalities, and let (D_C) be the dual of the linear program (L_C).

Let u be a m-vector satisfying (1). If there exists a p-vector v, $v \geqq 0$, $v \neq 0$, such that

$$(4) \qquad vC \leqq c - uA,$$

then (u, v) is a feasible solution to (D_C) and therefore $ub + ve_p$ is a lower bound on $z(D_C) = z(L_C)$, hence also on $z(P_C)$. We will say in this case that $ub + ve_p$ is a conditional lower bound on $z(P)$.

Now let z_U be a known upper bound on $z(P)$. If

$$(5) \qquad ve_p \geqq z_U - ub,$$

i.e., if the conditional lower bound on $z(P)$ exceeds or equals the upper bound z_U, then

$$z(P_C) \geqq z(L_C) \\ \geqq ub + ve_p \geqq z_U,$$

and hence every feasible solution to (P) better than the one associated with the bound z_U violates at least one of the inequalities (2), i.e., satisfies the disjunction

$$\bigvee_{i=1}^{p} \left(\sum_{j=1}^{n} c_{ij} x_j < 1 \right).$$

Since C is a 0–1 matrix and x is a 0–1 vector, this disjunction is the same as

$$\bigvee_{i=1}^{p} \left(\sum_{j=1}^{n} c_{ij} x_j = 0 \right).$$

If we denote

$$Q_i = \{ j \in N \mid c_{ij} = 1 \}, \quad i = 1, \ldots, p,$$

the result that we have just proved can be stated as follows.

Theorem 2.1. *Let z_U be a known upper bound on $z(P)$, and let u satisfy (1). If there exists a $p \times n$ matrix $C = (c_{ij})$, $c_{ij} \in \{0, 1\}$, $\forall i, j (1 \leq p \leq n)$, satisfying (3), and a p-vector $v \geq 0$, $v \neq 0$, satisfying (4) and (5), then every feasible solution x to (P) such that $cx < z_U$ satisfies the disjunction*

$$\text{(6)} \qquad \bigvee_{i=1}^{p} (x_j = 0, j \in Q_i).$$

We have stated Theorem 2.1 for the pure integer case in order to simplify the exposition, but it (and the rest of this section) is easily seen to carry over to the case of a mixed integer program. Indeed, if $N_1 \subset N$ is the set of integer-constrained variables, $N_1 \neq N$, let $c_{ij} = 0$, $\forall i = 1, \ldots, p$, $\forall j \in N \setminus N_1$, and the results are valid with some changes in the notation.

Note that the disjunction (6), though derived from a conditional bound, is an "unconditionally" valid constraint.

The first question that arises in connection with Theorem 2.1, is that of the existence of a matrix C and a vector v satisfying the above requirements.

Theorem 2.2. *Let z_U be a known upper bound on $z(P)$, let u be a feasible solution to (D), and let*

$$\text{(7)} \qquad s = c - uA.$$

Then there exists a pair C, v satisfying the requirements of Theorem 2.1 if and only if

$$\text{(8)} \qquad \sum_{j \in N} s_j \geq z_U - ub.$$

Proof. Let u and s satisfy (1) and (7). If (8) holds, then the pair C, v, defined by $C = I$ (the identity matrix of order N) and $v_j = s_j$, $j \in N$, satisfies (3), (4) and (5).

Conversely, if C and v satisfy (3), (4) and (5), then adding the inequalities (4) and substituting s for $c - uA$ yields

$$\sum_{j \in N} s_j \geq \sum_{i=1}^{p} v_i \left(\sum_{j \in N} c_{ij} \right)$$
$$\geq v e_p \qquad \text{[from (3)]}$$
$$\geq z_U - ub \quad \text{[from (5)].} \qquad\qquad \text{Q.E.D.}$$

Note that, if $p = 1$, i.e., if C has a single row, then the disjunction (6) becomes $x_j = 0$, $j \in Q_1$. Somewhat more generally, we have the following.

Remark 2.1. *Let z_U, u and s be as in Theorem 2.2, and define*

$$Q_0 = \{ j \in N \mid s_j \geq z_U - ub \}.$$

Then every feasible solution x to (P) such that $cx < z_U$ satisfies

$$x_j = 0, \quad j \in Q_0.$$

Thus, whenever $Q_0 \neq \emptyset$, the variables indexed by Q_0 can be set to zero permanently.

Example 1. Let (P) be an integer program (minimization problem) with 10 variables, and let $z_U = 35$ be the value of some known integer solution. Further, let u be a feasible solution to the dual of the associated linear program, with $ub = 27.9$, and let the reduced costs s_j associated with ub be

j	1	2	3	4	5	6	7	8	9	10
s_j	1.5	2.4	0	3.1	0	5.8	3.3	2.6	3.2	4.7

We have $z_U - ub = 35 - 27.9 = 7.1$. To construct a pair C, v satisfying the conditions of Theorem 2.1, we first choose $v_1 = s_1$, $v_2 = s_4$ and $v_3 = s_8$; then $v_1 + v_2 + v_3 = 7.2 \geq 7.1$. Next we choose the 1's among the elements c_{ij} of C in such a way that $v_1 c_{1j} + v_2 c_{2j} + v_3 c_{3j} \leq s_j$ for $j = 1, \ldots, 10$. (Neither v, nor C is of course unique.) C and v are shown below, where the blanks are 0's.

$$
C = \begin{array}{c}
\begin{array}{cccccccccc}
1 & 2 & 3 & 4 & 5 & 6 & 7 & 8 & 9 & 10
\end{array} \\
\begin{array}{|cccccccccc|}
\hline
1 & 1 & & & & & & & & 1 \\
& & & 1 & & 1 & 1 & & 1 & \\
& & & & & 1 & & 1 & & 1 \\
\hline
\end{array}
\end{array}
\qquad
v = \begin{array}{|c|}
\hline
1.5 \\
\hline
3.1 \\
\hline
2.6 \\
\hline
\end{array}
$$

From Theorem 2.1, every x satisfying the constraints of (P) and such that $cx < 35$ satisfies the disjunction

$$x_1 = x_2 = x_{10} = 0 \ \vee\ x_4 = x_6 = x_7 = x_9 = 0 \ \vee\ x_6 = x_8 = x_{10} = 0.$$

Given a feasible solution u to (D) whose associated reduced costs s satisfy (8), it is usually not difficult to find a pair C, v satisfying the conditions of Theorem 2.1; but the problem is to find one which defines a conveniently strong disjunction (6). The following procedure provides a general framework for doing this, which allows for many variants.

1. Choose a minimum-cardinality subset S of N, such that

$$(9) \qquad \sum_{j \in S} s_j \geq z_U - ub,$$

and order $S = \{j(1), \ldots, j(p)\}$ according to decreasing values of $s_{j(i)}$. Then set $v_i = s_{j(i)}$, $i = 1, \ldots, p$, and $c_{hj(i)} = 1$ for $h = i$, $c_{hj(i)} = 0$ for $h \neq i$, $i = 1, \ldots, p$.

2. For $j \in N \setminus S$, define recursively for $i = 1, \ldots, p$,

$$
c_{ij} = \begin{cases}
0 \text{ or } 1 & \text{if } \displaystyle\sum_{h=1}^{i-1} v_h c_{hj} + v_i \leq s_j \\
0 & \text{otherwise.}
\end{cases}
$$

The option of setting $c_{ij}=1$ or $c_{ij}=0$ in step 2 represents a choice between including i into Q_i, or leaving it available for inclusion into one or more sets Q_{i+k}. This can be decided by efficiency criteria, as will be seen later. It is easy to check that any pair C, v constructed by this procedure satisfies the requirements of Theorem 2.1.

A disjunction of the form (6) obtained from a conditional bound can be used to partition the feasible set in the framework of a branch and bound procedure, by creating p subproblems defined by the constraints

$$x_j=0, \quad j\in Q_1$$

$$\sum_{j\in Q_1} x_j\geqq 1; \quad x_j=0, \quad j\in Q_2$$

$$\cdots\cdots\cdots\cdots\cdots\cdots\cdots\cdots\cdots$$

$$\sum_{j\in Q_1} x_j\geqq 1, \quad \ldots, \quad \sum_{j\in Q_{p-1}} x_j\geqq 1; \quad x_j=0, \quad j\in Q_p.$$

For certain classes of integer programs, this way of branching seems highly efficient. It is currently being tested, for instance, in a new penalty method for solving traveling salesman problems [1], with excellent computational results.

Another way of using the disjunction (6) is to generate from it a family of valid inequalities. In the case of a set covering problem, these inequalities turn out to be of the set covering type, as shown in the next section.

In a broader context, the idea of deriving a valid ("unconditional") constraint from one or several conditional constraints may have many other applications. One of them appears in [7], where a properly chosen inequality is used to derive a bound from the fact that either the inequality or its complement must be satisfied by any feasible solution.

3. CUTTING PLANES FROM CONDITIONAL BOUNDS

From now on, we address ourselves to the set covering problem (SC) introduced in Section 1; i.e., $A=(a_{ij})$ with $a_{ij}\in\{0, 1\}$, $\forall i, j$, and $b=e$, where e is the m-vector of 1's. We will denote

$$N_i=\{j\in N\,|\,a_{ij}=1\}, \quad j\in M.$$

Theorem 3.1. *Suppose the conditions of Theorem 2.1 are satisfied, i.e., the disjunction (6) is a valid constraint for* (SC). *With each* $i\in\{1, \ldots, p\}$, *associate an index* $h(i)\in M$, *such that* $N_{h(i)}\cap Q_i\neq\emptyset$. *Then denoting*

$$W=\overset{y}{\underset{i=}{\cup}}\,[N_{h(i)}\backslash Q_i],$$

every cover x such that $cx<z_U$ satisfies

$$(10) \qquad\qquad \sum_{j\in W} x_j\geqq 1.$$

Proof. For $i=1, \ldots, p$, the i-th term of the disjunction (6) implies

$$\sum_{j \in N_h \setminus Q_i} x_j \geqq 1, \quad \forall h \in M.$$

Hence, for any choice of indices $h(i) \in M$, $i=1, \ldots, p$, (6) implies the disjunction

$$\bigvee_{i=1}^{p} \left(\sum_{j \in N_{h(i)} \setminus Q_i} x_j \geqq 1 \right),$$

which in turn implies (10).

Q.E.D.

The cutting planes of Theorem 3.1 are set covering inequalities, valid in the sense of being satisfied by every cover better than a given one. Since these properties are the same as those of the Bellmore–Ratliff cuts [3] obtained by the use of involutory bases, we next examine the relationship between the latter and our inequalities from conditional bounds. First, we show in the next theorem that the Bellmore–Ratliff inequalities are a subclass of the class of inequalities defined by Theorem 3.1. Then we show by way of example that the subclass in question is a proper one.

Theorem 3.2. *The Bellmore–Ratliff inequalities* [3] *are a subclass of the class* (10).

Proof. Let $\bar{x}$ be a prime cover, B an involutory basis associated with $\bar{x}$, and $c_j - c_B a_j$ the j-th reduced cost, where c_B is the m-vector whose i-th component is $c_{j(i)}$, if the basic variable associated with row i is (the structural variable) $x_{j(i)}$, and 0 if the basic variable associated with row i is a slack. (When B is an involutory basis, the reduced costs are known to be of the above form.) The Bellmore–Ratliff cut associated with $\bar{x}$ and B is then

$$(11) \qquad \qquad \sum_{j \in F} x_j \geqq 1,$$

where

$$F = \{ j \in N \mid c_j - c_B a_j < 0 \}.$$

To obtain this cut from a conditional bound, let $I_1 = \{ j(1), \ldots, j(p) \}$ be the index set of the structural basic variables, and set $u=0$, $s=c$, and $v_i = c_{j(i)}$, $j(i) \in I_1$. Then u satisfies (1) and v satisfies (5) (with equality) for $z_U = c\bar{x}$.

We now construct the matrix $C = (c_{ij})$ of Theorem 2.1 as follows. Let $h(i)$ be the row index associated with basic variable $x_{j(i)}$. Define

$$c_{ij} = \begin{cases} a_{h(i), j}, & j \in N \setminus F \\ 0, & j \in F \end{cases} \quad i=1, \ldots, p;$$

then

$$Q_i = N_{h(i)} \setminus F, \quad i=1, \ldots, p.$$

C trivially satisfies (3). To see that together with v it also satisfies (4), note that for $j \in N \backslash F$,

$$\sum_{i=1}^{p} v_i c_{ij} = \sum_{i=1}^{p} c_{j(i)} a_{h(i),j}$$
$$\leqq c_j - u a_j;$$

where the inequality follows from the fact that $c_j - \sum_{i=1}^{p} c_{j(i)} a_{h(i),j}$ is the j-th reduced cost (nonnegative for $j \in N \backslash F$), while $u = 0$. Further, for $j \in F$,

$$\sum_{i=1}^{p} v_i c_{ij} = 0$$
$$\leqq c_j - u a_j.$$

Thus C and v satisfy the conditions of Theorem 2.1. Applying Theorem 3.1, we now associate with each $i \in \{1, \ldots, p\}$ the row index $h(i)$, and define

$$W = \bigcup_{i=1}^{p} [N_{h(i)} \backslash Q_i]$$
$$= \bigcup_{i=1}^{p} [N_{h(i)} \cap F]$$

(from the definition of the sets Q_i)

$$= \left[\bigcup_{i=1}^{p} N_{h(i)} \right] \cap F.$$

On the other hand, from the definition of F, $j \in F$ implies $a_{h(i),j} = 1$ for some $i \in \{1, \ldots, p\}$, hence

$$F \subseteq \left[\bigcup_{i=1}^{r} N_{h(i)} \right].$$

Thus $W = F$, and the cut (10) of Theorem 3.1 is in this case identical with (11).

Q.E.D.

Thus the Bellmore–Ratliff cuts are a special case of the cuts (10). Furthermore, they are a proper subclass of the class (10), i.e., the conditional bound approach yields other inequalities besides those obtainable via involutory bases. Some of those other inequalities are considerably stronger, in the sense of having fewer positive coefficients. This is illustrated by the next example.

Example 2. Consider the set covering problem whose costs c_j and coefficient matrix A are shown in Table 1.

400

Table 1

	1	2	3	4	5	6	7	8	9	10	11	12	13	14	15	16	17	18	19	20
c_j	3	1	1	3	1	2	2	3	3	3	3	3	3	4	4	4	5	6	8	9
1						1							1				1	1		1
2	1								1						1	1	1	1		1
3		1						1		1									1	
4			1											1						1
5				1					1		1						1		1	
6					1						1			1		1				1
7							1								1		1		1	1
8												1	1						1	
9								1						1			1	1		1
10										1			1			1			1	
11						1						1						1	1	1

The 0–1 vector $\bar{x}$ whose support $\{2, 3, 5, 12, 13, 17\}$ is a cover, satisfying with equality all the inequalities except for 1 and 8, which are oversatisfied. To apply the Bellmore–Ratliff procedure, one associates with $\bar{x}$ an involutory basis. The variables x_2, x_3, x_5 can be basic only in rows 3, 4 and 6 respectively. Since rows 1 and 8 are slack, the variables x_{12} and x_{13} can be basic only is rows 11 and 10 respectively. Finally, x_{17} can be basic in any of the 4 rows 2, 5, 7, 9; and accordingly there are 4 involutory bases that can be associated with $\bar{x}$. We will denote them by B_2, B_5, B_7 and B_9, according as x_{17} is basic in row 2, 5, 7 or 9 respectively. The basis B_2 (after row permutations) is shown in Table 2.

Table 2

	2	3	5	12	13	17	25	27	29	21	28
3	1										
4		1									
6			1								
11				1							
10					1						
2						1					
5						1	-1				
7						1		-1			
9						1			-1		
1					1	1				-1	
8				1	1						-1

The variables $x_{21}, \ldots, x_{31}$ are slacks. The basis B_5 can be obtained from B_2 by interchanging rows 5 and 2, and replacing the slack variable x_{25} in row 5, by x_{22} in row 2. The other two bases can be obtained in an analogous way. The 4 cutting planes that can be obtained by the Bellmore–Ratliff procedure, depending on which basis is used, are

$$\begin{aligned}
x_1 + x_6 + x_9 + x_{10} + x_{15} + x_{16} + x_{18} + x_{20} &\geqq 1, \quad \text{from } B_2 \\
x_4 + x_6 + x_9 + x_{10} + x_{11} + x_{19} &\geqq 1, \quad \text{from } B_5 \\
x_6 + x_7 + x_{10} + x_{15} + x_{19} + x_{20} &\geqq 1, \quad \text{from } B_7 \\
x_6 + x_8 + x_{10} + x_{14} + x_{18} + x_{20} &\geqq 1, \quad \text{from } B_9.
\end{aligned}$$

On the other hand, using the conditional bound approach, we construct (by inspection or a heuristic) the dual vector

$$u = (0, 1, 1, 1, 1, 1, 2, 0, 1, 2, 2)$$

which, together with the associated reduced cost vector

$$s = (2, 0, 0, 2, 0, 0, 0, 1, 1, 0, 1, 1, 1, 1, 1, 0, 0, 2, 0, 1)$$

satisfies the condition (1).

The cover x whose support is $\{2, 3, 5, 12, 13, 17\}$ yields $z_U = c\bar{x} = 14$; and the dual vector u yields the lower bound $ue = 12$.

Since $z_U - ue = 2$, $Q_0 = \{j \in N \mid s_j \geqq 2\} = \{1, 4, 18\}$, and thus (Remark 2.1) every cover better than $\bar{x}$ satisfies $x_1 = x_4 = x_{18} = 0$. Hence we replace N by $N \setminus \{1, 4, 18\}$. Further, to apply Theorems 2.1 and 3.1, we set $p = 2$, with $v_1 = s_{12}$ and $v_2 = s_{13}$,

$$s_{12} + s_{13} = 2$$
$$\geqq z_U - ue.$$

Then, using (for instance) the conditional inequalities (and the matrix C) defined by

$$Q_1 = \{8, 12, 14, 20\}, \quad Q_2 = \{9, 11, 13, 15\}$$

we obtain the disjunction

$$x_8 = x_{12} = x_{14} = x_{20} = 0 \ \vee \ x_9 = x_{11} = x_{13} = x_{15} = 0.$$

Applying Theorem 3.1 with $h(1) = 11$, $h(2) = 10$, we have (using the newly defined set N)

$$N_{h(1)} \setminus Q_1 = \{6, 19\}$$
$$N_{h(2)} \setminus Q_2 = \{10, 16, 19\}$$

and thus

$$W = \{6, 10, 16, 19\}.$$

We have obtained the cut

$$x_6 + x_{10} + x_{16} + x_{19} \geqq 1$$

which has only 4 positive coefficients, whereas each of the involutory basis cuts has at least 6.

The above inequality cuts off $\bar{x}$. This is due to the way we chose the components of v and the row indices $h(i) \in M$, as will be shown in the next section. If we drop the requirement for a specified cover to be cut off, we can obtain inequalities which are "stronger" in the sense of having fewer positive coefficients. By a judicious choice of the column indices $j(i)$ for which we set $v_i = s_{j(i)}$, and the row indices $h(i)$, one can generate the cuts with the fewest possible positive coefficients. Thus, for instance, the choice $v_1 = s_{13}$, $v_2 = s_9$, and $h(1) = 8$, $h(2) = 5$, yields the valid inequality

$$x_{17} + x_{19} \geqq 1;$$

whereas $v_1 = s_{13}$, $v_2 = s_{14}$ and $h(1) = 8$, $h(2) = 4$ yields

$$x_3 + x_{19} \geqq 1.$$

The cutting planes derived by Bowman and Starr [4] by the use of a vector partial ordering are a special case of the Bellmore–Ratliff inequalities, hence they also belong to the family of cuts from conditional bounds.

4. SOME PROPERTIES OF CUTS FROM CONDITIONAL BOUNDS

In order to obtain a conditional bound, and thus to be able to generate the type of constraints discussed in the previous section, one needs a feasible solution u to (D), whose associated reduced costs satisfy (8). This requirement is easy to meet. The next theorem and its Corollary give a broad sufficient condition for u to satisfy them.

Let $S(\bar{x})$ denote the support of $\bar{x}$, i.e. $S(\bar{x}) = \{j \in N \,|\, \bar{x}_j = 1\}$.

Theorem 4.1. *Let $\bar{x}$ be a cover for* (SC) *and let $\bar{u}$ and $\bar{s}$ satisfy (1) and (7). If $\bar{u}$ also satisfies*

$$\bar{u}(A\bar{x} - e) = 0,$$

then

(12)
$$\sum_{j \in S(x)} \bar{s}_j = c\bar{x} - \bar{u}e.$$

Proof. Let

$$S^+ = \{j \in S(\bar{x}) \,|\, \bar{s}_j > 0\}$$

and consider the pair of dual linear programs

(L$_1$)
$$\min \{cx \,|\, Ax \geqq e,\ x_j \geqq 1,\ j \in S^+,\ x_j \geqq 0,\ j \in N \backslash S^+\}$$

and

(D$_1$)
$$\max \{ue + \sum_{j \in S^+} s_j \,|\, ua_j + s_j = c_j,\ j \in N;\ u \geqq 0,\ s \geqq 0\}.$$

Clearly, $\bar{x}$ is a feasible solution to (L$_1$), and $(\bar{u}, \bar{s})$ is a feasible solution to (D$_1$). Further, $\bar{x}$ and $(\bar{u}, \bar{s})$ satisfy the complementary slackness conditions

$$\begin{aligned}
(a^i x - 1) u_i &= 0, \quad i \in M \\
(x_j - 1) s_j &= 0, \quad j \in S^+ \\
x_j s_j &= 0, \quad j \in N \backslash S^+.
\end{aligned}$$

Hence $\bar{x}$ and $(\bar{u}, \bar{s})$ are optimal solutions to (L_1) and (D_1), respectively, which imply the equality of the two objective function values.

Q.E.D.

An immediate consequence of Theorem 4.1 is the following.
Let

$$T(\bar{x}) = \{i \in M \mid a^i \bar{x} = 1\}.$$

Corollary 4.1.1. *Let $\bar{x}$ be a cover, and let $\bar{u}$ and $\bar{s}$ satisfy (1) and (7). If $\bar{u}$ also satisfies*

$$(13) \qquad\qquad \bar{u}_i = 0, \quad \forall i \in M \backslash T(\bar{x}),$$

then there exists $\{j(1), \ldots, j(p)\} \subseteq S^+$ such that v defined by $v_i = \bar{s}_{j(i)}$, $i = 1, \ldots, p$, together with $\bar{u}$, satisfies (5).

Proof. If $\bar{u}$ satisfies (13), then together with $\bar{x}$ it satisfies the complementarity condition of Theorem 4.1, and therefore (12) holds. From (12), v_i as defined in the Corollary, together with $\bar{u}$, satisfies (5).
Q.E.D.

Thus, any feasible solution to (D) that has positive components only for $i \in T(\bar{x})$ produces a vector v which satisfies (5). A procedure of the type outlined in Section 2 can be used to find a matrix C which satisfies the other requirements of Theorem 2.1. Choosing appropriate rows $h(i) \in M$ will then produce a cut of the family (10) defined in Theorem 3.1.

From the point of view of having a finitely convergent procedure, it is essential that the inequalities generated be not only valid but also new, i.e., no cut should be generated twice.

Given two inequalities of the form (10) with associated index sets W_1 and W_2, the first inequality implies the second one if and only if $W_1 \subseteq W_2$. We will say that an inequality (10) is new, if it is not implied by any inequality of the current problem (SC).

Theorem 4.2. *The inequality*

$$(10) \qquad\qquad \sum_{j \in W} x_j \geq 1$$

of Theorem 3.1 is new if and only if $N \backslash W$ is the support of a cover.

Proof. If $N \backslash W$ is the support of a cover, say $\bar{x}$, then $\bar{x}$ violates (10), whereas it satisfies all the inequalities of the current problem (SC). Hence (10) is new. Conversely, if $N \backslash W$ does not define a cover, then there exists a row $i \in M$ such that $N_i \cap (N \backslash W) = \emptyset$, i.e., $N_i \subseteq W$. But then the cut (10) defined by W is either identical to, or implied by, the i-th inequality of (SC).
Q.E.D.

Theorem 4.2 gives a necessary and sufficient condition for an inequality (10) to cut off at least one cover. Next we give a sufficient condition for an inequality (10) to cut off a specified prime cover $\bar{x}$.

As before we will denote

$$M_j = \{i \in M \mid a_{ij} = 1\}, \quad j \in N.$$

Remark 4.1. *If $\bar{x}$ is a prime cover for* (SC), *then*

$$M_j \cap T(\bar{x}) \neq \emptyset, \quad \forall j \in S(\bar{x}).$$

Proof. Follows from the definition of $T(\bar{x})$ and that of a prime cover.

Theorem 4.3. *Let $\bar{x}$ be a prime cover for* (SC), *let z_U be a known upper bound on the value of* (SC), *and let u and s satisfy (1) and (7). Further, let*

$$(9) \qquad \sum_{j \ni S} s_j \geq z_U - ue$$

for some $S \subseteq S(\bar{x})$, $S = \{j(1), \ldots, j(p)\}$. Define

$$(14) \qquad v_i = s_{j(i)}, \quad i = 1, \ldots, p,$$

and let C be any $p \times n$ $0-1$ matrix satisfying conditions (3), (4) and

$$(15) \qquad c_{hj(i)} = \begin{cases} 1 & h = i \\ 0 & h \neq i \end{cases} \quad i = 1, \ldots, p.$$

Finally, let $Q_i = \{j \in N \mid c_{ij} = 1\}$ and

$$W = \bigcup_{i=1}^{p} [N_{h(i)} \backslash Q_i],$$

where

$$(16) \qquad h(i) \in T(\bar{x}) \cap M_{j(i)}, \quad i = 1, \ldots, p.$$

Then the inequality

$$(10) \qquad \sum_{j \in W} x_j \geq 1$$

is satisfied by every cover x such that $cx < z_U$, and cuts off $\bar{x}$.

Proof. By assumption, the pair u, s satisfies (1) and (7), while the matrix C satisfies (3) and, together with the vector v defined by (14), satisfies (4). From (9), u and v satisfy (5). The exisnce of such a pair C, v follows from (9) and Theorem 2.2. The additional requirement (15) on C can always be met. Hence the pair C, v defined in the Theorem exists and satisfies the requirements of Theorems 2.1 and 3.1.

Further, since $j(i) \in S(\bar{x})$, we have $T(\bar{x}) \cap M_{j(i)}$ $i=1, \ldots, p$ (Remark 4.1), i.e., there exists a set of row indices $h(i)$, $i=1, \ldots, p$, satisfying condition (16). From that condition, and the definition of $T(\bar{x})$, we have

$$S(\bar{x}) \cap N_{h(i)} = \{j(i)\}, \quad i=1, \ldots, p.$$

On the other hand, from (15), $j(i) \in Q_i$, $i=1, \ldots, p$. Hence

$$S(\bar{x}) \cap [N_{h(i)} \backslash Q_i] = \emptyset, \quad i=1, \ldots, p,$$

i.e. $S(\bar{x}) \cap W = \emptyset$ and the inequality (10) cuts off $\bar{x}$. Q.E.D.

The results of the last two sections can be used to generate a family of inequalities whose members are different from each other and from the initial constraints of (SC), and are satisfied by every cover whose value is smaller than z_U. Since the family of such cuts is of considerable size, one would like to be able to generate some of the "stronger" members, according to some reasonable measure of strength. This will be discussed in the next section. First, however, we examine another question related to the properties discussed above.

If the set Q_0 defined in Remark 2.1 is nonempty, then the variables indexed by Q_0 can be set to 0. Since the vector u (and the associated s) used in the definition of Q_0, is not required to have any other property but to satisfy (1), the question arises as to which vector u is more likely to produce a nonempty set Q_0, or more generally, a larger set Q_0. While this question is hard to answer in general, there is a clear answer for the case when the choice is between two vectors u^1, u^2 such that $u^1 \leq u^2$.

Theorem 4.4. *For $k=1, 2$, let u^k satisfy (1), and define*

$$s_j^k = c_j - \sum_{\in M} u_i^k a_{ij}, \quad j \in N,$$

and

$$Q_0^k = \{j \in N \mid s_j^k \geq z_U - u^k e\}.$$

Then $u^1 \leq u^2$ implies $Q_0^1 \subseteq Q_0^2$.

Proof. If $j \in Q_0^1$, then

$$s_j^1 = c_j - \sum_{i \in M} u_i^1 a_{ij}$$

$$\geq z_U - u^1 e,$$

i.e.,

$$c_j + \sum_{i \in M \backslash M_i} u_i^1 \geq z_U;$$

and if $u^1 \leq u^2$, the last inequality implies

$$c_j + \sum_{i \in M \backslash M_i} u_i^2 \geq z_U,$$

406

i.e.,

$$s_j^2 = c_j - \sum_{i \in M} u_i^2 a_{ij}$$

$$\geqq z_U - u^2 e,$$

hence $j \in Q_0^2$. Q.E.D

Thus, the set Q_0 defined with respect to a vector u satisfying (1) is always contained in the set Q_0 defined with respect to any vector $\hat{u}$ obtained from u by increasing some of its components (while maintaining $\hat{u}A \leqq c$). Therefore, in generating the set Q_0, one should always use a "maximal" dual vector u, i.e., one whose components cannot be increased without decreasing some component or violating $uA \leqq c$.

5. GENERATING CUTS

In this section we discuss a procedure for generating conveniently strong members of the family of cuts introduced above. Since all inequalities have a right hand side of 1 and left hand side coefficients of 0 or 1, we will use as a measure of strength the number of positive coefficients (the smaller the number, the stronger the inequality). The fact that an inequality A is stronger than B in this sense, i.e., has fewer positive coefficients, does not imply of course that A dominates B or implies B.

Procedure CUT generates a member of the subfamily characterized in Theorem 4.3, i.e., an inequality which is satisfied by every solution x such that $cx < z_U$, and which also cuts off a specified prime cover $\bar{x}$. The strength of an inequality (10), i.e., the size of the set W, depends on the integer p and the size of the sets $N_{h(i)} \backslash Q_i$, $i = 1$, $\ldots, p$, of Theorem 3.1. To have p conveniently small, the procedure sets $v_i = s_j$ for a sequence of indices $j(i)$, $i = 1, \ldots, p$, corresponding to the largest reduced costs s_j, $j \in S(\bar{x})$. Each row index $h(i)$, $i = 1, \ldots, p$, is of course chosen from $T(\bar{x}) \cap M_{j(i)}$, as prescribed by Theorem 4.3. Further, in order to have W as small as possible, the sequence of row indices $h(i)$ is chosen so as to come as close as possible at each step $k \in \{1, \ldots, p\}$ to minimize the set $W_k \backslash W_{k-1}$, where $W_0 = \emptyset$ and

$$W_k = \bigcup_{i=1}^{k} [N_{h(i)} \backslash Q_i], \quad k = 1, \ldots, p.$$

Since $|S| = 1$ implies (Remark 2.1) that some variable can be permanently set to zero, we assume that this has already been done for all variables indexed by Q_0, and thus there are no singleton sets S satisfying (9).

Let M and N be the row and column index sets of the current problem (SC), let $\bar{x}$ be a prime cover for (SC), and let $S(\bar{x}) = \{j \in N \mid \bar{x}_j = 1\}$, $T(\bar{x}) = \{i \in M \mid a^i \bar{x} = 1\}$. Further, let u and s satisfy (1), (7) and (9) for $S = S^+$, where

$$S^+ = \{j \in S(\bar{x}) \mid s_j > 0\}.$$

CUT

Step 0. Initialize $W_0=\emptyset$, $K_1=S^+$, $y_1=ue$, $s^1=s$. Let $t\leftarrow 1$ (iteration counter) and go to 1:

Step 1. Define

$$v_t=\min\left\{\max_{j\in K_t} s_j^t,\quad \min_{j\in K_t}\{s_j^t\mid s_j^t\geq z_U-y_t\}\right\},\quad K_t^*=\{j\in K_t\mid s_j^t=v_t\}$$

$$P_t=\{j\in N\mid s_j^t\geq v_t\},\quad M(t)=\bigcup_{j\in K_t^*} M_j.$$

Choose $i(t)$ such that

$$|N_{i(t)}\backslash(P_t\cup W_{t-1})|=\min_{i\in T(\bar{x})\cap M(t)}|N_i\backslash(P_t\cup W_{t-1})|$$

(breaking ties arbitrarily), and let

$$\{j(t)\}=N_{i(t)}\cap K_t^*.$$

Then define

$$W_t=W_{t-1}\cup[N_{i(t)}\backslash P_t],\quad y_{t+1}=y_t+v_t.$$

If $y_{t+1}\geq z_U$, go to 2. Otherwise, let

$$s_j^{t+1}=\begin{cases} s_j^t-v_t & j\in N_{i(t)}\cap P_t \\ s_j^t & \text{otherwise,}\end{cases}$$

$K_{t+1}=K_t\backslash\{i(t)\}$, set $t\leftarrow t+1$, and go to 1.

Step 2. Define $W=W_t$ and add to the constraints of (SC) the inequality

$$\sum_{j\in W} x_j\geq 1.$$

Theorem 5.1. *In $|S^+|$ iterations, procedure* CUT *generates an inequality satisfied by every cover x such that $cx<z_U$, and violated by $\bar{x}$.*

Proof. Since $K_1=S^+$ decreases by one element at each iteration, the procedure ends in $|S^+|$ iterations. Next we show that the inequality generated by the procedure satisfies the conditions of Theorem 4.3.

The vectors u, s satisfy (1) and (7). The vector v used in the procedure satisfied condition (14) of Theorem 4.3, while the matrix C used in the procedure (not stated explicitly) is defined by

$$c_{ij}=\begin{cases}1 & j\in Q_t=P_t\cap N_{i(t)} \\ 0 & \text{otherwise.}\end{cases}$$

This satisfies requirement (3) and also (15), since $\{j(t)\}=N_{i(t)}\cap K_t^*$ is unique [from $i(t)\in T(\bar{x})$] and $j(t)\neq j(t-1)$, $t=2,\ldots,p$. From the definition of the sets P_t, it follows that C and v satisfy (4). The stopping rule $y_{t+1}\geq z_U$ guarantees that the set $S=\{j(1),\ldots,j(p)\}$, where p is the number of iterations, satisfies (9). Finally, the choice of $i(t)$ in step 1 insures that condition (16) is met.

Q.E.D.

Note that the above inequality was obtained for a given value of z_U. If later the bound z_U is improved, i.e., replaced by $z_U' < z_U$, the cut can be strengthened. Namely, we have

Corollary 5.1.1. *Suppose*

$$(17) \qquad \sum_{j \in W_p} x_j \geqq 1$$

is a cut obtained in p iterations of CUT for an upper bound z_U, and subsequently z_U is replaced by $z_U' < z_U$. Then every cover x for which $cx < z_U'$ satisfies

$$(18) \qquad \sum_{j \in W_{p-k}} x_j \geqq 1,$$

where k is the greatest integer, if it exists, such that

$$(19) \qquad \sum_{i=1}^{k} v_{p+1-i} \leqq z_U - z_U',$$

or else $k = 0$.

Proof. At the end of p iterations of CUT, we have

$$(20) \qquad y_{p+1} = ue + \sum_{i=1}^{p} v_i \geqq z_U,$$

which implies that the inequality (17) is satisfied by every cover better than the one associated with z_U. Subtracting (19) from (20) yields

$$y_{p-k+1} = ue + \sum_{i=1}^{p-k} v_i \geqq z_U',$$

which implies that the inequality (18) is satisfied by every cover better than the one associated with z_U'.

The information required in order to be able to strengthen a cut generated by CUT 1 whenever z_U is improved, is the sequence of pairs

$$(W_1, v_1), \quad (W_2 \backslash W_1, v_2), \quad \dots, \quad (W_p \backslash W_{p-1}, v_p).$$

6. GENERATING PRIME COVERS

Any algorithm based on the cutting planes introduced in this paper needs to generate repeatedly prime covers for the current problem (SC), and feasible solutions to the dual (D) of the linear program associated with (SC). In this section we discuss some heuristics for generating reasonably good prime covers for (SC) and feasible solutions to (D), "good" in the sense of giving tight bounds.

The procedure PRIMAL 1, to be described below, is a kind of "greedy" algorithm, which selects at each step the locally most promising column, and in one pass produces a (usually remarkably good) prime cover $\bar{x}$.

As before, let M and N be the row and column index set of the current problem (SC), with $M_j=\{i\in M\,|\,a_{ij}=1\}$ and $N_i=\{j\in N\,|\,a_{ij}=1\}$. Further, let z_U be an upper bound and z_L a lower bound on the value of (SC).

PRIMAL 1

Step 0. Initialize $R_1=m$, $S_1=\emptyset$, $t\leftarrow 1$, and go to 1.

Step 1. If $R_t=\emptyset$, set $S=S_t$ and go to 2. Otherwise, define

$$I(t)=\{i\in R_t\,|\,|N_i|=\min_{h\in R_t}|N_h|\},\quad N_{I(t)}=\bigcup_{i\in I(t)}N_i$$

$$\text{and } k_j=|M_j\cap R_t|,\ j\in N.$$

If $N_{I(t)}=\emptyset$, stop; the cover corresponding to z_U is optimal. Otherwise, choose $j(t)\in N_{I(t)}$ such that

$$c_{j(t)}/k_{j(t)}=\min_{j\in N_{I(t)}}c_j/k_j$$

if this minimum is unique. If it is not, and it is attained for $j\in J(t)$, $|J(t)|>1$, choose $j(t)$ such that

$$|k_{j(t)}|=\max_{j\in J(t)}|k_j|$$

and if there are further ties, break them by maximizing $|M_j|$.

Then set $S_{t+1}=S_t\cup\{j(t)\}$, $R_{t+1}=R_t\backslash M_{j(t)}$, $t\leftarrow t+1$, and go to 1.

Step 2. Consider the elements $i\in S$ in order of decreasing costs c_i, and if

$$\bigcup_{j\in S\backslash\{i\}}M_j=M,$$

remove i from S. If all $i\in S$ have been considered, $\bar{x}$ defined by $\bar{x}_j=1$, $j\in S$, $\bar{x}_j=0$ otherwise, is a prime cover. If $c\bar{x}<z_U$, set $z_U=c\bar{x}$ and stop. If $z_U\leqq z_L$, $\bar{x}$ is optimal.

The logic behind the choice of $j(t)$ in Step 1 is that, the smaller the number of columns that can be used to cover a particular row, say i, the higher the price to be paid for not covering optimally row i is likely to be. Therefore the procedure first chooses a row $i(t)$ with a minimum number of 1's in it, then among the columns which can be used to cover row $i(t)$, it selects one with minimum cost per row covered. Ties are broken by choosing the column which covers the largest number of new rows (at the same unit cost), or, if there still is a tie, the column which covers the largest total number of rows. When all the rows have been covered ($R_t=\emptyset$), Step 2 checks whether the cover is prime, and makes it prime if necessary by dropping redundant columns in order of decreasing cost.

The above procedure generates a prime cover for the current problem (SC) by starting from scratch. Though this procedure is relatively cheap, it is nevertheless often desirable to have a procedure which starts with a partial cover rather than from scratch. For instance, if $S(\bar{x})\cap Q_0\neq\emptyset$, where $S(\bar{x})$ is the support of the current prime cover $\bar{x}$ and Q_0 is the set of Remark 2.1, i.e., if some variables which are at 1 in the current cover can be fixed at 0, it is desirable to find a new cover starting with, $S=S(\bar{x})\backslash Q_0$. Also, when $\bar{x}$ is a prime cover for the current problem (SC), but ceases to be a cover because of the addition to (SC) of new cuts, $\bar{x}$ can be used as a partial cover to start the procedure for finding a new cover. PRIMAL 2 is a modified version of PRIMAL 1, which differs from the latter only in the initialization step, where $S_1 = \emptyset$ is replaced by $S_1 = S(\bar{x})$, the support of the partial cover that we wish to start with.

Finally, it often happens that a good dual solution u is found, i.e., one which yields a high lower bound ue, but which, together with the associated reduced cost vector s, does not satisfy (9) for $S = S(\bar{x})$, where $\bar{x}$ is the current cover. If (9) cannot be satisfied for any set $S \subseteq N$, then of course we have no choice other than to modify u and s, i.e. decrease some components of u, which will then increase (usually by more) the sum of the reduced costs. If, however, u and s satisfy (9) for $S = N$, then it is worth trying to change the cover rather than the vector u; for if a cover (any cover) $\hat{x}$ can be found such that (9) holds for $S = S(\hat{x})$, then a cut can be generated whose strength depends only on u, s and z_U, but not on $c\hat{x}$. In other words, no matter how bad the new cover $\hat{x}$ is in terms of the associated objective function, if it makes (9) hold for u, s then it provides for a better starting point for a cut than could be obtained by keeping the old cover $\bar{x}$ and changing u and s.

The procedure PRIMAL 3 is meant for situations like this. It starts with the old cover $\bar{x}$, and successively introduces variables x_j such that $s_j > 0$. Every time such a variable is introduced, another variable x_h, such that $s_h = 0$, is dropped. Furthermore, x_j is chosen so as to cover at least one row not covered by other variables with positive reduced cost, and x_h is chosen so as to leave at least one row covered only by x_j. These choice rules are intended to keep the cover prime, if possible. When this procedure cannot be continued further, then the partial cover at hand is completed by using as few columns as possible, again in an attempt to keep the cover prime. The cover obtained in this way is then made prime, if necessary, by removing some variable(s) x_j, preferably such that $s_j = 0$.

As before, let $T(\bar{x}) = \{i \in M \mid a^i \bar{x} = 1\}$, $S^+ = \{j \in S(\bar{x}) \mid s_j > 0\}$, and denote $N^+ = \{j \in N \mid s_j > 0\}$. We assume that

$$\sum_{j \in N^+} s_j \geqq z_U - ue.$$

PRIMAL 3

Step 0. Initialize $S_1 = S(\bar{x})$, $K_1 = N^+ \backslash S^+$, $R_1 = M \backslash \bigcup_{j \in S^+} M_j$, and go to 1.

Step 1. If $R_t = \emptyset$, set $S = S_t$ and go to 3. If $R_t \neq \emptyset$, but $M_j \cap R_t = \emptyset$, $\forall j \in K_t$. Go to 2. If $R_t \neq 0$ and $M \cap R_t \neq \emptyset$ for some $j \in K_t$ but $s_j > 0$, $\forall j \in S_t$. Otherwise choose $j(t, 1) \in K_t$ and $j(t, 2) \in S_t \backslash N^+$ such that

$$|M_{j(t,1)} \cap R_t| = \max_{j \in K_t} |M_j \cap R_t|$$

$$|M_{j(t,2)} \cap M_{j(t,1)} \cap R_t| = \max_{j \in S_t \backslash N^+} |M_j \cap M_{j(t,1)} \cap R_t|.$$

Then

$$S_{t+1}=S_t\cup\{j(t,\,1)\}\backslash\{j(t,\,2)\},\quad K_{t+1}=K_t\backslash\{j(t,\,1)\},\quad R_{t+1}=R_t\backslash M_{j(t,\,1)},\quad t\leftarrow t+1,$$

and go to 1.

Step 1a. Choose $j(t,\,1)\in K_t$ such that $|M_{j(t,\,1)}\cap R_t|=\max\limits_{j\in K_t}|M_j\cap R_t|$ and set $S_{t+1}=S_t\cup\{j(t,\,1)\},\ K_{t+1}=K_t\backslash\{j(t,\,1)\},\ R_{t+1}=R_t\backslash M_{j(t,\,1)}$.

Step 2. If $R_t=\emptyset$, set $S=S_t$ and go to 3. If $R_t\neq\emptyset$, but, $M_j\cap R_t=\emptyset$, $\forall j\in N\backslash S_t$, stop; the cover corresponding to z_U is optimal. Otherwise, choose $j(t)$ such that

$$|M_{j(t)}\cap R_t|=\max_{j\in N\backslash S_t}|M_j\cap R_t|,$$

set

$$S_{t+1}=S_t\cup\{j(t)\},\quad R_{t+1}=R_t\backslash M_{j(t)},\quad t\leftarrow t+1,$$

and go to 2.

Step 3. Consider the elements $i\in S$ in order of increasing s_i^t and if

$$\bigcup_{j\in S\backslash\{i\}}M_j=M,$$

remove i from S. If all $i\in S$ have been considered, $\bar x$ defined by $\bar x_j=1$, $j\in S$, $\bar x_j=0$ otherwise, is a prime cover. If

$$\sum_{j\in S}s_j\geqq z_U-ue,$$

then s can be used to generate a cut. Otherwise u and s have to be modified.

Next we turn to the problem of generating "good" feasible solutions u to (D).

7. GENERATING FEASIBLE SOLUTIONS TO (D)

As already mentioned, the quality of a solution u to (D) is given by the quality of the lower bound ue on the value of (SC), provided that u and the associated reduced costs s_j satisfy (9) for $S=S(\bar x)$. Among two feasible solutions to (D) with the same value ue, the one with a higher value of $\sigma=\sum\limits_{j\in N}s_j$ is likely to produce a disjunction (6) with fewer terms and more elements in each term, i.e., a stronger cut. Hence the purpose of our heuristics will be to approximate as much as possible the lexicographic maximum of the two-component vector (ue,σ).

This is what the procedure DUAL 1, to be described below, tries to accomplish. It successively assigns values to components of u, which are maximal subject to the dual constraints and the earlier value-assignments. The order in which the values are assigned is crucial to the quality of the resulting solution, and DUAL 1 considers the rows of A in order of increasing N_i. Once a value has been assigned to each component of u, the set Q_0 of Remark 2.1 is identified and the corresponding variables are set to zero. If u and the associated s satisfy (9) for $S=S(\bar x)$, then s can be used to generate a cut. Otherwise either the cover must be changed as discussed in the previous section (see PRIMAL 3), or the vector u must be adjusted (see DUAL 3 below).

Let $S(\bar{x})$ be the support of a prime cover for the current problem (SC), and let z_U and z_L be an upper and a lower bound, respectively, on the value of (SC).

DUAL 1

Step 0. Initialize $R_1 = M$, $u^1 = 0$, $s^1 = c$, $t \leftarrow 1$, and go to 1.

Step 1. If $R_t = \emptyset$, go to 3. Otherwise choose $i(t)$ such that

$$N_{i(t)} = \min_{i \in R_t} |N_i|,$$

and set

$$u_i^{t+1} = \begin{cases} \min_{j \in N(t)} s_j^t & i = i(t) \\ u_i^t & \text{otherwise,} \end{cases} \qquad s_j^{t+1} = \begin{cases} s_j^t + u_i^t - u_i^{t+1} & j \in N_{i(t)} \\ s_j^t & \text{otherwise.} \end{cases}$$

Define

$$F_t = \{j \in N_{i(t)} \mid s_j^{t+1} = 0\}, \quad M(t) = \bigcup_{j \in F_t} M_j$$

set $R_{t+1} = R_t \setminus M(t)$, $t \leftarrow t+1$, and go to 1.

Step 2. If $u^t e > z_L$, set $z_L = u^t e$, store s^t in place of the vector s associated with the previous bound z_L, and stop. If $z_L \leqq z_U$, the cover associated with z_U is optimal.

As already mentioned, the choice of $i(t)$ is crucial for the efficiency of the procedure. Since the rule of Step 1 usually leads to ties, one may want to think of secondary criteria. Several such criteria have been tested. One which has been found to improve frequently the quality of the solutions is to give preference to $i \in T(\bar{x})$ over $i \in M \setminus T(\bar{x})$ in choosing $i(t)$ in Step 1.

While DUAL 1 is a computationally cheap one pass procedure, it starts from scratch every time a new pair of vectors u, s is to be found. Next we state a procedure (DUAL 2) which starts from a dual vector u and modifies it after the addition of a new cut and the finding of a new cover, by setting to 1 the variable associated with the new inequality, and adjusting the values of some other variables.

Let m be the index of the last inequality added to (SC) after generating the current dual vector $u = (u_1, u_{m-1})$ and the associated reduced cost vector s. Further, let $\bar{x}$ be a cover for the current (SC), $S(\bar{x})$ its support, and $T(\bar{x}) = \{i \in M \mid a^i \bar{x} = 1\}$.

DUAL 2

Step 0. Initialize

$$u_i^1 = \begin{cases} 1 & i = m \\ 0 & i \in [M \setminus T(\bar{x})] \\ u_i & \text{otherwise,} \end{cases}$$

$$s^1 = c - u^1 A, \quad \text{and} \quad K_1^- = \{j \in N \mid s_j^1 < 0\}.$$

the $t \leftarrow 1$ and go to 1a.

Step 1a. If $K_t^-=\emptyset$, set $K_t^0=\{j\in N \mid s_j^t=0\}$, $R_t=\{i\in M \mid N_i\cap K_t^0=\emptyset\}$, and go to 1. Otherwise, choose $i(t)$ such that

$$|N_{i(t)}\cap K_t^-|=\max_{i\in M\setminus\{m\}\mid u_i>0}|N_i\cap K_t^-|,$$

set

$$u_i^{t+1}=\begin{cases}u_i^t-1 & i=i(t)\\ u_i^t & \text{otherwise}\end{cases}\qquad s_j^{t+1}=\begin{cases}s_j^t+1 & j\in N_{i(t)}\\ s_j^t & \text{otherwise.}\end{cases}$$

Then set $K_{t+1}^-=\{j\in N\mid s_j^{t+1}<0\}$, $t\leftarrow t+1$, and go to 1a.

Steps 1–2 are the same as in DUAL 1.

Finally, we have to deal with the situation when a pair u, s has to be adjusted so as to satisfy (9) for some $S=S(\bar{x})$, where $\bar{x}$ is a given prime cover. The following procedure is meant to achieve this.

DUAL 3

Step 0. Initialize $u^1=u$, $s^1=s$, $U_t=\{i\in M\setminus T(\bar{x}) \mid u_i^1>0\}$. Set $t\leftarrow 1$ and go to 1.

Step 1. Choose $i(t)$ such that
$$|N_{i(t)}\cap S(\bar{x})|=\max_{i\in U_t}|N_i\cap S(\bar{x})|.$$

Then set

$$u_i^{t+1}=\begin{cases}u_i^t-1 & i=i(t)\\ u_i^t & \text{otherwise,}\end{cases}\qquad s_j^{t+1}=\begin{cases}u_j^t+1 & j\in N_{i(t)}\\ s_j^t & \text{otherwise.}\end{cases}$$

If u^{t+1} and s^{t+1} satisfy (9) for $S=S(\bar{x})$, stop: $s=s^{t+1}$ can be used in CUT 1. Otherwise, set $U_{t+1}=U_t\setminus\{i(t)\}$, $t\leftarrow t+1$, and go to 1.

The heuristics discussed here find reasonably good dual vectors u at a low computational cost. Another possibility is, of course, to find an optimal or near-optimal u by the simplex method or by subgradient optimization. While this is more costly, only computational testing can show whether the improvement in the bound ue is worth the extra computational effort.

8. ALGORITHMS

In this section we present two algorithms for solving set covering problems by cutting planes from conditional bounds.

They use as ingredients the procedures discussed in the last three sections (for generating prime covers, feasible dual vectors and valid cutting planes), and the following three subroutines for updating cuts and fixing variables whenever possible.

STRENGTHEN (to be used whenever PRIMAL replaces z_U^{old} by z_U^{new}).

In every cut $\sum_{j\in W_p} x_j\geq 1$, replace W_p by W_{p-k}, where k is the greatest integer such that $v_p+\ldots+v_{p-k+1}\leq z_U^{\text{old}}-z_U^{\text{new}}$.

414

TEST 1 (to be used after STRENGTHEN).

Let s be the reduced cost vector associated with z_L, and let $S(\bar{x})$ be the support of the last cover $\bar{x}$. Define

$$Q_0 = \{j \in N \mid s_j \geq z_U^{new} - z_L\},$$

set $x_j = 0$, $j \in Q_0$, and $N \leftarrow N \backslash Q_0$. If $N \subseteq S(\bar{x})$, the cover associated with z_U^{new} is optimal.

TEST 2 (to be used after DUAL).

Same as TEST 1, except for the fact that s, z_U^{new} and z_L are replaced by s', z_U and u' respectively, where u' is the last dual vector obtained, s' is the reduced cost vector associated with u', and z_U is the current upper bound.

Algorithm 1 starts by initializing the upper bound z_U at Σc_j and the lower bound z_L at 0. A typical iteration consists of the following sequence of steps.

A. Use PRIMAL 1 or 2 to generate a prime cover $\bar{x}$. If z_U is improved, go to C1; otherwise go to B1.

B1. Use DUAL 1 to generate a dual vector u and go to C2.

B2. Use DUAL 3 to adjust u and s and go to D.

C1. Apply TEST 1 to fix variables. If $N \subseteq S(\bar{x})$, stop. Otherwise, if $\bar{x}$ is still a cover, go to B1; else go to A.

C2. Apply TEST 2 to fix variables. If $N \subseteq S(\bar{x})$, stop. Otherwise, if $\bar{x}$ is still a cover and (9) holds for $S = S(\bar{x})$, go to D; if (9) does not hold for $S = S(\bar{x})$, go to B2; and if $\bar{x}$ is not any more a cover, go to A.

D. Use CUT to generate a valid inequality which cuts off the last cover. Add the new inequality to (SC) and go to A.

Algorithm 1 seems the simplest possible way to use the cuts from conditional bounds. Algorithm 2, stated below, is a more sophisticated procedure based on the same approach, which (a) strengthens the cuts whenever z_U is improved; (b) avoids starting from scratch every time a new cover or a new dual solution is generated by using the "updating" procedures PRIMAL 2 and DUAL 2, with a periodic return to PRIMAL 1 and DUAL 1; and (c) uses PRIMAL 3 whenever z_L is improved and (9) does not hold for the current cover, to find a prime cover $\bar{x}$ such that u and s satisfy (9) for $S = S(\bar{x})$. Since the chances for finding a better cover than the current best one are steadily diminishing during the procedure, the parameter α defining the number of iterations after which we return to PRIMAL 1 is doubled every time the latter procedure is applied; while the parameter β defining the frequency of return to DUAL 1 is kept constant.

After initializing z_U, z_L, α and β, a typical iteration of Algorithm 2 consists of some of the following steps.

A1. Use PRIMAL 1 to generate a prime cover $\bar{x}$, and double the value of α. If z_U is improved, go to C1; otherwise to B1.

A2. Use PRIMAL 2 to generate a prime cover $\bar{x}$. If z_U is improved, go to C1;

otherwise go to B2, or (if no cut was added after obtaining last vector u, or B2 was used β times since last use of B1) to B1.

A3. Use PRIMAL 3 to generate a prime cover $\bar{x}$. If the last u and s satisfying (9) for $S = S(\bar{x})$, go to D; otherwise go to B3.

B1. Use DUAL 1 to generate a dual vector u and go to C2.

B2. Use DUAL 2 to generate a dual vector u and go to C2.

B3. Use DUAL 3 to adjust u and s and go to D.

C1. Use STRENGTHEN to strengthen the cuts, and TEST 1 to fix variables. If $N \subseteq S(\bar{x})$, stop. Otherwise, if $\bar{x}$ is still a cover, go to B3; else go to A2.

C2. Use TEST 2 to fix variables. If $N \subseteq S(\bar{x})$, stop. Otherwise, if $\bar{x}$ is still a cover and (9) holds for $S = S(\bar{x})$, go to D; if (9) does not hold for $S = S(\bar{x})$, but holds for $S = N$, go to A3; and if (9) does not hold for $S = N$, go to B1; finally, if $\bar{x}$ is not any more a cover, go to A2.

D. Use CUT to generate an inequality which cuts off the last cover. If α cuts have been generated since the last use of PRIMAL 1 or 2, go to A1; otherwise go to A2.

We call an iteration of Algorithm 2 a sequence of steps, including C2, which results either in generating a cut or in fixing some variables.

9. NUMERICAL EXAMPLE

Consider the set covering problem whose cost vector c and coefficient matrix A are shown in Table 3, and which is obtained from the 32-variables example of [8] by removing 9 columns (12, 13, 17, 20, 21, 27, 28, 29, 30) dominated by sums of other columns.

Table 3

	1	2	3	4	5	6	7	8	9	10	11	12	13	14	15	16	17	18	19	20	21	22	23
c_j	1	1	1	1	1	1	1	1	2	2	2	2	2	3	3	3	4	4	5	5	5	8	9
A 1	1								1							1	1			1			1
2		1											1			1		1				1	1
3			1							1						1			1		1		1
4				1														1		1			1
5					1						1			1			1					1	
6						1													1		1		1
7							1													1		1	
8								1										1	1				1
9												1									1	1	
10										1										1		1	1
11												1						1	1				1
12															1	1	1		1		1	1	
13											1									1			1
14														1		1		1				1	
15												1	1		1								1

We apply Algorithm 2. To start with, we set $z_U = \sum_{j \in N} c_j = 67$, $z_L = 0$, $\alpha = 1$, $\beta = 2$.

Iteration 1.

A1. PRIMAL 1.

Step 0. $R_1 = \{1, \ldots, 15\}$, $S_1 = \emptyset$.

Step 1. $\min_{h \in R_1} |N_h|$ is attained for $h \in I(1) = \{7, 13\}$, $N_{I(1)} = \{7, 11, 20, 22, 23\}$. $\min_{j \in N_{I(1)}} c_j / |R_1 \cap M_j|$ is attained for $j(1) = 23$. $S_2 = \{23\}$, $R_2 = \{5, 7, 9, 12, 14\}$.

Applying Step 1 four more times, we determine the sequence of indices $j(2) = 7$, $j(3) = 8$, $j(4) = 16$, $j(5) = 5$, i.e., $S = \{5, 7, 8, 16, 23\}$.

Step 2. The vector $\bar{x}$ defined by $\bar{x}_j = 1$, $j \in S$, $\bar{x}_j = 0$, $j \in N \backslash S$, is a prime cover. The associated upper bound is $z_U = 15$, and $T(\bar{x}) = M \backslash \{1, 2, 3, 8\}$. We set $\alpha \leftarrow 2\alpha = 2$.

C1. STRENGTHEN. There is no cut yet to strengthen.

TEST 1. $Q_0 = \{j \in N \mid s_j \geq 15 - 0\} = \emptyset$ (here $s_j = c_j$, $j \in N$); we go to B1.

B1. DUAL 1.

Step 0. $R_1 = \{1, \ldots, 15\}$, $u^1 = 0$, $s^1 = c$.

Step 1. $\min_{i \in R_1 \cap T(\bar{x})} |N_i| = 3$, $i(1) = \{7\}$, $u_7^2 = 1$, $u_i^2 = u_i^1$, $i \neq 7$, and

$$s^2 = (1, 1, 1, 1, 1, 1, 0, 1, 2, 2, 2, 2, 2, 3, 3, 3, 4, 4, 5, 4, 5, 7, 9).$$
$$F_1 = \{7\}, \quad M(1) = M_7 = \{7\}, \quad R_2 = R_1 \backslash M(1) = \{1, \ldots, 6, 8, \ldots, 15\}.$$

Applying Step 1 seven more times we choose the sequence of indices

$$i(2) = 13, \quad i(3) = 4, \quad i(4) = 6, \quad i(5) = 9, \quad i(6) = 14, \quad i(7) = 10, \quad i(8) = 15,$$

and obtain the vectors

$$u^9 = (0, 0, 0, 1, 0, 1, 1, 0, 1, 1, 0, 0, 2, 3, 2),$$
$$s^9 = (1, 1, 1, 0, 1, 0, 0, 0, 0, 1, 0, 1, 2, 0, 1, 0, 4, 0, 2, 0, 1, 0, 2),$$

after which $R_9 = \emptyset$.

Step 2. $u^9 e = 12 > 0 = z_L$, hence we set $z_L = 12$ and store s^9.

C2. TEST 2. $Q_0 = \{j \in N \mid s_j^9 \geq 15 - 12\} = \{17\}$. We set $x_{17} = 0$, $N \leftarrow N \backslash \{17\}$. Since $Q_0 \cap S(\bar{x}) = \emptyset$, and $\sum_{j \in S(\bar{x})} s_j^9 = 3 \geq z_U - u^9 e = 3$, we go to D.

D. CUT. $T(\bar{x}) = M \backslash \{1, 2, 3, 8\}$, $S^+ = \{5, 23\}$, $s = s^9$.

Step 0. $W_0 = \emptyset$, $K_1 = \{5, 23\}$, $y_1 = 12$, $s^1 = s$.

Step 1. $v_1 = \min \{15 - 12, \max \{1, 2\}\} = 2$, $K_1^* = \{23\}$, $P_1 = \{13, 19, 23\}$, $M(1) = M_{23} = \{1, \ldots, 4, 6, 8, 10, 11, 13, 15\}$, and

$$_t N_{13} \backslash \{13, 19, 23\}| = \min_{i \in T(\bar{x}) \cap M(1)} |N_i \backslash \{13, 19, 23\}| = 2.$$

Hence $i(1)=13$, and $j(1)=N_{13}\cap K_1^*=\{23\}$. We have $W_1(13)=N_{13}\backslash P_1=\{11, 20\}$, $y_2=12+2=14<z_U=15$, and

$$s^2=(1, 1, 1, 0, 1, 0, 0, 0, 0, 1, 0, 1, 2, 0, 1, 0, *, 0, 2, 0, 1, 0, 0)$$

(here and in the following, stars replace the entries corresponding to variables that have been fixed). Further, $K_2=\{5, 23\}\backslash\{23\}=\{5\}$.

Step 1. $v_2=\min\{15-14, 1\}=1$; $K_2^*=\{5\}$; $P_2=\{1, 2, 3, 5, 10, 12, 13, 15, 19, 21\}$, $M(2)=\{5\}$. We have $i(2)=\{5\}$, $j(2)=\{5\}$, $W_2=\{11, 20\}\cup\{14, 22\}$, and $y_3=14+1\geqq$ $\geqq z_U=15$.

Step 2. We add to (SC) the cut

$$x_{11}+x_{14}+x_{20}+x_{22}\geqq 1$$

as the 16th inequality, and store the pairs

$$(W_1=\{11, 20\}, v_1=2), \quad (W_2\backslash W_1=\{14, 22\}, v_2=1).$$

Iteration 2.

A2. PRIMAL 2.
Step 0. $R_1=\{16\}$, $S_1=\{5, 7, 8, 16, 23\}$
Step 1. $I(1)=\{16\}$, $N_{I(1)}=\{11, 14, 20, 22\}$, $j(1)=11$.

$$S_2=\{5, 7, 8, 11, 16, 23\}, \quad R_2=\emptyset.$$

Step 2. The vector $\bar{x}$ whose support is $S=S_2\backslash\{5\}=\{7, 8, 11, 16, 23\}$, is a prime cover, with $c\bar{x}=16$ and $T(\bar{x})=M\backslash\{1, 2, 3, 8, 13\}$.

B2. DUAL 2.
Step 0. $m=16$. Initialize

$$u^1=(0, 0, 0, 1, 0, 1, 1, 0, 1, 1, 0, 0, 0, 3, 2, 1),$$

$s^1=(1, 1, 1, 0, 1, 0, 0, 0, 0, 1, 1, 1, 2, -1, 1, 0, *, 0, 2, 1, 1, -1, 4)$, $K_1^-=\{14, 22\}$.
Step 1a. $i(1)=\{14\}$, we set $u^2=(0, 0, 0, 1, 0, 1, 1, 0, 1, 1, 0, 0, 0, 2, 2, 1)$,

$s^2=(1, 1, 1, 0, 1, 0, 0, 0, 0, 1, 1, 1, 2, 0, 1, 1, *, 1, 2, 1, 1, 0, 4)$, $K_2^-=\emptyset$.
Step 1b. $K_2^0=\{4, 6, 7, 8, 9, 14, 22\}$, $R_2=\{3, 11, 13\}$.
Step 1. Since $R_2\cap T(\bar{x})=\{11\}$, we set $i(2)=11$, $u_{11}^3=\min\limits_{j\in N_{11}} s_j^2=1$, $u_i^3=u_i^2$, $i\neq 11$, and

$$s^3=(1, 1, 1, 0, 1, 0, 0, 0, 0, 1, 1, 0, 1, 0, 0, 1, *, 0, 1, 1, 1, 0, 3).$$
Further, $F_2=\{12, 15, 18\}$, $M(2)=\{2, 4, 8, 9, 11, 12, 14, 15\}$, and $R_3=\{3, 13\}$.
In two more iterations of Step 1, we set $i(3)=13$, $u_{13}^4=1$, and $i(4)=3$, $u_3^5=1$, and obtain the vectors $u^5=(0, 0, 1, 1, 0, 1, 1, 0, 1, 1, 1, 0, 1, 2, 2, 1)$ and

$$s^5=(1, 1, 0, 0, 1, 0, 0, 0, 0, 0, 0, 0, 1, 0, 0, 0, *, 0, 0, 0, 0, 0, 1),$$
at which point $R_5=\emptyset$.
Step 2. $u^5 e=13>12=z_L$, hence we set $z_L=13$ and store s_3^5 in place of s_9.

418

C2. TEST 2. $Q_0 = \{j \in N \mid s_j^5 \geqq 15 - 13\} = \emptyset$, $\sum\limits_{j \in S(\bar{x})} s_j^5 = 1 < 15 - 13$, hence we go to A3.

A3. PRIMAL 3.
Step 0. Initialize $S_1 = \{7, 8, 11, 16, 23\}$, $K_1 = \{1, 2, 5, 13\}$, $R_1 = \{5, 7, 9, 12, 14, 16\}$.
Step 1. $\max\limits_{j \in K_1} |M_j \cap R_1| = 1, j(1, 1) = 5; j(1, 2) = 11. S_2 = \{5, 7, 8, 16, 23\}, K_2 = \{1, 2, 13\}$,
$R_2 = \{7, 9, 12, 14, 16\}$.
Step 1. $\bigcup\limits_{j \in K_2} M_j \cap R_2 = \emptyset$; we set $R_2 = M \backslash \bigcup\limits_{j \in S_2} M_j = \{16\}$.
Step 2. $j(2) = \{20\}$, $S_3 = \{5, 7, 8, 16, 20, 23\}$, $R_3 = \emptyset$.
Step 3. $\bigcup\limits_{j \in S \backslash \{7\}} M_j = M$, $S = \{5, 8, 16, 20, 23\}$. Since $\sum\limits_{j \in S(\bar{x})} s_j^5 = 2 \geqq 15 - 13$, we go to D.

D. CUT generates the cut

$$x_6 + x_{11} + x_{14} + x_{19} + x_{21} + x_{22} \geqq 1$$

(which becomes the 17th inequality), and the pairs

$$(W_1 = \{6, 19, 21\}, v_1 = 1), \quad (W_2 \backslash W_1 = \{11, 14, 22\}, v_2 = 1).$$

Since $\alpha = 2$ and 2 cuts have been generated, we go to A1.

Iteration 3.
A1. PRIMAL 1 produces the prime cover $\bar{x}$ defined by $S(\bar{x}) = \{4, 6, 7, 8, 10, 11, 15, 16\}$, with $T(\bar{x}) = M \backslash \{3, 12, 17\}$ and $c\bar{x} = 14$. Since $14 < z_U = 15$, we set $z_U = 14$, $\alpha \leftarrow 2\alpha = 4$, and go to C1.

C1. STRENGTHEN. Since z_U was improved by $15 - 14 = 1$, and $v_p = v_2 = 1$ for both cuts 16 and 17, each of them can be strengthened by replacing W_2 with W_1. This yields

$$x_{11} + x_{20} \geqq 1 \quad \text{and} \quad x_6 + x_{19} + x_{21} \geqq 1$$

as the strengthened inequalities 16 and 17.

TEST 1. $Q_0 = \{j \in N \mid s_j^5 \geqq 14 - 13\} = \{1, 2, 5, 13, 23\}$. We set $x_1 = x_2 = x_5 = x_{13} = x_{23} = 0$, $N = \{3, 4, 6, \ldots, 12, 14, 15, 16, 18, \ldots, 22\}$, and since $\bar{x}$ is still a cover, we go to B1.
B1. DUAL 1 produces the vectors $u^{10} = (0, 2, 0, 1, 0, 1, 1, 1, 0, 1, 0, 1, 2, 0, 2, 0, 0)$ and $s^{10} = (*, *, 1, 0, *, 0, 0, 0, 0, 1, 0, 2, *, 3, 0, 0, *, 0, 0, 0, 1, 1, *)$, with $ue = 12$.
C2. TEST 2. $Q_0 = \{j \in N \mid s_j^{10} \geqq 14 - 12\} = \{12, 14\}$. We set $x_{12} = x_{14} = 0$, $N = \{3, 4, 6, \ldots, 11, 15, 16, 18, \ldots, 22\}$. Since $Q_0 \cap S(\bar{x}) = \emptyset$ and $\sum\limits_{j \in S(\bar{x})} s_j^{10} = 1 < 14 - 12$, but $\sum\limits_{j \in N} s_j^{10} = 4 \geqq 2$, we go to A3.

A3. PRIMAL 3. $S(\bar{x}) = \{4, 8, 11, 15, 16, 21, 22\}$, $c\bar{x} = 23$.
D. CUT. $x_6 + x_{19} + x_{20} \geqq 1$.
Iteration 4 uses the sequence of steps A2, B2, C2, D. Step C2 sets $x_{21} = 0$, while D generates the cut $x_{18} + x_{19} + x_{20} \geqq 1$.

Iteration 5. consists of A2, B2, C2, D, and produces the cut $x_8 + x_{19} \geqq 1$.
Iteration 6. A2, B1, C2, $x_{22} = 0$.
Iteration 7. A2, B1, C2, $x_9 = x_{18} = 0$.
Iteration 8. A2, B1. DUAL 1 produces the vector

$$u = (0, 3, 0, 1, 2, 1, 1, 0, 1, 2, 3, 0, 0, 0, 0, 0, 0, 0, 1, 0),$$

with $ue = 15$; we set $z_L = 15$, and since $z_L \geqq z_U = 14$, the cover associated with z_U is optimal. This is $\bar{x}$ such that $\bar{x}_j = 1$, $j = 4, 6, 7, 8, 10, 11, 15, 16$, $\bar{x}_j = 0$ otherwise.

10. COMPUTATIONAL EXPERIENCE

Several versions of the algorithms described in Section 8 were implemented and tested in the framework of a joint project with Andrew Ho, that is still under way and whose final results will be reported elsewhere. Preliminary results show the approach to be highly efficient on low-density problems. Several variants of the heuristics described in Sections 3, 4 and 5 were tested. Thus, for instance, the rule of choosing the smallest c_j/k_j in PRIMAL 1 was compared with other "greedy" rules, like choosing the smallest c_j, $c_j/\log_2 k_j$, or $c_j/k_j \log_2 k_j$, with the conclusion that alternating these rules between (or within) iterations seems better than using any single rule by itself. The results shown in Table 1 were obtained with a version of the code which returns to PRIMAL 1 every 7th iteration and to DUAL 1 every 4th iteration. PRIMAL 1 is enhanced to use intermittently min c_j/k_j and min $c_j/k_j \log_2 k_j$ as choice rules.

Table 1

Problem D Description			Algorithm 2		Bowman–Starr [4]		Salkin–Koncal [9]	Lemke–Salkin–Spielberg [8]
No.	m	n	# cuts	time [1]	# cuts	time [1]	time [1]	time [2]
1	104	133	15	3.9	120	2.48	5.7	424.0
2	200	300	16	11.1	49	5.34	15.9	461.3
3	200	413	28	15.1	365	15.81	26.7	625.9
4	200	500	55	28.5	3,000	>215	>120	803.5
5	50	450	6	1.8	464	16.2	>120	144.5
6	36	455	0	2.3	32	2.1	18.6	35.5
7	46	683	34	7.7	108	4.2	>120	56.9
8	50	905	15	6.3	241	11.77	>120	670.0

1) UNIVAC 1108 seconds
2) IBM 360/50 seconds

The test problems are those of Salkin and Koncal [9]. With the exception of problem 1, which is attributed in [9] to American Airlines, the remaining problems were

420

randomly generated by Salkin, with costs ranging between 0 and 99. The density of problem 1 is 0.07, that of the other problems ranges between 0.02 and 0.11.

The results obtained by our procedure (Algorithm 2) are compared with those of Bowman and Starr reported in [4], as well as those of Salkin and Koncal, and Lemke, Salkin and Spielberg, reported in [9]. The Bowman–Starr algorithm uses the cutting planes mentioned at the end of Section 2, in an enumerative framework. Salkin and Koncal [9] use Gomory's dual all integer cutting plane algorithm. The Lemke, Salkin and Spielberg algorithm [8] is a specialized implicit enumeration procedure with linear programming subproblems. While the Bowman–Starr and the Salkin–Koncal results, like ours, were obtained on a UNIVAC 1108 computer and hence are directly comparable, the Lemke–Salkin–Spielberg results refer to runs on an IBM 360/50 machine, believed to be 4–5 times slower (see [9], p. 192) than the UNIVAC 1108.

The number of cuts required to solve these problems is remarkably low. Surprisingly, large numbers of variables get fixed, often in the early stages of the runs. This and the pattern of successive bound-improvements are illustrated on a typical run shown in Table 2.

Table 2. Problem 1 (104×133)

Iteration	z_U	z_L	Variables left
1	1690	1637	105
2		1663	87
3	1686		
8	1682		
13	*1678*		
14		1667	76
17		1677	54
18		1717	

REFERENCES

[1] Balas, E. and Christofides, N.: A New Penalty Method for the Traveling Salesman Problem, Paper presented at the 9th International Mathematical Programming Symposium, Budapest, August 1976.

[2] Balas, E. and Padberg, M. W.: Set Partitioning, B. Roy (editor), *Combinatorial Programming: Methods and Applications,* Reidel Publishing Co., 1975, 205–258.

[3] Bellmore, M. and Ratliff, H. D.: Set Covering and Involutory Bases, *Management Science, 18* (1971), p. 194–206.

[4] Bowman, V. J. and Starr, J.: Set Covering by Ordinal Cuts. I: Linear Objective Functions, MSRR #321, Carnegie-Mellon University, June 1973.

[5] Christofides, N. and Korman, S.: A Computational Survey of Methods for the Set Covering Problem, Report No. 73/2, Imperial College of Science and Technology, April 1973.

[6] Garfinkel, R. and Nemhauser, G. L.: Optimal Set Covering: A Survey, A. M. Geoffrion (editor), *Perspectives on Optimization,* Addison–Wesley, 1972, pp. 164–183.

[7] B. Kovács, L. and Dienes, I.: Maximal Direction — Complete Paths and Their Application to a Geological Problem: Setting up Stratigraphic Units, Paper presented at the 9th International Mathematical Programming Symposium, Budapest, August 1976.

[8] Lemke, C. E., Salkin, H. M. and Spielberg, K.: Set Covering by Single Branch Enumeration with Linear Programming Subproblems, *Operations Research, 19* (1971), 998–1022.

[9] Salkin, H. M. and Koncal, R. D.: Set Covering by an All-Integer Algorithm: Computational Experience, *ACM Journal, 20* (1973), 189–193.

DECOMPOSITION OF MATHEMATICAL PROGRAMS BY MEANS OF THEOREMS OF ALTERNATIVE FOR LINEAR AND NONLINEAR SYSTEMS

G. CASTELLANI and F. GIANNESSI

(Venice, Italy)   (Pisa, Italy)

INTRODUCTION

The aim of this paper is first of all to review a decomposition method proposed for mixed integer programs [1]; it has been later extended to a wider class of mathematical programs, which include a linear program as a subproblem [2]. In spite of its potential power, such a method has not yet found a systematic application.

Here some recent applications of the method are described, which show its use both as a mathematical tool to decompose problems, and as a numerical way of treating problems of practical size.

A first application consists in a staffing problem coming from an airline [5], [7], which is formulated as an integer nonlinear program.

A second problem is the classical travelling salesman problem and some of its generalizations [3].

A third problem is the general linear complementarity problem and in particular an application to a problem of structural mechanics [4].

At last a theorem of alternative of the Farkas type for nonlinear systems is studied, and it is shown how to use it to decompose mathematical programs into nonlinear subproblems. This is accomplished also by means of a numerical example.

1. DECOMPOSITION OF MATHEMATICAL PROGRAMS BY MEANS OF A THEOREM OF ALTERNATIVE FOR LINEAR SYSTEMS

Consider the following problem

$$(1.1a) \qquad \min\,[z = c(y)x + d(y)]$$

under the constraints

$$(1.1b) \qquad A(y)x + B(y) \geqq 0$$

$$(1.1c) \qquad y \in Y,$$

where x and y are column-vectors with n and p components, respectively; $c : \mathbf{R}^p \to \mathbf{R}^n$; $d : \mathbf{R}^p \to \mathbf{R}$; $B : \mathbf{R}^p \to \mathbf{R}^m$; $A(y)$ is a real matrix, whose elements are functions of y; and $Y \subseteq \mathbf{R}^p$ is any given set.

Let $(\bar{x}, \bar{y})$ be any feasible solution of (1.1); set $\bar{z} \triangleq c(\bar{y})\bar{x} + d(\bar{y})$. We want to know a condition for a feasible solution (x, y) of (1.1) not to imply a value of the objective function $z = \bar{z}$; this happens if the system (in the unknown x, y)

$$(1.2) \qquad \begin{cases} c(y)x + d(y) < \bar{z} \\ A(y)x + B(y) \geqq 0 \\ y \in Y \end{cases}$$

is impossible. For any fixed $y \in Y$, (1.2) may be written this way

$$(1.3) \qquad \begin{cases} -c(y)x > d(y) - \bar{z} \\ -A(y)x \leqq B(y). \end{cases}$$

By a well-known theorem of alternative*, (1.3) is impossible, iff at least one of the following linear systems (in the unknown u)

$$(1.4\text{a}) \quad \begin{cases} uA(y) = c(y) \\ uB(y) \leqq d(y) - \bar{z}; \\ u \geqq 0 \end{cases} \qquad (1.4\text{b}) \quad \begin{cases} uA(y) = 0 \\ uB(y) < 0 \\ u \geqq 0 \end{cases}$$

is possible, where u is a p-row-vector.

Remark that (1.4a) at $y = \bar{y}$ has solutions iff $\bar{z}$ is the minimum of the problem

$$(1.5) \qquad P(\bar{y}) : \begin{cases} \min \, [c(\bar{y})x + d(\bar{y})] \\ A(\bar{y})x \geqq -B(\bar{y}) \end{cases}$$

and remark that the solutions of (1.4a) are the optimal solutions of the dual of (1.5), i.e. of the problem

$$(1.6) \qquad Q(\bar{y}) : \begin{cases} \max \, [-uB(\bar{y}) + d(\bar{y})] \\ uA(\bar{y}) = c(\bar{y}) \\ u \geqq 0. \end{cases}$$

In fact, whatever the feasible x and u may be, we have

$$-uB(\bar{y}) + d(\bar{y}) \leqq c(\bar{y})x + d(\bar{y}).$$

Then, iff x and u are respectively optimal for (1.5) and (1.6) the inequality

$$-uB(\bar{y}) + d(\bar{y}) \geqq c(\bar{y})x + d(\bar{y})$$

i.e.

$$uB(\bar{y}) \leqq d(\bar{y}) - \bar{z}$$

is satisfied.

---

* See for example [6], p. 32.

Remark also that (1.4b) has solutions, iff the feasible region of (1.1) is empty. In fact, when $y\in Y$, the feasible region of (1.1) is defined by the linear system (in the unknown x)

$$-A(y)x\leq B(y)$$

which, by a well-known theorem of alternative*, is impossible iff the following linear system (in the unknown u)

$$\begin{cases} -uA(y)=0 \\ uB(y)=-1 \\ u\geq 0 \end{cases}$$

is possible. Now it is enough to remark that the last system is equivalent to (1.4b).

If the feasible region of (1.1) is nonempty, whatever $y\in Y$ may be, we have obtained the following.

Theorem 1. *Let $\bar{u}$ and $\bar{z}$ be, respectively, an optimal solution and the maximum in in dual of the problem (1.1) at $y=\bar{y}\in Y$, i.e. in the problem (1.6). Then, to every $y\in Y$, which satisfies*

$$\begin{cases} \bar{u}A(y)=c(y) \\ \bar{u}B(y)\leq d(y)-\bar{z} \end{cases}$$

does not correspond in (1.1) a value $z<\bar{z}$, whatever the feasible solution x may be.

When Theorem 1 is used to decompose the problem (1.1), it is useful to have a lower bound of the minimum of (1.1). In fact, it may enable one to state that a feasible solution of (1.1) is optimal, even if Y has not been completely explored.

To this aim consider the set

$$U=\{u\in \mathbf{R}^m : uA(y)=c(y); u\geq 0\}$$

and assume, for sake of simplicity, that it is bounded.

Call z^* the minimum of (1.1). Then, the following relations

$$z^*= \min_{\substack{A(y)x+B(y)\geq 0 \\ y\in Y}} [c(y)x+d(y)]=\min_{y\in Y}\left\{d(y)+ \min_{A(y)x\geq -B(y)} [c(y)x]\right\}=$$

$$=\min_{y\in Y}\left\{d(y)+ \max_{\substack{uA(y)=c(y) \\ u\geq 0}} [-uB(y)]\right\}=\min_{y\in Y}\left\{d(y)+ \max_{u^i(y)\in \text{vert}.U(y)} [-u^i(y)B(y)]\right\}\geq$$

$$\geq\min_{y\in Y}\left\{d(y)+ \max_{u^i(y)\in S(y)} [-u^i(y)B(y)]\right\},$$

hold, where vert. $U(y)$ denotes the set of vertices of the polyhedron $U(y)$; and $S(y)\subseteq \subseteq U(y)$.

* See [6], p. 33.

The last problem offers a lower bound of z^*; however, its determination may be difficult, as the maximum depends on y. Then, a particularly interesting case is the one in which $S(y)$ is independent on y; if we set $S(y)=S$, we have

$$z^* \geq \min_{y \in Y} \left\{ d(y) + \max_{u^i \in S} [-u^i B(y)] \right\}.$$

Remark that, when U does not depend on y, in Theorem 1, the choice of $\bar{u}$ may be made in such a way that the above lower bound is referred to a prefixed region of Y, for instance an interval or any other desired set. The preceding remarks will now be clarified by the following numerical example. Consider the mixed integer linear program (M.I.L.P.)

$$\begin{cases} \min (z = 5x_1 + 2y_1 + 2y_2) \\ x_1 + 3y_1 + 2y_2 \geq 5 \\ 4x_1 - y_1 + y_2 \geq 7 \\ 2x_1 + y_1 - y_2 \geq 4 \\ x_1, y_1, y_2 \geq 0 \\ y_1, y_2 \text{ integer.} \end{cases}$$

Starting with $\bar{y}=(1, 1)$, the optimal solution of $Q(\bar{y})$ is $\bar{u}=\left(0, 0, \dfrac{5}{2}\right)$ and the common value of the extreme of $P(\bar{y})$ and $Q(\bar{y})$ is 4. Then the cut implied by Theorem 1 is

$$-y_1 + 9y_2 \geq 8$$

and we may restrict ourselves to consider the y's such that

(1.7)
$$-y_1 + 9y_2 \leq 7;$$

$y^*=(1, 0)$ is one of these; by repeating the preceding procedure, we consider an optimal solution of $Q(y^*)$; its optimal basic solutions are only

$$u^1 = (5, 0, 0) \ , \ u^2 = \left(0, \dfrac{5}{4}, 0\right).$$

The cuts implied by Theorem 1 and corresponding to u^1, u^2 are

$$13y_1 + 8y_2 \leq 13, \quad 13y_1 + 3y_2 \geq 13$$

respectively.

Then we may restrict ourselves to consider the y's such that

$$13y_1 + 8y_2 \geq 14, \quad 13y_1 + 3y_2 \leq 12.$$

As $\bar{y}$ implies $\bar{z}=14$ and y^* implies $z^*=12$, and as adding the last two inequalities to (1.7) we obtain an empty set over integer y, it follows that to y^* it corresponds an optimal solution of M.I.L.P., which is $(x_1=2, y_1=1, y_2=0)$. Now it will be shown how it is possible to obtain, in place of the preceding cuts, a cut parallel to an axis (in general a cut of a given family).

To this aim, remark that the cut implied by Theorem 1 is

$$(3u_1-u_2+u_3-2)y_1+(2u_1+u_2-u_3-2)y_2\leqq 5u_1+7u_2+4u_3-z.$$

It is parallel to y_1-axis, iff the following conditions

(1.8)
$$\begin{cases}3u_1-u_2+u_3-2=0\\2u_1+u_2-u_3-2\neq 0\end{cases}$$

hold.

Now we select an $\tilde{u}$ satisfying (1.8), for instance

$$\tilde{u}=\left(0,\ \frac{1}{6},\ \frac{13}{6}\right)$$

In order to apply Theorem 1 we look for a $\tilde{y}$ such that $\tilde{u}$ is an optimal solution of $Q(\tilde{y})$. This happens, iff $\tilde{y}$ is any point of the ray

(1.9)
$$y_1\geqq\frac{23}{30},\quad y_2=y_1-\frac{1}{3}.$$

Now the cut implied by Theorem 1 at $u=\tilde{u}$ is

$$0\tilde{y}_1-4\tilde{y}_2\leqq\frac{59}{6}-z$$

which becomes

$$z\leqq 4y_1+\frac{51}{6}\quad\left(y_1\geqq\frac{23}{30}\right),$$

if we take into account (1.9).

This inequality and the same Theorem 1 imply that $4y_1+\dfrac{51}{6}$ is a lower bound

for M.I.L.P. over the halfplane

$$y_2\geqq y_1-\frac{1}{3}\quad y_1\geqq\frac{23}{30}.$$

2. DECOMPOSITION OF AN OPTIMAL STAFFING PROBLEM

The results of the preceding section will now be applied to a problem of determining optimal shifts and minimum personnel requirements to meet given manpower loads at an airport.

Consider a generic day and divide it into 48 equal periods, each of half an hour, and assume to know the manpower load for every period.

Moreover, assume that, according to union rules, the policy of the airport management consists in forecasting three ordinary shifts and two supplementary shifts, each consisting of 17 consecutive periods. The starting times of the shifts are subjected to given constraints; for instance, the following:

(a) the ordinary shifts are connected so that the last period of the first (second) is the initial one of the second (third);

(b) the starting time of the first ordinary shift must be one of these: 6.30—7—7.30—8 a.m.

(c) the supplementary shifts may start at any period, but not at the starting times of the ordinary shifts.

If y_j, $j=1, 2, \ldots, 5$, denote the starting times of the shifts and if we put $y=(y_1, \ldots, y_5)$, only three elements of y are independent, that is y_1, y_4, y_5; while y_2, y_3 are determined, when y_1 has been fixed. Under our assumptions, y must belong to a given set Y, which has $4 \times \binom{48-3}{2} = 3960$ elements.

Denote by

r_i, the manpower load of period i, $i=1, \ldots, 48$;

x_j, the unknown number of persons to assign to shift j, $j=1, \ldots, 5$;

$$a_{ij}(y_j) = \begin{cases} 0, & \text{if period } i \text{ is not covered by shift } j, \\ 1, & \text{if period } i \text{ is covered by shift } j. \end{cases}$$

Let $A(y)$ be a 48×5 matrix, whose elements are $a_{ij}(y_j)$; r a 48-column vector of r_i;

$$e=(1, 1, 1, 1, 1); \quad x^T=(x_1, x_2, x_3, x_4, x_5).$$

Then, the preceding problem may be formulated as nonlinear integer program this way:

$$(2.1) \qquad \begin{cases} \min (z=ex) \\ A(y)x \geq r \\ x \geq 0 \text{ and integer} \\ y \in Y. \end{cases}$$

When y is regarded as fixed, and if the integer requirements on x are disregarded, (2.1) becomes a (real) linear program:

$$(2.2) \qquad P(y): \begin{cases} \min (z=ex) \\ A(y)x \geq r \\ x \geq 0. \end{cases}$$

In other words, (2.1), without the integer requirements on x, is a particular case of problem (1.1) and problem (2.2) a particular case of the linear program obtained from (1.1), when y has been fixed.

Remark that it is not restrictive to assume that (2.1) has feasible solutions, whatever $y \in Y$ may be. In fact, as r is a nonnegative vector, it is enough to assume that $A(y)$ has no null row, i.e. that each period may be covered by at least a shift.

We can now apply Theorem 1 of Section 1.

To this aim let us consider the dual of problem (2.2), i.e. the problem

$$(2.3) \qquad Q(y): \begin{cases} \max \ (ur) \\ uA(y) \leqq e \\ u \geqq 0, \end{cases}$$

where u is a 48-row-vector.

Theorem 1 becomes now.

Theorem 2. *Let $\bar{u}$ and $\bar{z}$ be respectively an optimal solution and the maximum of the dual of the problem (2.2), i.e. of the problem (2.3), at $y=\bar{y}\in Y$. Then, to every $y\in Y$ which satisfies*

$$\bar{u}A(y) \leqq e$$

does not correspond in (2.2) a value $z<\bar{z}$, whatever the feasible solution x of (2.2) may be.

A decomposition procedure for solving (2.1) can now be devised. First of all a $\bar{y}\in Y$ is fixed and (2.3) is solved, Theorem 2 is applied and a nonempty subset of Y is cut off. In the remaining part of Y, a new $\bar{y}$ is selected and the procedure is repeated, until an optimal solution of (2.1), without the integer requirements on x, is reached.

If this solution is integer, as it often happens, the problem (2.1) is completely solved; but if the solution is not integer, the following question must be answered: the $\bar{y}$ optimal for the real problem is optimal for the integer problem too or not?

If the answer were positive, it would be enough to solve the integer problem corresponding to the $\bar{y}$ optimal for the real problem.

Remark that, if the previous conditions, to which the starting times of the shift are subjected, are replaced by the following ones:

—each ordinary shift must begin at the end of the previous one*;
—the supplementary shifts may start at any period but not at the starting times of the ordinary shifts and each period may be covered by one supplementary shift at most,

then the matrix $A(y)$ is totally unimodular, whatever $y\in Y$ may be. In fact each row of $A(y)$ has two elements equal to 1 at most and the others equal to 0 and the columns may be divided into two sets I_1 and I_2, so that I_1 is formed by the three first columns corresponding to the ordinary shifts and I_2 is formed by the last two columns corresponding to the two supplementary shifts.

The two sets are such that, if in a row there are two 1, they are found in two columns, one of which belongs to the set I_1 (ordinary shifts) and the other one to the set I_2 (supplementary shifts).

By a well-known theorem it follows that $A(y)$ is totally unimodular and thus that the integer requirements on x may be disregarded in the problem (2.1).

---

* The number of periods of the shifts is now 16.

3. DECOMPOSITION OF THE TRAVELLING SALESMAN PROBLEM

Consider a connected and directed graph G, which can be assumed to be complete*. Let G have $n+1$ nodes, say

$$N_0, N_1, \ldots, N_n.$$

To every arc (i, j) of G a nonnegative real number (distance) is associated, call it c_{ij}, $i, j = 0, 1, \ldots, n$; call $C = (c_{ij})$ the distance matrix.

The travelling salesman problem (shortly, T.S.P.) we are considering consists in determining a circuit through the $n+1$ nodes (towns), having the minimum total distance, i.e. the sum of the distances of the arcs of the circuit. For the preceding problem it is easy to determine a feasible solution. A simple way is given by the "rule of the nearest unvisited node", according to which N_0 (or any other) is chosen as first node; a node N_{i_1}, having a minimum distance from N_0 among $N_1, \ldots, N_n$, as second node; N_{i_2} having a minimum distance from N_{i_1}, among the nodes not yet chosen, as third node, and so on.

Instead of considering c_{ij} as distances, we may consider

$$\bar{c}_{ij} \triangleq c_{ij} + c_{j0} - c_{i0}; \quad i = 0, 1, \ldots, n; \quad j = 1, \ldots, n$$
$$\bar{c}_{i0} \triangleq c_{i0} \quad i = 0, 1, \ldots, n$$

if we want to determine a "good" feasible solution.

When a T.S.P. is formulated as a linear assignment problem, it is well known that to an optimal solution of the latter does not necessarily correspond an Hamiltonian circuit of G. Now it will be shown that, by suitably modifying the objective function, an Hamiltonian circuit can always be obtained for an optimal solution.

Consider the binary values $\alpha_{ij} \in \{0, 1\} \triangleq \mathbf{B}$, $1 \leq i \neq j \leq n$, associated with the arcs, which are not loops; set $\alpha = (\alpha_{12}, \ldots, \alpha_{1n}, \ldots, \alpha_{n, n-1})$. Let M be a large enough real, and consider the generalized distances**

$$\hat{c}_{ij} = \hat{c}_{ij}(\alpha_{ij}) \triangleq \begin{cases} c_{ij} + (1 - \alpha_{ij})M, & \text{if } 1 \leq i \neq j \leq n \\ M, & \text{if } i = j = 0, 1, \ldots, n \\ c_{ij}, & \text{otherwise.} \end{cases}$$

Consider the problem

(3.1a)
$$\min \left\{ z(\alpha, x) = \sum_{i, j=0}^{n} \hat{c}_{ij}(\alpha_{ij}) x_{ij} \right\}$$

(3.1b)
$$\sum_{j=0}^{n} x_{ij} = 1, \quad i = 0, 1, \ldots, n$$

---

* When G is not complete, the non-existing arcs may be defined to have a lenght equal to that of the shortest path going from the initial node to the final one.

** It is assumed that G has no loops. Trivial variants are necessary when there are loops.

(3.1c)
$$\sum_{i=0}^{n} x_{ij}=1, \quad j=0, 1, \ldots, n$$

(3.1d)
$$x_{ij}\geqq 0, \qquad i, j=0, 1, \ldots, n.$$

Call it $P(\alpha)$ and $z(\alpha)$ and $x(\alpha)$ respectively the corresponding minimum and an optimal solution. Denote by R the feasible region of $P(\alpha)$, i.e. the polyhedron (3.1b, c, d).

Remark that an optimal solution of $P(\alpha)$ has $n+1$ elements equal to 1, and $n-1$ of them are x_{ij} having $1\leqq i\neq j\leqq n$. Thus, if $\alpha\in \mathbf{B}^{n(n-1)}$ has been chosen having more than $(n-1)^2$ elements equal to zero, then for at least a couple (i, j), with $1\leqq i\neq j\leqq n$, we have $\alpha_{ij}=0$ and in an optimal solution of $P(\alpha)$, x_{ij} is equal to 1, so that the minimum in $P(\alpha)$ is greater than or equal to M, and it does not correspond to a total distance of a path of G. For this reason we assume that α satisfies the constraint*

(3.2)
$$\sum_{1\leqq i\neq j\leqq n} \alpha_{ij}=n-1.$$

Even if (3.2) is satisfied, there is not necessarily a feasible solution of problem (3.1) with total distance less than M. In fact, if there are exactly $n-1$ of the α_{ij} equal to 1, and if they have the same row index or the same column index, then there is no feasible solution of (3.1) with total distance less than M. For this reason we are lead to consider the set of all $\alpha\in \mathbf{B}^{n(n-1)}$ which satisfy (3.2) and such that at least $n-1$ of the α_{ij}, which satisfy (3.2), correspond to intersections of different rows and different columns:

$$\alpha_{i_1 j_1}=\alpha_{i_2 j_2}=\ldots=\alpha_{i_{n-1} j_{n-1}}=1$$
$$i_{s+1}=j_s; \quad j_s\neq i_1, i_2, \ldots, i_s; \quad s=1, 2, \ldots, n-1.$$

Call this set H_1.

The following theorem holds

Theorem 3. $P(\alpha)$ *has feasible solutions with total distance less than* M, *iff* $\alpha\in H_1$.

Unfortunately, the set H_1 does not characterize the problems $P(\alpha)$ having optimal solutions corresponding to Hamiltonian circuits. To obtain this we consider the reals μ_i, $i=1, \ldots, n$, associated with the nodes $N_1, \ldots, N_n$ and we set

$$\mu(\alpha)\triangleq \left\{\mu\in R^n: \frac{n-1-(\mu_i-\mu_j)}{n}\geqq 0, \ 1\leqq i\neq j\leqq n; \text{ and } \frac{n-1-(\mu_i-\mu_j)}{n}\geqq 1, \text{ if } \alpha_{ij}=1\right\},$$

where $\mu=(\mu_1, \ldots, \mu_n)$.

Consider now the set of all $\alpha\in \mathbf{B}^{n(n-1)}$ and such that

(3.3)
$$\mu(\alpha)\neq\emptyset.$$

Call this set H_2.

The following theorem holds

---

* We assume that $n>2$.

Theorem 4. *An Hamiltonian circuit of G corresponds to every optimal solution of* $P(\alpha)$, *iff* $\alpha \in H_1 \cap H_2$.

By the preceding theorem, we can state that by solving (3.1) for every $\alpha \in H = H_1 \cap H_2$, i.e. by solving the problem

$$(3.4) \qquad\qquad \min_{\alpha \in H} \min_{x \in R} z(\alpha, x)$$

one obtains the minimum and an optimal solution of T.S.P.

As $P(\alpha)$ is a particular case of problem (1.1) of Section 1 and as its feasible region $R \neq \emptyset$, we can apply Theorem 1 to obtain a decomposition of $P(\alpha)$.

To this aim let us consider the dual of problem $P(\bar{a})$, $\bar{a} \in H$, i.e. the problem

$$(3.5) \qquad\qquad \max_{u A^r + v A^c \leq \hat{c}(\bar{a})} (u+v)e,$$

where A^r and A^c are, respectively, the matrices of the coefficients of the unknowns of equations (3.1b) and (3.1c); $\hat{c}(\bar{a})$ is the row-vector of the $\hat{c}_{ij}(\bar{a}_{ij})$; e a column-vector, whose elements are equal to 1; u and v two row-vectors of unknowns.

Theorem 1 becomes now

Theorem 5. *Let* $(\bar{u}, \bar{v})$ *and* $\bar{z}$ *be respectively an optimal solution and the maximum in the dual of the problem* $P(\bar{a})$, $\bar{a} \in H$, *i.e. in the problem (3.5). Then, to every* $\alpha \in H$ *and such that*

$$(3.6) \qquad\qquad \bar{u} A^r + \bar{v} A^c \leq \hat{c}(\alpha)$$

does not correspond in $P(\alpha)$ *a minimum* $z(\alpha) < \bar{z}$.

By this theorem we are now able to solve problem (3.4) by a decomposition procedure.

Let $x(\alpha)$ be an optimal solution of $P(\alpha)$. If every $\alpha \in H$ satisfies (3.6), then $x(\bar{a})$ is an optimal solution of T.S.P. too. If there is any $\tilde{\alpha} \in H$, which does not satisfy (3.6), then we can solve $P(\tilde{\alpha})$ and apply again Theorem 5 until every $\alpha \in H$ satisfies at least one of the systems (3.6). As every time Theorem 5 is applied, at least an $\alpha \in H$ is cut off, and, as H has a finite number of elements, the described algorithm is finite.

Remark now that, if γ_{ij} denotes the generic element of vector $\gamma = u A^r + v A^c$, (3.6) may equivalently be written this way

$$(3.7) \qquad\qquad \alpha_{ij} \leq 1 + \frac{c_{ij} - \bar{\gamma}_{ij}}{M}, \quad 1 \leq i \neq j \leq n.$$

Remark also that the preceding results can be obtained directly by a simple reasoning. However, the decomposition based on Theorem 5 is necessary if we want to obtain a lower bound at every step.

432

4. DECOMPOSITION OF A LINEAR COMPLEMENTARITY PROBLEM

Consider the following linear complementarity problem

$$\text{(4.1a)} \qquad \min \left[z(x) = c_{\mathrm{I}} x_{\mathrm{I}} + c_{\mathrm{II}} x_{\mathrm{II}} \right]$$

under the constraints

$$\text{(4.1b)} \qquad A_{\mathrm{I}} x_{\mathrm{I}} + A_{\mathrm{II}} x_{\mathrm{II}} = a$$

$$\text{(4.1c)} \qquad x_{\mathrm{I}},\, x_{\mathrm{II}} \geqq 0$$

$$\text{(4.1d)} \qquad x_{\mathrm{I}}^{T} x_{\mathrm{II}} = 0,$$

where A_{I}, A_{II} are square matrices of order m; c_{I}, c_{II} m-row-vectors; a, x_{I}, x_{II} m-column-vectors and where x_{I} and x_{II} are unknown. Call Q the problem (4.1).

Remark that the inner product (4.1d) implicates all the variables of Q; this is not necessary for the validity of what will be said.

Put

$$x = \begin{pmatrix} x_{\mathrm{I}} \\ x_{\mathrm{II}} \end{pmatrix}; \quad c = (c_{\mathrm{I}}, c_{\mathrm{II}}); \quad A = (A_{\mathrm{I}}, A_{\mathrm{II}}); \quad n = 2m;$$

$$R = \{x \in R^{n} :\ Ax = a;\ x \geqq 0;\ x_{\mathrm{I}}^{T} x_{\mathrm{II}} = 0\};$$

$$X = \{x \in R^{n} :\ Ax = a;\ x \geqq 0\};$$

$$B = \{0, 1\}; \quad \alpha = (\alpha_{1}, \ldots, \alpha_{m}) \in \mathbf{B}^{m}.$$

Now an equivalent formulation of Q will be defined, which enables one to decompose it into a finite sequence of linear programs. To this aim call M a real large enough number and put

$$c_{j}(\alpha_{j}) = \begin{cases} c_{j} + (1 - \alpha_{j})M, & \text{if } j' = 1, \ldots, m \\ c_{j} + \alpha_{j-m}M, & \text{if } j = m+1, \ldots, n, \end{cases}$$

$$c_{\mathrm{I}}(\alpha) = \left(c_{1}(\alpha_{1}), \ldots, c_{m}(\alpha_{m}) \right); \quad c_{\mathrm{II}}(\alpha) = \left(c_{m+1}(\alpha_{1}), \ldots, c_{n}(\alpha_{m}) \right)$$

$$c_{\mathrm{I}}(\alpha) = c_{\mathrm{I}} + (e - \alpha)M; \quad c_{\mathrm{II}}(\alpha) = c_{\mathrm{II}} + \alpha M; \quad c(\alpha) = \left(c_{\mathrm{I}}(\alpha), c_{\mathrm{II}}(\alpha) \right).$$

Consider the linear program

$$Q(\alpha) : \begin{cases} \min \left[z(x, \alpha) = c_{\mathrm{I}}(\alpha) x_{\mathrm{I}} + c_{\mathrm{II}}(\alpha) x_{\mathrm{II}} \right] \\ Ax = a \\ x \geqq 0. \end{cases}$$

The first natural question consists in determining when an element of R corresponds to an optimal solution of $Q(\alpha)$. To this aim consider the square matrix $B(\alpha)$ of order m, defined this way: $\forall j = 1, \ldots, m$ consider either the j-th column of A_{I} or the j-th column of A_{II} respectively according to either $\alpha_{j} = 0$ or $\alpha_{j} = 1$. Let y be a m-column vector; consider the sets

$$H = \left\{ \alpha \in \mathbf{B}^{m} : \{ y \in \mathbf{R}^{m} : B(\alpha)y = a;\ y \geqq 0 \} \neq \emptyset \right\};$$

$$J^{0}(\alpha) = \{ j = 1, \ldots, m : \alpha_{j} = 0 \}; \quad J^{1}(\alpha) = \{ j = 1, \ldots, m : \alpha_{j} = 1 \}.$$

It is easy to show [4] the following

Theorem 6. *Let X be nonempty. The optimal solutions x^* of Q, and the equality $z(x^*, \alpha) = z(x^*)$ holds, iff $\alpha \in H$.*

Theorem 7. *If $R \neq \emptyset$, then*

$$(4.2) \qquad \min_{x \in R} z(x) = \min_{\alpha \in H} \min_{x \in X} z(x, \alpha).$$

Remark that, if $X \neq \emptyset$, if $\alpha \in H$, and if det $B(\alpha) \neq 0$, then $Q(\alpha)$ has a trivial optimal solution, defined this way: it is $x_j = 0$ in correspondence of $\alpha_j = 0$; all the remaining x_j are given by the elements of the vector $B(\alpha)^{-1} \cdot a$.

We are n w ready to state a condition for an optimal solution of $Q(\alpha)$ to be an optimal solution of (4.1).

Assume $X \neq \emptyset$. Let $\bar{a} \in H$; and let $x(\bar{a})$ be an optimal solution of $Q(\bar{a})$ and $\bar{z} = z(x(\bar{a}),$ $\bar{a} = z(x(\bar{a}))$ the corresponding minimum.

We have

$$(4.3) \qquad \min_{x \in R} z(x) \leqq \bar{z},$$

and we want now to determine a subset of $\mathbf{B}^m$, to which corresponds a minimum of $Q(\alpha)$ greater than or equal to $\bar{z}$.

To this aim let us consider the dual of problem $Q(\alpha)$, i.e. the problem

$$(4.4) \qquad \begin{array}{c} \max ua \\ uA \leqq c(\bar{\alpha}), \end{array}$$

where u is a row-vector of unknowns.

Theorem 1 becomes now

Theorem 8. *Let $\bar{u}$ and $\bar{z}$ be respectively an optimal solution and the maximum in the dual of the problem $Q(\bar{a})$, i.e. in the problem (4.4).*

Then, to every $\alpha \in \mathbf{B}^m$, which satisfies the linear system

$$(4.5) \qquad \bar{u}A \leqq c(\alpha)$$

does not correspond a minimum of $Q(\alpha)$ less than $\bar{z}$.

Remark that (4.5) is equivalent to

$$(4.6) \qquad \frac{1}{M} \{\bar{u}A\}_{m+j} - \frac{c_{m+j}}{M} \leqq \alpha_j \leqq 1 + \frac{c_j}{M} - \frac{1}{M} \{\bar{u}A\}_j, \quad j = 1, 2, \ldots, m.$$

It follows that the set of α considered in the preceding theorem is the set of vertices of a face of the convex hull of $\mathbf{B}^m$, i.e. conv. $\mathbf{B}^m$.

Remark that (4.6) is possible on $\mathbf{B}^m$, as it is verified at least by $\alpha = \bar{a} \in \mathbf{B}^m$; thus,

$$c_j \langle \{\bar{u}A\}_j \Rightarrow \{\bar{u}A\}_{m+j};$$
$$c_{m+j} \langle \{\bar{u}A\}_{m+j} \Rightarrow \{\bar{u}A\}_j.$$

Then, if $\bar{u}A \leqq c$, it follows that $x(\bar{a})$ is an optimal solution of Q too. Let h be the number of negative elements of the vector $c - \bar{u}A$. The cut is of the kind

$$(4.7) \qquad\qquad \alpha_{i_1} = \delta_{i_1}; \; \ldots; \; \alpha_{i_h} = \delta_{i_h},$$

where $\delta_i \in \mathbf{B}$. (4.7) defines a $(m-h)$-dimensional face. It follows that the number of elements of conv. $\mathbf{B}^m$, which are cut off, is 2^{m-h}.

As $\bar{u}$ is an optimal solution of the dual of $Q(\bar{a})$, i.e. of the problem (4.4), we have

$$\bar{u} = c_B A_B^{-1},$$

where A_B is an optimal basis $Q(\bar{a})$, and c_B the subvector of c corresponding to the basic variables. (4.6) becomes now

$$\frac{1}{M}\{c_B A_B^{-1} A\}_{m+j} - \frac{c_{m+j}}{M} \leqq \alpha_j \leqq 1 + \frac{c_j}{M} - \frac{1}{M}\{c_B A_B^{-1} A\}_j.$$

If $\det B(\alpha) \neq 0$, then $A_B = B(\alpha)$. Now it naturally arises the problem of determining α, such that the number of negative elements of the vector $c - uA = c - c_B A_B = c - c_B B(\alpha)^{-1} A$ be a minimum.

A decomposition procedure for solving (4.1) can now be defined in the usual way.

The preceding procedure has been applied [4] to a particular case arising in the field of structural mechanics, which may be formulated as linear complementarity problem in the unknowns x, w, z:

$$(4.8) \qquad
\begin{aligned}
&\min ex \\
a \leqq\; &Cz - Cx \leqq b \\
&w = q + Mz - Mx \\
w \geqq 0,\; &z \geqq 0,\; x \geqq 0 \\
&w^T z = 0,
\end{aligned}$$

where C, M are square matrices of order m; and the vectors are defined according to the indicated matrix operations. Problem (4.8) can be put in the form (4.1), so that the preceding decomposition procedure may be applied.

Remark now that the preceding decomposition procedure is guided by the parameter α. To fix $\alpha \in H$ is equivalent to put m variables at zero level, in such a way that (4.1bcd) has solutions. This is not the only way to decompose the problem. For instance one can consider a set of m variables as the parameter for the decomposition procedure, and fix it at a nonnegative level, in such a way that the constraints (4.1bcd) have solutions on respect to the remaining set of m variables. This will be now explained by means of the particular case (4.8).

To this aim assume $\det M \neq 0$. Then, the 2-nd set of constraints of (4.8) becomes

$$(4.9) \qquad\qquad x = M^{-1}q + z - M^{-1}w,$$

so that (4.8) becomes

$$
(4.10) \qquad
\begin{cases}
eM^{-1}q + \min\,(ez - eM^{-1}w) \\
\quad -z + M^{-1}w \le M^{-1}q \\
\quad CM^{-1}w \le b + CM^{-1}q \\
\quad CM^{-1}w \ge a + CM^{-1}q \\
\quad z \ge 0, \quad w \ge 0 \\
\quad z^{T}w = 0
\end{cases}
$$

If $z > 0$ is fixed, then (4.10) becomes a linear program

$$
\mathcal{P}(z):
\begin{cases}
eM^{-1}q + ez + \min\,(-eM^{-1}w) \\
\quad M^{-1}w \le M^{-1}q + z \\
\quad CM^{-1}w \le b + CM^{-1}q \\
\quad CM^{-1}w \ge a + CM^{-1}q \\
\quad w_j = 0, \quad j \in J(z) \\
\quad w \ge 0
\end{cases}
$$

where

$$
J(z) \triangleq \{j : z_j > 0\}.
$$

Assume that $\bar{z} \ge 0$ has feen fixed in such a way that $\mathcal{P}(\bar{z})$ has optimal solutions; let $\bar{w}$ be one of them; and let $\bar{k}$ be the corresponding minimum, that is

$$
\bar{k} \triangleq eM^{-1}q + e\bar{z} - eM^{-1}\bar{w}.
$$

Consider now the dual of $\mathcal{P}(\bar{z})$, i.e. the problem

$$
(4.11) \qquad
\begin{cases}
eM^{-1}q + e\bar{z} + \max\,[-u(M^{-1}q + \bar{z}) - \mu(b + CM^{-1}q) + v(a + CM^{-1}q)] \\
\quad uM^{-1} + \mu CM^{-1} - vCM^{-1} + \varDelta z \ge + eM^{-1} \\
\quad u, \mu, v \ge 0
\end{cases}
$$

where

$$
\varDelta =
\begin{pmatrix}
\delta_1 & & & 0 \\
& \ddots & & \\
& & \ddots & \\
0 & & & \dot{\delta}
\end{pmatrix}
$$

and

$$
\delta_j = 0, \quad \text{if } j \notin J(\bar{z}), \quad \delta_j = 1 \quad \text{if } j \in J(\bar{z}).
$$

Theorem 8 becomes here

Theorem 9. *Let* $\bar{\xi} = (\bar{u}, \bar{\mu}, \bar{v}, \bar{\tau})$ *and* $\bar{k}$ *be respectively an optimal solution and the maximum in the dual of* $\mathcal{P}(\bar{z})$*, i.e. in the problem (4.11). Then, to every solution z of the system*

$$
\begin{cases}
(u - e)z \le eM^{-1}q - \bar{k} - [\bar{u} + (\bar{\mu} - \bar{v})C]M^{-1}q - \bar{\mu}b + \bar{v}a \\
z \ge 0
\end{cases}
$$

does not correspond a minimum of $\mathcal{P}(z)$ *less than* $\bar{k}$*.*

5. A THEOREM OF ALTERNATIVE FOR NONLINEAR SYSTEMS

In this section a theorem of alternative of the Farkas type for nonlinear systems is studied and it is shown how to apply it to the decomposition of nonlinear programs. A numerical example explains such an application.

Theorem 10. *Let f be an l-vector function and g an m-vector function, both real-valued and defined on a set $C \subseteq \mathbf{R}^n$. Then, the system*

$$(5.1) \qquad \begin{cases} f(x) > 0 \\ g(x) \geqq 0 \\ \bar{x} \in C \end{cases}$$

is impossible, iff there exist an l-vector function λ and an m-vector function μ, both real-valued and defined on C, with $\lambda(x) \geqq 0$, $\mu(x) \geqq 0$, $(\lambda(x), \mu(x)) \neq 0$, $\forall x \in C$, such that

$$(5.2) \qquad \lambda(y)f(y) + \mu(y)g(y) \leqq 0, \quad \forall y \in C,$$

where the inequality must be strongly verified $\forall y \in C$, such that $\lambda(y) = 0$.

Proof. Both systems (5.1) and (5.2) cannot have solutions. In fact, if $\bar{x}$ is a solution of (5.1), then, if $\lambda(\bar{x}) \neq 0$, $\lambda(\bar{x}) \geqq 0$ and $f(\bar{x}) > 0$ imply $\lambda(\bar{x})f(\bar{x}) > 0$; hence, as $\mu(\bar{x}) \geqq 0$ and $g(\bar{x}) \geqq 0$ imply $\mu(\bar{x})g(\bar{x}) \geqq 0$, it follows $\lambda(\bar{x})f(\bar{x}) + \mu(\bar{x})g(\bar{x}) > 0$ which contradicts (5.2). If $\lambda(\bar{x}) = 0$, (5.2) implies $\mu(\bar{x})g(\bar{x}) < 0$, while $\mu(\bar{x}) \geqq 0$ and $g(\bar{x}) \geqq 0$ imply $\mu(\bar{x})g(\bar{x}) \geqq 0$; a contradiction is again obtained. To complete the proof it is enough to show that the feasibility of (5.1) implies the impossibility of (5.2). To this end consider the sets

$$H = \{(y^l, y^m) : y^l > 0; y^m > 0\};$$
$$K(x) = \{(y^l, y^m) : y^l \leqq f(x); y^m \leqq g(x)\};$$

where y^l and y^m are real vectors having l and m elements, respectively. The impossibility of (5.1) implies $K(x) \cap H = \emptyset$, $\forall x \in C$. Then, the convexity of $K(x)$ and H and a well-known theorem of separation for convex sets imply the existence of a closed half-space containing $K(x)$, while H is contained in the opposite (open) half-space. Hence, there exist an l-vector λ and a m-vector μ, with $\lambda \geqq 0$, $\mu \geqq 0$, $(\lambda, \mu) \neq 0$, such that, $\forall (y^l, y^m) \in K(x)$, we have $\lambda y^l + \mu y^m \leqq 0$ if $\lambda \neq 0$, or $\mu y^m < 0$ if $\lambda = 0$ (as H is open at $y^m = 0$). As no assumption has been made on C, it follows that there exist functions $\lambda(x)$ and $\mu(x)$ such that (5.2) is fulfilled. This completes the proof.

Remark. When f and g are concave and C is convex, then function $\lambda(x)$ and $\mu(x)$ can be assumed to be constant with respect to x, as it is easily seen by looking at the second part of the above proof. In this case Theorem 10 becomes a well-known statement.

437

By means of a numerical example it will be shown how to use the preceding theorem for decomposing a nonlinear programming problem.

Consider the following mixed-integer quadratic program:

$$(5.3a) \qquad \min \; (z = x_1^2 + x_2^2 + y_1 + y_2)$$

under the constraints

$$(5.3b) \qquad \begin{cases} 2x_1 + x_2 - y_1 - 2y_2 \geq 1 \\ x_1 + 2x_2 - 2y_1 - y_2 \geq 1 \\ x_1, \, x_2 \geq 0 \\ y_1, \, y_2 \geq 0 \text{ and integer.} \end{cases}$$

Put $\bar{y} = (\bar{y}_1 = \bar{y}_2 = 0)$. Then, (5.3) becomes a (real) convex quadratic program, whose optimal solution is

$$\bar{x} = \left(\bar{x}_1 = \bar{x}_2 = \frac{1}{3} \right); \quad \bar{z} = \frac{2}{9}.$$

We want to know if $(\bar{x}, \bar{y})$ is an optimal solution of (5.3) or not. More precisely, we want to know if there exist y_s, which imply, for the corresponding convex quadratic program, a minimum less than $\bar{z}$. This does not happen, iff the system

$$(5.4) \qquad \begin{cases} f(x) \triangleq \dfrac{2}{9} - (x_1^2 + x_2^2 + y_1 + y_2) > 0 \\ g_1(x) \triangleq 2x_1 + x_2 - y_1 - 2y_2 - 1 \geq 0 \\ g_2(x) \triangleq x_1 + 2x_2 - 2y_1 - y_2 - 7 \geq 0 \\ x \in C \triangleq R_+^2 \cap \{ x \in R^2 : \|x\| \leq \gamma \} \end{cases}$$

is impossible (where γ is positive and large enough).

As the assumption of Theorem 10 is satisfied, such a theorem can be applied to the present case (5.4) is identified with (5.1); then (5.2) becomes

$$(5.5) \qquad \begin{cases} (2\mu_1 + \mu_2)x_1 - \lambda x_1^2 + (\mu_1 + 2\mu_2)x_2 - \lambda x_2^2 + \left[\dfrac{2}{9}\lambda - \mu_1 - \mu_2 - (\lambda + \mu_1 + 2\mu_2)y_1 - \right. \\ \left. \qquad\qquad\qquad - (\lambda + 2\mu_1 + \mu_2)y_2 \right] \leq 0, \quad \forall x \in C \\ \lambda, \, \mu_1, \, \mu_2 \geq 0; \; (\lambda, \mu_1, \mu_2) \neq 0 \end{cases}$$

where λ, μ_1, μ_2 can be assumed to be constant as f, g_1, g_2 are now concave and C is convex.

Now we have to state if (5.5) has solution or not. To this aim consider first the case $\lambda > 0$, where the quadratic form, which appears in (5.5), results to be negative definite. Its maximum value is obtained at the point

$$\left(x_1 = \frac{2\mu_1 + \mu_2}{2\lambda}, \quad x_2 = \frac{\mu_1 + 2\mu_2}{2\lambda} \right),$$

438

and it is equal to

$$\frac{(2\mu_1+\mu_2)^2}{4\lambda}+\frac{(\mu_1+2\mu_2)^2}{4\lambda}+\frac{2}{9}\lambda-\mu_1-\mu_2=$$

$$=\frac{5}{4\lambda}\mu_1^2+\frac{2}{\lambda}\mu_1\mu_2+\frac{5}{4\lambda}\mu_2^2-\mu_1-\mu_2+\frac{2}{9}\lambda.$$

The last expression is in turn a positive definite quadratic form in μ_1, μ_2, whose minimum value is attained at the point

$$\left(\mu_1=\mu_2=\frac{2\lambda}{9}\right)$$

and it equals zero. It follows that (5.5) has solutions.

Now it is not necessary to analyze the case $\lambda=0$. As we have shown that (5.5) has solutions without any assumption on y_1, y_2, it follows that $(\bar{x}, \bar{y})$ is an optimal solution of (5.3).

REFERENCES

[1] Benders, J. F.: Partitioning procedures for solving mixed-variables programming problems, *Numerische Mathematik* 4 (1962), 238–252.

[2] Geoffrion, A. M.: Generalized Benders decomposition. Proceedings of the Symposium on Nonlinear Programming, Mathematics Research Center, University of Wisconsin, Madison, May 4–6 1970.

[3] Giannessi, F.: A decomposition method for the travelling salesman problem and its applications, Paper of the Dept. of Operations Research, Univ. of Pisa, 1976 (to appear).

[4] Giannessi, F. Jurina, L. and Maier, G.: Optimal excavation profile for a pipeline freely resting on the seafloor, *Engineering Structures* 1 (1979), January.

[5] Khanna: Optimal staffing at airline terminals, (private comunication).

[6] Mangasarian, O. L.: *Nonlinear programming*, McGraw-Hill Co., New York, 1969.

[7] Natali, L., Nicoletti, B. and Scalas, M. R.: Optimal staffing at an airport: a generalized set covering problem, Paper of the Dept. of Operations Research, Univ. of Pisa, 1976.

MAXIMUM TRANSITIVE PATHS
AND THEIR APPLICATION TO A GEOLOGICAL PROBLEM:
SETTING UP STRATIGRAPHIC UNITS

I. DIENES and L. B. KOVÁCS

(Budapest, Hungary)

1. INTRODUCTION

The present paper is mainly devoted to the examination and determination of maximal transitive paths of a given directed graph. The notion of transitive path is a counterpart of a complete subgraph of an undirected graph. The use of this new notion has become necessary in connection with our geological research shortly outlined in Section 2. The mathematical formulation of our basic problem is followed by the basic properties of transitive paths. Sections 5–7 consist of further properties and some procedures promoting the solution of the basic problem, i.e. a relaxation by maximal complete subgraphs, introducing the connection numbers for the pair of vertices as a kind of "likeness being together in a good solution" and a special decomposition for large problems. Section 8 shows the exact (heuristic) filtering for eliminating vertices which cannot (not very likely to) be in any optimal solution. Section 9 is a short outline of the branch-and-bound procedure used for solving the basic problem. Finally, Section 10 is a short summary of the goals and the present experiments with the data of Dorog basin.

The relaxation of our basic problem is the determining of maximum complete subgraphs of given undirected graph (see Section 6) which may be solved directly [3, 4] or as a set covering problem [2, 10, 12, 13, 14, 15, 16]. It should be mentioned however, that many procedures and ideas, described in the present paper, may also be used to accelerate the solution procedure for many well-structured large scale discrete problems, e.g., for the complete subgraph problem itself, especially if only the maximum element solutions are needed.

2. DESCRIPTION OF THE GEOLOGICAL PROBLEM

The knowledge of extent of rock bodies is a prerequisite for the solution of the majority of geological problems, e.g.:

—the estimation of tonnage of ore reserves

—determination of approximate position of tectonic faults, limiting mining operations
—production control.

Formalized stratigraphy provides a good way for the exact formulation of particular problems in this area. The problem of finding a maximum set of ordered rock bodies plays an important role in the above-mentioned research and also in other geological tasks [7]. This problem may be described as follows:

Any (suitably defined) part of the Earth is called rock body. A set of rock bodies is called stratigraphic unit of first order. There are several precedence relations among rock bodies. In the present research we use the following one:

A rock body T_i precedes a rock body T_j in investigation area T in sense "32" (serial number of ordering relation) if

$$(*) \qquad\qquad T \cap T_i \cap T \neq \emptyset$$

and

$$P_{x,y}(T) = \{(x, y) \mid \exists z : (x, y, z) \in T\}$$

$$\inf \{z \mid (x, y, z) \in T_i\} \geq \sup \{z \mid (x, y, z) \in T_j\}$$

for any

$$(x, y) \in P_{x,y}(T \cap T_i \cap T_j).$$

This precedence is denoted by T_i "32" T_j.

If the relation $(*)$ does not hold, then T_i and T_j are not comparable in the investigation area T.

Assume, that a set of geological bodies $\{T_i\}_{i \in I}$ have been defined. Then our aim is to find the maximum number of ordered rock bodies, i.e., a set $J \subset I$ of maximum number of elements that is ordered in the following sense:

Let us denote the ordering defined on the set J by ϱ. Then for any $i \neq j$, $i, j \in J$ the relation $i\varrho j$ should imply, that either T_i "32" T_j or T_i and T_j not comparable according to the relation "32".

Notes. 1. Any other relation may also be used instead of "32".

2. As the investigation area is known only in the bore-holes already examined, in the above definition area T may be substituted by the union of bore-holes:

3. MATHEMATICAL FORMULATION OF THE PROBLEM

It is given an $n \times n$ prohibition matrix D, the elements of which

$$d_{ij} = 0 \quad \text{or} \quad 1 \quad \text{for any} \quad 1 \leq i \leq n, \quad 1 \leq j \leq n$$
$$d_{jj} = 0 \qquad\qquad \text{for any} \quad 1 \leq j \leq n$$

Problem i. The longest ordered series of positive integers (subscripts)

$$(j_1, j_2, \ldots, j_k)$$

is to be found, subject to the following conditions:

$$(1a) \qquad j_1, j_2, \ldots, j_k \in \{1, 2, \ldots, n\}$$
$$(1b) \qquad j_r \neq j_s \quad \text{if} \quad r \neq s$$
$$(1c) \qquad d_{j_r j_s} = 1 \quad \text{for any} \quad r, s : 1 \leq r < s \leq k.$$

Note. $d_{ij} = 1$ if the supposition, that rock body i precedes rock body j, is not contradictory to the geological data obtained from the bore-holes (i.e. either rock body i precedes rock body j in each of the common boreholes or they have no common bore-holes at all), otherwise $d_{ij} = 0$.

Problem i may be reformulated as a graph theoretic problem if a new notion is introduced.

Let us consider a directed graph

$$G = (V, E),$$

where

$V = \{v_1, v_2, \ldots, v_n\}$ is the set of vertices and
$E = \{e_1, e_2, \ldots, e_m\}$ is the set of directed edges. (arcs)

As usually denoted in the literature an *elementary path* of a directed graph G is a set of connecting arcs

$$((v_{j_1}, v_{j_2}), (v_{j_2}, v_{j_3}), \ldots, (v_{j_{k-1}}, v_{j_k})),$$

where

$$v_{j_1}, \ldots, v_{j_k} \in V : (v_{j_1}, v_{j_2}), \ldots, (v_{j_{k-1}}, v_{j_k}) \in E,$$

such that the vertices

$$v_{j_1}, \ldots, v_{j_k}$$

are all different. The number of vertices; k is called the *v-length* of the path. The elementary path may be simply denoted by the ordered set of edges it goes through:

$$(v_{j_1}, v_{j_2}, \ldots, v_{j_k}).$$

Definition 1. An elementary path

$$v_{j_1}, v_{j_2}, \ldots, v_{j_k}$$

of a directed graph G is called *transitive path* if

$$(v_{j_r}, v_{j_s}) \in E$$

for any integers r, s satisfying the inequalities

$$1 \leq r < s \leq k.$$

Definition 2. A transitive path

$$(v_{j_1}, v_{j_2}, \ldots, v_{j_k})$$

of a directed graph G is called *maximal transitive path* if it cannot be augmented, i.e., there exists no transitive path of $k+1$ vertices in graph G containing the above k vertices.

Definition 3. A transitive path

$$v_{j_1}, v_{j_2}, \ldots, v_{j_k}$$

of a directed graph G is called *maximum transitive path* if it contains the maximum number of vertices, i.e., there exists no transitive path of $k+1$ vertices in graph G.

Now we can define.

Problem ii. The following directed graph is given:

$$G = (V, E).$$

A maximum transitive path is to be found in graph G.

Problem ii is equivalent to Problem i if graph G is defined as follows:

$$
\begin{aligned}
(2) \qquad & V = \{v_1, v_2, \ldots, v_n\} \\
& E = \{(v_j, v_k) \,|\, v_j, v_k \in V \text{ and } d_{jk} = 1\}.
\end{aligned}
$$

In the rest of the paper it will be supposed, that the graph G is given by the defining equation (2), thus the solution of this problem will be discussed.

4. BASIC PROPERTIES OF TRANSITIVE PATHS

We often suppose, that transitive paths of v-length L or longer are to be determined. There might be several reasons:

1. Shorter ones are not interesting at all or they are already known from previous calculations.
2. This statement is used as an indirect supposition. (See Lemma 6.)

The following lemmata will give some properties of transitive paths and thus they will provide a good base for algorithms solving Problem ii.

Lemma 1. *No transitive path of v-length L (or longer) can contain vertex v_j if*

$$(3) \qquad \sum_{i=1}^{n} (d_{ij} + d_{ji} - d_{ij}d_{ji}) \leqq L - 2.$$

Proof. The expression

$$d_{ij} + d_{ji} - d_{ij}d_{ji}$$

444

is equal to 1 iff at least one of the numbers d_{ij} and d_{ji} is equal to 1, i.e., vertices v_i and v_j may stand in the same transitive path. In other words, the left hand side of the inequality (3) is the number of vertices which may stand in a transitive path together with vertex v_j and this proves the lemma.

This lemma may be used iteratively for eliminating vertices, because if several vertices were already left out, then it is possible, that inequality (2) will be satisfied also for further vertices.

To state the next lemmata we need to define and calculate some counting numbers for each vertex v_j

$$(4) \qquad b_j = \sum_{i=1}^{n} d_{ij}$$

$$(5) \qquad a_j = \sum_{k=1}^{n} d_{jk}$$

$$(6) \qquad h_j = \sum_{i=1}^{n} (d_{ij} + d_{ji} - d_{ij}d_{ji}).$$

Their meaning is the following: the number of vertices that may stand in a path together with vertex v_j before this element b_j, after this element a_j or anywhere in the path h_j. It should be noted, that

$$(7) \qquad h_j = a_j + b_j - \sum_{i=1}^{n} d_{ij}d_{ji}$$

which may be shown either formally from the defining equations (4)–(6) or thinking over the meaning of numbers a_j, b_j, h_j and that the product $d_{ij}d_{ji}$ is equal to 1 iff vertex v_i may stand both before and after vertex v_j in the same transitive path.

Let us denote the optimum of Problem ii, i.e., the length of maximum transitive path of graph G by $L_{\max}$, Obviously

$$(8) \qquad L_{\max} = 1 + \max_{j} h_j$$

but a much better bound may be obtained by

Lemma 2. *Supposing, that the vertices are indexed such that*

$$h_1 \geqq h_2 \geqq \ldots \geqq h_n$$

and denoting

$$h'_j = \min(j-1, h_j) \quad (j = 1, 2, \ldots, n)$$

if $h_2 \geqq 1$, then the following inequality holds:

$$(9) \qquad L_{\max} \leqq 1 + \max_{j} h'_j.$$

Denote

$$h'_k = \max_{j} h'_j.$$

Let us suppose indirectly, that

$$L_{\max}\geq 2+h'_k.$$

This inequality means, that there is a transitive path containing at least $2+h'_k$ vertices, i.e., there are at least $2+h'_k$ vertices v_j which can stand together with at least $1+h'_k$ other vertices in a path. In other words

$$h_j\geq 1+h'_k\geq k$$

for at least

$$2+h'_k\geq 1+k$$

vertices. According to the ordering of vetices this means, that

$$h_{k+1}\geq k,$$

which implies, that

$$h_{k+1}=k>\min\,(k-1,\,h_k)=h'_k.$$

This inequality is a contradiction to the maximality of h'_k, which proves the lemma.

The following lemma gives a necessary condition for the existence of a transitive L-path:

Lemma 3. *If there exist a transitive path of length L, then there must exist L vertices having count numbers*

$$(b_{j_k},\,a_{j_k})\quad (k=1,2,\ldots,L)$$

such that

(10)
$$b_{j_k}\geq k-1\quad k=1,2,\ldots,L$$
$$a_{j_k}\geq L-k+1.$$

Proof. Considering any transitive path of v-length L the k-th vertex has $k-1$ preceding vertices and $L-k$ vertices which follow it, thus the corresponding counts must exceed $k-1$ and $L-k$ respectively.

Note. In general, it is difficult to check if these necessary conditions are met for a given number L.

It is possible to find weaker necessary conditions which are easier to check, e.g.:

Corollary 1. *A necessary condition for the existence of a transitive path of v-length L is the following: For any integer $k(0\leq k\leq L)$ the sets*

(11)
$$\{v_j\,|\,v_j\in V,\,b_j\geq k\}\quad and \quad \{v_j\,|\,v_j\in V,\,a_j\geq L-k\}$$

must contain at least $L-k$ and k elements respectively.

For obtaining better and better upper bounds for $L_{\max}$, the following lemma is very important. Let us consider an arbitrary set of vertices $S\subset V$, and let us suppose that we have a procedure for obtaining an upper bound $L(S)$ of the v-length of maximum transitive paths on the vertex set S. This maximum length is denoted by $L_{\max}(S)$.

Introducing the notation

$$S_k = \{v_j \mid v_j \in V, h_j \leqq k-1\},$$
(12)

where h_j is defined by the equation (6), we can state.

Lemma 4. *For any positive integer*

$$k < L(S)$$
(13)

if

$$L(S-S_k) < L(S),$$
(14)

then

$$\hat{L}(S) = \max\{k, L(S-S_k)\} = L_{\max}(S).$$
(15)

Proof. Let us suppose (as a conditional constraint—see the note after the proof), that

$$L_{\max}(S) = k+1.$$
(16)

No matter if the supposition (16) holds or not we can draw some conclusions. If the inequality (16) holds, then obviously all vertices which can stand only in a transitive path of v-length k or shorter (defined on the vertex set S) may be left out from further consideration. This results in the new upper bound

$$L(S-S_k) \geqq L_{\max}S.$$
(17)

On the other hand, if the supposition (16) is false, then we obtain, that

$$L_{\max}(S) \leqq k.$$
(18)

In both cases the new upper bound (15) is proved to be correct and this is a better one than $L(S)$ because of the inequalities (13) and (14).

Obviously, this lemma may be applied consecutively for different values of k, choosing a good strategy. We may start with high or low values of k or we may apply a bisection procedure. The choice of appropriate k may depend on the sets S_k.

Note. The idea of conditional constraints is the substance of this lemma and it is a very useful tool in solving large scale structured integer programming problems. It can be shortly described as follows: A constraint which helps to improve the state of the algorithm is added to the constraint set in such a way, that the opposite of the constraint is also helpful. The name of conditional constraint was given by E. Balas [1]. We have discovered the same idea independently and presented it on the same symposium. The idea is closely related to that of Generalized Branch-and-Bound. Balas' idea was formulated for a more general problem. The idea in the present paper is used only for a special problem, but it was applied in a different way. Before solving any of the subproblems a universal bound is obtained for both branches, better than the one obtained before introducing the conditional constraint.

5. RELAXATION

A good relaxation of Problem ii may be obtained in the following way. Let us consider the undirected graph

$$\bar{G}=(V,\bar{E})$$

which is obtained from the graph $G=(V,E)$ defined in Section 2 by disegrading the directions, i.e.,

$$\bar{E}=\{(v_i,v_j)\,|\,(v_i,v_j)\in E \text{ and/or } (v_j,v_i)\in E\},$$

where the edges (v_i,v_j) and (v_j,v_i) are considered identical in the graph $\bar{G}$. Now let us consider a maximal complete subgraph

$$\bar{G}_N=(N,\bar{E}_N)$$

of the graph $\bar{G}=(V,\bar{E})$. This means, that

(i) $\bar{G}_N$ is a subgraph of $\bar{G}$:

$$N\subseteq V, \quad \bar{E}_N=\{(v_i,v_j)\,|\,v_i,v_j\in N,(v_i,v_j)\in\bar{E}\}$$

(ii) $\bar{G}_N$ is a complete graph:
$$\text{for any } v_i,v_j\in N(v_i,v_j)\in\bar{E}_N$$

(iii) $\bar{G}_N$ is a maximal complete subgraph:
$$\text{for any } N\subset M\subseteq V \text{ the subgraph } \bar{G}_M$$
is not complete.

The maximal complete subgraph $\bar{G}_N$ is called *maximum* complete, if

$$M\subseteq V \quad \text{and} \quad |M|>|N|$$

implies, that $\bar{G}_M$ is not complete, i.e., $\bar{G}_N$ is a complete subgraph containing the maximum number of elements. Now we can state the relation between maximum complete subgraphs and maximum transitive paths.

Lemma 5. *If a maximum complete subgraph $\bar{G}_N=(N,\bar{E}_N)$ of the undirected graph $\bar{G}=(V,\bar{E})$ has the property, that the directed subgraph $G_N=(N,E_N)$ contains no circuit, then the vertex set*

$$N=\{v_{j_1},v_{j_2},\ldots,v_{j_k}\}$$

can be ordered so that it gives a maximum transitive path of the directed graph $G=(V,E)$, i.e., an optimal solution of Problem ii.

Proof. Obviously, if

$$(v_{r_1},v_{r_2},\ldots,v_{r_t})$$

is a transitive path then the graph $G_M=(M,E_M)$

with

$$M=\{v_{r_1}, v_{r_2}, \ldots, v_{r_t}\}$$

is a complete subgraph of graph G. Therefore, in order to prove the lemma, it is sufficient to give a transitive path containing k vertices, because it is given a maximum complete subgraph of k elements. For this purpose, let us consider the set N. As there is no circuit in graph G_N there is at least one vertex without negative arc. Arc of type $(v_i, v_j)\in E_N$ is called negative arc for the vertex v_j, similarly $v_j, v_k E_N$ is a positive arc for the vertex v_j. Let us denote this vertex by v_{r_1}. The procedure may be repeated for the vertex set $N-v_{r_1}$ etc. The resulting path is

$$(v_{r_1}, v_{r_2}, \ldots, v_{r_k}).$$

From the procedure, it follows, that there is no arc of type

$$(v_{r_q}, v_{r_p})\in E \quad \text{for any} \quad p<q.$$

But graph $\bar{G}_N$ is complete, thus

$$(v_{r_p}, v_{r_q})\in E \quad \text{for any} \quad p<q,$$

which means, that the above path is transitive and the proof of the lemma is finished.

There are further possibilities to improve this result from the algorithmic point of view. In the worst case, the circuit may be broken by leaving out a vertex. This way good feasible solutions may be obtained, which accelerates the algorithm by all means.

Note. If all circuits of subgraph $G_N=(N, E_N)$ may be eliminated by leaving out an arc set $F\subset E_N$ in such a way, that

$$(19) \qquad \bar{G}_N=(N, \overline{E_N-F})$$

remains a complete graph, then the statement of the lemma still holds. (Thus arc elimination must be tried before vertex elimination.)

In the next section a measurement is introduced for the pairs of vertices: how "good" this pair may be together.

6. CONNECTION NUMBERS FOR PAIRS OF VERTICES

For two different vertices the more number of further vertices they may together stand for the better. Therefore we introduce the "connection number" of vertices v_i and v_j by the help of the following sets:

$$(20) \qquad S(v_i, v_j)=\{v_k \mid v_k\in V, d_{ik}+d_{ki}\geq 1, d_{jk}+d_{kj}\geq 1\}$$

$$(21) \qquad S_L(v_i, v_j)=\{v_k \mid v_k\in S(v_i, v_j) \text{ and } h_k\geq L-1\},$$

where L is a lower bound for $L_{\max}$ (a transitive path of v-length $L-1$ is already obtained or it is not interesting for some other reasons). Now the connection number of vertices v_i and v_j, if transitive paths of v-length L or longer are to be obtained:

$$(22) \qquad f_{ij}(L) = \begin{cases} |S_L(v_i, v_j)| & \text{if } d_{ij} + d_{ji} \geq 1 \\ 0 & \text{otherwise.} \end{cases}$$

The following lemma gives an additional tool for iterative filtering, i.e., for eliminating vertices:

Lemma 6. *If the set*

$$(23) \qquad R_k = \{v_i \mid v_i \in V, f_{ik}(L) \geq L-2\}$$

contains less than $L-1$ vertices, then no transitive path of v-length L (or longer) may contain vertex v_k.

Proof. If the condition of the lemma holds, then there are at most $L-2$ vertices for which the following two statements are true (formulated for a vertex v_i):

1° There may be a transitive path of v-length L containing vertices v_i and v_k.
2° There are at least $L-2$ vertices which may stand together with v_i and v_k in a transitive path of v-length L.

Conditions 1° and 2° are necessary for the existence of a transitive path of v-length L containing vertices v_k and v_i. As the number of such vertices v_i is at most $L-2$, together with v_k it is only $L-1$, no transitive path of v-length L may be obtained with vertex v_k, the lemma is proved.

7. DECOMPOSITION OF LARGE PROBLEMS

If Problem ii is large, i.e., if V consists of several hundred or even more vertices, then it is very difficult to find a maximum transitive path. In this case the problem may be decomposed into subproblems of the same type. This procedure may be outlined as follows.

Decomposition procedure

STEP 1. Decompose the graph $G = (V, E)$ into subgraphs $G_k = (N_k, E_k)$ using "connection numbers", where

$$V = \bigcup_{k=1}^{r} N_k \quad (N_i \cap N_j = \emptyset, i \neq j)$$

and the subgraph G_k is the one induced by the set of vertices N_k in the original graph G. The following two kinds of subsets N_L are preferred:

(a) $L_{\max}(N_k) \ll |N_k|$ ($L_{\max}(N_k) = 1$ or 2 if possible)
(b) $L_{\max}(N_k)$ large relative to $|N_k|$
    ($L_{\max}(N_k)$ is close to $L_{\max}$ if possible
    $L_{\max}(N_k)$ denotes the v-length of maximum transitive path in graph G_k.

STEP 2. Determine maximum complete subgraphs of the undirected graphs $\bar{G}_k$ $k = 1, \ldots, r$ and denote the number of vertices in these maximum complete subgraphs by n_k.

STEP 3. Determine the relationship between the subproblems: $m_j(i, k)$; $(k = 1, 2, \ldots, n_i)$ denotes the maximum number of elements that might be taken from set N_j if k element is taken from set N_i. (Good upper bound should be obtained.)

STEP 4. Apply an enumeration procedure for the new variables x_k, denoting the number of vertices taken from the set N_k.

This procedure may also be combined with filtering as it can be seen in the second part of the next section.

8. FILTERING

This word is used in the present paper for eliminating vertices that cannot be or not very likely to be in any optimal solution of Problem ii, depending on whether we consider an exact or a heuristic procedure. Obviously, the results of Lemmata 1–6 may be used and combined with the procedures given below.

8.1. Filtering with connection numbers—a heuristic tool

It is important to obtain good solutions as early as possible, because this may result in the elimination of vertices and in better upper bounds for the optimum. On the other hand, as soon as the present upper bound is reached by any feasible solution, the latter is also optimal. Now let us see the outline of a simple, but useful heuristic procedure.

(Transitive paths of v-length at least L are to be obtained):

STEP 1. Choose a "good vertex" v_{j_1}, e.g., the one with the largest connection number. Let $r = 1$.

STEP 2. Eliminate all vertices v_k, for which

$$f_{j,k}(L) < L - 2$$

and repeat this step with recalculated f_{ij} numbers until no more elimination is possible.

STEP 3. Choose a further "good vertex" $v_{j_{r+1}}$ from among the remaining ones (e.g. by the help of the last corrected connection numbers).

Repeat Steps 2 and 3 until no more vertex remains.

Then the vertices obtained by the procedure

$$(v_{j_1}, v_{j_2}, \ldots, v_{j_r})$$

give a maximal complete subgraph of the undirected graph $\bar{G}$ (which is not necessarily maximum complete) and a feasible solution of Problem ii may be obtained by breaking, the circuits (directed cycles) of graph G if there is any. (See Section 4.)

8.2. *An exact filtering with connection numbers in the decomposition*

In the case of decomposition described in Section 6, there is a further possibility for filtering. Using the notations of Section 6, the statement of Lemma 6 may be improved. (The set R_k is defined by the relation (23).)

Lemma 7. *If*

$$(24) \qquad t_p = \sum_{k=1}^{r} \min \{n_k, {}_1R_p \cap N_k|\} < L - 1,$$

then no transitive path of v-length L may contain vertex v_p.

Proof. It is sufficient to show, that the number of vertices that may stand together with vertex v_p in a transitive L-path is at most t_p. It was already shown in Lemma 6, that R_p is such an upper bound.

The set R_p may be decomposed:

$$R_p = \sum_{k=1}^{r} (R_p \cap N_k).$$

But at most n_k vertices of the set N_k may stand in the same transitive path according to the definition of the numbers n_k (Step 2 of Section 6) which completes the proof.

This result may be even further improved by using the relations of the vertex groups N_k, defined in Step 3.

9. OUTLINE OF THE BRANCH-AND-BOUND ALGORITHM

The substance of the algorithm for solving Problem (ii) is the gradual improvement of lower bounds L_w (feasible solutions) and upper bounds L_u. Any time these two bounds are equal to each other the algorithm is terminated.

452

Now we shall give some of the basic properties of the branch-and-bound algorithm:

1° Lower bounds and good feasible solutions can be obtained by procedures similar to the one described in Section 7.1.

2° Upper bounds may be determined after iterative filtering by Lemmata 1–3, 6, 7.

3° Both bounds may be improved by their alternate calculations and the conditional constraint of Lemma 4. The latter permits also bisectioning between lower and upper bounds.

4° Mainly depth-first approach should be used to obtain new (and better) feasible solutions as soon as possible.

5° Relaxation of Section 4 is used and the relaxed problem is heuristically solved and utilized to provide further good solutions.

6° Further heuristic subprocedures are used, e.g., subsets of good vertices are chosen—according to the evaluations by connection numbers and the best transitive path is augmented by the temporarily deleted elements if possible.

7° For larger problems the decomposition of Section 6 and the filtering of Section 7.2 is applied.

10. EXPERIMENTS WITH THE DATA OF DOROG BASIN

During the first century of brown coal mining several data have become known which were summarized by L. Gidai [9].

The aims of our investigations in Dorog basin are

(i) To test an improved methodology for making geological maps and cross sections by computer plotting.

(ii) To find a set of ordered rock bodies for better orientation of mining.

(iii) To obtain a maximum detailed formalized cronostratigraphic scale and respective subdivision.

A data file containing full faunal lists of the Eocene in 47 bore-holes from the Dorog basin and its enviroment have been formed on a CDC 3300 computer. The first and last appearance along the bore-holes of each of the more than 1000 fossile taxa defined the rock bodies used for further computations.

The total number of stored data exeeded 20,000.

Up to now only experimental results are available. Using the most frequent 150 rock bodies a solution consisting of 8 rock bodies resulted from the computations for the entire area. Another set of 8 rock bodies was obtained for a subregion under stronger constraints. Another set of 252 rock bodies-defined in a different way resulted in an ordered set of 26 rock bodies.

Results proved to be useful for the solution of Problem i (see Dienes and Kovács [7, 8]). Further investigation is devoted to the solution of Problems (ii) and (iii) and also to an improved solution of Problem (i) by larger sets of rock bodies.

REFERENCES

[1] Balas, E.: Set Covering With Cutting Planes from Conditional bounds, presented on IX. International Symposium on Mathematical Programming, 23–27 August, 1976. Printed as Management Science Research Report No. 399. Carnegie-Mellon University Pittsburgh, Pennsylvania 15213.

[2] Balas, E. and Padberg, M. V.: On the set Covering Problem, *Operations Research* 20 (1972), 1152–1161.

[3] Bron, C. and Kerbosch, J.: Algorithm 457-Finding All Cliques of an Undirected Graph, *Comm of ACM* 16 (1973), 575–577.

[4] Christofides, N.: *Graph Theory—An Algoritmic Approach*, Academic Press, 1975.

[5] Dienes, I.: Subdivision of a geological body into ordered parts 1974. In: Matematika és Számítástechnika a nyersanyagkutatásban MFt. kiadv. Szerk.: Dienes I. (Mathematics and Computational Techniques in the Search for Raw Materials, editor: I. Dienes).

[6] Dienes, I.: Proposal for Formalisation of Stratigraphic Nomenclature, *Journal of IAMG* (in prep.).

[7] Dienes, I. and Kovács, L. B.: Formalized Eocene stratigraphy of the Dorog Basin, Transdanubia, Hungary, *Acta Geologica* (in prep.).

[8] Dienes, I. and Kovács, L. B.: An algorithm for setting up optimal chronostratigraph c scales and plotting haemera tables, *Computers and Geosciences* (in prep.).

[9] Gidai, L.: L'Eocéne de la region de Dorog, *Ann. Inst. Geol. Publ. Hung. LV*. (1973), 1.

10] Kovács, L. B.: A New Solution for the Set Covering Problem, Proceedings of the 5th IFIP Conference on Optimization Techniques, Rome 1973, *Lecture Notes in Computer Science* 4–5, Springer 1974.

[11] Kovács, L. B.: A Computer Program System for the Solution of Pure Linear Discrete Problemns in *Colloquia Mathematica Societatis János Bolyai* 12. *Progress in Operations Research*, Eger, (Hungary), 1974. ed. A. Prékopa, North-Holland, 1975, pp. 573–589.

[12] Kovács, L. B.: *Discrete Programming*, North-Holland (to appear).

[13] Lemke, C. E., Salkin, H. M. and Spielberg, K.: Set Covering by Single Branch Enumeration with Linear Programming Subproblems, *Operations Research*, 19 (1971), 998–1022.

[14] Marsten, R. E.: An Algorithm for Large Set Partitioning Problems; *Management Science*, 20 (1974), 779–787.

[15] Salkin, H. M. and Koncal, R. D.: Set Covering by an All Integer Algorithm: Computational Experience. *ACM Journal*, 20 (1973), 189–193.

[16] Thiriez, H. M.: The Set Covering Problem: A Group Theoretic Approach, *RAIRO*, 5 (1971), 83–104.

AN ALGORITHM FOR INTEGER CONVEX PROGRAMMING

CS. FABIAN

(Bucharest, Romania)

INTRODUCTION

The difficulty in solving the integer programming problems and the nonlinear programming ones explains the more reduced attention given to the tackling of integer nonlinear problems.

The first work in this respect is owed to Kelley [9] and has appeared in the period in which Gomory's [6] method also appeared, the first method for integer linear problems. Kelley's method for linear objective functions and convex constraints is iterative and it proposes, at each iteration, the utilization of Gomory's method for solving an integer linear programming problem.

Lawler's and Bell's [8] work has appeared later, a lexicographical method for problems resulting from the difference of monotonous functions in each variable. The Korte, Krelle and Oberhofer [10] method for some nonlinear problems is also a lexicographical searching method, which is combined with the "branch-and-bound" method and which has proved more efficient [10].

Further, we mention works for particular cases of integer nonlinear programming, such as those of Künzi and Oettli [12] for integer quadratic programming, Hammer and Rudeanu [7] for pseudoboolean problems.

Special integer nonlinear problems are also the "bottleneck" problems, quadratic assignation and the travelling salesman problem for which particular methods are known [5].

Worth mentioning are also the attempts to reduce a more complicated problem to a simpler one by the transformation method. For example, the reducing of discrete linear problems to continuous quadratic problems [11], or the transformation of quadratic problems in (0, 1) into linear problems in variables (0, 1) [4].

The present study gives an algorithm for solving integer, *bounded,* convex problems based on Kelley's cutting plane method [9] and on the lexicographical searching method of Korte, Krelle and Oberhofer [10].

At the end of the study, we present the results of the numerical test of the method on the computer by using some test examples from literature (Lawler and Bell [8]).

FORMULATION OF THE PROBLEM AND NOTATIONS

Consider the integer convex programming problem:

$$\min f(x)$$
$$(1) \qquad g_i(x) \leq 0, \quad i = 1, \ldots, m$$
$$0 \leq x \leq d,$$
$$x, \quad \text{integer,}$$

where $f(x)$ and $g_i(x)$, $i = 1, \ldots, m$ are differentiable convex functions with bounded partial derivates and $x = (x_1, \ldots, x_n)$ and $d = (d_1, \ldots, d_n)$.

By using notation $g_0(x) = f(x) - x_{n+1}$ problem (1) can be written:

$$\min x_{n+1}$$
$$(2) \qquad g_i(x) \leq 0, \quad i = 0, \ldots, m$$
$$0 \leq x \leq d,$$
$$x, \quad \text{integer.}$$

Let G be the feasible domain of problem (2). It is obvious that G is bounded and it is contained in a P^0 bounded polyhedron. Let $x^e = (x_1^e, \ldots, x_n^e)$ be a point exterior to domain G and denote by S^0 an upper bound of the minimum of problem (1). With these elements we pass to the presentation of the algorithm.

ALGORITHM

The proposed algorithm requires the knowing of a point $x^e \notin G$ with the aid of which a polyhedron P^0 [12] is constructed, which contains G.

Further, search lexicographically for the first feasible solution [2] $x^0 \in P^0$ which is used in constructing an "internal" cutting plane if $x^0 \in G$ and in constructing an "external" cutting plane if $x^0 \notin G$ with the constructed cutting planes, polyhedron P^1 is obtained out of P^0.

The search for the first feasible solutions in a lexicographical order of an intermediate problem is made by setting out from the first feasible lexicographical solution of the previous problem [2], so that, in fact, we have to solve a single integer linear problem, a problem which changes by adding more constraints or by changing a right-hand-side.

The proposed algorithm is finite, because P^0 is bounded and thus has a finite number of points with integer coordinates and at each iteration is cut by the cutting planes constructed at least one point with integer coordinates.

The steps of the algorithm are:

Step 1.—Initializations.

$i:=0$ (for the counting of iterations)
$r:=0$ (1 if the problem has and 0 if the problem has no solution)

$$x_j^e:=d_j+1, \quad j=1, \ldots, n \text{ (construction of point exterior to } G)$$

$$P^0:=\{x \mid x \triangledown g_i(x^e) \leqq x^e \triangledown g_i(x^e)-g_i(x^e), i=0, \ldots, m, 0\leqq x \leqq d\}$$

(construction of the bounded polyhedron containing G).

Step 2.—Searching for the first feasible solution of problem (1) in a lexicographical order

$$
\begin{aligned}
z^i:=&\min x_{n+1}\\
&x \in P^i\\
&0 \leqq x \leqq d,\\
&x \text{ integer.}
\end{aligned}
$$

(3)

If (3) has no solution go to Step 4 otherwise let x^i be the solution of problem (3) and go to Step 3.

Step 3.—Constructing the cutting planes.
If $x^i \in G$, then put $r:=1$, $z^*:=z^i$, $x^*:=x^i$ and after it is put $i:=i+1$ and, go to Step 2.
If $x^i \notin G$, then construct

$$P^{i+1}:=P^i \cap \{x \mid x \triangledown g_j(x^i) \leqq x^i \triangledown g_j(x^i)-g_j(x^i)\},$$

where:

$$g_j(x^i):= \max_{k=0,\ldots,m} g_k(x^i), \quad \text{put} \quad i:=i+1$$

go to Step 2.

Step 4.—Ending test stop.
If $r=0$, then problem (1) has no integer solutions; otherwise (x^*, z^*) is the optimum integer solution of problem (1).

OBSERVATIONS

1. Problems (3) are not solved optimally at each iteration as proposed in Kelley [9], but a feasible solution is used in constructing the cutting planes.

2. Two types of cutting planes called "internal" and "external" are constructed.

3. The problems are not solved with Gomory's [6] method, but with the lexicographical research method of Korte, Krelle and Oberhofer [10].

4. The lexicographical searching method is modified for solving problems (3) which partly have the structure of Benders's "master" problems and for which the modifications are made in [2].

The given algorithm was programmed in the Fortran IV language and run on an IBM 370/145 computer of the Economic Cybernetics Department, Bucharest.

The test problems used are those given by Lawler and Bell [8] by modifying problems 2, 3, 4, 5 which contained nonconvex constraints and which were eliminated.

The characteristic data of the problems are presented in

Table 1

No.	Problem	n	m	$f(x^*)$	$d_1 \ldots d_n$	Remarks
1.	L-B-1	5	6	8	3, 3, 3, 3, 3	$f(x)$ quadratic
2.	L-B-2	7	5	8	7, 7, 7, 15, 15, 7, 7	$f(x)$ bi-linear
3.	L-B-3	7	5	8	7, 7, 7, 15, 15, 7, 7	$f(x)$ linear
4.	L-B-4	7	5	42	7, 7, 7, 15, 15, 7, 7	$f(x)$ bi-linear
5.	L-B-5	7	5	16	7, 7, 7, 15, 15, 7, 7	$f(x)$ quadratic

Table 2

No.	Problem	KECON no. it. sec.		KEKO no. it. sec.		KEKOF no. it. sec.	
1.	L-B-1	12	5.07	4	3.78	5	2.64
2.	L-B-2	1	2.57	3	23.30	1	0.79
3.	L-B-3	1	2.22	1	4.78	0	1.64
4.	L-B-4	1	2.62	2	7.34	1	1.38
5.	L-B-5	179	67.42	16	220.06	41	21.88

Table 2 contains the results obtained with the algorithm proposed (KEKOF) compared to the results of Kelley's algorithm for the case of continuous variables (KECON) and to the results of Kelley's algorithm for the discrete case when Gomory's algorithm is replaced by the lexicographical searching method (KEKO).

REFERENCES

[1] Benders, J. F.: Partitioning Procedures for Solving Mixed-Variables Programming Problems, *Numerische Mathematik,* 4 (1962), 238–252.

[2] Fabian, Cs.: Verknüpfung des Dekompositionsprinzips vor Benders mit dem Prinzip des lexikographischen Suchens, Dissertation, 1975, University of Bonn.

[3] Fabian, Cs.: Some Algorithms for Integer Nonlinear Programming, Economic Computation and Economic Cybernetics Studies and Research, no. 3 (1976).

[4] Forgó, F.: Some Transformations of Integer Programming Problems, Department of Mathematics, Karl Marx, University of Economics Paper 1974—1. Budapest.

[5] Garfinkel, S. R. and Nemhauser, L. G.: *Integer Programming,* John Wiley and Sons, New York–London–Sydney–Toronto, 1972.

[6] Gomory, R. E.: Algorithm for the Mixed Integer Problems, Rand Paper no. RM-2597 no. P-1885, 1960.

[7] Hammer, P. L. and Rudeanu, S.: *Boolean Methods in Operations Research and Related Areas.* Springer-Verlag, Berlin, 1968.

[8] Lawler, E. L. and Bell, M. D.: A Method for Solving Discrete Optimization Problems, *Operations Research*, 14, no. 6 (1966), 1098–1112.

[9] Kelley, J. E.: The Cutting Plane Method for Solving Convex Programs, *SIAM* 8 (1960), 208–218.

[10] Korte, B., Kelle, W. and Oberhofer, W.: Ein lexikographischer Suchalgorithms zur Lösung allgemeiner ganzzahliger Programierungsaufgaben, *Unternehmensforschung*, 13 (1969), 73–98, 171–192, 14 (1970), 228–234.

[11] Korte, B.: Über eine Klasse kombinatorischer Extremalprobleme, Bericht no. 7206 (1972), Bonn.

[12] Künzi, H. P. and Oetti, W.: *Nichtlineare Optimierung: Neuere Verfahren, Bibliographie*, Springer-Verlag, 1969.

A SUB-OPTIMAL ALGORITHM TO SOLVE A LARGE SCALE 0–1 PROGRAMMING PROBLEM*

J. A. FERLAND and M. FLORIAN

(Montreal, Canada)

1. INTRODUCTION

This paper is concerned with the following problem:

$$\text{Min } y_0$$

$$\text{subject to } y_0 \geqq \sum_{j=1}^{n} \sum_{i=1}^{I_j} b_k(i,j)x(i,j), \quad k=1, 2, \ldots, m,$$

$$(1) \qquad \sum_{i=1}^{I_j} x(i,j)=1, \qquad j=1, 2, \ldots, n,$$

$$x(i,j)=0 \text{ or } 1, \qquad i=1, 2, \ldots, I_j, j=1, 2, \ldots, n.$$

This problem appears often as a sub-problem during the application of Bender's decomposition. We tried to use the Lagrangean relaxation approach [4] to solve this problem. Unfortunately, due mainly to a lack of a good strategy to move optimally in the feasible region of the Lagrangean dual of (1), the numerical results were very disappointing.

In this paper we develop a sub-optimal algorithm to solve (1). The algorithm is sub-optimal, because it does not necessarily generate an optimal solution. On the other hand, it yields a good feasible solution together with an upper bound on the distance from the optimal solution. The algorithm is very simple, because it involves only a scanning process of each block of variables and the application of a simple procedure to reduce the objective function. The process is repeated with different initial solutions; these are generated using the information about the relative costs of the variables.

In Section 2, we present the scanning process that determines a "pseudo-local minimum" of (1) associated with an initial feasible solution. The structure of this

* This research was supported by the National Research Council of Canada, Grants No. RD-96 and No. A 7406.

problem is such that it is very easy to determine the relative costs of the variables, given a basis associated with a "pseudo-local minimum". This analysis is discussed in Section 3. In Section 4, we use duality theory to determine a bound on the distance of a "pseudo-local minimum" from the optimal solution. Next, in Section 5, we analyse two strategies to move away from a "pseudo-local minimum" and restart the scanning process with a new initial feasible solution. Computational results are reported in Section 6. We shall attempt to show that this method is suitable for the solution of (1) in the application of Bender's decomposition since it produces nearly optimal solutions in small time requirements.

2. PSEUDO-LOCAL MINIMUM

Problem (1) can be summarized to be that choosing exactly one variable equal to 1 in each block in order to minimize y_0 subject to the m constraints

$$y_0 \geq \sum_{j=1}^{n} \sum_{i=1}^{I_j} b_k(i,j)x(i,j).$$

The basic strategy of the algorithm to be developed is the following: fix the variables of all the blocks but one, and try to reduce the value of y_0 by working in this block only. The process is repeated with all blocks taken sequentially. When no further improvement is possible, the solution on hand is (called) a "pseudo-local minimum". This solution is not necessarily optimal as it will be illustrated at the end of this Section.

Given an initial solution of (1), an algorithm that determines a "pseudo-local minimum" may be stated as follows. Let $p(j)$ be the index of the variable equal to 1 in the j-th block:

Step 0. Evaluate y_0 $\left(\text{i.e. } y_0 = \underset{1 \leq k \leq m}{\text{Max}} \left\{ \sum_{j=1}^{n} b_k(p(j),j) \right\} \right)$. Then

$$\text{set} \quad \bar{y}_0 = y_0, \quad F1 = 1, \quad \text{and} \quad \bar{j} = 1.$$

Step 1. Let $x(i,j)$ be fixed for all $i=1, 2, \ldots, I_j$ and all $j \neq \bar{j}$. Set $i=0$.

Step 2. If $i < I_{\bar{j}}$, then set $i=i+1$ and move to Step 3. Otherwise, move to Step 4.

Step 3. Evaluate y_0 with $x(i,\bar{j})=1$. If $y_0 < \bar{y}_0$, then set $\bar{y}_0 = y_0$, $F_1 = \bar{j}$, and $p(\bar{j})=i$. Return to Step 2.

Step 4. If $\bar{j} < n$, then set $\bar{j}=\bar{j}+1$. Otherwise, $\bar{j}=1$. If $F_1 \neq \bar{j}$, then return to Step 1. Otherwise terminate.

Unfortunately, the solution obtained by this algorithm is not optimal as shown by the following example:

462

$$\text{Min } y_0$$

$$\text{subject to} \quad y_0 \geq x(1, 1) + 4x(2, 1) + 2x(1, 2) + 3x(1, 3)$$
$$y_0 \geq 2x(1, 1) + 3x(1, 2) + x(1, 3) + 4x(2, 3)$$
$$y_0 \geq 3x(1, 1) + x(1, 2) + 4x(2, 2) + 2x(1, 3)$$
$$x(1, 1) + x(2, 1) = 1, \quad x(1, 2) + x(2, 2) = 1, \quad x(1, 3) + (2, 3) = 1,$$
$$x(i, j) = 0 \text{ or } 1, \quad i = 1, 2, \quad j = 1, 2, 3.$$

Suppose $p(1) = p(2) = p(3) = 1$ (i.e. $x(1, 1) = x(1, 2) = x(1, 3) = 1$ and $x(2, 1) = x(2, 2) = x(2, 3) = 0$) for the initial solution. Then $y_0 = \bar{y}_0 = 6$. Since the value of y_0 increases to 9 whenever $x(2, 1) = x(1, 2) = x(1, 3) = 1$, or $x(1, 1) = x(2, 2) = x(1, 3) = 1$, or $x(1, 1) = x(1, 2) = x(2, 3) = 1$, it follows that the algorithm terminates with $\bar{y}_0 = 6$. On the other hand, the feasible solution $x(2, 1) = x(2, 2) = x(2, 3) = 1$ has a value of $y_0 = 4$.

If this method were to produce acceptable results, we would need a method that will enable a move away from this "pseudo-local minimum" if this is too far from the optimal value. The analysis in the next Sections presents a technique to accomplish this.

3. RELATIVE COSTS

The structure of the matrix of constraints is such that the relative costs for the variables associated with a basic feasible solution for the problem with relaxed integrality constraints of $x(i, j)$ constructed from a "pseudo-local minimum" are very easy to compute, and this computation does not require any matrix inversion.

Suppose that $p(j)$ is the index of the variable equal to 1 in block j. Furthermore, assume that

$$\bar{y}_0 = \sum_{j=1}^{n} b_1(p(j), j)$$

(if it is not the case, then a simple permutation of the indices k can be done to insure that the assumption is satisfied). Then a basis associated with this "pseudo-local minimum" has the following structure

$$
\begin{array}{c}
m \text{ rows} \left\{ \begin{array}{c} \\ \\ \\ \\ \end{array} \right. \\[2em]
n \text{ rows} \left\{ \begin{array}{c} \\ \\ \\ \\ \end{array} \right.
\end{array}
\begin{bmatrix}
\begin{array}{ccccc|cccc}
 & y_0 & \text{slack variables} & & & & \text{variables } x(i,j) & & \\
1 & 0 & 0 & \cdots & 0 & -b_1(p(1), 1) & -b_1(p(2), 2) & \cdots & -b_1(p(n), n) \\
1 & -1 & 0 & \cdots & 0 & -b_2(p(1), 1) & -b_2(p(2), 2) & \cdots & -b_2(p(n), n) \\
\vdots & \vdots & \vdots & \cdots & \vdots & \vdots & \vdots & \cdots & \vdots \\
1 & 0 & 0 & \cdots & -1 & -b_m(p(1), 1) & -b_m(p(2), 2) & \cdots & -b_m(p(n), n) \\
\hline
 & & & & & 1 & & & \\
 & & & & & & 1 & & \\
 & & 0 & & & & & \ddots & \\
 & & & & & & & & 1 \\
\end{array}
\end{bmatrix}
$$

or more simply

$$A = \begin{bmatrix} J & B \\ 0 & I \end{bmatrix},$$

where the $m \times m$ matrix J and the $m \times n$ matrix B are appropriately defined. In general, for matrix A having this structure, if J^{-1} exists, then

$$A^{-1} = \begin{bmatrix} J^{-1} & -J^{-1}B \\ 0 & I \end{bmatrix}.$$

Furthermore, since

$$J = \begin{bmatrix} 1 & 0 & 0 & \cdots & 0 \\ 1 & -1 & 0 & \cdots & 0 \\ \cdot & \cdot & \cdot & \cdots & \cdot \\ \cdot & \cdot & \cdot & \cdots & \cdot \\ \cdot & \cdot & \cdot & \cdots & \cdot \\ 1 & 0 & 0 & \cdots & -1 \end{bmatrix}$$

it is easy to verify that $J^{-1} = J$. Hence

$$A^{-1} = \begin{bmatrix} J & -JB \\ 0 & I \end{bmatrix}.$$

Using standard results from linear programming theory [2], it follows that the vector Π of simplex multipliers is simply the first row of A^{-1}:

$$\Pi = [1, 0, 0, \ldots, 0, b_1(p(1), 1), b_1(p(2), 2), \ldots, b_1(p(n), n)].$$

Hence the relative cost $\bar{c}(i, j)$ of variable $x(i, j)$ is

$$\bar{c}(i,j) = b_1(i,j) - b_1(p(j), j), \quad i = 1, 2, \ldots, I_j, \quad j = 1, 2, \ldots, n.$$

These relative costs will be useful to implement some strategies to move away from a "pseudo-local minimum" to a new initial solution.

4. UPPER BOUND ON THE ERROR

Given a "pseudo-local minimum", an analysis using duality theory can be performed to determine an upper bound on its distance from optimality.

The Lagrangean dual of (1) with respect to the first m constraints is

$$(2) \qquad \underset{\lambda \geq 0}{\text{Max}} \left[\underset{x \in X \, y_0 \in R}{\text{Min}} \left\{ \left(1 - \sum_{k=1}^{m} \lambda_k\right) y_0 + \sum_{k=1}^{m} \sum_{j=1}^{n} \sum_{i=1}^{I_j} \lambda_k b_k(i,j) x(i,j) \right\} \right],$$

where $X = \left\{ x \mid \sum_{i=1}^{I_j} x(i,j) = 1, \, x(i,j) = 0 \text{ or } 1, \, i = 1, 2, \ldots, I_j, j = 1, 2, \ldots, n \right\}$, and R is the set of all real numbers. Since y_0 is not restricted in sign, it follows that there exists at least one optimal solution of (2) such that

$$\sum_{k=1}^{m} \lambda_k = 1.$$

Hence, if we denote

$$\Lambda=\left\{\lambda\mid\sum_{k=1}^{m}\lambda_k=1,\ \ \lambda_k\geqq0,\ k=1,2,\ldots,m\right\},$$

then an equivalent problem to (2) is

$$(3)\qquad\operatorname*{Max}_{\lambda\in\Lambda}\left[\operatorname*{Min}_{x\in X}\left\{\sum_{k=1}^{m}\sum_{j=1}^{n}\sum_{i=1}^{l_j}\lambda_k b_k(i,j)x(i,j)\right\}\right].$$

Suppose that $p(j)$ is the index of the variable equal to 1 in block j for the "pseudo-local minimum" on hand, and that $\bar{y}_0$ is its value. Furthermore, define $\bar{K}$ as the set of indices k such that

$$\bar{y}_0=\sum_{j=1}^{n}b_k(p(j),j).$$

Define

$$\bar{\lambda}_k=\begin{cases}0 & \text{if }\ k\notin\bar{K}\\[4pt]\dfrac{1}{|\bar{K}|} & \text{if }\ k\in\bar{K},\end{cases}$$

where $|\bar{K}|$ is the number of elements in $\bar{K}$. For all j, let $v(j)$ be the index of the variable for which

$$\sum_{k\in\bar{K}}b_k(v(j),j)=\operatorname*{Min}_{1\leq i\leq l_j}\left\{\sum_{k\in\bar{K}}b_k(i,j)\right\}.$$

Then $\bar{\lambda}$, $x(v(j),j)=1$ and $x(i,j)=0$ for all $i\neq v(j)$ and for all $j=1,2,\ldots,n$, is a feasible solution of (3). A standard result of duality theory indicates that

$$\bar{y}_0\geqq y_0^*\geqq\frac{1}{|\bar{K}|}\sum_{j=1}^{n}\sum_{k\in\bar{K}}b_k(v(j),j).$$

where y_0^* is the optimal value of (1).

Hence

$$\bar{y}_0-\frac{1}{|\bar{K}|}\sum_{k\in\bar{K}}\sum_{j=1}^{n}b_k(v(j),j)$$

is an upper bound on the distance of $\bar{y}_0$ from y_0^*.

5. STRATEGIES OF SOLUTION

Two strategies to determine a new initial solution for the algorithm of Section 2 will be compared.

FIRST STRATEGY: In each block j, the variable with the *smallest* relative cost takes the value 1.

SECOND STRATEGY: In each block j, the variable with the *largest* relative cost takes the value 1.

The logic behind strategy 1 is identical to the logic behind the simplex algorithm: given the solution on hand we move in the direction of steepest descent. With the second strategy we move in a direction in which it seems the most probable that we will not obtain the same "pseudo-local minimum".

The complete algorithm to solve (1) with the first strategy is now stated; the appropriate modifications to implement it with the second strategy are indicated in parentheses.

Step 0. Initialization

Let $p(j)=\Pi(j)=1$ for $j=1, 2, \ldots, n$. Set YSOL=COUNT=0, FLI=0, and YD=$-\infty$.

Step 1. "Pseudo-local minimum"

With the initial solution on hand, determine a "pseudo-local minimum". Let $\bar{y}_0$ be its value and $p(j)$ the index of the variable equal to 1 in block j, $1 \le j \le n$.

Step 2. Best feasible solution

If COUNT=0, then YSOL=$\bar{y}_0$, $\Pi(j)=p(j)$ for all $j=1, 2, \ldots, n$, COUNT=1 and move to Step 3.

If $\bar{y}_0 >$ YSOL, then COUNT=COUNT+1, FL1=0, and move to Step 3.

If $\bar{y}_0=$ YSOL and $\Pi(j)=p(j)$ for all $j=1, 2, \ldots, n$, then COUNT=COUNT+1, FL1=1, and move to Step 3.

If $\bar{y}_0 \le$ YSOL and $\Pi(j) \ne p(j)$ for at least one j, then YSOL=$\bar{y}_0$, $\Pi(j)=p(j)$ for all $j=1, 2, \ldots, n$, COUNT=1, FL1=0, and move to Step 3.

Step 3. Dual feasible solution

Determine the set $\bar{K}$ of indices k such that

$$\bar{y}_0 = \sum_{j=1}^{n} b_k(p(j), j).$$

If

$$YD < \frac{1}{|\bar{K}|} \sum_{j=1}^{n} \left\{ \operatorname*{Min}_{1 \le i \le I_j} \sum_{k \in K} b_k(i, j) \right\},$$

then

$$YD = \frac{1}{|\bar{K}|} \sum_{j=1}^{n} \left\{ \operatorname*{Min}_{1 \le i \le I_j} \sum_{k \in K} b_k(i, j) \right\}.$$

Step 4. Sub-optimality test

If FL1=1, move to Step 5.

If $\dfrac{\text{YSOL} - \text{YD}}{\text{YSOL}} <$ P1, then terminate because the maximum relative distance from optimality is small enough.

If COUNT$\geq$P2, terminate, because this solution has been the best one for the last P2 iterations.
Set $k=0$.

Step 5. Set $k=k+1$.

Step 6. If $k\in\bar{K}$, then move to Step 7.

If $k\notin\bar{K}$ and $k<m$, then return to Step 5. Otherwise terminate, because the solution on hand is optimal or the best solution possible to generate using this strategy.

Step 7. Relative costs

(a) Set $j=0$ and C0$=0$.
(b) $j=j+1$
(c) Determine $b_k(\mu_1(j),j)=\underset{1\leq i\leq l_j}{\text{Min}}\ b_k(i,j)$

$\quad$ and $b_k(\mu_2(j),j)=\underset{1\leq i\leq l_j}{\text{Max}}\ b_k(i,j)$.

If $b_k(\mu_1(j),j)<b_k(p(j),j)$, then C0$=$C0$+1$.
If $j<n$, repeat (b), otherwise move to Step 8.

Step 8. Optimality check or new initial solution

If C0$=0$ terminate, because the solution on hand is optimal.
If C0>0, set $p(j)=\mu_1(j)$ $(p(j)=\mu_2(j))$, and repeat Step 1.

Notice that optimality of the solution on hand can be detected at Step 6 or Step 8 and that parameters P1 and P2 specified by the user serve to measure the sub-optimality of the solution on hand.

6. NUMERICAL RESULTS

The algorithm of Section 5 was implemented with both strategies on the Cyber 74 of the Computer Center at the University of Montreal.

Two series of problems were generated at random. The first series was used to compare the first strategy, the second strategy, and a $0-1$ integer programming code, DSZ1IP, developed in [3] and found in [1]. The problems of the second series were solved using the first strategy, the second strategy, and the random strategy where the new basic variables are determined randomly each time Step 1 of the algorithm is repeated, respectively. The results are summarized in Table 1 and Table 2 where each entry is the average over 5 problems generated at random having the same characteristics.

Table 1

Problems			DSZ1IP		First strategy		Second strategy	
m	ΣI_j	j	Sol.	Time (Sec.)	Sol.	Time (Sec.)	Sol.	Time (Sec.)
10	25	5	191.2	11.940	194.8	0.207	192.2	0.334
10	25	10	—	—	408.6	0.313	407	0.380
15	25	5	204.2	16.834	210.4	0.318	209.6	0.302
10	35	5	176.8	19.113	182.8	0.373	181.2	0.274
20	50	10	—	—	389.8	1.433	389.8	1.748
30	70	10	—	—	415.2	2.867	418	3.127
30	80	10	—	—	424.2	3.097	430.2	3.536
50	150	20	—	—	845	17.386	864	18.424

P1 = 0.01, P2 = 5.

Table 2

Problems			First strategy			Second strategy			Random strategy		
m	ΣI_j	j	Sol.	YD	Time (Sec.)	Sol.	YD	Time (Sec.)	Sol.	YD	Time (Sec.)
10	25	5	199.6	62	.237	199.2	60.6	.258	208.4	67.8	.209
10	25	10	420.2	241.2	.269	417.4	248.4	.262	429.2	270	.225
10	25	15	650	508	.260	655.4	526.6	.274	681.8	511.4	.234
10	35	5	201.4	43.2	.332	204.4	49.4	.348	204.0	45.2	.299
10	40	5	193.2	36.4	.388	194.0	43.8	.378	198.4	53.4	.344
10	50	10	345	104.6	.485	354.8	126.8	.447	343.2	97.4	.486
15	50	10	396	95.2	.749	404.2	125.2	.676	408	94.6	.736
20	50	10	415.6	89.8	.956	419.6	96.6	.902	416	106.8	.901
25	50	10	444.8	108.4	1.074	456.6	167.6	1.078	450.4	112.4	1.123
25	70	10	422.2	82.0	1.466	424.2	118.0	1.513	420.0	89.8	1.515
25	90	10	414	62.6	1.923	421.8	56.8	1.997	409.2	70.0	1.922
50	150	20	868.6	145.2	6.281	862.2	189.0	6.535	864.8	133.2	6.132

P1 = 0.01, P2 = 5.

In Table 1, note that a dash indicates that it was not possible to solve some of the problems generated in a reasonnable amount of time. Furthermore, when it was possible to solve the problems exactly with DSZ1IP we observe that the relative error with both stragegies is always less than 4%. On the other hand, the gain in the solution times is very significant. Hence the results are very encouraging.

It is not that easy to determine which of the tree strategies (first, second, or random) is the best. Indeed, their relative efficiency as far as precision and execution time are concerned are comparable.

The results in Table 1 and Table 2 indicate that execution time is proportional to the number of variables and also proportional to the number of constraints.

It is to be noticed that the maximum relative distance $\left(\dfrac{\text{YSOL}-\text{YD}}{\text{YSOL}}\right)$ significantly small to stop the algorithm. Hence this parameter is not very relevant in the algorithm.

Finally, it is worth noticing that a problem having 300 variables and 75 constraints was solved in 87.31 sec with the first strategy.

7. CONCLUSION

It is well known that, in the application of Bender's decomposition, it is not necessary to solve the relaxed master problem to optimality at each step. Rather, an appropriate method that yields efficiently good solution may be used at least at the beginning of the algorithm. The method given in this paper is very appropriate for doing this when the relaxed master has the form of (1). Indeed, as mentioned in Section 6, the method yields good solutions by requiring only very short solution times.

Acknowledgement

We wish to express our gratitude to Guy Lapalme and Réjean Lessard who have coded this algorithm and run the tests.

REFERENCES

[1] Cohen, C. and Stein, J.: Multi Purpose Optimization System, User's Guide, Vogelback Computing Center, Northwester University, Illinois (May 1974).

[2] Dantzig, G. B.: *Linear Programming and Extensions* (Princeton University Press, Princeton, New Jersey, 1963).

[3] Lemke, C. and Spielberg, K.: Direct Search 0–1 Integer Programming, DSZ1IP, Contributed Program Library, IBM New York Scientific Center 360 D-15, 2.001 (May 1966).

[4] Geoffrion, A. M.: Lagrangean Relaxation for Integer Programming, *Mathematical Programming Study 2*, (1974), 82–114.

DOUBLE-RELAXATION OPTIMALITY CONDITIONS
FOR INTEGER PROGRAMMING

J. M. FLEISHER and R. R. MEYER

(Madison, Wi., USA)

1. INTRODUCTION

A bounded* variable mixed integer program can be written in the form

(P) Maximize cx
  subject to $Ax \leqq b$
       $0 \leqq x \leqq d$
  x_j integer, $j \in I \subseteq N = \{1, 2, \ldots, n\}$

where A is an $m \times n$ matrix, c, d, and $x \in {}^{**}R^n$, $b \in R^m$, and I contains the set of indices corresponding to integer variables ($I = N$ for pure integer programs). Given $x^* \in R^n$, conditions are sought which will ensure x^* solves a problem of the form (P) with particular data. Having established sufficient optimality conditions, it is possible to construct a test problem of the form (P) with a known optimum x^* by selecting data such that the optimality conditions are satisfied.

Rosen [7] discusses construction of nonlinear programming problems whereby satisfaction of the Kuhn–Tucker conditions [5] implies optimality. Such optimality conditions are not generally applicable to mixed integer programming problems which have disjoint feasible regions.

Various sufficient optimality conditions for problems of the form (P) exist but are not amenable to constructing test problems. The relaxation criterion [2] states that

---

* Most mixed integer programming formulations of physical models have bounded feasible regions and most integer programming codes require that upper bounds on the variables be specified. For these reasons, bounds on the variables are included in this report; an analogous theory could be developed with the bounds excluded.

** Vectors may be row vectors or column vectors. If A is an $m \times n$ matrix, x and $y \in R^n$ and $v \in R^m$, then xy will denote $\sum_{j=1}^{n} x_j y_j$ and vAx will denote $\sum_{i=1}^{m} \sum_{j=1}^{n} v_i A_{ij} x_j$. It is also assumed throughout the body of this report that all data elements are rational.

if x^* is optimal for any problem whose feasible region contains the feasible region of (P) and whose objective function is the objective function of (P), then x^* solves (P) if and only if x^* is feasible for (P). This criterion forms the basis for branch and bound and cutting plane algorithms [2]. Other optimality conditions based upon coefficients obtained from appropriate transformations of variables are given in reference [3].

2. OPTIMALITY CONDITIONS

In order to develop sufficient optimality conditions for (P), it is convenient to introduce a problem obtained from (P) by a transformation of variables.

Lemma 1. *Consider the mixed integer program*

(Q) *Maximize* $\quad cy$
$\qquad$ *subject to* $\quad Ay \leqq s^*$
$$-x^* \leqq y \leqq d - x^*$$
$$y_i \ \text{integer}, \quad i \in I \subset N = \{1, 2, \dots, n\}$$

where A is an $m \times n$ matrix, c, d, and $y \in R^n$, $x^* \in R^n_+(I, d)^*$, and $s^* \in R^m_+$. Then $y^* = 0$ solves (Q) if and only if $x = x^*$ solves (P) with $b = Ax^* + s^*$.

Proof. Let $x = y + x^*$ and assume $b = Ax^* + s^*$. Then $\langle Ay \leqq s^* \rangle \Leftrightarrow \langle A(x - x^*) \leqq s^* \rangle \Leftrightarrow \langle Ax^* \leqq b \rangle$; $\langle -x^* \leqq y \leqq d - x^* \rangle \Leftrightarrow \langle -x^* \leqq x - x^* \leqq d - x^* \rangle \Leftrightarrow \langle 0 \leqq x \leqq d \rangle$; $\langle y_i$ integer, $i \in I \rangle \Leftrightarrow \langle y_i + x_i^*$ integer, $i \in I \rangle \Leftrightarrow \langle x_i$ integer, $i \in I \rangle$. Thus, there is a 1–1 correspondence between the feasible regions of (P) and (Q). Also, $\langle cy$ is maximized$\rangle \Leftrightarrow \langle cy + cx^*$ is maximized$\rangle \Leftrightarrow \langle cx$ is maximized$\rangle$.

Hence, problems (P) and (Q) are equivalent so that $\langle y^* = 0$ solves (Q)$\rangle \Leftrightarrow \langle y^* + x^* = x^*$ solves (P)$\rangle$. ■

In what follows, A, b, c, and d will refer to data for (P) and m, n, and I will denote respectively the number of constraints, number of variables, and index set of integer variables for (P).

Lemma 2. *Let $x^* \in R^n_+(I, d)$, s^*, $u^0 \in R^m_+$, and v^0, $w^0 \in R^n_+$. If*

(1) $$c = A^T u^0 - v^0 + w^0,$$

(2) $$b = Ax^* + s^*,$$

(3) $$s^* u^0 + x^* v^0 + (d - x^*) w^0 = 0,$$

then x^* solves (P).

$*\ R^n_+$ will denote $\{x \in R^n \mid x \geqq 0\}$ and $R^n_+ (I, d)$ will denote $\{x \in R^n \mid 0 \leqq x \leqq d$ and x_i integer for $i \in I\}$.

472

Proof. Since $s^* \geqq 0$ and $0 \leqq x^* \leqq d$, $y^* = 0$ is feasible for (Q). For y feasible for (Q) we have

(4) $$Ay \leqq s^*$$

(5) $$-y \leqq x^*,$$

(6) $$y \leqq d - x^*.$$

Multiplying (4) by u^0, (5) by v^0, (6) by w^0 and summing yields

(7) $$cy = u^0 Ay - v^0 y + w^0 y \leqq s^* v^0 + x^* v^0 + (d - x^*) w^0 = 0$$

using (1) and (3). Since $cy \leqq 0$, $y^* = 0$ solves (Q).

By (2) and Lemma 1, it follows that x^* solves (P). ∎

Note that the triple (u^0, v^0, w^0) is a feasible solution to the *dual* of the linear program (CLQ) obtained by deleting the integrality requirements of (Q), and that $0 = s^* u^0 + x^* v^0 + (d - x^*) w^0$ is the objective value of the dual of (CLQ) at (u^0, v^0, w^0). By the duality theory of linear programming, 0 is an upper bound on the optimal objective value of (CLQ), and thus must also be an upper bound on the optimal objective value of (Q). Since $y^* = 0$ is feasible for (Q), it must therefore also be optimal. This line of argument may be thought of as an alternative method of proof for the lemma. The same approach is also generalized to obtain stronger results in Section 4.

The conditions used in Lemma 2 are the Kuhn–Tucker conditions for linear programming [5]. No use is made of the integrality requirements in (P) and as may be seen from the alternate proof, the Kuhn–Tucker conditions are not satisfied at the optimum x^* unless the solution to (CLP), the continuous relaxation of (P), is also x^*. It is thus desirable to have more general optimality conditions which make use of the integrality requirements.

Lemma 3 below is a generalization of Lemma 2 which provides a set of sufficient optimality criteria that may hold when the strong criteria of Lemma 2 do not hold. Lemma 3 makes use of a generalization to rationals of the number theoretic concept of *greatest common divisor (gcd)*. Specifically, the *generalized greatest common divisor* of n rationals, $c_1, c_2, \ldots, c_n$ (assumed not all 0's), denoted as $ggcd (c_1, c_2, \ldots, c_n)$, is defined to be the minimum of $\sum_{j=1}^{n} c_j z_j$ subject to $\sum_{j=1}^{n} c_j z_j > 0$ and z_j integer, $j = 1$, $2, \ldots, n$. It is shown in the appendix that this definition is, in some sense, the dual of the usual definition of the *gcd*, and that its value is the usual *gcd* when the arguments c_j are all integers. The appendix also shows that the *ggcd* is well defined and positive when the arguments c_j are rationals (not all 0), and gives a number of important properties of this function. If the arguments c_j are integers, the *ggcd* may be efficiently computed by the Euclidean algorithm in at most $5 [\log_{10} c_p] + n + 3$ iterations where $c_p = \min_{c_i \neq 0} \{|c_i|\}$ as shown in [1].

Lemma 3. *Let $x^* \in R_+^n(I, d)$, s^*, $v^0 \in R_+^m$, and v^0, $w^0 \in R_+^n$. If*

$$(8) \qquad\qquad c = A^T u^0 - v^0 + w^0$$

$$(9) \qquad\qquad b = Ax^* + s^*$$

$$(10) \qquad \delta_0 = s^* v^0 + x^* v^0 + (d - x^*) w^0 < \gamma_0 \quad where \quad \gamma_0 = ggcd(c_1, c_2, \ldots, c_n)$$

$$(11) \qquad and \qquad j \notin I \Rightarrow c_j = 0 \; (continuous \; variables),$$

then x^* solves (P).

Proof. Following the same argument used in Lemma 2, $y^* = 0$ is feasible for (Q), and for any y feasible for (Q) we have

$$(12) \qquad cy = u^0 Ay - v^0 y + w^0 y \leq s^* u^0 + x^* v^0 + (d - x^*) w^0 < \gamma_0$$

using (8) and (10). By (11), $cy = \sum_{j \in I} c_j y_j$ and $\gamma_0 = ggcd(\{c_j\})$. Suppose $y^* = 0$ is not optimal for (Q). Then, there exists a feasible y for (Q) with $cy > 0$. This would imply $0 < cy = \sum_{j \in I} c_j y_j < \gamma_0 = ggcd(\{c_j\})$, in contradiction to then definition of the *ggcd* function, whereby $\gamma_0 = \min \sum_{j \in I} c_j y_j$ over $\sum_{j \in I} c_j y_j > 0$. Thus, $y^* = 0$ necessarily solves (Q) and by (9) and Lemma 1 it follows that x^* solves (P). ∎

Corollary. *Let $x^* \in R_+^n(I, d)$, s^*, $u^0 \in R_+^m$, and v^0, $w^0 \in R_+^n$. If c is an integer vector and*

$$(13) \qquad\qquad c = A^T u^0 - v^0 + w^0,$$

$$(14) \qquad\qquad b = Ax^* + s^*,$$

$$(15) \qquad \varepsilon_0 = s^* u^0 + x^* v^0 + (d - x^*) w^0 < 1,$$

$$(16) \qquad\qquad j \notin I \Rightarrow c_j = 0,$$

then x^* solves (P).

Lemma 3 is a generalization of Lemma 2 in that the primal solution pair (x^*, s^*) and the dual solution trio (u^0, v^0, w^0) are required to be complementary in Lemma 2, but in Lemma 3 the quantity δ_0 is allowed to assume a positive value less than γ_0. Equation (10) states that the primal solution pair (x^*, s^*) and the dual solution trio (u^0, v^0, w^0) need to be "not too far from being complementary". When (10) holds we will say that the solutions are δ_0-quasicomplementary.

Note that the continuous relaxation of (Q) may have an optimal objective value as large as δ_0. Thus, the gap between the optimal objective value of (P) and the optimal objective value of (CLP) may be as large as δ_0 where $\delta_0 < \gamma_0$ instead of 0 as in Lemma 2. The corollary thus states that the gap between the optimal objective value of (P) and the optimal objective value of (CLP) must be a nonnegative value less than 1 when

474

c is an integer vector. If c is an integer vector and $c_1, c_2, \ldots, c_n$ are not relatively prime but have a greatest common divisor of $\mu_0 \geqq 2$, the corollary still holds but (15) can be replaced by the weaker condition $\varepsilon_0 < \mu_0$: i.e., the gap between the optimal objective value of (P) and the optimal objective value of (CLP) may be any nonnegative value less than μ_0.

A further generalization is obtained by representing c as a nonnegative linear combination of p vectors $c^{(1)}, c^{(2)}, \ldots, c^{(p)}$ and dividing them into two groups. The set T will denote the set of indices k such that $j \notin I \Rightarrow c_j^{(k)} = 0$, group I will contain those $c^{(k)}s$ such that $k \in T$, and group II will contain those $c^{(k)}s$ such that $k \notin T$. Thus, group I contains those $c^{(k)}s$ where only integer variables may have nonzero costs $c_j^{(k)}$ and group II contains those $c^{(k)}s$ where some continuous variables have nonzero cost $c_j^{(k)}$. For the $c^{(k)}s$ in group I, we require that a dual solution trio $(u^{(k)}, v^{(k)}, w^{(k)})$ be δ_k-quasicomplementary with the primal solution pair (x^*, s^*) where $\delta_k < \gamma_k = ggcd(c_1^{(k)}, c_2^{(k)}, \ldots, c_n^{(k)})$ and apply Lemma 3; for the $c^{(k)}s$ in group II, we require that a dual solution trio $(u^{(k)}, v^{(k)}, w^{(k)})$ be complementary with the primal solution pair (x^*, s^*) and apply Lemma 2. Alternatively, let (P_k) denote problem (P) with $c = c^{(k)}$ and (CLP_k) denote the continuous relaxation of (P_k). Then if $k \in T$, we require that the gap between the optimal objective value of (P_k) and the optimal objective value of (CLP_k) have an upper bound of $\delta_k < \gamma_k$ and apply Lemma 3; if $k \notin T$, we require that the optimal objective values of (P_k) and (CLP_k) be equal and apply Lemma 2. The main sufficient optimality criteria for mixed integer programming problems now follows:

(SOC)   **Theorem 1.** *(Sufficient Optimality Criteria)*
Let $x^* \in R_+^n(I, d)$, $s^* \in R_+^m$,

$$u^{(k)} \in R_+^m, \quad k = 1, 2, \ldots, p,$$

$$v^{(k)}, w^{(k)} \in R_+^n, \quad k = 1, 2, \ldots, p,$$

$$\lambda_k \geqq 0, \quad k = 1, 2, \ldots, p.$$

If

$$(17) \qquad c^{(k)} = A^T u^{(k)} - v^{(k)} + w^{(k)}, \quad k = 1, 2, \ldots, p,$$

$$(18) \qquad c = \sum_{k=1}^{p} \lambda_k c^{(k)},$$

$$(19) \qquad b = Ax^* + s^*,$$

$$(20) \qquad k \in T \Rightarrow \delta_k = s^* u^{(k)} + x^* v^{(k)} + (d - x^*) w^{(k)} < \gamma_k$$

$$\text{where } \gamma_k = ggcd(c_1^{(k)}, c_2^{(k)}, \ldots, c_n^{(k)}),$$

$$(21) \qquad k \notin T \Rightarrow s^* u^{(k)} + x^* v^{(k)} + (d - x^*) w^{(k)} = 0,$$

then x^* *solves* (P).

Proof. By (19), $s^* \geqq 0$, and $0 \leqq x^* \leqq d$, x^* is feasible for (P). If $k \in T$ and $c = c^{(k)}$, then by Lemma 3 x^* solves (P). If $k \notin T$ and $c = c^{(k)}$, then by Lemma 2 x^* solves (P).

Hence, $c^{(k)} x \leqq c^{(k)} x^*$, $k = 1, 2, \ldots, p$, for any feasible point x of (P). Since $\lambda_k \geqq 0$, $k = 1, 2, \ldots, p$, it follows that

$$cx = \sum_{k=1}^{p} \lambda_k c^{(k)} x \leqq \sum_{k=1}^{p} \lambda_k c^{(k)} x^* = cx^*$$

for any feasible point x of (P) establishing the optimality of x^* for (P). ∎

(SOC 1) **Corollary 1.** *Let $x^* \in R_+^n(I, d)$, $s^* \in R_+^m$,*
$$u^{(k)} \in R_+^m, \quad k = 1, 2, \ldots, p,$$
$$v^{(k)}, w^{(k)} \in R_+^n, \quad k = 1, 2, \ldots, p,$$
$$\mu_k \geqq 0, \quad k = 1, 2, \ldots, p,$$
and $c^{(k)}$ be an integer vector, $k = 1, 2, \ldots, p$. If

$$(22) \qquad c^{(k)} = A^T u^{(k)} - v^{(k)} + w^{(k)}, \quad k = 1, 2, \ldots, p,$$

$$(23) \qquad c = \sum_{k=1}^{p} \mu_k c^{(k)},$$

$$(24) \qquad b = Ax^* + s^*,$$

$$(25) \qquad k \in T \Rightarrow \varepsilon_k = s^* u^{(k)} + x^* v^{(k)} + (d - x^*) w^{(k)} < 1,$$

$$(26) \qquad k \notin T \Rightarrow s^* u^{(k)} + x^* v^{(k)} + (d - x^*) w^{(k)} = 0,$$

then x^* solves (P).

The optimality conditions in (SOC) are more general than the conditions in Lemma 3. Lemma 3 requires that only integer variables may have nonzero costs c_j and that the gap between the optimal objective value of (P) and the optimal objective value of (CLP) be less than $\gamma_0 = ggcd\,(c_1, c_2, \ldots, c_n)$. However, if c can be expressed as a nonnegative linear combination of $c^{(k)}$s where each problem (P_k) is such Lemma 2 or Lemma 3 holds, continuous variables in (P) may have nonzero costs c_j and the gap between the optimal objective value of (P) and the optimal objective value of (CLP) may be considerably larger than γ_0. Such cases are exhibited in examples 1 and 2 of Section 3 where the conditions in (SOC) hold and the conditions in Lemma 3 certainly do not hold.

Although (SOC) is more general than the Kuhn–Tucker conditions and Lemma 3, it is not a necessary condition for x^* to be an optimum of (P). That is, given an optimum x^* to (P) it may not be possible to find p nonnegative scalars $\lambda_1, \lambda_2, \ldots, \lambda_p$, and p dual solution trios, $(u^{(k)}, v^{(k)}, w^{(k)})$, $k = 1, 2, \ldots, p$, such that the conditions in (SOC) hold. Such a case is exhibited in example 3 of Section 3. In Section 4 it will be shown that Lemma 3 may be extended to provide *necessary* and *sufficient* conditions for optimality, but that these necessary and sufficient conditions are not as easily applied.

The following corollaries are immediate since bounded variable mixed integer programming problems can always be expressed in the form (P) [2].

(SOC 2) **Corollary 2.** *Consider the problem*

(P2)   *Maximize* cx

   subject to $*A_ix \leqq b_i, \quad i \in Q_1$

   $A_ix \geqq b_i, \quad i \in Q_2$

   $A_ix = b_i, \quad i \in Q_3$

   $0 \leqq x \leqq d$

   x_i *integer,* $\qquad i \in I \subset N = \{1, 2, \ldots, n\},$

where A is an $m \times n$ matrix, c, d, and $x \in R^n$, $b \in R^m$, and $Q_1 \oplus Q_2 \oplus Q_3 = M = \{1, 2, \ldots, m\}$. Let $x^* \in R_+^n(I, d)$, $s^* \in R_+^m$,

$$u^{(k)} \in R^m, \quad k = 1, 2, \ldots, p,$$
$$v^{(k)}, w^{(k)} \in R_+^n, \quad k = 1, 2, \ldots, p,$$
$$\lambda_k \geqq 0, \quad k = 1, 2, \ldots, p.$$

If

(27)
$$c_j^{(k)} = \sum_{i \in Q_1 \cup Q_3} A_{ij} u_i^{(k)} - \sum_{i \in Q_2} A_{ij} u_i^{(k)} - v_j^{(k)} + w_j^{(k)}, \quad \begin{aligned} k &= 1, 2, \ldots, p, \\ j &= 1, 2, \ldots, n, \end{aligned}$$

28)
$$c = \sum_{k=1}^{p} \lambda_k c^{(k)},$$

(29)
$$b_i = \begin{cases} A_ix^* + s_i^* & \text{if} \quad i \in Q_1, \\ A_ix^* - s_i^* & \text{if} \quad i \in Q_2, \\ A_ix^* & \text{if} \quad i \in Q_3, \end{cases}$$

(30)
$$i \in Q_1 \cup Q_2 \Rightarrow u_i^{(k)} \geqq 0,$$

(31)
$$k \in T \Rightarrow \delta_k = \sum_{i \in Q_1 \cup Q_2} s_i^* u_i^{(k)} + x^* v^{(k)} + (d - x^*) w^{(k)} < \gamma_k$$

where $\gamma_k = ggcd(c_1^{(k)}, c_2^{(k)}, \ldots, c_n^{(k)})$,

(32)
$$k \notin T \Rightarrow \sum_{j \in Q_1 \cup Q_2} s_i^* u_i^{(k)} + x^* v^{(k)} + (d - x^*) w^{(k)} = 0,$$

then x^* solves (P2). Note that for the equality constraints, the corresponding $u_i^{(k)}$ multipliers are unrestricted in sign.

(SOC 3) **Corollary 3.** *Suppose (P3) is the same problem as (P2) except that the objective is to minimize. If all of the assumptions of (SOC 2) hold except that the algebraic signs in (27) are reversed, then x^* solves (P3).*

---

* A_i denotes the i-th row of A.

(SOC 4) **Corollary 4.** *Suppose (P4) is the same problem as (P2) except that there are no explicit upper bounds on the variables (i.e., no d vector). If all of the assumptions of (SOC 2) hold except that the $w^{(k)}$ vectors along with the terms in (27), (31), and (32) containing $w^{(k)}$ are omitted, then x^* solves (P4).*

3. EXAMPLES

Example 1. In this example, the vectors d, x^*, s^*, $u^{(k)}$, $v^{(k)}$, $w^{(k)}$, the scalars λ_k and γ_k, and the fourth row of A were specified a priori; the remainder of A, b, and c were then selected in such a manner that (SOC) would hold.

Maximize
subject to

$$33x_1 + 5x_2 + 48x_3 + 20x_4 + 20x_5$$
$$10x_1 + 3x_2 + 7x_3 + 4x_4 + 2x_5 \leqq 52$$
$$8x_1 + 7x_2 + 12x_3 + 6x_4 + 10x_5 \leqq 103$$
$$4x_1 + 0x_2 + 15x_3 + 14x_4 + 8x_5 \leqq 116$$
$$x_1 + x_2 + x_3 + x_4 + x_5 = 10$$

$$0 \leqq (x_1, x_2, x_3, x_4, x_5) \leqq (5, 10, 7, 9, 10)$$
$$x_1, x_2, x_3 \quad \text{integer.}$$

Solution: Maximum objective value $= 325$,

$$(x_1^*, x_2^*, x_3^*, x_4^*, x_5^*) = (0, 1, 5, 1.5, 2.5) \quad \text{and}$$

$$(s_1^*, s_2^*, s_3^*, s_4^*) = (3, 2, 0, 0).$$

We have in the notation of (SOC 2),

$$A = \begin{bmatrix} 10 & 3 & 7 & 4 & 2 \\ 8 & 7 & 12 & 6 & 10 \\ 4 & 0 & 15 & 14 & 8 \\ 1 & 1 & 1 & 1 & 1 \end{bmatrix}, \quad \begin{aligned} b &= (52, 103, 116, 10), \\ c &= (33, 5, 48, 20, 20), \\ d &= (5, 10, 7, 9, 10), \end{aligned}$$

$$I = \{1, 2, 3\}, \quad Q_1 = \{1, 2, 3\}, \quad Q_2 = \emptyset, \quad Q_3 = \{4\}.$$

Let $\lambda_1 = \lambda_2 = \lambda_3 = 1$,

$$\begin{aligned}
u^{(1)} \ (2, 1, 0, -13), \quad & v^{(1)} = (0, 0, 0, 1, 1), \quad & w^{(1)} = (0, 0, 0, 0, 0), \\
u^{(2)} \ (1, 2, 1, -30), \quad & v^{(2)} = (0, 2, 1, 0, 0), \quad & w^{(2)} = (0, 0, 0, 0, 0), \\
u^{(3)} = (0, 0, 0, \ \ 20), \quad & v^{(3)} = (0, 0, 0, 0, 0), \quad & w^{(3)} = (0, 0, 0, 0, 0).
\end{aligned}$$

Note that the $u_4^{(k)}$s are unrestricted in sign since the fourth constraint is an equality. Computation using (27) yields

$$c^{(1)}=(13, 0, 13, 0, 0), \qquad \gamma_1=13,$$
$$c^{(2)}=(0, -15, 15, 0, 0), \quad \gamma_2=15,$$
$$c^{(3)}=(20, 20, 20, 20, 20), \ \gamma_3=20,$$

and
$$\sum_{k=1}^{3} \lambda_k c^{(k)}=(33, 5, 48, 20, 20)=c$$

whence (28) holds. Computation using (29) yields
$$b=(52, 103, 116, 10).$$

Since only some $u_4^{(k)}<0$, (30) holds. For this example, T in $(31)=\{1, 2\}$ and

$$\delta_1= \sum_{i=1}^{3} s_i^* u_i^{(1)}+x^* v^{(1)}+(d-x^*)w^{(1)}=12<13=\gamma_1,$$

$$\delta_2= \sum_{i=1}^{3} s_1^* u_i^{(2)}+x^* v^{(2)}+(d-x^*)w^{(2)}=14<15=\gamma_2,$$

whence (31) holds. Finally, (32) holds since

$$\sum_{i=1}^{3} s_i^* u_i^{(3)}+x^* v^{(3)}+(d-x^*)w^{(3)}=0.$$

Therefore, the conditions of (SOC 2) are satisfied. For this example, a solution to the continuous relaxation of the problem is $x^0=(0.82142, 0.00000, 4.62502, 1.15177, 3.40178)$ with an objective value of 340.179. Thus, the gap between the optimal objective value of the problem and the optimal objective value of its continuous relaxation is 15.179. Note that $\gamma_0=ggcd(c_1, c_2, \ldots, c_5)=1$, so the conditions of Lemma 3 do not hold.

Example 2. The following class of problems is discussed in reference [4] as an example of a class of difficult test problems:

Maximize
$$-x_1$$
subject to
$$2px_1-qx_2=p$$
$$x_1, x_2 \geqq 0, \quad \text{integer,}$$

where $(p, q)\geqq(1, 3)$ integer, and $gcd(2p, q)=1$.

Solution: Maximum objective value$=-\frac{1}{2}(q+1)$, $(x_1^*, x_2^*)=\left(\frac{1}{2}(q+1), p\right)$, and $s_1^*=0$.

Since no upper bounds are given, assume $(x_1, x_2)\leqq(p+q, p+q)$. We have

$$A=[2p,-q], \quad b=p,$$
$$c=(-1, 0), \quad d=(p+q, p+q),$$
$$I=\{1, 2\}, \quad Q_1=Q_2=\emptyset, \quad Q_3=\{1\}.$$

Since $gcd(2p, q) = 1$, there exists integers μ_1, μ_2 such that $2p\mu_1 + q\mu_2 = 1$ [8].

Let $\lambda_1 = \lambda_2 = 1$,

$$u^{(1)} = (\mu_1), \quad v^{(1)} = (1, 0), \quad w^{(1)} = (0, 0),$$
$$u^{(2)} = (-\mu_1), \quad v^{(2)} = (0, 0), \quad w^{(2)} = (0, 0).$$

Computation using (27) yields

$$c^{(1)} = (2\mu_1 p - 1, -\mu_1 q) = (-\mu_2 q, -\mu_1 q), \quad \gamma_1 \geqq q,$$
$$c^{(2)} = (-2\mu_1 p, \mu_1 q), \quad \gamma_2 \geqq \min(\mu_1, 1),$$

and $c^{(1)} + c^{(2)} = (-1, 0) = c$ whence (28) holds. Since $b = 2px_1^* - qx_2^*$, (29) holds. With one equality constraint, (30) holds trivially. For this example, T in $(31) = \{1, 2\}$, and

$$\delta_1 = s^* u^{(1)} + x^* v^{(1)} + (d - x^*) w^{(1)} = \frac{1}{2}(q + 1) < q \leqq \gamma_1,$$

$$\delta_2 = s^* u^{(2)} + x^* v^{(2)} + (d - x^*) w^{(2)} = 0 < \min(\mu_1, 1) \leqq \gamma_2,$$

whence (31) holds and (32) holds vacuously.

Therefore, the conditions of (SOC 2) are satisfied.

Jeroslow and Kortanek [4] show that the solutions to this class of problems require an arbitrarily large number of cuts using the Gomory algorithm as p and q become large. What is even more interesting is that the solution to the continuous relaxation of the problem is always $(x_1^0, x_2^0) = \left(\frac{1}{2}, 0\right)$ with an objective value of $-\frac{1}{2}$, regardless of p and q. Thus, as q becomes large, so does the differential between the optimal objective value of the problem and the optimal objective value of its continuous relaxation. The gap is $\frac{q}{2}$ and $\gamma_0 = ggcd(c_1, c_2) = 1$.

Example 3. The following example shows that the optimality conditions (SOC) are not necessary conditions for x^* to be optimal for (P).

Maximize $\qquad\qquad\qquad\qquad 10x_1 + 9x_2$

subject to $\qquad\qquad\qquad\quad\; x_1 + 6x_2 \leqq 18,$

$$5x_1 + 2x_2 \leqq 18,$$

$$0 \leqq x_1, \quad x_2 \leqq 10, \quad x_1, \quad x_2 \text{ integer.}$$

Solution: Maximum objective value $= 39$, $(x_1^*, x_2^*) = (3, 1)$ and $(s_1^*, s_2^*) = (9, 1)$. We have in the notation of (SOC),

$$A = \begin{bmatrix} 1 & 6 \\ 5 & 2 \end{bmatrix}, \quad b = (18, 18), \quad c = (10, 9), \quad d = (10, 10).$$

A graph of the problem is Fig. 1. The feasible region consists of those integer lattice points corresponding to heavy dots.

For this problem, we cannot find a set of $\lambda_k' s \geqq 0$ and dual solution trios $(u^{(k)}, v^{(k)}, w^{(k)}) \geqq 0$ such that the conditions in (SOC) hold.

480

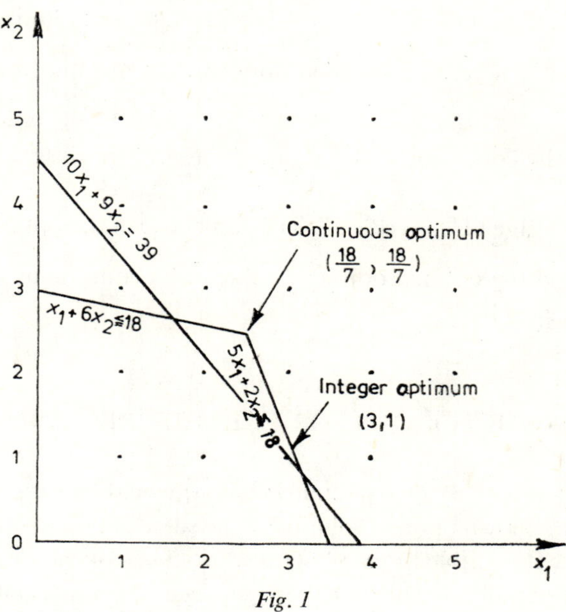

Fig. 1

First note that $c_1^{(k)} < 0$ cannot hold, because $\gamma_k = ggcd(c_1^{(k)}, c_2^{(k)})$ would be no larger than $|c_1^{(k)}|$ and δ_k in (20) would have to be at least $3|c_1^{(k)}|$ for $(u^{(k)}, v^{(k)}, w^{(k)}) \geqq 0$. This would mean (20) would be violated. Similarly, $c_2^{(k)} < 0$ cannot hold, because $\gamma_k = ggcd(c_1^{(k)}, c_2^{(k)}) \leqq |c_2^{(k)}|$ and δ_k in (20) would have to be at least $|c_2^{(k)}|$ for $(u^{(k)}, v^{(k)}, w^{(k)}) \geqq 0$ and again (20) would be violated. Thus, if (SOC) were to hold, $(c_1^{(k)}, c_2^{(k)}) \geqq 0$ is required.

In order for $c = (10, 9) = \sum_{k=1}^{P} \lambda_k(c_1^{(k)}, c_2^{(k)})$ where $\lambda_k, c_1^{(k)}, c_2^{(k)} \geqq 0$ to hold, we must be able to find $c^{(1)}$ and $c^{(2)}$ (possibly $c^{(1)} = c^{(2)}$) such that

$$(33) \qquad -9c_1^{(1)} + 10c_2^{(1)} \geqq 0, \quad c_1^{(1)} + c_2^{(1)} > 0, \quad (c_1^{(1)}, c_2^{(1)}) \geqq 0$$

$$(34) \qquad -9c_1^{(2)} + 10c_2^{(2)} \leqq 0, \quad c_1^{(2)} + c_2^{(2)} > 0, \quad (c_1^{(2)}, c_2^{(2)}) \geqq 0.$$

If (33) cannot be satisfied, $-9c_1 + 10c_2 < 0$ (or $c = 0$) and if (34) cannot be satisfied, $-9c_1 + 10c_2 > 0$ (or $c = 0$) but for $c = (10, 9)$, $-9c_1 + 10c_2 = 0$ (and $c \neq 0$).

Thus, applying (SOC 1), we must be able to find *integers* $c_1^{(k)}$ and $c_2^{(k)}$ such that

$$(35) \qquad \delta_k = 9u_1^{(k)} + u_2^{(k)} + 3v_1^{(k)} + v_2^{(k)} + 7w_1^{(k)} + 9w_2^{(k)} < 1$$

$$(36) \qquad \begin{aligned} u_1^{(k)} + 5u_2^{(k)} - v_1^{(k)} + w_1^{(k)} - c_1^{(k)} &= 0 \\ 6u_1^{(k)} + 2u_2^{(k)} - v_2^{(k)} + w_2^{(k)} - c_2^{(k)} &= 0 \\ u_1^{(k)}, u_2^{(k)}, v_1^{(k)}, v_2^{(k)}, w_1^{(k)}, w_2^{(k)} &\geqq 0 \end{aligned}$$

and (33) holds if $k = 1$; (34) holds if $k = 2$.

For $k=2$, the minimum of δ_k in (35) subject to (34) and (36) is $\frac{8}{7}>1$ so that (35) cannot be satisfied. Therefore, the conditions in (SOC) cannot be satisfied for this example.

A solution to the continuous relaxation of the problem is $x^0=\left(\dfrac{18}{7}, \dfrac{18}{7}\right)$ with an optimal objective value of $\dfrac{342}{7}\cong 48.857$. The gap between the optimal objective value of the problem and the optimal objective value of its continuous relaxation is 9.857 and $\gamma_0=ggcd(c_1, c_2)=1$.

4. DIRECTIONS FOR FURTHER RESEARCH

In order to test an integer programming code, it would be very useful to construct test problems with known optimal solutions. The sufficient optimality criteria derived in Section 2 make it possible to construct test problems of the form (P) by selecting data such that the conditions of (SOC) are satisfied. Alternate optima are possible: however, the maximum objective value of (P) must be cx^* so we would always know how close to being optimal a solution obtained from a code would be. Of particular interest is the construction of test problems for problems with physical interpretations.

This line of approach also suggests characterizing the class of problems that can be constructed using (SOC), and thereby determining a class of problems for which the conditions of (SOC) are also *necessary* for optimality (i.e., they would be necessary in the sense that they must hold at some optimal solution).

Clearly, from Example 3 of Section 3, these conditions are *not necessary* for all problems of the form (P), so another area for further research lies in further analysis and sharpening of the optimality conditions. In particular, from the proof of Lemma 3, it is easily seen that the *conclusion* of the Lemma holds if the optimal objective value $m(F)$ of the problem

$$
\begin{array}{lll}
(37) & \text{Maximize} & cy \\
& \text{subject to} & y\in F,
\end{array}
$$

where F is any set containing F^*, the feasible set of (Q), is *strictly less than* the optimal value $M(F')$ of the problem

$$
\begin{array}{lll}
(38) & \text{Minimize} & cy \\
& \text{subject to} & cy>0 \\
& & y\in F',
\end{array}
$$

where F' also contains F^*.

482

Theorem 2. *If $m(F) < M(F')$ for some pair of sets F, F' such that $F \supseteq F^*$, $F' \supseteq F^*$, then x^* solves* (P). *A necessary and sufficient condition for x^* to solve* (P) *is that $m(F^*) < < M(F^*)$.*

Proof. Suppose $y^* = 0$ does not solve (Q). Then there exists a $\bar{y} \in F^*$ such that $c\bar{y} > 0$. Since $\bar{y} \in F$ and $\bar{y} \in F'$, $\bar{y}$ is feasible for both (37) and (38), from which we have $c\bar{y} \leq \leq m(F) < M(F') \leq c\bar{y}$, a contradiction. Thus, $y^* = 0$ must solve (Q) whence x^* solves (P).

To see that this condition is also necessary when $F = F' = F^*$, suppose that x^* solves (P). Then $y^* = 0$ solves (Q), so $m(F^*) = 0$. On the other hand, (38) must be infeasible, and by convention, $M(F^*) = + \infty$. ∎

Lemma 3, in fact, corresponds to the case in which F is the continuous relaxation of F^* and F' is the relaxation of F^* obtained by discarding all constraints other than integrality. From a computational point of view, these relaxations are particularly convenient since the problems (37) and (38) may then be solved relatively easily. On the other hand, weaker sufficient conditions may be obtained by choosing F and F' closer to F^*. For example, the bounds on the integer variables might be retained in F'. The possibility of using tighter relaxations is currently under study. Note that an extension of (SOC) may be obtained by using Theorem 2 in place of Lemma 3.

APPENDIX

PRIMAL AND DUAL FORMULATIONS

OF THE *ggcd* FUNCTION

We will show that the optimization problem used in the definition of the *ggcd* of $c_1, c_2, \ldots, c_n$ as given in Section 2, namely,

(A1) Minimize
$$\sum_{j=1}^{n} c_j z_j$$

  subject to
$$\sum_{j=1}^{n} c_j z_j > 0$$
$$z_j \text{ integer}, \qquad j = 1, 2, \ldots, n$$

may be thought of as the "dual" of the following problem, which corresponds to the usual definition of the greatest common divisor:

(A2) Maximize
$$x$$

  subject to
$$\frac{c_j}{x} = y_j, \quad j = 1, 2, \ldots, n$$
$$y_j \text{ integer}, \qquad j = 1, 2, \ldots, n.$$

Theorem A1. *If $c_1, c_2, \ldots, c_n$ are rational numbers not all 0, then the optimal values of the problems* (A1) *and* (A2) *exist and are equal.*

Proof. We will first show that (A1) and (A2) both have optimal solutions. Let each c_j be written in the form $c_j = p_j/q_j$, where p_j and q_j are integers with $q_j > 0$,

and let $q = \prod_{j=1}^{n} q_j$.

Then the constraints $cz > 0$, z integer are satisfied if and only if the constraints $(qc)z > 0$, z integer are satisfied. Since qc is an integer vector, this is equivalent to $(qc)z \geq 1$, z integer or $cz \geq q^{-1}$, z integer. Hence, the problem (A1) is equivalent to the problem

(A3) Minimize $\qquad\qquad cz$

$\qquad$ subject to $\qquad\qquad cz \geq q^{-1}$

$\qquad\qquad\qquad\qquad\qquad z \quad$ integer.

Since (A3) is an integer programming problem with rational data, (A3) has an optimal solution if it is not unbounded or infeasible [6]. Since (A3) requires $cz \geq q^{-1}$, it is not unbounded, and since $c \neq 0$, it is easily seen to have feasible solutions. Thus, (A3) has an optimal solution z^* and an optimal objective value $\alpha > 0$.

To see that (A2) has an optimal solution, note that (A2) is equivalent to the problem

(A4) Maximize $\qquad\qquad\qquad x$

$\qquad$ subject to $\qquad\qquad\quad \dfrac{c_j}{x} = y_j, \quad j = 1, 2, \ldots, n,$

$\qquad\qquad\qquad\qquad\quad y_j, \text{ integer}, \qquad j = 1, 2, \ldots, n,$

$\qquad\qquad\qquad\qquad\quad q^{-1} \leq x \leq \min_{c_j \neq 0} \{|c_j|\}$

$\qquad\qquad\qquad\qquad\quad \dfrac{c_j}{\min_{c_j \neq 0} \{|c_j|\}} \leq y_j \leq c_j q^{-1}.$

Since the objective function of (A4) is continuous and its feasible set is compact, (A4) and hence (A2) has an optimal solution (x^*, y^*) whose optimal objective value is $x^* > 0$.

We will now show that $\alpha = x^*$. We have $\alpha = cz^* = (x^* y^*)z^* = x^*(y^* z^*)$ so that $y^* z^* = \dfrac{\alpha}{x^*} > 0$. Since $y^* z^*$ is integer, $y^* z^* \geq 1$ and thus $\alpha \geq x^*$. The components of y^* must be relatively prime for if they had a common factor $\mu \geq 2$, $(\mu x^*, \mu^{-1} y^*)$ would be feasible for (A2) with an objective value of $\mu x^* > x^*$, contradicting the optimality of (x^*, y^*). Thus, there exists an integer vector v^* such that $v^* y^* = 1$ [8]. Thus, $cv^* = (x^* y^*)v^* = x^*(y^* v^*) = x^*$. Since v^* is feasible for (A1), $x^* \geq \alpha$ and thus $x^* = \alpha$. ∎

Note that if $c_1 = c_2 = \ldots = c_n = 0$, then (A1) is infeasible and (A2) is unbounded, so in this case the "optimal" values may be considered to be $+\infty$.

If some of the c_j are *irrational*, then the optimal objective values are equal if (A2) has a feasible solution ((A1) will have a feasible solution as long as the c_js are not all 0). For if (A2) has a feasible solution it has a feasible solution of the form $(\bar{x}, \bar{y})$, where $\bar{x}>0$. We may show in a manner analagous to the proof of Theorem A1 that (A2) has an optimal solution. Furthermore, since feasibility of $(\bar{x}, \bar{y})$ implies $c=\bar{x}\bar{y}$, problem (A1) is equivalent to

(A5)   Minimize $\quad\quad\quad\quad\quad\quad cz$

       subject to $\quad\quad\quad\quad\quad\quad \bar{y}z>0$

$$z\quad \text{integer.}$$

Again it follows from the same argument used in the proof of the preceding theorem that (A5) has an optimal solution. Given the existence of optimal solutions to both problems, the proof of equality of the optimal values may be carried through unchanged.

On the other hand, if (A2) is infeasible, there exists, a k such that $c_k \neq 0$ and a k' such that $c_{k'}/c_k$ is irrational. The problem (A1) is thus equivalent to

(A6)   Minimize $\quad\quad\quad\quad \hat{c}z/(sgn(c_k)c_k)$

       subject to $\quad\quad\quad\quad \hat{c}z>0$

$$z\quad \text{integer,}$$

where $\hat{c}=c/(sgn(c_k)c_k)$. Since $\hat{c}$ has both irrational and non-zero rational components, it follows from arguments presented in [6] that (A6) has no optimal solution since the imfimum of the objective values is 0. An example in which (A2) is infeasible is given by $c_1=1$ and $c_2=\sqrt{2}$ for if (A2) were feasible, we would have $\sqrt{2}=\dfrac{y_1}{y_2}$, contradicting the irrationality of $\sqrt{2}$, since y_1 and y_2 are nonzero integers. These results are summarized in the following theorem:

Theorem A2. *If the problems* (A1) *and* (A2) *are both feasible, then they both have optimal solutions and their optimal objective values are equal. Problem* (A1) *is infeasible if and only if $c=0$, in which case the optimal objective values of both problems may be taken to be $+\infty$. If* (A2) *is infeasible, then* (A1) *has no optimal solution and the infimum of the objective function values of* (A1) *is* 0.

As a consequence of the preceding "strong duality" theorem, we may conclude that if $\bar{z}$ is feasible for (A1) and $(\bar{x}, \bar{y})$ is feasible for (A2), then $c\bar{z}\geq\bar{x}$. In fact, an even stronger relationship holds between objective function values for feasible solutions, as the following theorem indicates.

Theorem A3. *If $\bar{z}$ is feasible for* (A1) *and* $(\bar{x}, \bar{y})$ *is feasible for* (A2)*, then $c\bar{z}=k\bar{x}$, where k is a non-zero integer.*

Proof. We have $c\bar{z}=(\bar{x}\bar{y})\bar{z}=\bar{x}(\bar{y}\bar{z})=k\bar{x}$, where $k=\bar{y}\bar{z}$. Since $c\bar{z}>0$ and $\bar{y}\bar{z}$ is integer, the result follows. ∎

Given an $(\bar{x}, \bar{y})$ which is feasible for (A2), the following theorem gives a necessary and sufficient criterion for $(\bar{x}, \bar{y})$ to be optimal for (A2), and hence for $\bar{x}$ to be the optimal objective value of (A1):

Theorem A4. *If $(\bar{x}, \bar{y})$ is feasible for* (A2), *it is optimal for* (A2) *(and thus $\bar{x}$ is the optimal objective value of* (A1)*) if and only if the components in $\bar{y}$ are relatively prime and $\bar{x} > 0$.*

Proof. Assume $c \neq 0$, otherwise $\bar{y} = 0$, so certainly the components in $\bar{y}$ are not relatively prime. Then (A2) has an optimal solution (x^*, y^*) and we have

$$
x^* = \left\langle
\begin{array}{c}
\text{Min. } cz \\
\text{subject to } cz > 0 \\
z \text{ integer}
\end{array}
\right\rangle
= \left\langle
\begin{array}{c}
\text{Min. } (\bar{x}\bar{y})z \\
\text{subject to } (\bar{x}\bar{y})z > 0 \\
y, z \text{ integer}
\end{array}
\right\rangle
= \left\langle
\begin{array}{c}
\bar{x} \text{ Min. } \bar{y}z \\
\text{subject to } \bar{y}z > 0 \\
z \text{ integer}
\end{array}
\right\rangle =
$$

$= \bar{x}\, gcd(\bar{y}_1, \bar{y}_2, \ldots, \bar{y}_n) = \bar{x}$ if and only if $gcd(\bar{y}_1, \bar{y}_2, \ldots, \bar{y}_n) = 1$: i.e., the components in $\bar{y}$ are relatively prime. ∎

The theorem states even more: namely, if $(\bar{x}, \bar{y})$ is feasible for (A2) and $\bar{y} \neq 0$ then (x^*, y^*), the optimal solution for (A2), has x^* equal to $\bar{x}$ multiplied by the greatest common divisor of the components in $\bar{y}$.

As a final observation, it might be noted that the *continuous* relaxations of (A1) and (A2) have no optimal solutions. If $c \neq 0$, the continuous relaxation of (A1) has 0 as the infimum of its objective values, and the continuous relaxation of (A2) has $+\infty$ as the supremum of its objective values.

PROPERTIES OF THE GENERALIZED GREATEST COMMON DIVISOR FUNCTION

Properties of the generalized greatest common divisor function, *ggcd*, are listed below. In what follows, all data are rational numbers. The proofs of those properties which are immediate consequences of the previous theorems are omitted.

1. If $c_1, c_2, \ldots, c_n$ are not all 0, then $ggcd(c_1, c_2, \ldots, c_n, \alpha) = ggcd(ggcd(c_1, c_2, \ldots, c_n), \alpha)$.

2. $ggcd(c_1, c_2, \ldots, c_n) = ggcd(c_{\sigma_1}, c_{\sigma_2}, \ldots, c_{\sigma_n})$, where $\{\sigma_1, \sigma_2, \ldots, \sigma_n\}$ is any permutation of $\{1, 2, \ldots, n\}$: i.e., *ggcd* is a symmetric function.

3. If $c_1, c_2, \ldots, c_n$, not all 0, are of the form $c_j = \dfrac{p_j}{q_j}$, where p_j and q_j are relatively prime integers with $q_j > 0$, $j = 1, 2, \ldots, n$, then

$$
ggcd(c_1, c_2, \ldots, c_n) = \frac{gcd\,(p_1, p_2, \ldots, p_n)}{2\, cm\,(q_1, q_2, \ldots, q_n)},
$$

where $2\, cm$ denotes, the least common multiple.

4. If $\alpha \neq 0$, $ggcd(\alpha) = |\alpha|$.

5. If $\alpha = 0$, $c_1, c_2, \ldots,$ or c_n, then $ggcd(c_1, c_2, \ldots, c_n, \alpha) = ggcd(c_1, c_2, \ldots, c_n)$.

6. $ggcd(c_1, c_2, \ldots, c_n) = ggcd(|c_1|, |c_2|, \ldots, |c_n|)$.

7. If $\alpha \neq 0$, $ggcd(\alpha c_1, \alpha c_2, \ldots, \alpha c_n) = |\alpha|\, ggcd(c_1, c_2, \ldots, c_n)$.

8. $ggcd(c_1, c_2, \ldots, c_n, \alpha) = \dfrac{1}{\mu}\, ggcd(c_1, c_2, \ldots, c_n)$, where μ is a positive integer. In particular, $ggcd(c_1, c_2, \ldots, c_n, \alpha) \leqq ggcd(c_1, c_2, \ldots, c_n)$.

9. For each $c_k \neq 0$, there exists a positive integer μ such that

$$ggcd(c_1, c_2, \ldots, c_n) = \frac{1}{\mu} |c_k|.$$

In particular,

$$ggcd(c_1, c_2, \ldots, c_n) \leq \operatorname*{Min}_{c_j \neq 0} |c_j|.$$

10. $ggcd(c_1, c_2, \ldots, c_n)$ is integer if and only if $c_1, c_2, \ldots, c_n$ are integers not all 0.

11. If $c_1, c_2, \ldots, c_n$ are integers, $ggcd(c_1, c_2, \ldots, c_n, 1) = 1$.

12. If $ggcd(c_1, c_2, \ldots, c_n) = \mu\gamma$, where μ is any positive integer, then $ggcd(c_1, c_2, \ldots,$

$$c_n) = \gamma \gcd\left(\frac{c_1}{\gamma}, \frac{c_2}{\gamma}, \ldots, \frac{c_n}{\gamma}\right) \text{ since } \frac{c_j}{\gamma} \text{ is integer, } j = 1, 2, \ldots, n.$$ Thus, if we

know the c_j's are integral multiples of some number γ, the $ggcd$ can be computed by scaling the arguments and computing the number theoretic greatest common divisor function.

13. The $ggcd$ function is not continuous. (Let $\alpha_n = 3 \sum\limits_{k=1}^{n} 10^{-k} = \underbrace{..333...3}_{n \text{ places}}$. Then

$$ggcd\left(\alpha_n, \frac{1}{2}\right) = 10^{-n}, \lim_{n \to \infty} \alpha_n = \frac{1}{3} \text{ and } \lim_{n \to \infty} 10^{-n} = 0, \text{ but } ggcd\left(\frac{1}{3}, \frac{1}{2}\right) = \frac{1}{6}).$$ Thus

small errors in approximating fractional values of the arguments of the $ggcd$ function can greatly change the value of the $ggcd$ function.

Proofs of properties 1 and 3, which are not as immediate as the remaining properties, are given below.

Proof of Property 1. Property 1 gives a means for computing the $ggcd$ of n arguments in terms of $n-1$ calculations of the $ggcd$ of 2 arguments. If $c_1, c_2, \ldots, c_n$ are rational numbers not all 0 and α is rational, then

$$ggcd(c_1, c_2, \ldots, c_n) = \gamma_1 = \operatorname{Max} \gamma$$
$$\text{subject to} \quad c_j = \gamma z_j, \quad j = 1, 2, \ldots, n,$$
$$z_j \text{ integer}$$

where the z_j's are relatively prime integers, and

$$ggcd(ggcd(c_1, c_2, \ldots, c_n), \alpha) = \gamma_2 = \operatorname{Max} \gamma$$
$$\text{subject to } \gamma_1 = \gamma y_1$$
$$\alpha = \gamma y_2,$$
$$y_1 \text{ integer}$$
$$y_2 \text{ integer}$$

where y_1 and y_2 are relatively prime integers, by Theorem A4.

Thus,

$$c_j = \gamma_1 z_j^* = \gamma_2 (y_1^* z_j^*) = \gamma_2 x_j, \quad j = 1, 2, \ldots, n,$$
$$\alpha = \gamma_2 y_2^*,$$

where $x_j = y_1^* z_j^*$ are integers.

Then $x_1, x_2, \ldots, x_n, y_2^*$ are necessarily relatively prime for if they had a common prime factor of $\mu \geqq 2$, since $z_1^*, z_2^*, \ldots, z_n^*$ are relatively prime and thus have no common prime factors, y_1^* and y_2^* would have to have a prime factor of $\mu \geqq 2$ but y_1^* and y_2^* are relatively prime. Since $x_1, x_2, \ldots, x_n, y_2^*$ are relatively prime, by Theorem A4, $\gamma_2 = ggcd(c_1, c_2, \ldots, c_n, \alpha)$. ■

Proof of Property 3. Property 3 gives an explicit form for the generalized greatest common divisor function in terms of number theoretic functions whose arguments are integer. Any rational number c_j can be expressed uniquely (except for algebraic signs) as $\dfrac{p_j}{q_j}$ where p_j and q_j are relatively prime integers (note $p_j = 0$ implies $q_j = +1$). Property 3 states that the generalized greatest common divisor of n rational numbers $\{c_j\}$ is equal to the greatest common divisor of the numerators $\{p_j\}$ divided by the least common multiple of the denominators $\{q_i\}$.

Lemma. *If a, b, c, d are four integers with $gcd(a, b) = gcd(c, d) = 1$, then $gcd(ad, bc) = gcd(a, c)gcd(b, d)$.*

Suppose one or more of a, b, c, or d is zero. There are four cases to consider:

Case	a	b	c	d	$gcd(ad, bc)$		$gcd(a, c)gcd(b, d)$
1	0	± 1	c	d	$|c|$	$=$	$|c| \cdot 1$
2	± 1	0	c	d	$|d|$	$=$	$|d| \cdot 1$
3	a	b	0	± 1	$|a|$	$=$	$|a| \cdot 1$
4	a	b	± 1	0	$|b|$	$=$	$|b| \cdot 1$

Thus, the lemma holds if any of a, b, c, or d is zero. If all of a, b, c and d are non-zero we can write

$$(A7) \qquad |a| = \prod_{p\ \text{prime}} p^{\alpha_p}, \quad |b| = \prod_{p\ \text{prime}} p^{\beta_p}, \quad |c| = \prod_{p\ \text{prime}} p^{\gamma_p}, \quad |d| = \prod_{p\ \text{prime}} p^{\delta_p}$$

where $\alpha_p, \beta_p, \gamma_p, \delta_p \geqq 0$, integer, and $\mathrm{Min}\,(\alpha_p, \beta_p) = \mathrm{Min}\,(\gamma_p, \delta_p) = 0$.

We thus have

$$(A8) \qquad gcd(ad, bc) = \prod_{p\ \text{prime}} p^{\mathrm{Min}\,(\alpha_p + \delta_p,\ \beta_p + \gamma_p)}$$

$$(A9) \qquad gcd(a, c)gcd(b, d) = \prod_{p\ \text{prime}} p^{\mathrm{Min}\,(\alpha_p, \gamma_p) + \mathrm{Min}\,(\beta_p, \delta_p)}.$$

To show that $(A8) = (A9)$, it suffices to show that $\mathrm{Min}\,(\alpha_p + \delta_p, \beta_p + \gamma_p) = \mathrm{Min}\,(\alpha_p, \gamma_p) + \mathrm{Min}\,(\beta_p, \delta_p)$ when $\alpha_p, \beta_p, \gamma_p, \delta_p$ are nonnegative integers with $\mathrm{Min}\,(\alpha_p, \beta_p) = \mathrm{Min}\,(\gamma_p, \delta_p) = 0$.

There are four cases to consider:

Case	α_p	β_p	γ_p	δ_p	$\mathrm{Min}\,(\alpha_p + \delta_p, \beta_p + \gamma_p)$		$\mathrm{Min}\,(\alpha_p, \gamma_p) + \mathrm{Min}\,(\beta_p, \delta_p)$	
1	0	β_p	0	δ_p	$\mathrm{Min}\,(\delta_p, \beta_p)$	$=$	0	$+ \mathrm{Min}\,(\beta_p, \delta_p)$
2	0	β_p	γ_p	0	0	$=$	0	$+\ 0$
3	α_p	0	0	δ_p	0	$=$	0	$+\ 0$
4	α_p	0	γ_p	0	$\mathrm{Min}\,(\alpha_p, \gamma_p)$	$= \mathrm{Min}\,(\alpha_p, \gamma_p) +$	0	

Thus (A8)=(A9), so the lemma holds if all of a, b, c, and d are non-zero. This completes the proof of the lemma. ∎

The proof of property 3 now follows.

For $n=1$, the result is trivial.

For $n=2$, we have

$$ggcd(c_1, c_2)=ggcd\left(\frac{p_1}{q_1}, \frac{p_2}{q_2}\right)=\frac{1}{q_1 q_2}\, gcd(p_1 q_2, p_2 q_1)=$$

$$=(\text{by lemma})\,\frac{1}{q_1 q_2}\, gcd(p_1, p_2)gcd(q_1, q_2)=\frac{gcd(p_1, p_2)}{lcm\ (q_1, q_2)}$$

$$\left(\text{since } lcm\ (a, b)=\frac{ab}{gcd(a, b)}\right).$$

Assume property 3 holds for $n=N\geqq 2$. Then for $n=N+1$, we have

$$ggcd(c_1, c_2, \ldots, c_{N+1})=ggcd(ggcd(c_1, x_2, \ldots, c_N), c_{N+1})=$$

(A10)
$$ggcd\left(\frac{gcd(p_1, p_2, \ldots, p_N)}{lcm\ (q_1, q_2, \ldots, q_N)}, \frac{p_{N+1}}{q_{N+1}}\right).$$

Note that $gcd(p_1, p_2, \ldots, p_N)$ and $1\text{ cm }(q_1, q_2, \ldots, q_N)$ must be relatively prime otherwise, some p_j and q_j would have a common factor and thus not be relatively prime. Therefore, (A10) equals

$$\frac{gcd(gcd(p_1, p_2, \ldots, p_N), p_{N+1})}{lcm\ (lcm\ (q_1, q_2, \ldots, q_N), q_{N+1})}=\frac{gcd(p_1, p_2, \ldots, p_{N+1})}{lcm\ (q_1, q_2, \ldots, q_{N+1})}.$$

Thus, property 3 holds for any number of arguments, n. ∎

REFERENCES

[1] Bradley, G. H.: Algorithm and Bound for the Greatest Common Divisor of n Integers, *Communications of the ACM*, 13 (1970), 433–436.

[2] Garfinkel, R. S. and Nemhauser, G. L.: *Integer Programming*, John Wiley & Sons, N. Y., 1972.

[3] Greenberg, H.: *Integer Programming*, Academic Press, N. Y., 1971.

[4] Jeroslow, R. G. and Kortanek, K. O.: On an Algorithm of Gomory, *Siam Journal of Applied Math.*, 21 (1971), 55–59.

[5] Mangasarian, O. L.: *Nonlinear Programming*, McGraw-Hill Book Co., N. Y., 1969.

[6] Meyer, R. R.: On Existence of Optimal Solutions to Integer and Mixed-Integer Programming Problems, *Mathematical Programming*, 7, (1974), 223–235.

[7] Rosen, J. B. and Suzuki, S.: Construction of Nonlinear Programming Test Problems. *Communications of the ACM*, 8, (1965), 113.

[8] Stark, H. M.: *An Introduction to Number Theory*, Markham Publishing Co., Chicago, 1970.

MAINTENANCE SCHEDULING

MONIQUE GUIGNARD and K. SPIELBERG

(Stanford, U.S.A.)

1. INTRODUCTION

Maintenance scheduling problems have been formulated as mixed integer problems for quite a while (see [1, 2] for examples and for other references), but the practical solution of such problems is still not attempted widely, undoubtedly due to their large dimensionality and difficulty. Important exceptions appear to be in the utility industry, as suggested by [3, 4, 5].

One may well wish to take a new look at the situation. Computers and production mixed integer codes are more powerful than ever, and there have also been some advances in the technique of solving mixed integer problems.

We report on work in two areas. They are both, to some extent, related to opportunities opening up with the new "extended control language" of IBM's MPSX–MIP/370 [6, 7, 8], which permits an unprecedented level of *user interaction* with a robust mixed integer programming system. Within such a framework, some large practical problems of special structure may become amenable to solution for the first time.

We first consider certain relaxations of the problem, such as "wraparound" or abandonment of integrality for the start of the maintenance period. We were encouraged to try the latter approach by the success of the work on investment planning reported in [9]. It became apparent later that for our problem this approach amounts to an application of type 2 (S2) "special ordered sets", as described in [10]. However, the absence of an *explicit* S2 feature in our system led us to use equivalent constraint formulations which may have their own utility.

Secondly, we considered the possibility of inserting an enumerative module into the mixed integer production code (via the extended control language of MPSX/370), possibly as a *heuristic aid*. In the course of this investigation, we found that our enumerative code, written and tested as an experimental program in APL, could handle rather large problems quite effectively, at least for the case of ready feasibility (problem not too constrained; this appears to be the most likely case in practice), or for the case of problem infeasibility (which could arise often within a branch and code approach to the overall problem).

The enumeration module could be heuristic within a larger production code treating a somewhat more complex practical situation. It might serve as an efficient tool for producing good feasible solutions (or solution candidates), while the production code would provide all the flexibility, generality and robustness of a well-designed large system.

2. SOME MAINTENANCE PROBLEMS

Consider a set of

machines i, $i=1, 2, \ldots, M$

which are to be removed from service (for maintenance) at the start of *one*

time period t, $t=1, 2, \ldots, T$

over a time horizon of T periods.

Let a *(starting) variable* $x(i, t)=1(0)$

if and only if maintenance of machine i starts at the beginning of period t.

It will be understood that maintenance for a given machine i requires a given number (integral, unless stated otherwise) of contiguous time periods, denoted by $D(i)$.

We shall consider two scheduling situations:

2.1. Scheduling of Labor Requirements

Let $L(i)$ be a vector of length $D(i)$, specifying the labor requirement in each subunit of the maintenance period. If one denotes by $R(t)$ the vector of total labor requirements in periods t, $t=1, 2, \ldots, T$, then one may wish to consider the problems:

P1:
$$\min R$$
$$R \geqq R(t), \quad t=1, 2, \ldots, T$$
$$R(t) = \Sigma_i L(i) \cdot \xi(i, t), \quad t=1, 2, \ldots, T$$

where the *(removal) variable* $\xi(i, t)=1(0)$

if and only if machine i is removed from service in period t.

Note: (i, t) does not appear to be useful for the solution of the problem. The variable $x(i, t)$ is used instead and the constraints are enforced in terms of entries corresponding to the $L(i)$ in the constraint matrix (see example below).

P2:
$$\min \sum_{t=1}^{T} R(t) - \sum_{i=1}^{M} L(i) \cdot \xi(i, t)$$
$$\sum L(i) \cdot \xi(i, t) \leqq R(t) \quad \text{for all } t.$$

492

Note: We have used P2 only for given $R(t)$, in which case the objective function is a constant and the algorithm becomes one of a search for feasibility. Actually, this renders the problem difficult (dually degenerate) and the imposition of a more realistic objective function would appear to be helpful.

The constraints of problems P1 and P2 can be represented in terms of the $x(i, t)$, as seen from the following small.

Example: $M=5, T=8$

i	$D(i)$	$L(i)$		
1	3	4	6	3
2	3	3	2	5
3	3	7	1	1
4	3	1	3	6
5	3	8	9	2

	$i=1$				$i=2$			$i=3$				$i=4$					$i=5$			
	11	12	13	14	21	22	23	32	33	34	35	42	43	44	45	46	53	54	55	
	1	1	1	1																$=1$
					1	1	1													$=1$
								1	1	1	1									$=1$
												1	1	1	1	1				$=1$
																	1	1	1	$=1$
	4				3															$\leq R1$
	6	4			2	3		7				1								$\leq R2$
	3	6	4		5	2	3	1	7			3	1				8			$\leq R3$
		3	6	4		5	2	1	1	7		6	3	1			9	8		$\leq R4$
			3	6			5		1	1	7		6	3	1		2	9	8	$\leq R5$
				3						1	1			6	3	1		2	9	$\leq R6$
											1				6	3			2	$\leq R7$
																6				$\leq R8$

2.2. Scheduling to Maximize Reserve

We now consider a problem for which the *output* of the machines is of paramount importance, such as in the utility industry. Let $a(i)$ stand for the *capacity* of machine i, and let the $L(i)$ be vectors of length $D(i)$ having equal coefficients $a(i)$. Let $R(t)$ then be an upper bound on the amount of capacity which may be removed in time period t.

The objective is to stay away from the upper bounds as much as possible, i.e. to leave the system with as much "reserve" as possible (in some sense).

Denote a constraint set in terms of the $x(i, t)$, as in the example but with all non-zero entries of the $L(i)$ equal to $a(i)$,

$$A \cdot x = b$$

or in structured form

$$M \cdot x = e$$

$$Q \cdot x = R$$

with M representing the multiple choice constraints, so that e is a vector of 1's.
Then the integer program becomes:

P3:
$$\max$$

$$M \cdot x = e$$

$$Q \cdot x + \lambda = R.$$

It is possible to restrict the problem further, in a number of meaningful ways. Certain of the $x(i, t)$ can be set to 0 (the corresponding starting points are "prohibited") by deletion of the column, others can be set to 1 (starting point "fixed"), again be deletion of the column and modification of the right hand side.

Less obviously, one may enforce *precedence* requirements of various types, or *conflict* restrictions which prohibit the simultaneous removal of machines belonging to prespecified classes of machines (e.g., the machines within a certain geographical configuration may constitute a conflict class).

For instance, one may avoid the concurrent outage of machines $i = 1$ and $i = 2$ by adjoining to the matrix $A = (M, Q)$ another matrix of the same size as Q, with coefficients 1 in the place of the $a(i)$ for $i = 1$ and $i = 2$, all other coefficients being 0 and the associated right hand side having coefficients of 1.

3. RELAXED VERSIONS OF THE PROBLEMS

Problem P2, for the data of the example is not feasible with values of the $R(t)$ uniformly ≤ 13. A first attempt at relaxing the problem might be to permit
. all starting points for machine such that completion is guaranteed within the time horizon T.
In addition to the above one might permit *wraparound*, i.e. one would allow
. all starting points within the time horizon, even if that requires the completion of the maintenance period at the "beginning" of the overall period, i.e. at $t = 1$, etc. (the interpretation being that one permits maintenance to be finished "next year").

494

In the example at hand, one might consider all the columns listed below:

$$
\begin{array}{cccccccc}
11 & 12 & 13 & 14 & 15 & 16 & 17 & 18 \\
4 & & & & & & 3 & 6 \\
6 & 4 & & & & & & 3 \\
3 & 6 & 4 & & & & & \\
 & 3 & 6 & 4 & & & & \\
 & & 3 & 6 & 4 & & & \\
 & & & 3 & 6 & 4 & & \\
 & & & & 3 & 6 & 4 & \\
 & & & & & 3 & 6 & 4 \\
\end{array}
$$

Even with wraparound, the example has no solution for $R(t) \leqq 13$.

More interesting is the consideration of *adjacent fractions* $x(i, t)$:

$$
x(i, t) = \varepsilon
$$
$$
x(i, t+1) = 1 - \varepsilon.
$$

Interpretation:

Maintenance starts in period t, ε units from the end of the period (e.g. 5 day week, 52 weeks, $\varepsilon = .2$ ..service starts on Friday of week t).

We impose uniform values $R = R(1) = \ldots = R(t)$. Problem P2 then has many solutions with $R \geqq 8.125$. Two representative solutions are:

$$
R = 8.5
$$

$$
x(1, 4) = 1, \quad x(2, 1) = 1, \quad x(3, 3) = .45833, \quad x(3, 4) = .54167, \quad x(4, 4) = .25,
$$
$$
x(4, 5) = .75, x(5, 7) = .5, \quad x(5, 8) = .5
$$

$$
R = 8.23867
$$

$$
x(1, 6) = .47697, \quad x(1, 7) = .52303, \quad x(2, 1) = .14803, \quad x(2, 8) = .85197,
$$
$$
x(3, 6) = .63816, \quad x(3, 7) = .36184, \quad x(4, 1) = 1, \quad x(5, 3) = .125,
$$
$$
x(5, 4) = .875 \text{ (wraparound)}
$$

3.2. Implementation in MPSX/370

We first implemented the adjacent fractional value formulation (S2 sets) by introducing a new set of integer variables $y(i, t)$, designating the $y(i, t)$ as special zero-one variables (see property below), by relaxing the $x(i, t)$ to be continuous, and by imposing the constraints:

Y-Formulation

$$
x(i, 1) \leqq y(i, 1) \leqq [x(i, T)] + x(i, 1) + x(i, 2)
$$
$$
x(i, 2) \leqq y(i, 2) \leqq x(i, 1) \ + x(i, 2) + x(i, 3)
$$

$$
\cdots\cdots\cdots\cdots\cdots\cdots\cdots
$$

$$
x(i, T-1) \leqq y(i, T-1) \leqq x(i, T-2) + x(i, T-1) + x(i, T)
$$
$$
x(i, T) \quad \leqq y(i, T) \quad \leqq x(i, T-1) + x(i, T) \quad + [x(i, 1)]
$$

The square brackets are applicable to the case with wraparound.

Property. The $y(i, t)$ have exactly two 1's in adjacent positions.

Example. $(1, 1, 0, \ldots, 0)$, $(0, 1, 1, 0, \ldots, 0)$, $\ldots$, $(0, \ldots, 0, 1, 1)$.

Also: The constraints have the (advantageous) form typical of "weakly linked" problems (see [11]), or of "generalized variable upper bounds" [12].

After developing the y-formulation, we found the following, simpler z-formulation in [1].

Z-Formulation

$$x(i, 1) \leqq [z(i, T)] + z(i, 1)$$
$$x(i, 2) \leqq z(i, 1)\ + z(i, 2)$$

$$\cdots\cdots\cdots\cdots$$

$$x(i, T) \leqq z(i, T-1) + z(i, T)$$
$$\Sigma_i z(i, t) = 1 \quad \text{for all } t \text{ (multiple choice)}$$

$x(i, t)$.. continuous, $z(i, t)$..$(0, 1)$ integer.

We used both formulations with MPSX–MIP/370, and found them of similar utility for relatively small problems.

3.3. A small Sample Problem

Consider a problem P3 for maximizing reserve, with the data:

$M = 4, \quad T = 8$
$D(i) = 2 \quad 4 \quad 1 \quad 2$
$A(i) = 200 \quad 200 \quad 90 \quad 300$
$R(t) = 541 \quad 225 \quad 514 \quad 511 \quad 234 \quad 483 \quad 386 \quad 495.$

In Fig. 1 we give the machine output (of an experimental APL program; see next section). Its value (maximal value for minimal reserve) is 34.

BEST POSS. SLACK, BEST SLACK, CURR. SLACK 34 34 34
SCHEDULED NO. OF MACHINES IS 4 OUT OF 4
SCHEDULE, OUTAGE, RESERVE-SLACK, RESERVE

SCHEDULE	OUTAGE	RESERVE-SLACK	RESERVE
	0	507	541
	0	191	225
✱ ○	400	80	114
✱ ○	400	77	111
○	200	0	34
○	200	249	283
□	300	52	86
▽ □	390	61	95

LAMBDA = 34   *Fig. 1*

The printout gives a plot of the schedules for the 4 machines, with the time-axis pointing downward and the machine-axis pointing to the right. For example, the second column represents the outage of machine 2 in the periods 3, 4, 5 and 6, with its capacity of 200 removed in each of these periods.

The first numerical column represents the sum of the capacities removed for the printed schedule. The second numerical column is the difference between the $R(t)$ and the first column. The last column, finally, gives the remaining reserves for each t, the minimal value corresponding to the objective function of the problem.

It is clear that this problem has many degrees of freedom. One could find many alternative, equivalent solutions. The defaults of the program have been set such as to move the schedules as much as possible to the end of the time-horizon, and this is reflected in the fact that the printed solution has the first two time periods free (which may be an advantage: i.e., the slackeness of the system can be exploited for secondary benefits).

In Fig. 2 we give an optimal schedule corresponding to relaxation via wraparound. It can be seen that the schedules for the first two machines occupy the first and last periods of the overall interval. The objective function value is drastically increased. It is, of course, inadvisable to extrapolate such a favorable result to larger problems. But for realistic problems, even small improvements in the minimal reserve may be significant.

BEST POSS. SLACK, BEST SLACK, CURR. SLACK 186 85 85
SCHEDULED NO. OF MACHINES IS 4 OUT OF 4
SCHEDULE, OUTAGE, RESERVE-SLACK, RESERVE

* O	400	56	141
	0	140	225
□	300	129	214
□	300	126	211
	0	149	234
O ▽	290	108	193
O	200	101	186
* O	400	0	85

LAMBDA = 85

Fig. 2

Finally, we present the results for integrality relaxations, obtained with MPSX–MIP/370, in Fig. 3. There is another substantial improvement in objective function value. It may well justify the inconvenience of having to go to split-schedules (represented by the ε).

FRACTIONAL   (MPSX–MIP/370)

K	1	2	3	4	R	Σ	λ SLACK
$D(K)$	2	4	1	2			
$A(K)$	200	200	90	300			
1					541	200	341
2					225	$.356 \times 300 = 107$	118
3					514	$300 + 200 \times .42 = 384$	130
4					511	$200 + 300 \times .644 = 393$	118
5					234	$200 \times .58 = 116$	118
6					483	$200 + 90 = 290$	193
7					386	200	186
8					495	200	295
ε	.42..	—	—	.356			
$1-\varepsilon$	.58..	—	—	.644		$\lambda_{\max} = 118$	

4. AN ENUMERATIVE SCHEDULING SCHEME

A maintenance scheduling problem such as P3, and more complex variants of it, can today be solved with general purpose mixed-integer codes. An industrial user is probably well-advised to go that route, to avoid the necessity of redesigning his code whenever the model changes. But the problems become rather large. A 40 machine, 52 time-period problem expands into an integer program with 2081 integer variables and more than 100 constraints. We have therefore developed an enumerative procedure which handles moderately sized problems very well and could be used as an effective heuristic aid (built into an all-purpose code as an add-on; e.g., into MPSX–MIP/370 via the extended control language) for larger models.

4.1. The Enumeration

We have an experimental APL enumeration program (not available outside of IBM) with the following features:

1. All constraints are treated *implicitly*, i.e. are generated as needed from the data: $M, T, A(i), D(i), R(t)$, and a P by M *conflict matrix* which represents P conflicts, one per row.
2. The method is a modified additive algorithm ([13], also [14]), starting at the 0 origin and scheduling outages one by one. The machines are either preordered

498

absolutely in terms of descending $A(i).D(i)$, or ordered in terms of conflict classes, the above ordering being respected within each class. A dynamic ordering could be imposed in the sense of *minimal preferred variable* inequalities [15], but has not been found necessary so far.

3. The book-keeping is implemented through binary matrices of sizes M by T, and therefore is modest in terms of storage requirement.

4. A row of the conflict matrix has entries of 1 for all machines in potential conflict with each other (the conflicts are assumed to impose a partitioning; this could easily be relaxed). If machine k', of maintenance duration d' has been scheduled ("fixed") to start maintenance at time t' in the search, and another machine (k, d, t) is considered subsequently (at larger "levels") of the search, then an entire string of starting points is prohibited for machine k (over successor nodes of the enumeration tree).

The prohibited string is determined by (wraparound not being considered here):

$$\text{starting point:} \quad ts0 = t' + 1 - d$$
$$ts = \max\{1, ts0\}, \qquad \text{and}$$

$$\text{length:} \qquad L = d' + d - 2 \qquad \text{if} \quad ts0 > 0$$
$$= (d' + d - 2) - (1 - ts0) \quad \text{if} \quad ts0 \leqq 0.$$

5. Upper bounds on λ are easily computable and may be used to stop the search when one has an incumbent solution close enough to the bound.

4.2. Results for Some Large Problems

We give two experimental results. In the first case we consider 50 machines parti tioned into 6 conflict classes plusone non-conflict class (indicated in the plot by changing character representation from class to class). It can be seen that the problem is quite tightly constrained by virtue of the conflicts. In the first conflict all but three time periods must be used if there is to be no conflict. As a consequence, the search for feasibility and then optimality is not too easy, and optimality was not attained after about 5 minutes (on an APL time sharing system). But there were many solutions and the plotted schedule is within 119 units of optimality.

In the second plot we have a 40 machine, 52 time period problem with no conflicts, it is clearly not very tightly constrained, and the depicted solution is found and attested optimal within a few minutes. It can again be seen, that the defaults of the system tend to create a schedule which leaves initial periods unused. In this case the entire maintenance requirement could be compressed into 44 time periods only.

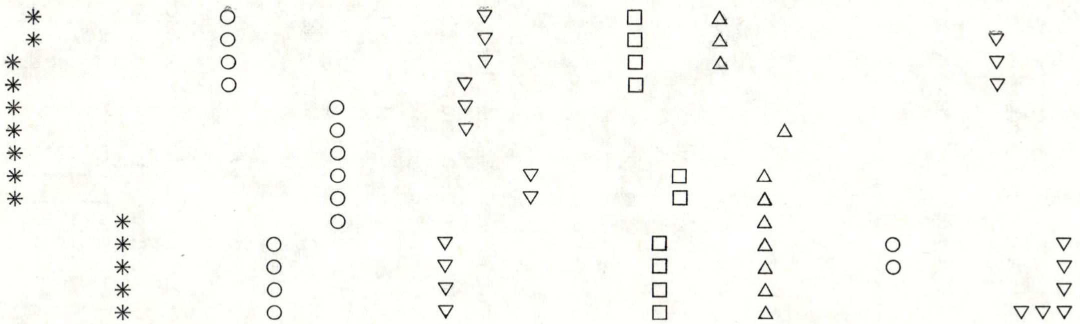

500

50 MACHINES
52 TIME PERIODS

ITERATION   100
STOPPED AT   500

SOLUTION   913
BEST POSS.:   1032

592	275
695	228
781	820
661	58
385	26
480	1507
198	97
518	731
518	703
157	32
692	843
692	751
533	94
735	88

BEST SLACK, CURRENT SLACK 913 913
SCHEDULED NO. OF MACHINES IS 50 OUT OF 50
SCHEDULE, OUTAGE, RESERVE-SLACK

NO CONFLICTS

221	1742
223	1674
254	1437
254	313
317	746
317	700
166	29
201	616
370	771
676	1065
850	19
977	470
1237	490
1390	29
1390	199
1390	11
1416	55
227	0
463	1298
463	574
644	271
56	63
253	1162
402	563
507	816
425	928
369	18
498	1323
498	1221
498	1271
512	563
216	51
299	590
497	1470
368	51
558	323
558	61
592	1159

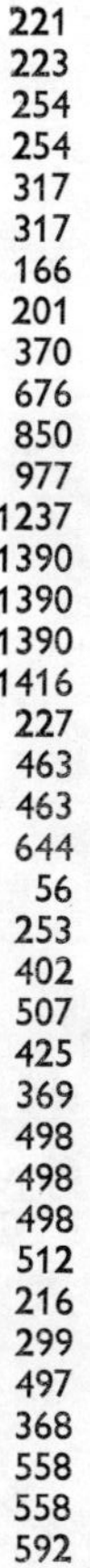

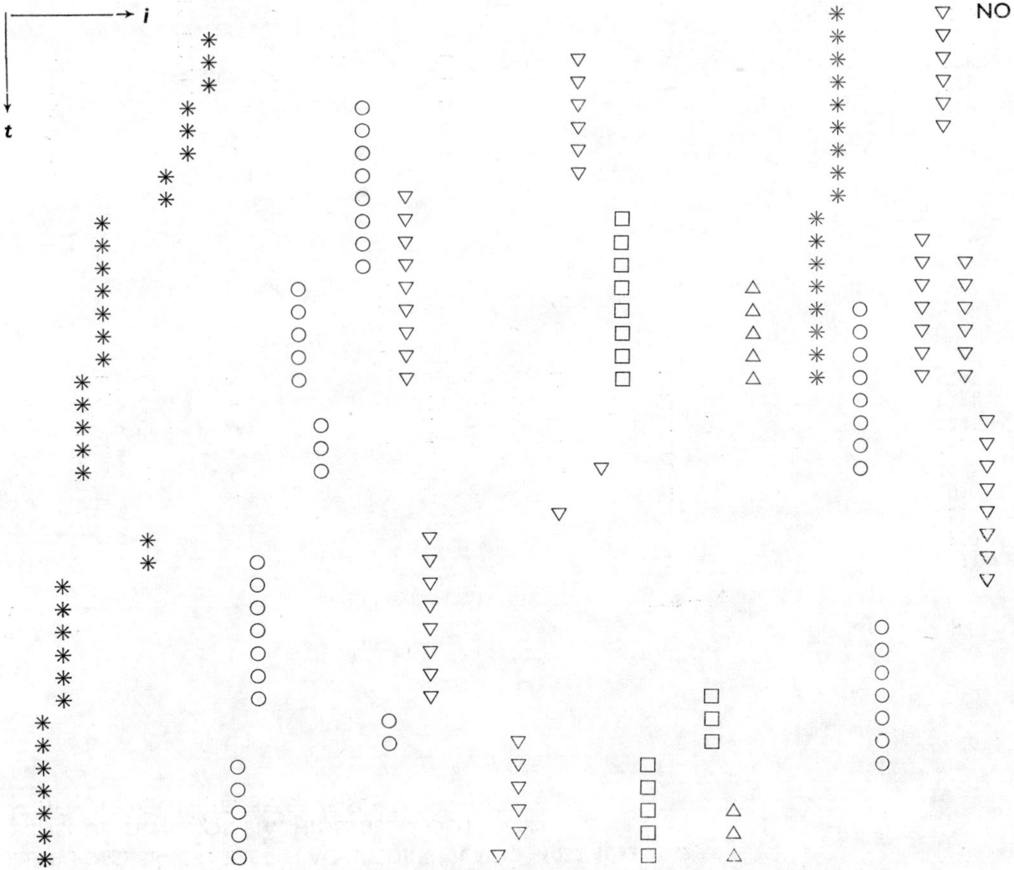

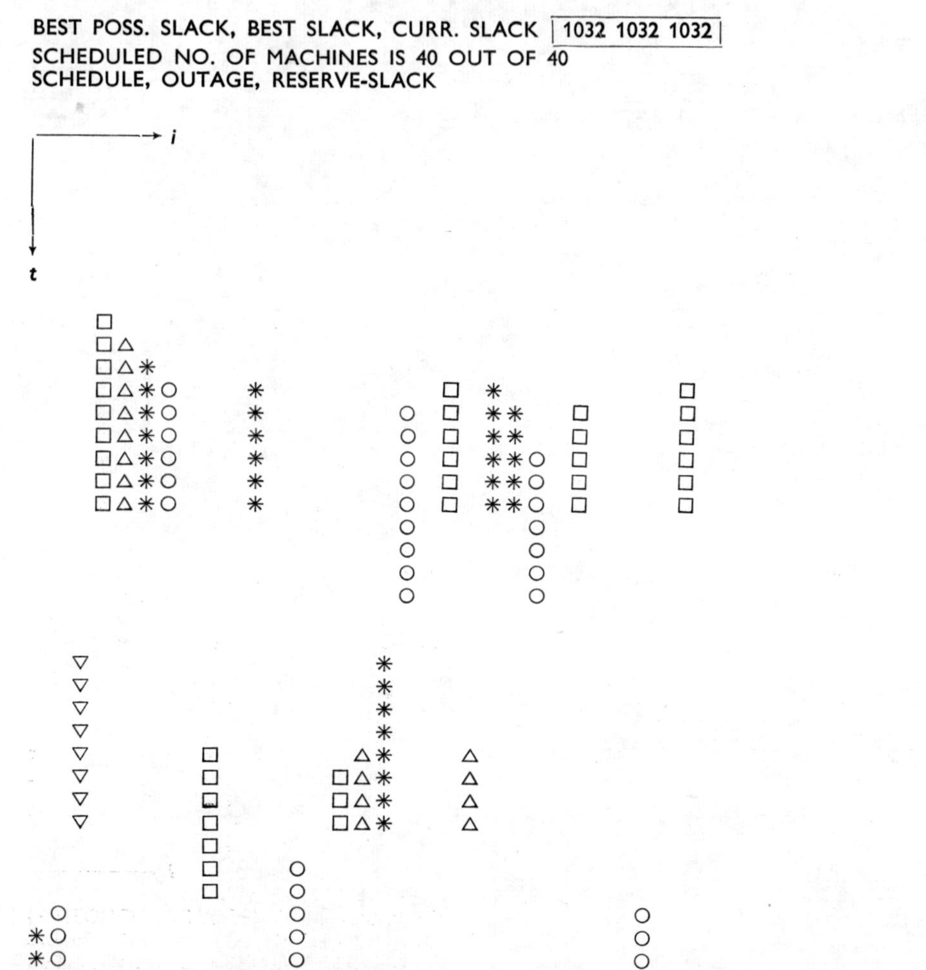

BEST POSS. SLACK, BEST SLACK, CURR. SLACK   1032 1032 1032
SCHEDULED NO. OF MACHINES IS 40 OUT OF 40
SCHEDULE, OUTAGE, RESERVE-SLACK
i
t
0   1844
0   1778
0   1572
0   448
0   944
0   898
0   76
0   698
171   851
359   1263
533   217
1010   318
1193   415
1193   107
1241   229
1241   41
1241   111
97   11
97   1545
97   821
97   699
0   0
0   1296
249   597
249   955
249   985
249   19
598   1104
778   822
778   872
778   178
135   13
200   570
200   1648
297   3
497   265
497   3
SAVED TIME PERIODS
(DEFAULTS FOR TIES)

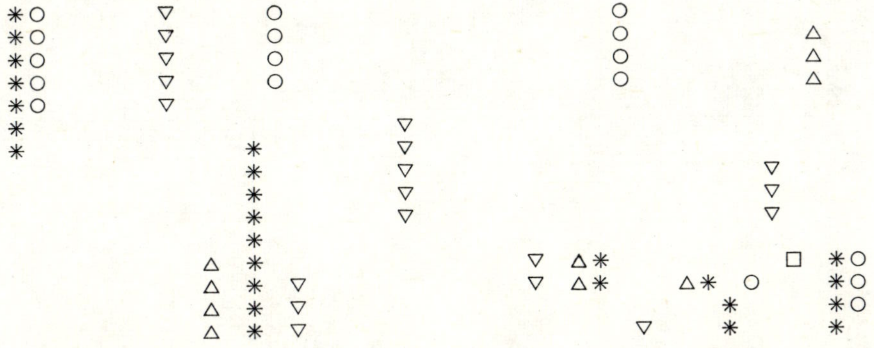

691   941
705   43
705   99
705   777
592   8
283   9
353   1515
171   5
171   959
171   931
70   0
752   664
1317   7
508   0
703   1

TERMINATION AT IT=42

40 MACHINES
52 TIME PERIODS

REFERENCES

[1] Wagner, H. M.: *Principles of Operations Research,* Prentice Hall, Englewood Cliffs, N. J., 1969.

[2] Wagner, H. M., Giglio, R. J., Glaser, R. G.: Preventive Maintenance Scheduling by Math. Programming, *Mgmt. Sc.,* 10 (1964), 2, Jan.

[3] Kingston, P. L., Lipton, S. L., Stojka, J.: Mixed Integer Programming, Models for Generator Maitenance Scheduling, Working Paper, Niagara Mohawk Power Co. (to be presented at the ORSA/TIMS Meeting, Miami, No. 1976).

[4] Dopaz, J. F., Merrill, H. M.: Optimal Generator Maintenance Scheduling using Integer Programming, IEEE Trans. Power Appl. & Systems, Vol PAS-94, No. 5, Sept. 1975.

[5] Merrill, H. M.: Power Plant Maintenance Scheduling with Integer Programming, IEEE Tutorial Course, 76 CH1107-2-PWR, 1976.

[6] IBM Math. Progr. System Extended/370, (MPSX/370), Program Ref. Manual SH19–1095, Mixed Integer Progr./370 (MIP/370), Program Ref. Manual SH19–1099, Dec. 1974.

[7] Benichou, M., Gauthier, J. M., Hentges, G., Ribiere, G.: The Efficient Solution of Large Scale Linear Programming Problems. Some Algorithmic Techniques and Computational Results., 9th Int. Symp. on Math. Progr., Budapest, 1976.

[8] Slate, L., Spielberg, K.: The Extended Control Language of MPSX/370 (and possible applications), IBM Scientific Marketing, White Plains, N. Y., 1976.

[9] Chen, R., Crowder, H., Johnson, E. L.: An Integer Programming Formulation of the Installment Scheduling Problem, IBM Th. J. Watson Research Center, Yorktown Heights, N. Y., 1976.

[10] Beale, E. M. L., Tomlin, J. A.: Special Facilities in a General Math. Progr. System for Nonconvex Problems using Ordered Sets of Variables, *Proc. 5th Int. Conf. on Operational Research* (ed. J. Lawrence), Tavistock Publ., London 1970.

[11] Balinski, M. L., Spielberg, K.: *Methods for Integer Programming: Algebraic, Combinatorial and Enumerative,* in *Progr. in OR,* Vol III (ed. J. Aronofsky), J. Wiley & Sons, N. Y. 1969.

[12] Schrage, L.: Implicit Representation of Generalized Upper Bunds in Linear Programs, Report 7543, Dept. Econ. Grad. School of Bus., Univ. of Chicago, Oct. 1975.

[13] Balas, E.: An Additive Algorithm for Solving Linear Programs with Zero-One Variables, *J. ORSA,* 13 (1965).

[14] Lemke, C. E., Spielberg, K.: Direct Search Zero-One and Mixed Integer Programming, *J. ORSA,* 15 (1967).

[15] Guignard, M., Spielberg, K.: The State Enumeration Method for Mixed Zero-One Programming, IBM Phil. Sc. Center Report, #320 3024, 1971.

THEORETICAL AND SIMULATION STUDIES ON COMPUTATIONAL PERFORMANCE OF BRANCH-AND-BOUND ALGORITHMS

TOSHIHIDE IBARAKI

(Kyoto, Japan)

INTRODUCTION

Branch-and-bound is a computational principle which can be applied to a wide variety of combinatorial optimization problems such as the integer programming problem, various scheduling problems and network problems. In the first half of this paper, the recent theoretical results on the computational performance are summarized through the use of general model of branch-and-bound algorithms which utilize the lower bound test, the upper bound test and the dominance test. It is investigated how strengthening each test contributes to the improvement of the computational efficiency, and how the introduction of an allowance function (i.e., being satisfied with a suboptimal solution rather than the exact optimal solution) affects the total computation steps. Since only weak properties can be proved theoretically, however, some other properties, which are more useful in practical applications, are then investigated by simulation in the second half of this paper. Although this experiment is rather preliminary, the obtained results appear to be stable and meaningful conclusions may be drawn. As an example, a decision rule for selecting appropriate search strategies in branch-and-bound algorithms under various situations is proposed in the final section.

1. CONSTRUCTION OF BRANCH-AND-BOUND ALGORITHMS

Five constituents of branch-and-bound algorithms are first introduced in Sections 1.1–1.5, and a formal description of a branch-and-bound algorithm is given. We assume that the reader is familiar with basic notions of branch-and-bound. The branch-and-bound algorithms discussed here are quite general since the dominance test is incorporated as well as the ordinary lower bound test and upper bound test.

The decomposition process (when all possible decompositions are performed) of a given minimization problem P_0 is represented by a *finite rooted tree* $\mathcal{B}=(\mathcal{P}, \mathcal{E})$, where $\mathcal{P}$ is a set of nodes with root P_0 (corresponding to the given problem) and $\mathcal{E}$ is a set of arcs. Each node $P_i \in \mathcal{P}$ represents a partial problem P_i generated from P_0. $(P_i, P_j) \in \mathcal{E}$ iff P_j is generated from P_i by decomposition; P_j is called a *son* of P_i. P_j is a *descendant* of P_i (i.e., P_i is an *ancestor* of P_j) if $P_j = P_i$ or there exist $P_{i_1}, P_{i_2}, \ldots, P_{i_s}$ ($s \geq 0$) such that $(P_i, P_{i_1}) \in \mathcal{E}$, $(P_{i_1}, P_{i_2}) \in \mathcal{E}, \ldots, (P_{i_s}, P_j) \in \mathcal{E}$. A descendant (ancestor) P_j of P_i is *proper* if $P_j \neq P_i$. A set of nodes $\mathcal{A} \subset \mathcal{P}$ is *independent* if no node in $\mathcal{A}$ is a proper ancestor of the other. The set of *leaf nodes* in $\mathcal{P}$ is denoted $\mathcal{T}$. The *depth* of $P_i \in \mathcal{P}$, denoted $d(P_i)$, is the length of the path from P_0 to P_i.

Let $f: \mathcal{P} \to E \cup \{\infty\}$ denote the optimal values of partial problems, where E is the set of real numbers; $f(P_i) = \infty$ if P_i is infeasible. f satisfies

$$f(P_i) = \min \{f(P_j) \mid (P_i, P_j) \in \mathcal{E}\}$$

and hence

$$f(P_i) \leq f(P_j) \quad \text{for} \quad (P_i, P_j) \in \mathcal{E}.$$

Let $O(P_i)$ denote the set of optimal solutions of P_i. $O(P_i)$ satisfies

$$O(P_i) = \cup \{O(P_j) \mid f(P_i) = f(P_j), (P_i, P_j) \in \mathcal{E}\}.$$

$(\mathcal{B}, O, f)$ (O is sometimes omitted) is called the *branching structure* of P_0.

1.2 Lower Bounding Function g

In executing a branch-and-bound algorithm, $f(P_i)$ is usually not known, but a *lower bounding function* $g(P_i)$ is evaluated for each generated P_i. $g: \mathcal{P} \to E \cup \{\infty\}$ satisfies the following conditions.

(a) $\qquad\qquad\qquad\qquad g(P_i) \leq f(P_i) \quad \text{for} \quad P \in \mathcal{P}$
(b) $\qquad\qquad\qquad\qquad g(P_i) = f(P_i) \quad \text{for} \quad P_i \in \mathcal{T}$
(c) $\qquad\qquad\qquad\qquad g(P_i) \leq g(P_j) \quad \text{for} \quad (P_i, P_j) \in \mathcal{E}.$

$\mathcal{G}$ denotes the set of nodes P_i which are incidentally solved or $O(P_i) \cap O(P_0) = \emptyset$ is concluded, in the computing process of g. It satisfies

(A)* $\qquad\qquad\qquad\qquad g(P_i) = f(P_i) \quad \text{for} \quad P_i \in \mathcal{G}$
(B) $\qquad\qquad\qquad\qquad\qquad \mathcal{G} \supset \mathcal{T}$
(C) $\qquad\qquad P_i \in \mathcal{G} \quad \text{implies} \quad P_j \in \mathcal{G} \quad \text{for} \quad (P_i, P_j) \in \mathcal{E}.$

* When $O(P_i) \cap O(P_0) = \emptyset$ is concluded, the value $g(P_i)$ itself is not relevant to the computation process of a branch-and-bound algorithm. Thus condition (A) is assumed for simplicity even though it is not actually satisfied.

For each P_i tested in a branch-and-bound algorithm, a good feasible solution may sometimes be obtained by simple calculation. This gives rise to an *upper bounding function* $u : \mathcal{P} \to E \cup \{\infty\}$ satisfying the following conditions.

(I) $\qquad\qquad\qquad\qquad\qquad\qquad u(P_i) \geqq f(P_i)$ for $P_i \in \mathcal{P}$

(II) $\qquad\qquad\qquad\qquad\qquad\qquad u(P_i) = f(P_i)$ for $P_i \in \mathcal{G}$.

$u(P_i) = \infty$ if no feasible solution is obtained, or if the calculation of $u(P_i)$ is not attempted.

$u = \infty$ denotes that $u(P_i)$ is never computed, and $u = u(P_0)$ denotes that $u(P_i)$ is computed only for P_0 (the given problem). Note however that $u(P_i) = f(P_i)$ for $P_i \in \mathcal{G}$ are assumed even in these cases.

1.4 Dominance Relation D

Many of the existing branch-and-bound algorithms also incorporate a test based on a *dominance relation D*. *D* is a *partial ordering* on $\mathcal{P}$ satisfying the following conditions.

(1) *When all optimal solutions of P_0 are sought:*

(i) $P_i D P_j \wedge P_i \neq P_j$ implies $f(P_i) < f(P_j)$.

(ii) $P_i D P_j \wedge P_i \neq P_j$ implies that there exists a descendant $P_{i'}$ of P_i satisfying $P_{i'} D P_{j'}$ for each descendant $P_{j'}$ of P_j.

(2) *When only a single optimal solution of P_0 is sought:*

(i) $P_i D P_j \wedge P_i \neq P_j$ implies $f(P_i) \leqq f(P_j)$ and that P_i is not a proper descendant of P_j.

(ii) $P_i D P_j \wedge P_i \neq P_j$ implies that there exists a descendant $P_{i'}$ of P_i satisfying $P_{i'} D P_{j'}$ for each descendant $P_{j'}$ of P_j.

(iii) There exists no sequence of nodes $P_{i_1}, P_{i_2}, \ldots, P_{i_{k+1}}$ ($k \geqq 2$ and $P_{i_1}, P_{i_2}, \ldots, P_{i_k}$ are distinct), generated during computation, such that P_{i_s} is a proper descendant of $P_{i_{s+1}}$ or $P_{i_s} D P_{i_{s+1}} \wedge f(P_{i_s}) = f(P_{i_{s+1}})$, for $s = 1, 2, \ldots, k$, and $P_{i_{k+1}} = P_{i_1}$.

$D = I$ (identity relation) denotes that the test based on a dominance relation is not effective.

1.5 Search Strategies

The order of partial problems P_i tested in a branch-and-bound algorithm is specified by a *search function* $s : \Pi \to \mathcal{P}$ such that $s(\mathcal{A}) \in \mathcal{A}$ for $\mathcal{A} \in \Pi$, where Π denotes the family of independent subsets of $\mathcal{P}$. Various types of search functions are used in existing algorithms.

s is a *heuristic search function* based on *heuristic function* $h: \mathcal{P} \to E$ if

$$h(s(\mathcal{A})) = \min \{h(P_i) \mid P_i \in \mathcal{A}\}$$

holds for $\mathcal{A} \in \Pi$. In this case, s is denoted s_h. It is usually assumed that $h(P_i) \neq h(P_j)$ for $P_i \neq P_j$ by using an appropriate tie breaking rule if necessary. In particular, $s = s_g$ is called *best-bound search function*.

Let

$$\bar{N}(\mathcal{A}) = \{P_i \in \mathcal{A} \mid d(P_i) = \max \{d(P_j) \mid P_j \in \mathcal{A}\}\}.$$

Then the *depth-first search function* based on h, denoted $\bar{s}_h$, satisfies

$$h(\bar{s}_h(\mathcal{A})) = \min \{h(P_i) \mid P_i \in \bar{N}(\mathcal{A})\}$$
$$\bar{s}_h(\mathcal{A}) \in \bar{N}(\mathcal{A}).$$

On the other hand, the *breadth-first search function* based on h, denoted $\tilde{s}$ satisfies

$$h(\tilde{s}_h(\mathcal{A})) = \min \{h(P_i) \mid P_i \in \tilde{N}(\mathcal{A})\}$$
$$\tilde{s}_h(\mathcal{A}) \in \tilde{N}(\mathcal{A}),$$

where

$$\tilde{N}(\mathcal{A}) = \{P_i \in \mathcal{A} \mid d(P_i) = \min \{d(P_j) \mid P_j \in \mathcal{A}\}\}.$$

It is known that a heuristic search function s_h is most general among the above search strategies, in the sense that the other three can be viewed as s_h with special h.

A *depth-m search function* based on h, denoted $s_h(m)$, is a further generalization of the above search functions. In depth-m search, current active nodes are stored in an ordered linear list L, and $s_h(m)$ selects the last node in L. Initially L stores only P_0, and then it is modified as follows.

(i) If the selected node P_i is decomposed into k sons, the last $m-1$ nodes in L (after eliminating P_i) and the k generated nodes are rearranged in the decreasing order of h and stored in the last $m+k-1$ positions of L.

(ii) If the selected node P_i is terminated by some test, the last m nodes in L (after eliminating P_i) is rearranged in the decreasing order of h.

The detailed description and its properties are found in [8]. $s_h(1)$ is equivalent to depth-first search function $\bar{s}_h$, and $s_h(\infty)$ is equivalent to heuristic search function s_h. By changing m form 1 to ∞, $s_h(m)$ can realize a wide spectrum of intermediate search functions. $s_h(m)$ is not generally a heuristic search function.

1.6 Description of Branch-and-Bound Algorithm

In the following, $\mathcal{M} \subset \mathcal{P}$ denotes the set of nodes currently generated. A node in $\mathcal{M}$ is *active* if it is yet neither tested nor decomposed into smaller partial problems. $\mathcal{A}$ denotes the set of current active nodes. $\mathcal{A}$ is always an independent set. O denotes the set of best feasible solutions currently available. z, called the *incumbent value*,

508

denotes the best upper bound of $f(P_0)$. In general, $z \leq f(O)$ (i.e., $f(x)$ for $x \in O$), but $z = f(O)$ holds if $u = \infty$ or if the computation has terminated. It is assumed that $O(P_i)$ is obtained as a by-product of testing P_i, in case $P_i \in \mathcal{G}$ and $f(P_i) \leq z$.

(1) Branch-and-bound algorithm $A = ((\mathcal{B}, O, f), (\mathcal{G}, g, u), D, s)$: All optimal solutions of P_0

A1 (Initialize): $\mathcal{A} \leftarrow \{P_0\}$, $\mathcal{N} \leftarrow \{P_0\}$, $z \leftarrow \infty$, $O \leftarrow \emptyset$.

A2 (Search): If $\mathcal{A} = \emptyset$, go to A9; else $P_i \leftarrow s(\mathcal{A})$, $z \leftarrow \min[z, u(P_i)]$ and go to A3.

A3 (Test by $\mathcal{G}$): If $P_i \in \mathcal{G}$, go to A7; else go to A4.

A4 (Lower bound test): If $g(P_i) > z$, go to A8; else go to A5.

A5 (Dominance test): If $\exists P_k (\neq P_i) \in \mathcal{N}$ satisfying $P_k D P_i$, then go to A8; else go to A6.

A6 (Decompose): Generate sons $P_{i_1}, P_{i_2}, \ldots, P_{i_k}$ of P_i. Return to A2 after letting $\mathcal{A} \leftarrow \mathcal{A} \cup \{P_{i_1}, \ldots, P_{i_k}\} - \{P_i\}$ and $\mathcal{N} \leftarrow \mathcal{N} \cup \{P_{i_1}, \ldots, P_{i_k}\}$.

A7 (Improve): Go to A8 after letting

$$O \leftarrow \begin{cases} O(P_i) & \text{if } f(P_i) < f(O) \\ O \cup O(P_i) & \text{if } f(P_i) = f(O) \\ O & \text{otherwise.} \end{cases}$$

A8 (Terminate P_i): $\mathcal{A} \leftarrow \mathcal{A} - \{P_i\}$, and return to A2.

A9 (Halt): Halt. $O = O(P_0)$ and $z = f(P_0)$ hold.

(2) Branch-and-bound algorithm $A = ((\mathcal{B}, f), (\mathcal{G}, g, u), D, s)$: A single optimal solution of P_0

It is assumed that a feasible solution x of P_i satisfying $f(x) = u(P_i)$ is obtained in computing $u(P_i)$ if $u(P_i) < \infty$. In particular, an optimal solution of P_i is obtained if $P_i \in \mathcal{G}$. In the following algorithm, O stores at most one solution x, and $z = f(x)$ always holds during computation.

The algorithm description in this case is obtained by changing the above A2, A3, A4 and A9. Thus only these steps are given below. A7 may be eliminated since it is never executed.

A2 (Search): If $\mathcal{A} = \emptyset$, go to A9; else $P_i \leftarrow s(\mathcal{A})$, $z \leftarrow \min[z, u(P_i)]$,

$$O \leftarrow \begin{cases} O & \text{if } u(P_i) \geq z \\ \{x\}, & \text{otherwise,} \end{cases}$$

where x is a feasible solution of P_i with $f(x) = u(P_i)$, and go to A3.

A3 (Test by $\mathcal{G}$): If $P_i \in \mathcal{G}$, go to A8; else go to A4.

A4 (Lower bound test): If $g(P_i) \geq z$, go to A8; else go to A5.

A9 (Halt): Halt. x stored in O is an optimal solution of P_0 (i.e., $f(x) = f(P_0)$), and z is its value.

The finiteness and correctness of the above two types of algorithms (or special cases of them) may be found in references such as [1, 2, 5, 6, 7, 16, 18, 19, 20, 22].

509

The following parameters may be useful to measure the computational efficiency of a branch-and-bound algorithm A.

$T(A)$: The number of nodes decomposed in A6 prior to termination in A9.

$B(A)$: The number of nodes decomposed in A6 prior to the last modification of O occurred in A7 (or A2 when a single optimal solution is sought).

$F(A)$: The number of nodes decomposed in A6 prior to the first modification of O (i.e., when O becomes $O \neq \emptyset$).

$M(A)$: The maximum size of $|\mathcal{A}|$ attained during computation of A.

It is obvious that $T(A)$ is relevant to the total computation time of A, and $B(A)$, $F(A)$ are relevant to the quality of solutions stored in O when the computation is cut off before the normal termination in A9, due to the insufficiency of the available computer time. $M(A)$ represents the memory space required to execute A.

In the subsequent discussion, subscripts a and s are sometimes added, e.g., A_a, A_s, $T_a(A)$, $B_s(A)$, to distinguish the case of all optimal solutions and the case of a single optimal solution, respectively; no subscript is added if it is not necessary to distinguish them.

2. THEORETICAL RESULTS

2.1 Theoretical Bounds on $T(A)$

Given $A = ((\mathcal{B}, O, f), (\mathcal{G}, g, u), D, s)$, where $D = I$, $T(A)$ is bounded below as follows ([20, 12] and possibly in others).

$$T_a(A) \geq |\mathcal{F} - \mathcal{G}|$$
$$T_s(A) \geq |\mathcal{H} - \mathcal{G}|$$
$$\mathcal{F} = \{P_i \in \mathcal{P} \mid g(P_i) \leq f(P_0)\}$$
$$\mathcal{H} = \{P_i \in \mathcal{P} \mid g(P_i) < f(P_0)\}.$$

It is interesting to see that these bounds are actually attained if best-bound search is used [18, 20], i.e., $A = ((\mathcal{B}, O, f), (\mathcal{G}, g, u), D = I, s_g)$ satisfies

$$T_a(A) = |\mathcal{F} - \mathcal{G}|$$
$$T_s(A) \leq |\mathcal{F} - \mathcal{G}|.$$

(Note that $|\mathcal{F} - \mathcal{G}| \geq |\mathcal{H} - \mathcal{G}|$ holds but the difference is usually negligible.) Best-bound search, however, is not recommended from the view point of $B(A)$ and $F(A)$, since

$$B_a(A) = T_a(A)$$

holds in general, and furthermore

$$F_a(A) = T_a(A), \quad B_s(A) = F_s(A) = T_s(A)$$

510

holds if $u=\infty$. Namely (good) feasible solutions are stored in O at the last stage of the entire computation. This property of best-bound search is true even if D is taken into account.

For general case $D\neq I$, it seems difficult to derive similar bounds unless D is somewhat restricted. Now call that D is *consistent with heuristic function h* if P_iDP_j implies $h(P_k)<h(P_j)$ for any proper ancestor P_k of P_i. In addition, D is called to be *consistent with lower bounding function g* if P_iDP_j implies $g(P_i)<g(P_j)$ $(g(P_i)\leq f(P_j)$ if a single optimal solution is sought). D is consistent with h in the following important special cases.

(a) $h(P_k)<h(P_l)$ if P_l is a son of P_k, and $P_iDP_j\wedge P_i\neq P_j$ implies $h(P_i)<h(P_j)$.

(b) $h=g$ (i.e., best-bound search) and D is consistent with g (and $g(P_k)<g(P_l)$ for a son P_l of P_k if a single optimal solution is sought).

(c) $s=s_h$ is a breadth-first search function, and P_iDP_j implies $d(P_i)\leq d(P_j)$.

Under the above two consistency assumptions (in case of (c), D may not necessarily be consistent with g), we have [12]

$$|\mathcal{F}\cap\mathcal{P}_D-\mathcal{G}|\leq T_a(A)\leq|\mathcal{P}_D-\mathcal{G}|$$
$$|\mathcal{H}\cap\mathcal{P}_D-\mathcal{G}|\leq T_s(A)\leq|\mathcal{P}_D-\mathcal{G}|,$$

where

$$\mathcal{P}_D=\{P_i\in\mathcal{P}\,|\,P_i \text{ is minimal with respect to partial ordering } D\}.$$

Furthermore, $T_a(A)=|\mathcal{F}\cap\mathcal{P}_D-\mathcal{G}|$ holds if $s=s_g$.

The upper bound $|\mathcal{P}_D-\mathcal{G}|$ is sometimes used to find very efficient branch-and-bound algorithms for some problems (i.e., when $\mathcal{P}_D-\mathcal{G}$ is a very small subset of $\mathcal{P}$) [12].

2.2 Role of Heuristic Function h

In this section, it is first discussed what is the theoretical goal in designing a heuristic function h. h is called *nonmisleading* if $h(P_i)<h(P_j)$ implies $f(P_i)\leq f(P_j)$ for $P_i, P_j\in\mathcal{P}$. A branch-and-bound algorithm A with a nonmisleading h is most efficient in both senses of $T(A)$ and $B(A)$, i.e.,

$$T_a(A)=|\mathcal{F}\cap\mathcal{P}_D-\mathcal{G}|$$
$$T_s(A)\leq|\mathcal{H}\cap\mathcal{P}_D-\mathcal{G}|+|\mathcal{J}\cap\mathcal{K}\cap\mathcal{P}_D-\mathcal{G}|$$
$$B_s(A)\leq B_a(A)\leq|\mathcal{J}\cap\mathcal{P}_D-\mathcal{G}|$$
$$\mathcal{K}=\{P_i\in\mathcal{P}\,|\,g(P_i)=f(P_0)\}$$
$$\mathcal{J}=\{P_i\in\mathcal{P}\,|\,f(P_i)=f(P_0)\}$$

if D is consistent with g and h, or $D=I$ (note $\mathcal{P}_{D=I}|=\mathcal{P}$) [4, 9].

It is of course not reasonable to assume that a nonmisleading h is easily obtainable since it requires almost complete knowledge of f which is usually unknown. When general heuristic search is concerned, however, it is known as shown below that

$T(A)$, $B(A)$ become smaller if h is closer to nonmisleading. If h is almost nonmisleading, the computational result is also close to the nonmisleading case (i.e., T, B are stable with respect to h). Thus we should try to design a heuristic function which is as close as possible to nonmisleading.

Now let $\mathcal{P}=\{P_{i_1}, P_{i_2}, \ldots, P_{i_m}\}$ and let h satisfy

$$h(P_{i_1}) < h(P_{i_2}) < \ldots < h(P_{i_m}).$$

h' is said to be *contiguous* to h if h' satisfies for some $p(1 \leqq p \leqq m-1)$ that

$$h'(P_{i_1}) < \ldots = h'(P_{i_{p-1}}) < h'(P_{i_{p+1}}) < h'(P_{i_p})$$
$$< h'(P_{i_{p+2}}) < \ldots < h'(P_{i_m}).$$

Denote $h' \triangleright h$ if h and h' are contiguous and furthermore $f(P_{i_{p+1}}) > f(P_{i_p})$ or $f(P_{i_{p+1}}) = f(P_{i_p}) \wedge (P_{i_{p+1}} \in \mathcal{G} \Rightarrow P_{i_p} \in \mathcal{G})$. Then branch-and-bound algorithms $A' = A(h')$ and $A = A(h)$ with $D = I$ and $u = \infty$ (or $u = z(P_0)$) satisfy

$$T_a(A) \leqq T_a(A') \leqq T_a(A) + 1$$
$$T_s(A) \leqq T_s(A') \leqq T_s(A) + 1$$
$$B_s(A) \leqq B_s(A') \leqq B_s(A) + 1$$

if $h' \triangleright h$. In addition, for any h, there is a sequence of h_i such that

$$h = h_0 \triangleright h_1 \triangleright h_2 \triangleright \ldots \triangleright h_s = h^*,$$

where h^* is nonmisleading [9]. In other words, improving h in the direction of $\triangleright$ always results in more efficient algorithms. It is also possible to extend this result to algorithms using D which is consistent with both g and h.

It should be emphasized here that the above results cannot be extended to the class of depth-first search functions; a depth-first search function $\bar{s}_h$ with h closer to nonmisleading may sometimes give larger $T(A)$ and $B(A)$.

A breadth-first search function (viewed as a heuristic search function s_h) is not desirable from the above view point since it is usually very far from nonmisleading. This is not the case, however, if a strong dominance relation consistent with h is available.

2.3 Power of Lower Bounding $\mathcal{G}$ and g

We consider here whether computational efficiency always becomes higher if $\mathcal{G}$ and g are improved. Let

$$g_1 \geqq g_2 \Leftrightarrow \mathcal{G}_1 \supset \mathcal{G}_2 \wedge g_1(P_i) \geqq g_2(P_i) \quad \text{for} \quad P_i \in \mathcal{P}.$$

Table 1 shows whether g_1 with $g_1 \geqq g_2$ improves T and B under various assumptions on D and search strategies. The results for depth-first search and breadth-first search using D consistent with g are proved in [16], in which it is also shown that $T_s(A_1) \leqq \leqq T_s(A_2)$ may not hold for best-bound search. [7] treats the case of best-bound search with $D = I$ (and some additional assumptions). Other results are found in [14].

512

Table 1. Computational Efficiency of $A_1 = A(g_1)$ and $A_2 = A(g_2)$ with $g_1 \geq g_2$
[— in the following entries indicates that a monotonic change in T (or B) does not necessarily hold]

Search strategies / D	General	Consistent with g
Heuristic (general)	— —	$T(A_1) \leq T(A_2)$ $B(A_1) \leq B(A_2)$
Heuristic (nonmisleading)	$B_a(A_1) = B_a(A_2)$ $B_s(A_1) \leq B_s(A_2)$	$T(A_1) \leq T(A_2)$ $B(A_1) \leq B(A_2)$
Depth-first Breadth-first	— —	$T(A_1) \leq T(A_2)$ $B(A_1) \leq B(A_2)$
Best-bound[a]	— — — —	$T_a(A_1) \leq T_a(A_2)$ — (b) $B_a(A_1) \leq B_a(A_2) + |\mathcal{K}^* - \mathcal{G}|$ [c] — (d)

(a) It should be noted that search functions also change by improving g, in case of best-bound search.

(b) $T_s(A_1) \leq T_s(A_2) + |\mathcal{K}^* - \mathcal{G}_1|$ if $D = I$

(c) $\mathcal{K}^* = \{P_i \in \mathcal{P} \,|\, g_1(P_i) = g_2(P_i) = f(P_0)\}$

(d) $B_s(A_1) \leq B_s(A_2) + |\mathcal{K}^* - \mathcal{G}_1|$ if $D = I$ and $u = \infty$.

2.4 Power of Upper Bounding Function u

Denote $u_1 \leq u_2$ if $u_1(P_i) \leq u_2(P_i)$ for all $P_i \in \mathcal{P}$. It is summarized in Table 2 whether $u_1 \leq u_2$ implies $T(A_1) \leq T(A_2)$ and $B(A_1) \leq B(A_2)$. The cases of best-bound search, depth-first search and breadth-first search using D consistent with g are treated in [16] under the assumption $u = u(P_0)$. Other results are found in [14].

Table 2. Computational efficiency of $A_1 = A(u_1)$ and $A_2 = A(u_2)$ with $u_1 \leq u_2$.
[— denotes that a monotonic change in T (or B) does not necessarily hold]

Search strategies / D	General	Consistent with g
Heuristic (general)	— —	$T(A_1) \leq T(A_2)$ $B(A_1) \leq B(A_2)$
Heuristic (nonmisleading)	$T_a(A_1) = T_a(A_2)$ — $B_a(A_1) = B_a(A_2)$ $B_s(A_1) \leq B_s(A_2)$	$T_a(A_1) = T_a(A_2)$ $T_s(A_1) \leq T_s(A_2)$ $B_a(A_1) = B_a(A_2)$ $B_s(A_1) \leq B_s(A_2)$
Depth-first Breadth-first	— —	$T(A_1) \leq T(A_2)$ $B(A_1) \leq B(A_2)$
Best-bound	$T_a(A_1) = T_a(A_2)$ $T_s(A_1) \leq T_s(A_2)$ $B_a(A_1) = B_a(A_2)$ $B_s(A_1) \leq B_s(A_2)$	$T_a(A_1) = T_a(A_2)$ $T_s(A_1) \leq T_s(A_2)$ $B_a(A_1) = B_a(A_2)$ $B_s(A_1) \geq B_s(A_2)$

Let D and D' be dominance relations. D' is said to be *stronger* than D, denoted $D' \supset D$, if $P_i D P_j$ implies $P_i D' P_j$. This was first defined in [16] and it was shown that a stronger D does not necessarily provide a more efficient algorithm, using the definition of D which is consistent with g, but may not satisfy conditions (ii), (iii) of Section 1.4. Based on the definition of Section 1.4, however, [10] found four special cases of branch-and-bound algorithms in which monotonicity with respect to D is guaranteed.

(1) Let $A_1 = A(D_1)$ and $A_2 = A(D_2)$, $D_1 \supset D_2$, be branch-and-bound algorithms with a nonmisleading heuristic search function. Then

$$T_a(A_1) \leqq T_a(A_2), \quad B_a(A_1) \leqq B_a(A_2)$$
$$T_s(A_1) \leqq T_s(A_2) + |\mathcal{J} \cap \mathcal{K} - \mathcal{G}|$$
$$B_s(A_1) \leqq B_s(A_2) + |\mathcal{J} \cap \mathcal{K} - \mathcal{G}|,$$

where $\mathcal{J}$ and $\mathcal{K}$ were defined in Section 2.2.

(2) If A_1 and A_2 use best-bound search, then $T_a(A_1) \leqq T_a(A_2)$, $B_a(A_1) \leqq B_a(A_2)$, $T_s(A_1) \leqq T_s(A_2) + |\mathcal{K} - \mathcal{G}|$, $B_s(A_1) \leqq B_s(A_2) + |\mathcal{K} - \mathcal{G}|$.

(3) Let A_1 and A_2 use breadth-first search, and assume that branching structured satisfies $P_i \in \mathcal{T} \Leftrightarrow d(P_i) = n$ for some $n > 0$, and $\mathcal{G} = \mathcal{T}$. Further assume that D_1 and D_2 satisfy

$$P_i D_1 P_j (P_i D_2 P_j) \quad \text{implies} \quad d(P_i) \leqq d(P_j),$$

and $u = \infty$. Then $T(A_1) \leqq T(A_2)$ and $B(A_1) \leqq B(A_2)$ hold.

(4) Let A_1 and A_2 use depth-first search, and assume that A5 (Dominance test) is modified as follows.

A5: If there exists $P_k (\neq P_i) \in \mathcal{N} - \mathcal{A}$ such that $P_k D P_i$, go to A8; else go to A6.

In addition, assume that $P_i D P_j$ and $P_i \neq P_j$ imply that P_j is not a descendant of P_i. Then $T(A_1) \leqq T(A_2)$ and $B(A_1) \leqq B(A_2)$ hold.

2.6 Approximate Branch-and-Bound Algorithms

When sufficient computer time is not available, it is sometimes attempted to improve the efficiency at the sacrifice of obtaining an optimal solution; A4 (Lower bound test) is strengthened beyond its limit (it is assumed that only a single optimal solution of P_0 is sought):

A4: If $g(P_i) \geqq z - \varepsilon(z)$, go to A8; else go to A5.

$\varepsilon(z)$ is called an *allowance function* and satisfies the following conditions.

(i) $\varepsilon(z) \geqq 0$ for $z \in E$.
(ii) $z_1 \leqq z_2$ implies $z_1 - \varepsilon(z_1) \leqq z_2 - \varepsilon(z_2)$.

Two types of allowance functions are often used [18, 17].

(a) $\varepsilon(z)=\varepsilon$, a nonnegative constant. This guarantees that a suboptimal solution with its *absolute error* from $f(P_0)$ at most ε.

(b) $\varepsilon(z)=r\varepsilon$, where r is a constant satisfying $0\leq r\leq 1$, and in this case z is assumed nonnegative. This guarantees that the exact optimal solution is within $r\times 100\%$ *relative error* from the final suboptimal value.

Now denote $\varepsilon_2\leq\varepsilon_1$ iff $\varepsilon_2(z)\leq\varepsilon_1(z)$ for $z\in E$. Although it seems intuitively obvious that a larger ε implies smaller T and B, the results are also dependent on search strategies and D. They are summarized in Table 3.

Table 3. Computational efficiency of $A_1=A(\varepsilon_1)$ and $A_2=A(\varepsilon_2)$
with allowance functions $\varepsilon_2\leq\varepsilon_1$

Problem	D	$T(A_1)\leq T(A_2)$ when $\varepsilon_2\leq\varepsilon_1$			$T(A_1)\leq T(A_2)$ when $\varepsilon_2=0$		
Search strategies		General	Consistent with g	I	General	Consistent with g	I
Heuristic (general)		No	No	No	No	Yes	Yes
Heuristic (nonmisleading)		No	Yes	Yes	No	Yes	Yes
Depth-first		No	No	No	No	Yes	Yes
Best-bound Breadth-first[a]		Yes [$T(A)$ is not dependent on ε]					

(a) For breadth-first search, it is further assumed that $u=\infty$ and $\mathcal{B}, \mathcal{G}$ satisfy the conditions in (3) of Section 2.5.

2.7 Memory Space Requirement

From the viewpoint of memory space, depth-first search is most economical since it is known that

$$M(A)\leq n(k-1)+1$$

always holds, where n is the height of $\mathcal{B}$ and k is the number of sons generated by one decomposition (k is assumed to be constant in the following discussion).

For other search strategies, such as heuristic search, best-bound search and breadth-first search, it seems difficult to derive exact formulae of $M(A)$ in terms of n and k since $M(A)$ is dependent on individual problems to be solved. It is empirically known, however, that $M(A)$ grows exponentially as a function of n. Thus it could be a serious problem to execute the entire computation within the available memory space.

In this respect, depth-*m* search mentioned in Section 1.5 is worth considering, since it can provide search strategies located in between depth-first search and heuristic (best-bound) search by adjusting *m*, and furthermore $M(A)$ is bounded by $O(n^m)$ [8]. Thus employing an appropriate *m*, it is possible to restrict the required memory space within the limit.

In practice, the following strategy is often taken: Use heuristic (best-bound) search as long as at least $n(k-1)+1$ memory locations are left; otherwise switch it to depth-first search.

In view of the above properties of depth-*m* search, it would also be natural to use the following strategy: Classify the computation into a certain number of categories according to how many memory locations are left, and use depth-*m* search with *m* defined for each category (i.e., a larger *m* if more memory locations are left). This strategy may give a better performance since sudden change in search function can be avoided.

Finally, it is noted that the measure $M(A)$ is dependent on other constituents of branch-and-bound algorithms in a manner similar to T and B. In each case discussed previously, in which $T(A_1) \leq T(A_2)$ and $B(A_1) \leq B(A_2)$ are proved for A_1 and A_2, it is possible to show that $M(A_1) \leq M(A_2)$ holds.

3. SIMULATION STUDY AND ITS IMPLICATION

3.1 Simulation Model

Since only weak properties can be theoretically clarified, it seems important to investigate more detailed computational behavior by simulation. For this purpose, models of branch-and-bound algorithms are constructed by assigning random numbers to $f(P_i)$, $g(P_i)$, $h(P_i)$ for $P_i \in \mathcal{M}$. It is briefly sketched below; see [21] for details.

$\mathcal{B}$: Each node (except a leaf node) has exactly two sons. Node P_i is leaf iff $d(P_i)=n$, i.e., the height of $\mathcal{B}$ is *n*. *n* is set to 30 throughout the experiment unless otherwise stated

$f: f(P_0)=1$. For two sons P_{i_1} and P_{i_2} of P_i, $f(P_{i_1})=f(P_i)$ and $f(P_{i_2})$ is randomly determined by

$f(P_{i_2})=f(P_i)+x$, where $x \geq 0$ is a random variable with exponential probability density function $\lambda e^{-\lambda x}$ (λ is a given constant).

$g: g(P_0)=0$. For a son P_j of P_i, $g(P_j)$ is randomly determined by a probability density function defined over $[g(P_i), f(P_j)]$ (see [21] for details), so that

$$g(P_i) \leq g(P_j) \leq f(P_j)$$

always holds.

Note here that the left most path in $\mathcal{B}$, denoted $P_0, P_1, \ldots, P_n$, satisfies $f(P_0)=$ $=f(P_1)=\ldots=f(P_n)$, i.e., P_n provides optimal solutions of P_0. The probability density function of g used above guarantees

$$E\big(g(P_i)\big)=f(P_0)-1+\frac{i}{n}, \quad i=0, 1, \ldots, n.$$

It is also noted that, since $f(P_0)-g(P_0)$ is normalized to 1, parameter λ specifies the accuracy of lower bounding function g; a small λ implies an accurate g.

$h: h(P_i)$ is a random variable normally distributed around $f(P_i)$ with the standard deviation

$$(n-i)\sigma \quad (\sigma: \text{constant}).$$

A small σ of course implies that h is an accurate eatimate of f.

Throughout the subsequent discussion, dominance relation D is not used, and $u=\infty$ is assumed unless otherwise stated.

3.2 Exponential Growth of $\bar{T}(n)$

Let $\bar{T}(n)$ denote the average of $T(A)$ applied to A's which treat $\mathcal{B}$'s with height n. The behavior of $\bar{T}(n)$ conveys an important information on the maximum size of problems which can be practically solved. $\bar{B}(n)$, $\bar{F}(n)$, $\bar{M}(n)$ are similarly defined.

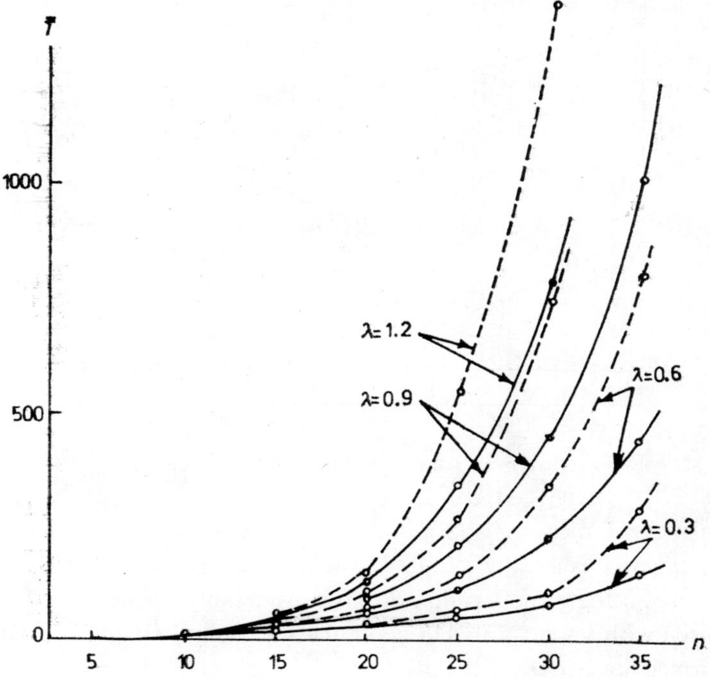

Fig. 1. Exponential growth of $\bar{T}(n)$. (The search is guided by $h=g$. ——— denotes best-bound search, and – – – denotes depth-first search.)

Fig. 1 shows how parameter $\bar{T}(n)$ changes as n increases, for $m=1$ and ∞ when the search is guided by g. (Note that $\bar{T}(n)=\bar{B}(n)=\bar{F}(n)=\bar{M}(n)-1$ if $m=\infty$ (best-bound search).) $\bar{T}(n)$ clearly increases exponentially. This is not special to this particular model, but there is theoretical evidence that most branch-and-bound algorithms have this tendency [12]. For certain practical algorithms, the exponential growth is in fact proved [15, 20].

3.3 Effects of m

Fig. 2 shows how parameters $\bar{T}, \bar{B}, \bar{F}, \bar{M}$ depend on m of depth-m search. Broken curves show the results when the search is guided by $h=g$; best-bound search results when $m=\infty$. Note that $\bar{T}(g)=\bar{B}(g)=\bar{F}(g)=\bar{M}(g)-1$ always holds if $m=\infty$. On the other hand, solid curves represent the cases of general heuristic functions h; heuristic search results when $m=\infty$.

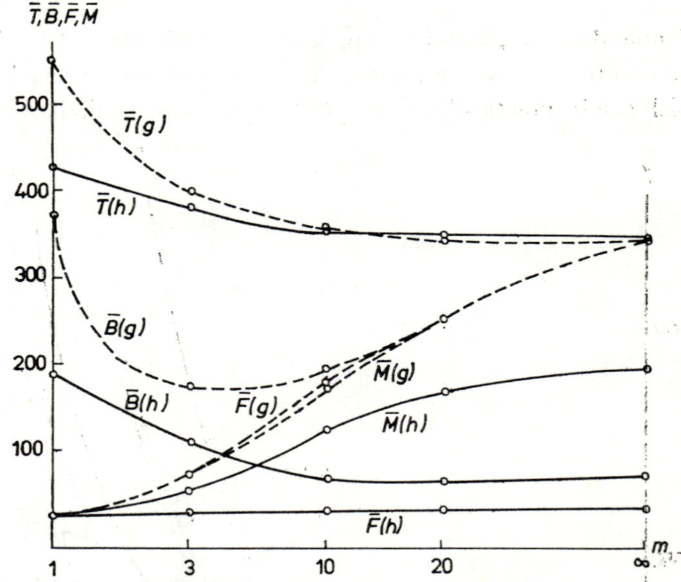

Fig. 2. Effects of m on parameters $\bar{T}, \bar{B}, \bar{F}$ and $\bar{M}$. ($\lambda=0.75, \sigma=0.3$)

In terms of the T measure a larger m always provides better performance, while larger memory space is usually required. The most significant difference between two types of searches using g and h is that $\bar{B}(h)$ is considerably smaller than $\bar{B}(g)$, implying that the search with h tends to yield good feasible solutions quickly. This tendency is also confirmed in various practical algorithms, e.g., see [3]. It may be also interesting to see that $\bar{B}(g)$ is not monotone with respect to m; the smallest $\bar{B}(g)$ is attained somewhere around $m=5$ in this experiment. Although m with the smallest $\bar{B}(g)$ varies

518

with problem parameter λ, this characteristic is usually observed (see Fig. 3). From the viewpoint of $\overline{F}$ measure, the search with h is much preferred to the search with g, since $\overline{F}(h)$ is smaller than $\overline{F}(g)$ and moreover is almost constant. Finally, it is noted, though $\overline{T}(g)$ with $m=\infty$ is the smallest possible $\overline{T}$ among all branch-and-bound algorithms (as noted in Section 2.1), $\overline{T}(h)$ with $m=\infty$ is very close to $\overline{T}(g)$. This indicates that heuristic search with relatively good heuristic function h is almost as efficient as best-bound search, in terms of the T count.

In conclusion, it may be said that the search with h is usually preferred to the search with g, from the viewpoint of both computation time and required memory space.

3.4 Effects of g

As noted in Section 3.1, the quality of lower bounding function g is controlled by parameter λ in our simulation model. (A small λ implies an accurate g.) Fig. 3 shows

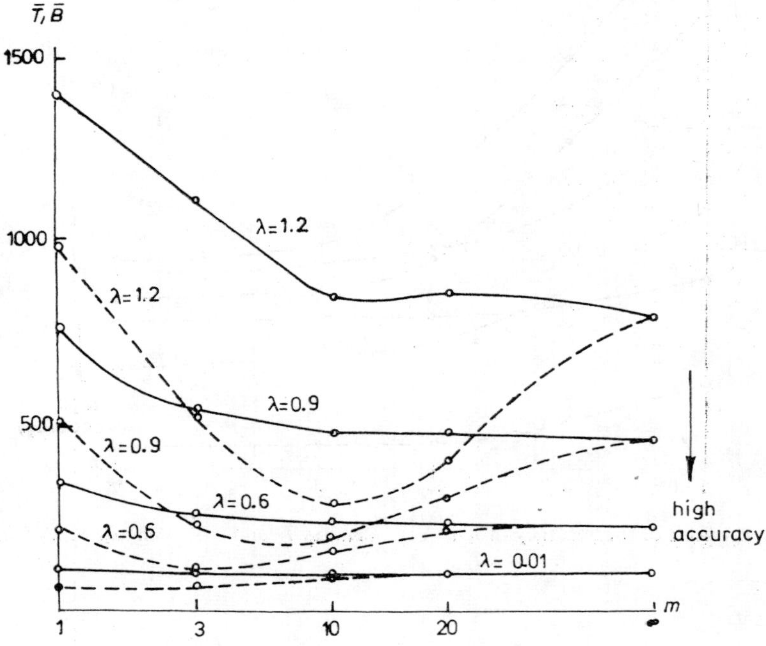

Fig. 3. Effects of the accuracy of g on $\overline{T}$ and B. ($---$ denotes $\overline{B}$, and $\underline{\qquad}$ denotes $\overline{T}$)

how $\overline{T}$ and $\overline{B}$ depend on λ when the search is guided by $h=g$. From these results, it may be strongly recommended to find an accurate lower bounding function, in practical applications, provided of course that computing $g(P_i)$ does not require excessive computation time. It may also be interesting to see that the influence of m becomes rather small when a good lower bounding function g is used.

The quality of heuristic function h is controlled by parameter σ. Fig. 4 shows for various m how $\bar{T}$ and $\bar{B}$ depend on σ. It may concluded that performance of algorithms (in particular $\bar{B}$) can be considerably improved by using h which is an accurate estimation of f. Thus the efforts in improving h would be worthwhile.

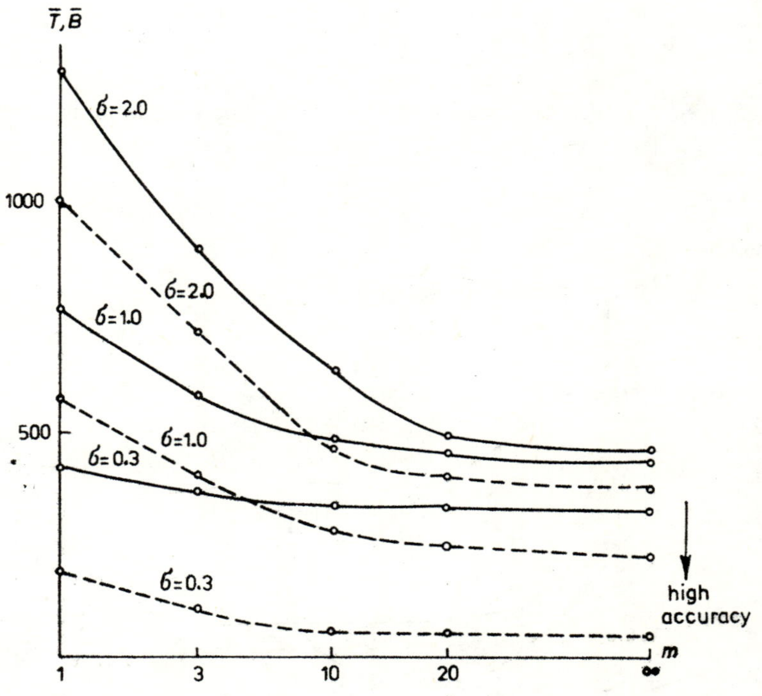

Fig. 4. Effects of the accuracy of h. (——— denotes $\bar{T}$, and – – – denotes $\bar{B}$. $\lambda=0.75$)

3.6 Effects of u

Fig. 5 (a), (b) illustrates the results when upper bounding function $u=u(P_0)$ is introduced. Although $\bar{T}$ and $\bar{B}$ become smaller when $u(P_0)$ is improved, the effect does not seem to be drastic (especially when the search with h is employed). Thus it may not be worth spending too much computation time on $u(P_0)$, unless very small m (e.g., $m=1\sim3$) is used.

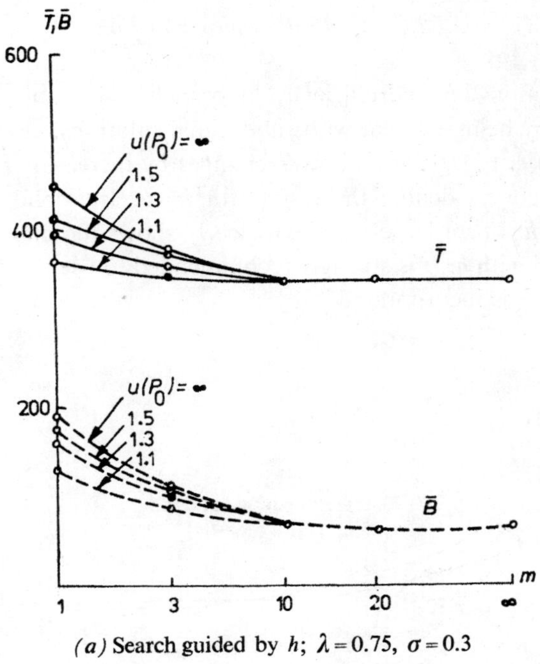

(a) Search guided by h; $\lambda=0.75$, $\sigma=0.3$

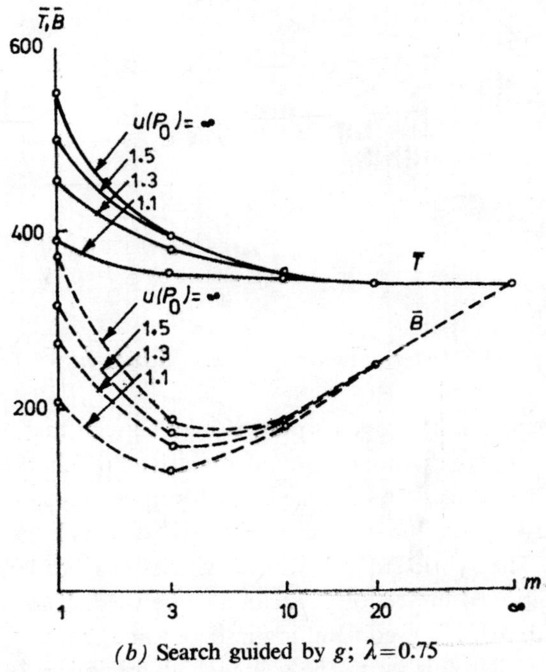

(b) Search guided by g; $\lambda=0.75$

Fig. 5. Effects of $u(P_0)$ on $\bar{T}$ and $\bar{B}$. $\left(f(P_0)=1.0 \text{ in all cases}\right)$

521

3.7 Effects of Allowance Function ε

As noted in Section 2.6, the efficiency of branch-and-bound algorithms can be further improved by being content with suboptimal solutions, i.e., by introducing an allowance function ε. Fig. 6 shows how $\bar{T}$ changes by increasing r of $\varepsilon = rz$ (see Section 2.6). Broken curves denote the results for $h=g$, and solid curves denote the results for general h. For a large m, Fig. 6 clearly shows the difference between two searches with g and with h. It is strongly recommended to use the search with h when allowance function ε is incorporated.

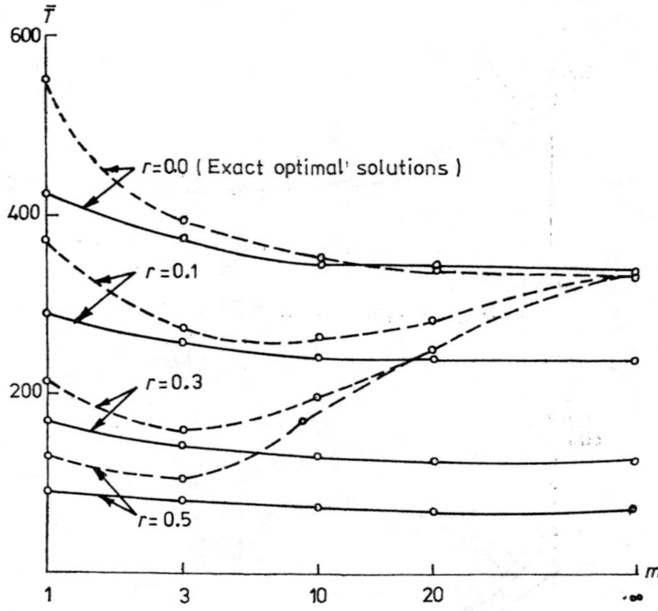

Fig. 6. Effects of allowance function $\varepsilon = rz$ on $\bar{T}$. (——— denotes when the search is guided by h, and – – – denotes when the search is guided by g. $\lambda = 0.75$, $\sigma = 0.3$.)

3.8 Determination of Search Strategies

As a result of the simulation given so far, it may be seen that appropriate search strategies vary according to the situations under which a branch-and-bound algorithm is executed. Fig. 7 summarizes the suggested search strategies under various situations.

Notes: (a) The largest possible m implies that m is taken as large as possible within memory limitation. The hybrid strategy which switches m according to the remaining memory space, mentioned in Section 2.7, can also be used. If the available memory space is very limited, $m=1$ is used resulting in depth-first search.

(b) $m=m^*$ implies that m is set to the value which minimizes $\bar{B}$ under the present lower bounding function g; see Fig. 3.

522

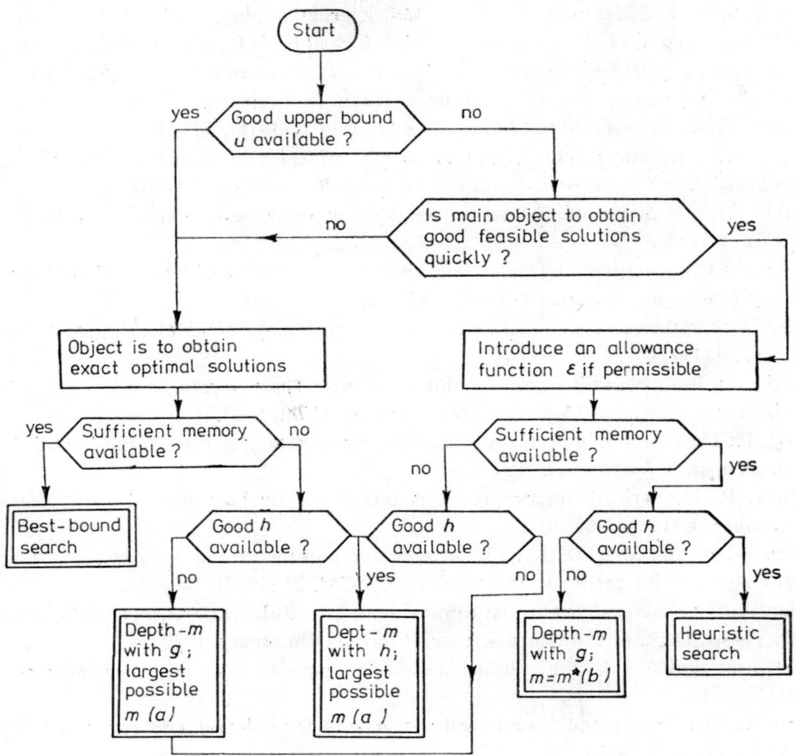

Fig. 7. Determination of an appropriate search strategy under various situations

Acknowledgement

The author wishes to thank Professors H. Mine and T. Hasegawa of Kyoto University for their comments, and Mr. T. Ohkawa for implementing the simulation program discussed in Chapter 3.

REFERENCES

[1] Agin, N.: Optimum seeking with branch and bound, *Management Science,* 13 (1966), B176–B185.

[2] Balas, E.: A note on the branch-and-bound principle, *Operations Research,* 16 (1968), 442–445.

[3] Forrest, J. J. H., Hirst, J. P. H. and Tomlin, J. A.: Practical solution of large mixed integer programming problems with UMPIRE, *Management Science,* 20 (1974), 736–773.

[4] Fox, B. L. and Schrage, L. E.: The values of various strategies in branch-and-bound, Technical Report, Graduate School of Business, University of Chicago, 1972.

[5] Geoffrion, A. M. and Marsten, R. E.: Integer programming algorithms: A framework and state-of-the-art survey, *Management Science,* 18 (1972), 465–491.

523

[6] Golomb, S. W. and Baumert, L. D.: Backtrack programming, *J. ACM,* 12 (1965), 516–524.

[7] Hart, P. E., Nilsson, N. J. and Raphael, B.: A formal basis for the heuristic determination of minimal cost paths, *IEEE Trans. System Science and Cybernetics,* SSC-4, (1968), 100–107.

[8] Ibaraki, T.: A generalization of depth-first search in branch-and-bound algorithms, 1974; *International J. of Computer and Information Sciences* 7, (1978), 315–343.

[9] Ibaraki, T.: Theoretical comparisons of search strategies in branch-and-bound algorithms, *International J. of Computer and Information Sciences,* 5 (1976), 315–344.

[10] Ibaraki, T.: The power of dominance relations in branch-and-bound algorithms, *J. ACM,* 24 (1977), 264–279.

[11] Ibaraki, T.: Computational efficiency of approximate branch-and-bound algorithms, *Mathematics of Operations Research,* 1 (1976), 287–298.

[12] Ibaraki, T.: On the computational efficiency of branch-and-bound algorithms, *J. OR Society of Japan,* 20 (1977), 16–35.

[13] Ibaraki, T.: Branch-and-bound procedure and state-space representation of combinatorial optimization problems, *Information and Control* 36 (1978), 1–27.

[14] Ibaraki, T.: The power of upper and lower bounding operations in branch-and-bound algorithms, Working Paper, *J. Kyoto Univ.* (1977).

[15] Jeroslow, R. G.: Trivial integer programs unsolvable by branch-and-bound, *Mathematical Programming,* 6 (1974), 105–109.

[16] Kohler, W. H. and Steiglitz, K.: Characterization and theoretical comparison of branch-and-bound algorithms for permutation problems, *J. ACM,* 21 (1974), 140–156.

[17] Kohler, W. H.: Exact and approximate branch-and-bound algorithms for permutation problems, Ph. D. Thesis, Dept. of Computer Science, Princeton University, 1972.

[18] Lawler, E. L. and Wood, D. E.: Branch-and-bound method: A survey, *Operations Research,* 14 (1966), 699–719.

[19] Mitten, L. G.: Branch-and-bound method: general formulation and properties, *Operations Research,* 18 (1970), 24–34.

[20] Nilsson, N. J.: *Problem-Solving in Artificial Intelligence,* McGraw-Hill, New York, 1971.

[21] Ohkawa, T.: Simulation study on computational behavior of branch-and-algorithms, Master Thesis, Department of Applied Mathematics and Physics, Faculty of Engineering, Kyoto University, 1976.

[22] Rinnooy Kan, A. H. G.: On Mitten's axioms for branch-and-bound, *Operations Research,* 24 (1976), 1176–1178.

TIGHTER EQUIVALENT FORMULATIONS
OF INTEGER PROGRAMMING PROBLEMS

I. KALISZEWSKI and S. WALUKIEWICZ

(Warsaw, Poland)

1. INTRODUCTION

For a given formulation of an integer programming problem (IP) there are many other formulations (called equivalent formulations) with exactly the same integer feasible solutions. Computational results reported in [2], [9], [19], [20] indicate that an IP is much easier to solve or at least a good near-optimal solutions is obtained relatively quickly if the feasible region to the corresponding linear programming problem (LP) is smaller, i.e., if the continuous formulation of a given IP is tighter. In [2], [9], [19] and [20] the equivalence of given and tighter formulations is established using logical analysis. This method can be applied only to special IP's mostly to IP's with $0-1$ coefficients matrix. Paper [6] gives the general method for constructing an equivalent inequality by the reduction of coefficients of a given inequality and [19] uses this method in constructions of tighter equivalent formulations.

In this paper we show how to obtain tighter equivalent formulations of a given IP using the dynamic programming procedure for the rotation of a given constraint in any direction without adding or eliminating any integer feasible solution [13]. In Section 2 we define a strongest cut as a constraint that cannot be rotated by means of above-mentioned procedure and introduce so-called best direction of such a rotation. An IP in which every constraint is a strongest cut is called an almost linear (integer programming) problem. In Section 3 we prove that a finite IP can be transformed to an equivalent almost linear problem with the same number of variables and the number of constraints increased at most by one. A linear optimum to the corresponding LP is often integer for such problems. If such a case does not happen we construct a cut that cuts off a linear optimum but does not cut off any feasible solution and rotate this cut in its best direction. We repeat adding and rotating new cuts until a linear optimum is integer. We show that all Gomory cuts and many other ones can be rotated.

In our computational experiments we use the method of integer form. We study an influence of the rotation on the number of cuts needed and on the total solution time.

Different direction of rotation are investigated. Results of this experiments are in Section 4, while Section 5 containes the results of a solution of Problem 1 from [19] using the IBM MIP/370 system.

We will consider only finite IP's, i.e., problems with finite feasible solutions set. Without loss of generality, such problems can be formulated as

(IP)
$$f(x^*)=\max \sum_{j=1}^{n} c_j x_j$$

subject to

$$g_i(x)= \sum_{j=1}^{n} a_{ij}x_j \leq a_{i0}, \quad i=1, \ldots, m$$

$$x_j=0 \quad \text{or} \quad 1, \quad j \in N=\{1, \ldots, n\}$$

where all data are integers. Replacing $x_j=0$ or 1 by $0 \leq x_j \leq 1, j=1, \ldots, n$, we obtain LP associated with IP. We denote by $f(x_{\mathrm{LP}})$ the optimal objective value to LP.

2. STRONGEST CUTS

Since we rotate every constraint separately using an information contained in the objective function, then we have to consider the binary knapsack problem. It is well known that a constraint of an IP can be transformed to an equivalent knapsack-type constraint

(1)
$$g(x)= \sum_{j=1}^{n} a_j x_j \leq a_0,$$

where

$$0 < a_j \leq a_0, \quad \sum_{j=1}^{n} a_j > a_0.$$

We also assume that

(2)
$$\frac{c_1}{a_1} \geq \frac{c_2}{a_2} \geq \ldots \geq \frac{c_n}{a_n}.$$

Then the optimal objective value to linear knapsack problem is [7]

(3)
$$f(x_{\mathrm{LP}})=f\left(1, 1, \ldots, 1, \frac{1}{a_{t+1}}\left(a_0- \sum_{j=1}^{t} a_j\right), 0, 0, \ldots, 0\right).$$

Kianfar [13] described a method for the rotation of $g(x)=a_0$ in any direction without adding or eliminating any integer points of the feasible solution set, in such a way that the final hyperplane

(4)
$$\hat{g}(x)= \hat{a}_1 x_1 + \hat{a}_2 x_2 + \ldots + \hat{a}_n x_n = a_0$$

passes through at least as many integer points as $g(x)=a_0$. It is easy to see that we

526

will have a tighter formulations of a given knapsack problem if $\hat{g}(x)=a_0$ will contain as many feasible solutions as possible.

If the hyperplane $g(x)=a_0$ cannot be rotated we call the corresponding inequality $g(x)\leq a_0$ a strongest cut or a strongest constraint. Let S be a feasible solutions set to the binary knapsack constraint (1). Then $g(x)\leq a_0$ is the strongest cut if

$$(5) \qquad \underset{i\in N}{\forall}\ \underset{x\in S}{\exists}\ \sum_{j=1}^{n} a_j x_j = a_0 \quad \text{and} \quad x_i = 1.$$

In this section we give a description of rotation procedure [13] for constructing a strongest cut with slight improvements.

We have

$$a_r x_r \leq a_0 - \sum_{j\in(N-\{r\})} a_j x_j = a_0 - b_r.$$

Note that $b_r \leq a_0 - a_r$ for any $x\in S$ such that $x_r = 1$. We define

$$b_r^* = \max_{x\in S} \sum_{j\in(N-\{r\})} a_j x_j \quad \text{and such that} \quad x_r = 1.$$

We may compute b_r^* as the maximum element of the set

$$B_r = \left\{ b \mid b = \sum_{j\in J} a_j,\ J\subset(N-\{r\}),\ b\leq a_0 - a_r \right\},$$

where J is a subset of $N-\{r\}$.

If $b_r^* < a_0 - a_r$, then the strongest coefficient

$$\hat{a}_r = a_0 - b_r^*$$

is greater than a_r, and the hyperplane

$$\sum_{j\in(N-\{r\})} a_j x_j + \hat{a}_r x_r = a_0$$

passes through at least one more binary point than $g(x)=a_0$.

Let now $\hat{g}(x)\leq a_0$ be a strongest cut and $\hat{S}$ be a feasible solution set to it. A strongest cut has the following properties:

1. $\hat{S}=S$.

2. $a_j \leq \hat{a}_j \leq a_0$ and hence $\dfrac{c_j}{\hat{a}_j} \leq \dfrac{c_j}{a_j}$, $j\in N$.

3. If $g(\bar{x})=a_0$ and $\bar{x}\in S$, then $\hat{g}(\bar{x})=a_0$.

4. If $(a_0 - a_r)\in B_r$, then $\hat{a}_r = a_r$.

5. Let $K=\{x\in S\mid g(x)=a_0\}$ and $\hat{K}=\{x\in S\mid \hat{g}(x)=a_0\}$, then $|\hat{K}|\geq|K|$ and can be proved [18] that $|\hat{K}|\geq 2$.

In the rotation procedure (we call it Procedure R) we construct the set B_r sequentially by dynamic programming for $r=1, 2, \ldots, n$ (see also [13]). If during the construction of B_r we find $(a_0 - a_r) \in B_r$ we set $\hat{a}_r = a_r$ and go to the construction of B_{r+1}. We denote

$$a_m = \min_{j \in N} a_j.$$

Procedure R

Step 1: (Initialization, $r = 1$). Set $B_1^n = \{0, a_n\}$ and compute

$$(6) \qquad B_1^i = B_1^{i+1} \cup \{b + a_i \mid b \in B_1^{i+1},\, b + a_i \le a_0 - a_m\}$$

for $i = n-1, \ldots, 2$. Set $B_1^1 = B_1^2$ and $B_1 = \{b \mid b \in B_1^1,\, b \le a_0 - a_1\}$. Compute b_1^* and $a_1^* = a_0 - b_1^*$.

Step 2: ($2 \le r \le n-1$). Set $r = r+1$. If $r = n$, then go to STEP 3. Otherwise

$$(7) \qquad B_r^r = \{b \mid b \in B_1^{r+1},\, b \le a_0 - a_r\}.$$

For $i = r-1, \ldots, 1$ compute

$$(8) \qquad B_r^i = B_r^{i+1} \cup \{b + \hat{a}_i \mid b \in B_r^{i+1},\, b + \hat{a}_i \le a_0 - a_r\}.$$

Set $B_r = B_r^1$, compute b_r^* and $\hat{a}_r = a_0 - b_r^*$.

Step 3: (Termination, $r = n$). Set $B_n^0 = \{0\}$, $i = 1$ and compute

$$(9) \qquad B_n^i = B_n^{i-1} \cup \{b + \hat{a}_i \mid b \in B_n^{i-1},\, b + \hat{a}_i \le a_0 - a_n\}$$

for $i = 1, \ldots, n-1$. Set $B_n = B_n^{n-1}$, compute b_n^* and $\hat{a}_n = a_0 - b_n^*$. Stop.

Our improvements consist in using $(a_0 - a_m)$ and $(a_0 - a_r)$ in (6)–(9), while in [13] the bound $b \le a_0$ was used in these expressions.

As a result of Procedure R we obtain a strongest cut, i.e., a hyperplane (4) having the property (5). It can be shown [13] that Procedure R requires in the worst-case example $0(n^2 a_0)$ additions and comparisons to rotate a constraint (1) or to establish that it is a strongest cut.

Procedure R gives a strongest cut for a given sequencing of the variables in (1). Now we consider the problem of determining best sequence and since there is a one-to-one correspondence between a given sequence of variables and the direction of rotation of $g(x) = a_0$ we say that $g(x) = a_0$ is rotated in the direction $s = (s_1, \ldots, s_n)$ if $\hat{a}_{s_1}$ is computed first, then $\hat{a}_{s_2}$ and so on up to $\hat{a}_{s_n}$.

In (18) the following theorem is proved.

Theorem 1. *If during rotation a coefficient $\hat{a}$ is computed twice as the h-th and as the i-th, then for $h < i$ we have*

$$\frac{c_h}{\hat{a}_h} \le \frac{c_i}{\hat{a}_i}.$$

Therefore, if we rotate $g(x) = a_0$ in the direction $s_0 = (1, 2, \ldots, n)$, we in general case, decrease $f(x_{\mathrm{LP}}, s_0)$—the optimal objective value for the knapsack problem

528

with the rotation in this direction constraint. For this reasons we call s_0 the best direction for the rotation (of a knapsack constraint). We note that in [13] $g(x)=a_0$ is rotated in $s=(n, n-1, \ldots, 1)$.

Let $F(\mathrm{LP})$ be a feasible solution set to the linear knapsack problem. From the above we have

Corollary 1. *If $g(x)\leq a_0$ is not a strongest cut, then after the rotation in the direction s_0 we have*

$$f(x^*)\leq f(x_{\mathrm{LP}}, s_0)\leq f(x_{\mathrm{LP}}) \quad \text{and} \quad F(\mathrm{LP}(s_0))\subsetneq F(\mathrm{LP}).$$

In other words, after the rotation the feasible solution set to $\mathrm{LP}(s_0)$ is strictly contained in $F(\mathrm{LP})$.

3. ALMOST LINEAR PROBLEMS

In this section we consider a straightsforward generalization of our result for a finite IP. An IP, in which every constraint is a strongest cut, is called an almost linear (integer programming) problem. If at least one constraint of an IP is not a strongest cut, then after the rotation of this constraint in its best direction, we have (see Corollary 1)

$$f(x^*)\leq f(x_{\mathrm{LP}}, s_0)\leq f(x_{\mathrm{LP}}) \quad \text{and} \quad F(\mathrm{LP}(s_0))\subsetneq F(\mathrm{LP}).$$

Obviously, the best direction for different constraints may not be the same and we rotate all constraints that are not strongest cuts. Therefore s_0 is, in general, a matrix. The above consideration allow us to propose an algorithm for solving IP's as it is outlined in the Introduction.

Theorem 2. *A finite IP can be transformed to an equivalent almost linear problem with the same number of variables and the number of constraints increased at most by one.*

Proof. For an inequality constrained IP the rotation does not increase the number of variables or number of constraints. Consider an equality constrained IP

$$f(x)=\max \sum_{j=1}^{n} a_j x_j$$

subject to

$$\sum_{j=1}^{n} a_{ij}x_j=a_{i0}, \quad i=1, \ldots, m, \quad x_j=0 \text{ or } 1, \quad j=1, \ldots, n.$$

It is well known that such an IP is equivalent to an inequality constrained IP

$$f(x^*)=\max \sum_{j=1}^{n} c_j x_j$$

subject to

$$\sum_{j=1}^{n} a_{ij}x_j \leqq a_{i0}, \quad i=1, \ldots, m,$$

$$\sum_{j=1}^{n} \left(\sum_{i=1}^{m} a_{ij} \right) x_j \geqq \sum_{i=1}^{m} a_{i0},$$

$$x_j=0 \quad \text{or} \quad 1, \quad j=1, 2, \ldots, n,$$

and this problem has $m+1$ constraints.

From the definition of almost linear problems we have

Theorem 3. *An IP with $0-1$ coefficient matrix is an almost linear problem. In particular, set packing, set covering and set partitioning problems are almost linear problems.*

In [8] an equivalence between a finite IP and set partitioning or set packing problems in which each column has no more than three ones is established. But during this transformation the number of variables increased to an $O\left(\sum_{i=1}^{m} \sum_{j=1}^{n} a_{ij} \right)$ while the number of constraints remains the same. Also a finite IP can be transformed to an equivalent set covering problems [10], but then the number of constraints increases exponentially in the worst-case example while the number of variables remains the same.

Consider the n-city traveling salesman problem, where c_{ij} is the distance from i to j. This problem can be formulated as follows:

$$(10) \qquad f(x^*)=\min \sum_{i=1}^{n} \sum_{j=1}^{n} c_{ij}x_{ij}$$

subject to

$$(11) \qquad \sum_{j=1}^{n} x_{ij}=1 \quad \text{for} \quad i=1, \ldots, n,$$

$$(12) \qquad \sum_{i=1}^{n} x_{ij}=1 \quad \text{for} \quad j=1, \ldots, n$$

and

$$(13) \qquad \sum_{i \in V} \sum_{j \in \overline{V}} x_{ij} \geqq 1,$$

where V is a nonempty subset of $\{1, \ldots, n\}$, $\overline{V}=\{1, \ldots, n\}-V$ and $x_{ij}=1(0)$ if the salesman (does not) travels from i to j. This problem has $2n+2^n-2$ constraints. From Theorem 3 we have

530

Theorem 4. *The formulation (10)–(13) of the traveling salesman problem is an almost linear problem.*

Using the same notation we give Tucker's formulations of the traveling salesman problem (14)

$$(14) \qquad f(x^*) = \min \sum_{i=1}^{n} \sum_{j=1}^{n} c_{ij} x_{ij}$$

subject to

$$(15) \qquad \sum_{j=1}^{n} x_{ij} = 1 \quad \text{for all} \quad i, \text{ except } i = i_0$$

$$(16) \qquad \sum_{i=1}^{n} x_{ij} = 1 \quad \text{for all} \quad j, \text{ except } j = i_0$$

$$(17) \qquad u_i - u_j + n x_{ij} \le n - 1 \quad \text{for all} \quad i, j, \quad i, j \ne i_0$$

where u_i, u_j are real variables, but, without loss of generality, it can be shown that we may consider these variables as integer ones not greater than n. One of possible interpretation of u_i, u_j is an order in which the salesman visits city i and j. Then (17) is tight not for every pair of cities. Therefore we have proved

Theorem 5. *The formulation (14)–(17) of the traveling salesman problem is not an almost linear problem.*

The fact that (13) is a stronger cut than (17) was mentioned in [3]. This formulation has only $n(n-1)$ constraints.

4. COMPUTATIONAL EXPERIENCE

Recently, there has been interest in constructing equivalent inequalities that are in a certain sense "stronger" than the original ones. In [6] an equivalent knapsack inequality is constructed and it can be shown that Corollary 1 does not hold for such an equivalent inequality, and it can be rotated (see, for instance, inequality (9) on p. 275 of [6]).

It is known [5], [12], that procedures for aggregating integer constraints make the feasible solution set to the associated LP problem large. Therefore these knapsack constraints can be rotated.

In [4] a deeper cut is constructed by reducing a_0 (in our notation) with respect to a chosen subset of constraints. The cut $g(x) \le a_0$ is the deepest one if and only if $g(x) = a_0$ passes through at least one feasible solution. The standard cut associated with dual (method of integer form) and dual-all-integer algorithms can be made deeper. In

Table 1. Results for R.a.D. problem 1 with $n=6$ and $m=17$ from (15)

		Given formulation	Constraints rotated in $(1, 2, \ldots, n)$	Constraint rotated in its best directions
No cut rotated	$f(x_{\mathrm{LP}})$	41.340	39.792	39.585
	$f(x^*)$	38.0	38.0	38.0
	Simplex iterations	341	97	18
	Cuts	47	10	3
	Total CPU time in msec.	82.700	23.780	5.660
Cuts rotated in direction $(1, \ldots, n)$	Simplex iterations	51	26	12
	Cuts	9	5	1
	Rotated cuts	5	5	1
	Total CPU time in msec.	13.600	7.500	3.159
	Total rotation time in msec.	400	300	40
Cut rotated in its best direction	Simplex iterations	25	16	12
	Cuts	5	2	1
	Rotated cuts	4	1	1
	Total CPU time in msec.	8.420	4.960	3.380
	Total rotation time in msec.	$\geqq 360$	$\geqq 220$	$\geqq 60$
Rotated constraints			9	9
Maximal number of coefficients changed in a constraint			3	3
Total number of coefficients changed			16	16
Maximal rotation time of a constraint in msec.			60	60
Total rotation time in msec.			280	480

general, these deeper cuts be rotated (see, for instance, example 2(a) of [4]). It can be shown that the cuts associated with the primal-all-integer algorithm can also be rotated. In [11] and [16] cuts are constructed using the group theory approach and generalized Lagrange multiplier method. From the numerical examples in these references we can see that these cuts can be rotated as well.

In our computational experience we study the method of integer form. We use the triangularization method of realization of the simplex algorithm [1], since this approach guarantees more numerical stable algorithm than usually used revised simplex method, although it is slover. We do not use the bounded simplex algorithm, therefore we add additional constraints $x_j \leqq 1$ for $j=1, \ldots, n$.

Suppose we have a basic representation of the LP corresponding to an IP given by

$$(18) \qquad x_{B_i} = y_{i0} - \sum_{j \in R} y_{ij} x_j, \quad i=0, \ldots, m$$

where R is the index set of nonbasic variables, and for $i=0$ we have the row of the objective function. Since all data in an IP are integer we have one more constraint

(19)
$$f(x^*)\leqq [y_{i0}]-\sum_{j\in R} y_{ij}x_j,$$

where $[x]$ is the integer part of x. Therefore now $m=m+n+1$.

Using the above notations the cut of the method of integer form is

(20)
$$\sum_{j\in R} f_{ij}x_j\geqq f_{i0},$$

where f_{ij} is the fractional part of y_{ij}. As a source row we choose the row with the largest f_{i0}. We use retriangularization before adding a new cut in order to decrease the influence of rundoff errors. The other way of dealing with such errors, we have not tested yet, is the following. Since $f_{ij}\geqq 0$ we may increase f_{ij} to, say, 2–3 digit number and after rotate (20) is its best direction. For the same reason we may decrease f_{i0}.

We solve examples taken from [15] and [17]. Preliminary results for runs on the IBM 360/50 are given in Tables 1–3.

Table 2. Results for fixed charge problem 2 from (17) with $n=22$, $m=17$

		Given formulation	Constraints rotated in $(1, 2, \ldots, n)$	Constraint rotated in its best directions
No cut rotated	$f(x_{LP})$	9.612	9.508	9.508
	$f(x^*)$	no found	8.0	8.0
	Simplex iterations	>518	50	50
	Cuts	>39	8	8
	Total CPU time in msec.	>271.000	24.340	24.180
Cuts rotated in direction $(1, \ldots, n)$	Simplex iterations	493	183	183
	Cuts	27	17	17
	Rotated cuts	5	10	10
	Total CPU time in msec.	259.520	80.360	80.640
	Total rotation time in msec.	60.620	4 320	4 340
Cut rotated in its best direction	Simplex iterations	>426	50	50
	Cuts	>22	8	8
	Rotated cuts	$\geqq 5$	3	3
	Total CPU time in msec.	>252.040	25.660	25.820
	Total rotation time in msec.	$\geqq 57.480$	560	500
Rotated constraints			1	1
Maximal number of coefficients changed in a constraint			1	1
Total number of coefficients changed			1	1
Maximal rotation time of a constraint in msec.			240	180
Total rotation time in msec.			640	600

Table 3. Results for fixed charge problem 4 from (17) with $m=21$ and $n=16$

		Given formulation	Constraints rotated in $(1, 2, \ldots, n)$	Constraint rotated in ist best directions
No cut rotated	$f(x_{\mathrm{LP}})$	9.219	9.219	9.219
	$f(x^*)$	8.0	8.0	8.0
	Simplex iterations	52	52	52
	Cuts	11	6	6
	Total CPU time in msec.	26.419	22.480	24.180
Cuts rotated in direction $(1, \ldots, n)$	Simplex iterations	>218	>271	>271
	Cuts	>22	>24	>24
	Rotated cuts	$\geqq 8$	$\geqq 10$	$\geqq 10$
	Total CPU time in msec.	>127.780	>169.660	>170.480
	Total rotation time in msec.	27.960	34.360	34.540
Cut rotated in its best direction	Simplex iterations	40	52	52
	Cuts	8	6	6
	Rotated cuts	1	0	0
	Total CPU time in msec.	19.940	21.040	21.000
	Total rotation time in msec.	60	—	—
Rotated constraints			2	2
Maximal number of coefficients changed in a constraint			1	1
Total number of coefficients changed			2	2
Maximal rotation time of a constraint in msec.			260	140
Total rotation time in msec.			680	560

5. SOLUTION OF PROBLEM 1 FROM [19]

From data sent us by H. P. Williams we write a problem with 8 equality constraints, 23 binary variables and 16 continuous variables. In [19] the first formulation of this problem has 9 constraints and 40 integer variables, but one constraint in this formulation is a sum of two previous constraints. One can find even more compact formulations of this problem, but we do not study them, since we want to compare our result with those from [19]. We use the IBM MIP/370 system on the IBM 370/140 computer.

The problem was solved in the way similar to [19]. We obtain tighter formulation by rotating 6 inequality constraints corresponding to given three equality constraints. The rotation was done on the IBM 360/50 computer. The total rotation time was 1.300 msec.

Although the solution time of the tighter formulation is in this example greater than for the given formulation, good near-optimal solution can be find easier for the tighter formulation.

Table 4. Given formulation

	CPU time since MIXSTART in min.	Simplex iterations since SETUP	Node number	Functional value
Continuous optimum		39	1	0.0
First integer solution	0.05	108	23	11.256
Optimal integersolution	0.35	431	114	2.821
Optimality proved	2.22	2480	567	

Table 5. Tighter formulation

	CPU time since MIXSTART in min.	Simplex iterations since SETUP	Node number	Functional value
Continuous optimum		52	1	0.0
First integer solution	0.07	134	24	7.331
Optimal integer solution	1.25	1218	317	2.821
Optimality proved	2.90	2812	633	

6. CONCLUSIONS

So far dynamic programming and linear programming have been used as the separate method for solving IPs. In this paper it is shown how both dynamic programming (rotation) and linear programming (cutting plane methods, branch-and-bound methods) may be used simultaneously in solving IPs. Preliminary computational results suggest that it is reasonable to consider the best direction of the rotation and an IP as an almost linear problem.

REFERENCES

[1] Bartels, R. H., Stoer, J., Zenger, Ch.: A Realization of the Simplex Method Based on Triangular Decompositions, In *"Linear Algebra"*, Vol. II. Eds. J. H. Wilkinson and C. Reinsh.

[2] Beale, E. M. L., Tomlin, J. A.: An Integer Programming Approach to Class of Combinatorial Problems, *Mathematical Programming 3* (1972), 339–344.

[3] Bellmore, M., Nemhauser, G. L.: The Traveling Salesman Problem. A Survey, *Operations Research 16* (1968), 538–558.

[4] Bowman, V. J., Nemhauser, G. L.: Deep Cuts in Integer Programming, *Opsearch 8* (1971), 89–111.

[5] Bradley G. H.: Transformation of Integer Programming Problems to Knapsack Problems, *Discrete Mathematics 1* (1971), 29–46.

[6] Bradley, G. H., Hammer, P. L., Wolsey, L.: Coefficient Reduction for Inequalities in 0–1 Variables, *Mathematical Programming 7* (1974), 263–282.

[7] Dantzig, G. B.: Discrete-Variable Extremum Problems, *Operations Research 5* (1957), 266–277.

[8] Garfinkel, R. S., Nemhauser, G. L.: A Survey of Integer Programming Emphasing Computation and Relations among Models, In *"Mathematical Programming"*. Eds. T. C. Hu and S. M. Robinson. London 1973 Academic Press, pp. 77–156.

[9] Geoffrion, A. M., Graves, G. W.: Multicommodity Distribution System Design by Benders Decomposition, *Management Science 20* (1974), 822–844.

[10] Granot, F., Hammer, P. L.: On the Use of Bolean Functions in 0–1 Programming, Operations Research Mimeograph 70 Technion, Haifa Sept. 70.

[11] Johnson, E. L.: On the Group Problem for Mixed Integer Programming, *Mathematical Programming Study 2* (1974), 137–179.

[12] Kendal, K. E., Zionts, S.: Solving Integer Programming Problems by Aggregating Constraints, Working Paper No. 155. School of Management State University of New York at Buffalo, Nov. 1972.

[13] Kianfar, F.: Stronger Inequalities for 0–1 Integer Programming, *Operations Research 19* (1971), 1373–1392.

[14] Miller, G. T., Tucker, A. W., Zemlin, R. A.: Integer Programming Formulations and Traveling Salesman Problem, *J. of ACM 7* (1960), 326–329.

[15] Petersen, C. C.: Computational Experience with Variants of the Balas Algorithm Applied to the Selection of R.a.D Projects, *Management Science 13* (1967), 736–750.

[16] Shapiro, J. F.: Generalized Lagrange Multiplier in Integer Programming, *Operations Research 19* (1971), 69–76.

[17] Trauth, C. A., Woolsey, E.: Integer Linear Programming. A Study in Computational Efficiency, *Management Science 15* (1969), 481–493.

[18] Walukiewicz, S.: Prawie liniowe zadania programowania dyskretnego, Warszawa 1975. Prace IOK, Seria B z. 23.

[19] Williams, H. P.: Experiments in the Formulation of Integer Programming Problems, *Mathematical Programming Study 2* (1974), 180–190.

[20] Williams, H. P.: The Reformulation of Two Mixed Integer Programming Problems, Research Report 75–11, University of Sussex, Oct. 1975.

THE "TUBE-PASSING PROBLEM" AND THE TRAVELLING SALESMAN PROBLEM

D. G. LIESEGANG

(Cologne, FRG)

1. INTRODUCTION

Given some tubes, each with two openings, one can formulate the problem of finding the shortest circuit passing once through each tube. One can state the problem as follows:

Given an undirected graph $G=(V, E)$ with vertex set $V=\{1, \ldots, n\}$, the set of edges $E=\{\{(i,j), (j, i)\}/i\in V, j\in V\}$ and the weight $c_{ij}=c_{ji}\geqq 0$ for each edge $\{(i,j), (j, i)\}\in E$ and a subgraph $G'=(V, E')$ with $E'\subset E$ and with degree $d(i)\leqq 1$ for each vertex $i\in V$, find a Hamiltonian circuit on G using all the edges of G' with minimal total weight.

The motivation to deal with this problem arose in the following context. Considering the branch-and-bound method of Little et al. [1963], which can be used highly effectively up to 150 cities for asymmetric travelling salesman problems (see for instance Liesegang [1974]), one must admit that the results obtained for symmetric problems are very poor. By each step of the branching procedure a directed arc (i, k) is chosen and the problem is reduced to a smaller t.s.p with the edges i and k melted in a new edge $\boxed{ik}$ with entrance i and exit k. The symmetric structure of the problem—if it was existant in the beginning—is lost. In addition, the solution of the initial assignment problem produces too many short cycles. Here, the symmetry is neglected too.

Thus, for symmetric t.s.p's a conclusion for a solution strategy might be:

(1) try to save symmetry during the reduction phase
(2) chose undirected *links* between cities during the branching procedure instead of directed arcs.

Hence we are led to the tube-passing problem stated in the beginning.

2. FRAMEWORK OF THE BRANCH-AND-BOUND PROCEDURE FOR SOLVING THE TUBE-PASSING PROBLEM

The tube-passing problem can be restated algebraically. Given a permutation $(p(1), \ldots, p(n))$ of $(1, \ldots, n)$ representing the tubes and a triangular distance matrix (c_{ik}), $1 \leq k \leq i \leq n$ with $c_{ik} \geq 0$ minimize

$$(2.1) \qquad f(x) = \sum_{1 \leq k \leq i \leq n} c_{ik} x_{ik}$$

subject to

$$(2.2) \qquad x_{ik} \in \{0, 1\} \quad 1 \leq k \leq i \leq n$$

$$(2.3) \qquad x_{ik} = 1 \quad \text{if} \quad p(i) = k$$

$$(2.4) \qquad \sum_{(i,k) \in Q_j} x_{ij} = 2, \quad j = 1, \ldots, n$$

(2.5) x corresponds to a connected subgraph.

By the class Q_j are meant all the possible links with city j:

$$Q_j := \{(j, i)/1 \leq i < j\} \cup \{(i, j)/j < i \leq n\}.$$

If we put $c_{ik} \sim \infty$ for $p(i) = k$, then (2.3) and (2.4) can be replaced by

$$(2.6) \qquad \sum_{(i,k) \in Q_j} x_{ik} = \begin{cases} 2 & \text{if } p(j) = j \\ 1 & \text{else} \end{cases}$$

and (2.5) changes to

(2.7) x combined with the set of tubes corresponds to a connected subgraph.

Using condition (2.6) the distance matrix can be reduced similar to the algorithm of Little et al. [1963] until in each class Q_j is at least one zero. For classes Q_j with $p(j) = j$ the reduction can be extended. There can be at least two zero or non-positive elements. If a class has outstanding many non-positive elements, one can try to find a further reduction by readding a constant to this class and by reducing the crossing classes.

(2.1) and (2.6) represent a b-matching problem and the reductions correspond to the task of finding a feasible dual solution (Lenstra [1976], pp. 66–67).

Starting from the reduced matrix, the branching is performed by fixing the non-positive link (i, k) which yields the most immediate additional costs, if it is not chosen. This alternative cost is the sum of the second or third smallest elements of the two classes Q_i and Q_k according to being part of a tube or not.

As general procedure a depth-first search with backtracking was carried out.

538

3. AN EXAMPLE

Let us consider the following problem (which may arise after the congress at Budapest):

Given the 7 cities Amsterdam, Berlin, Budapest, Paris, Rome, Vienna and Zurich, you want to find the minimum tour under the additional constraints that the links Vienna–Zurich and Paris–Amsterdam are a part of the tour (by sight-seeing reasons).

The distance matrix (in 10 km) is given in Fig. 1.

	AMS	BER	BUD	PAR	ROM	WIE
BER	69					
BUD	144	92				
PAR	[48]	110	156			
ROM	177	153	135	144		
WIE	118	66	26	130	125	
ZUR	88	85	101	58	89	[75]
	xPAR	BER	BUD	xAMS	ROM	xZUR	xWIE

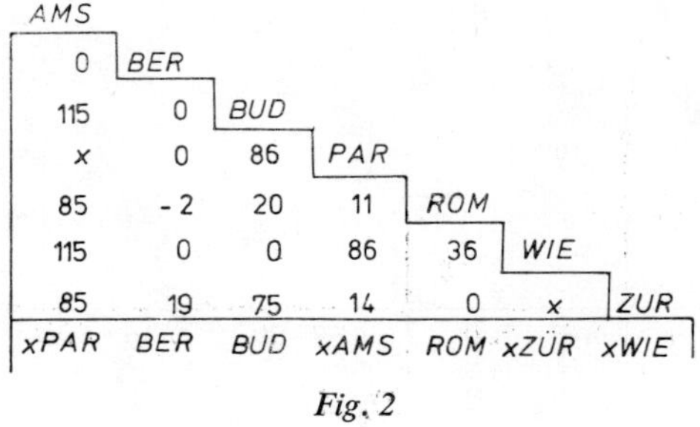

Fig. 1

The permutation vector is indicated below.

At first a number of reduction steps are performed by subtracting constants u_i from the classes $Q_i(i=1, \ldots, 7)$.

The reduction r_i of the objective function depends on whether a city is part of a tube or not. This reduction is based on (2.6). We get:

1. AMS, $u_1=69$, $r_1=69$; 2. BER, $u_2=66$, $r_2=66$;
3. BUD, $u_3=26$, $r_3=26+26$; 4. PAR, $u_4=44$, $r_4=44$;
5. ROM, $u_5=89$, $r_5=89+87$; 6. AMS, $u_1'=66$, $r_1'=0$.

In the distance matrix negative elements might occur. They are treated in the following manner: the value of this link is zero, if chosen in a tour; but if this link is not chosen, the objective function has to be increased by the positive value of this negative element.

If a negative element occurs in a class Q_j with $p(j)\neq j$, then we can add the same

	AMS	BER	BUD	PAR	ROM	WIE
BER	0					
BUD	115	0				
PAR	x	0	86			
ROM	85	-2	20	11		
WIE	115	0	0	86	36	
ZUR	85	19	75	14	0	x
	xPAR	BER	BUD	xAMS	ROM	xZÜR	xWIE

Fig. 2

positive value to this class without changing the objective function. This is the case in Step 6.

We obtain the distance matrix as shown in Fig. 2.

As Berlin has too many zero links, we try a further reduction by the steps:

7. BER, $u_2' = -100$, $r_2' = -100-98$;
8. AMS, $u_1'' = 85$, $r_1'' = 85$;
9. BUD, $u_3' = 20$, $r_3' = 20$;
10. PAR, $u_4' = 11$, $r_4' = 11$;
11. BER, $u_2'' = 80$, $r_2'' = 80+15$.

The sum of all reductions is 420. The starting point for the branch-and-bound procedure is the following reduced distance matrix with three additional rows for the smallest elements of the related classes in order to facilitate the calculation of the lower bounds (Fig. 3).

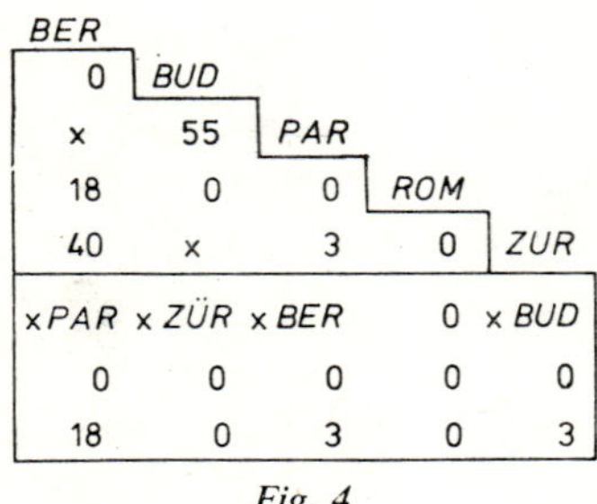

Fig. 3

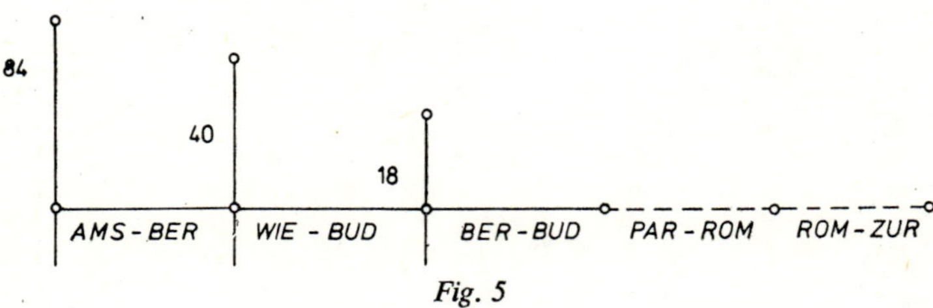

Fig. 4

Fig. 5

We get the search tree of Fig. 5, which shows that the first solution found is already the unique optimal solution. Fig. 4 shows the state of distance matrix after the execution of the first two branching steps.

4. RESULTS AND FURTHER DEVELOPMENTS

The procedure was written in FORTRAN IV and mainly used to try symmetric t.s.p's. There are no storage difficulties, because the changes from one node to the next of the search tree are executed on the original distance matrix, whilst the operations are stored in a "memory". During the retreat, operations are done in the inverse direction and then "forgotten". Only 30 K are needed for treating 80-cities problems. The problems were randomly generated by arbitrarily chosing a pair of coordinates

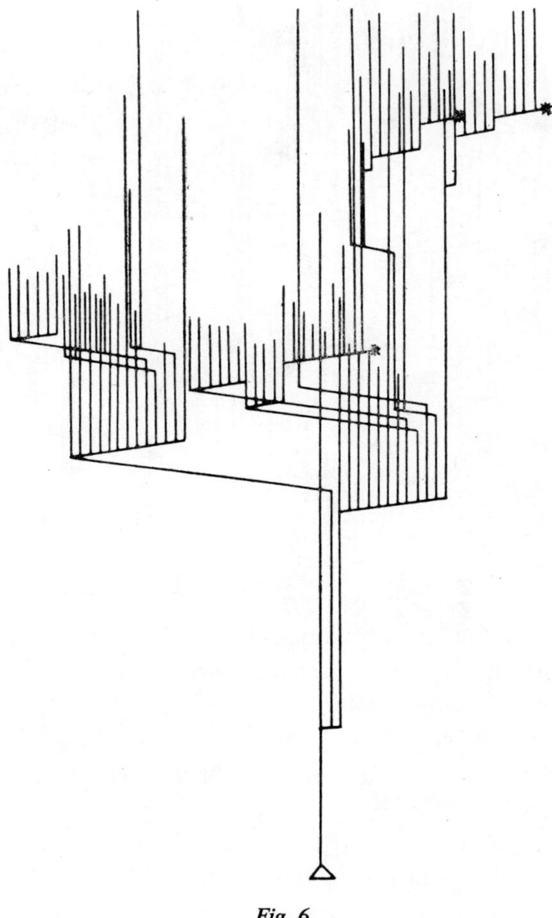

Fig. 6

out of (1000×1000) and then calculating the Euclidean distances. Each problem is to be identified and can be regenerated by its seed number IX (see Liesegang [1974]).

The basic reduction in the problems examined from 10 to 80 cities was on the average 15% better than the related assignment problem solution. The time for the reduction is relatively stable for each problem size and varied from 0.01 to 2 seconds (on a CDC 700).

Since the reduced matrix offers valuable insights and is easy to calculate, it may be the starting point even for other problems with more constraints, as for instance for the vehicle scheduling problem.

But even though the reduction is very promising, the optimal solution could only easily been attained up to 25 cities within 1 to 5 seconds. A typical search tree of a 25-cities problem is shown in Fig. 6. For bigger problems the search trees had the structure shown in Fig. 7.

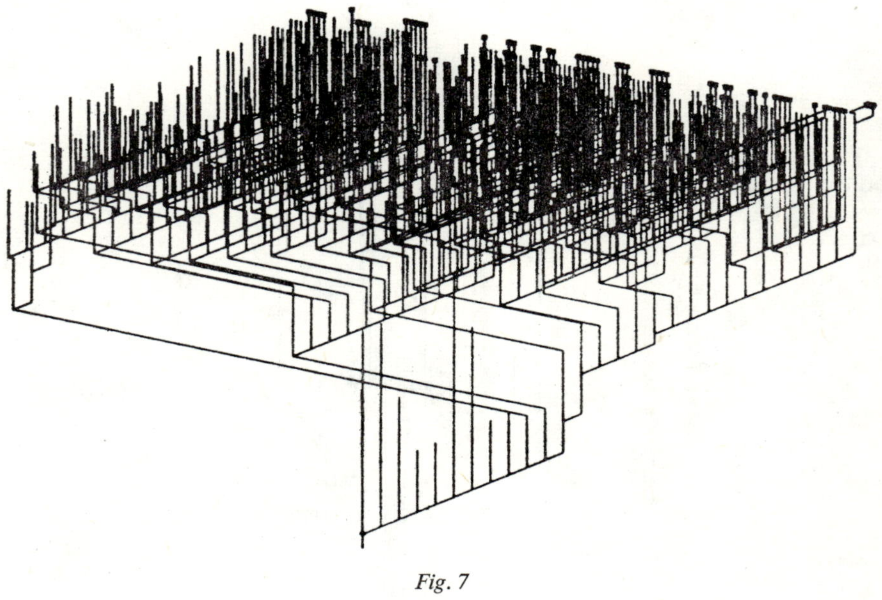

Fig. 7

Nevertheless there seem to be substantially little difference in the difficulty of solving a 40- or 80-cities problem. This is indicated by using approximative methods derived from the original branch-and-bound procedure.

The reason for this difficulty is the fact, that the constraint of connectivity is neglected calculating the lower bounds of the subproblems. Regarding the graphs of zero and negative arcs of a 10-, 20-, 40- and 80-cities problem with seed number 11, 21, 41 and 81 respectively (see Figs 8, 9, 10, 11; the zero links are dashed) it is obvious that the connectivity constraint is heavily violated for the 40-cities problem, because we get 4 distinct components. But, it was a second experience that the number of

542

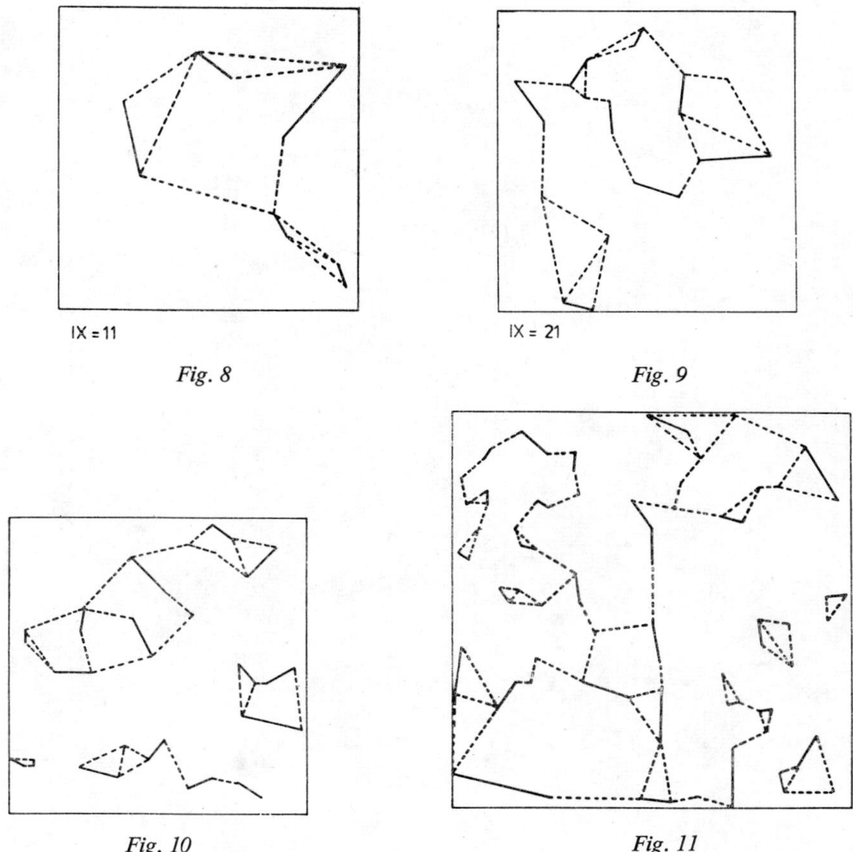

Fig. 8

Fig. 9

Fig. 10

Fig. 11

connectivity components of the "non-positive graph" does not essentially increase further: the number of components for 80-cities problems varied from 3 to 6.

Thus, further investigation will be made to include the constraints of connectivity into the evaluation of lower bounds. One promising way seems to be to determine the minimum spanning tree on the non-positive graph.

Another improvement is still left by the fact, that the tour must be 2-connected, which means that after destroying an arbitrary link of the tour, the remaining graph must be connected still.

REFERENCES

Lenstra, J. K.: Sequencing by Enumerative Methods, Dissertation, Amsterdam, 1976.
Liesegang, D. G.: Möglichkeiten zur wirkungsvollen Gestaltung von Branch-and-Bound Verfahren, Dissertation, Köln, 1974.
Little, J. D. C., Murty, K. G., Sweeny, D. W., Karel, C.: An Algorithm for the Travelling-Salesman Problem. *Operations Research* 11 (1963), No. 6. November.

OPTIMAL CUTTING ALGORITHM FOR RECTANGLE ELEMENTS

M. T. MISZCZYŃSKI and K. R. STRZELEC

(Łódź, Poland)

1. INTRODUCTION

Two-dimensional cutting stock problem ignoring guillotine methods is stated and the heuristic algorithm of finding optimal layouts of rectangles is presented in the paper.

2. PROBLEM FORMULATION

We start the presentation of the algorithm by calling attention to the following remark.

If one takes a rectangle and cuts it into a certain number of rectangular pieces and then tries to put them all together in order to obtain the original rectangle, one realizes that it could be done in many different ways.

This remark states the groundwork for the following achievements

Let

S_j — number of j-th stock rectangles to be supplied, $j \in \{1, \ldots, k\}$

D_i — number of i-th rectangles to be ordered, $i \in \{1, \ldots, m\}$

l_i — length of R_i rectangle

w_i — width of R_i rectangle.

Ignoring the guillotine cutting methods consider the problem stated as follows:

How can one portion the set of ordered elements into subsets (max number $2^m - 1$) and assign (simultaneously) these subsets to the supplied elements?

Cutting problem optimization is meant as a bi-partial process: making the cutting patterns from one side and an optimal choice from the other.

Here, in the context of the given question, the cutting patterns generating process is mentioned as the following PADD procedure:

1. Portion the set of ordered elements into subsets.
2. Assign the supplied elements to each subset.

3. Determine lattice points and calculate the minimal waste for each supplied element together with assigned subset.

4. Decide how many elements (ordered) to choose and what lattice points have to be active.

Now we explain the above steps more precisely. Consider a supplied element as an area bounded both in x and y directions (Fig. 1).

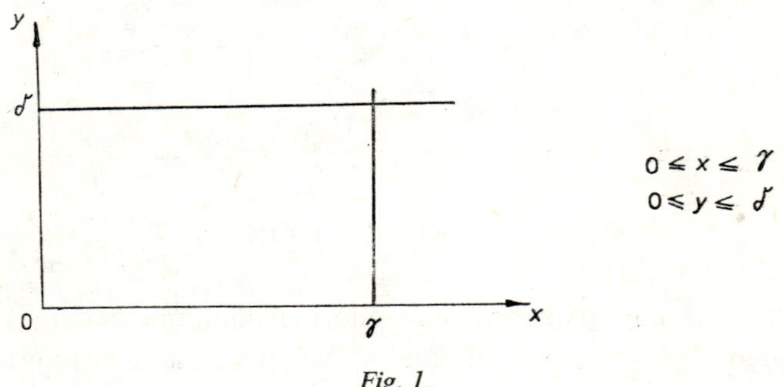

Fig. 1

Notice that, if we assume the elements are not rotated, only one point suffices to determine a position of the rectangle. Secondly, any element of the layout may be treated as belonging to two respective stripes determined respectively by its length and width. We refer here only to theoretical stripes rather than guillotine cuttings in which the stripes really exist.

Hence, in the case when the demand subsets are already known, we may define all theoretically *feasible stripes* corresponding to any supplied rectangle.

The procedure of obtaining all theoretically feasible stripes for a given rectangle can be demonstrated by the following example.

Example. Let the set of orders contain elements of two types (Fig. 2):

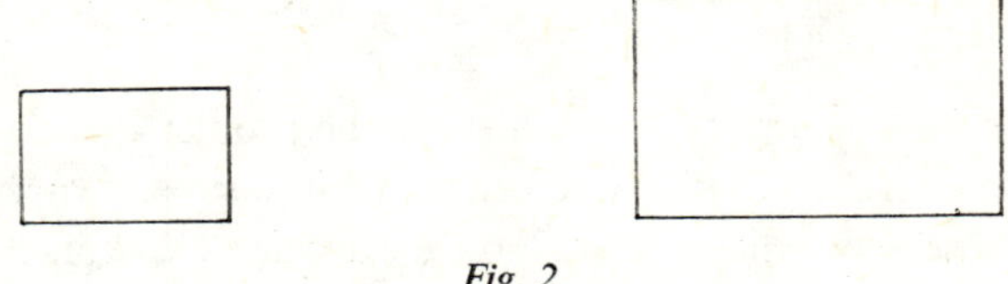

Fig. 2

If we suppose that the stripes adjoin each other and the first stripe corresponding to the width of the elements adjoins the $0x$ axis and that corresponding to the length, adjoins the $0y$ axis we obtain Fig. 3.

The dashed stripes consist of what may be called *waste rectangle elements*. In accordance to dimensions of ordered elements this area states the necessary waste. Of

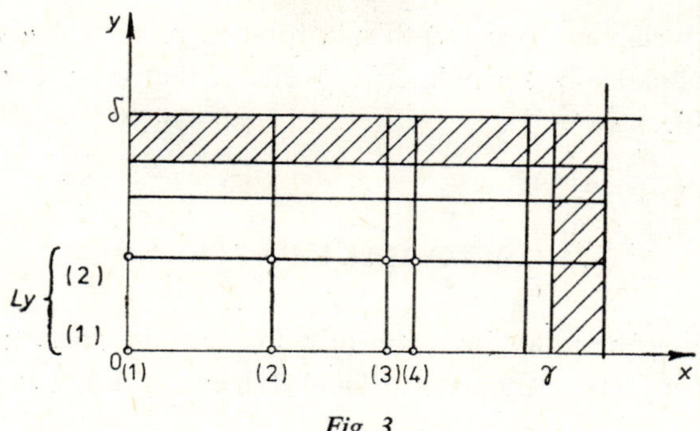

Fig. 3

course one may think of the stripes in another way and then the location of the necessary waste will be different. But in all cases the total area of necessary waste will be the same. Therefore only the lattice points within the light area are worth considering. Denote these points by $P_{(i)(j)}$. We will call them active or passive on whether the origin of the rectangle element covers the point or not.

Any layout may be presented as a *lattice points matrix* (LPM):

$$\text{LPM} = [l_{(i)(j)}]$$
$$l_{(i)(j)} \in \{0, 1, \ldots, m\}.$$

If an entry of the LPM matrix is equal to zero it means that the respective point is passive. To the active points we correspond a number which denotes the type of the element of the layout. An optimization problem has been established as the following zero-one programming problem:

$$\text{Maximize } z = \sum_{k=1}^{m} x_{j/i}^k w_k$$

$$\text{subject to } \sum_{k=1}^{m} x_{j/i}^k \leqq 1 \text{ for each } j \in L_y$$

$$\left(\sum_{k=1}^{m} x_{j_1/i}^k\right)\left(\sum_{k=1}^{m} x_{j_2/i}^k\right) = 0 \text{ for each pair } (j_1, j_2) \text{ such that } |j_1 - j_2| < w_k$$

$$\sum_{j \in L_y} \sum_{k=1}^{m} x_{j/i}^k w_k \leqq \delta$$

$$x_{j/i}^k l_k \leqq \gamma \text{ for each } k \in \{1, \ldots, m\} \text{ and for each } j \in L_y$$

$$x_{j/i}^k \in \{0, 1\}.$$

We solve several such problems. The first one for $i = (1)$, the following respectively to the growing extent in x direction. Parameter may be the same for all subproblems or we may introduce a string of gammas:

$$0 < \gamma_{(1)} < \gamma_{(2)} < \ldots < \gamma_{(n)} = \gamma$$

and include its (i)-th element to the (i)-th subproblem. It is worthwhile to emphasize that the problem may also be formulated conversely, that is with respect to growing extent in y direction.

3. HOW TO MAKE PATTERNS

From the foregoing the following main principles result. The steps of the optimal cutting algorithm for rectangle elements (OCARE) are as follows:

1. Determine the sets L_x and L_y in respect to the determined demand subset together with the supplied element.
2. Calculate the necessary waste and determine the LPM matrix according to cartesian product of L_x and L_y. Subset zero into all entries.
3. Formulate the zero-one programming problem. For $i=(1)$ find a solution. This process proceeds as follows:
3.1. Select the largest element (in the sense of length) not exceeding the parameter $\gamma_{(i)}$. Put it into the proper stripe.
3.2. Arrange all the smaller elements in decreasing order.
3.3. Find, sequentially, the proper or the most proper (i.e. in the sense of difference between dimensions) stripe for a given element. Put the chosen element into its stripe. Repeat as long as the restrictions of $0-1$ programming problem are still fulfilled.
4. When Step 3 is over, the first break has been obtained. Check all segment lines and cancel those lines to which any one element does not fit. Calculate the waste and add it into the actual value.
5. Make sure that parameter γ has not been reached. If not, formulate the following optimization problem and repeat Steps 3 and 4. In another case the algorithm is terminated.

4. HOW TO USE THE ALGORITHM

Recalling the definition of cutting problem optimization we may distinguish three possible alternatives for solving the problem:

(1) to check all feasible patterns (acceptable only for very small problems),
(2) to generate patterns seeking the feasible layouts randomly and then, to solve the linear programming problem,
(3) to solve the problem with the help of heuristics.

Now we present a heuristic method used for finding solutions. The method contains four steps.

Let, additionally,

x_q — number of repetitions of the q-th cutting pattern, $q \in \{1, \ldots, N\}$

d_{iq} — number of the i-th elements in the q-th cutting pattern.

Step 1. Use the OCARE algorithm for each $j \in \{1, \ldots, k\}$ together with $S_j > 0$ and for each $i \in \{1, \ldots, m\}$ together with $D_i > 0$. If it is impossible the procedure is terminated. Choose the pattern in which the percentage waste is the smallest. Let this pattern be connected with j_0-th stock rectangle.

Step 2. Calculate the average percentage waste. If the difference between the percentage waste in the last pattern and the current average percentage waste is less than ε per cent, then go to the next step, otherwise act as it is described in Step 4.

Step 3. Determine the number of repetitions of the last slitting pattern as follows:

$$x_q = \min \left\{ \min_{1 \leq i \leq m} \{E[D_i/d_{iq}]\}, \, S_{jo} \right\} \text{ for } d_{iq} > 0.$$

Assign the number of stock rectangles of the type which has been chosen according to $S_j \leftarrow S_j - x_q$ and the number of ordered rectangles to

$$D_i \leftarrow D_i - x_q d_{iq} \quad \text{for each} \quad i \in \{1, \ldots, m\}.$$

Return to Step 1.

Step 4. For these ordered elements for which $D_i > 0$ set infinity demand D_i'. For the other elements set $D_i' = 0$. Then act as it was described in Step 1, using D_i' instead of D_i. Calculate the number of repetitions of the chosen pattern as follows:

$$x_q = \min \left\{ \max \left\{ \min_{1 \leq i \leq m} \{E[D_i/d_{iq}]\}, \, 1 \right\}, \, S_{jo} \right\} \text{ for } d_{iq} > 0.$$

Assign the supply and demand quantities as described in the formulas of Step 3. Return to Step 1.

5. COMPUTATIONAL EXPERIENCE

The OCARE algorithm has been proved by a computer experiment. Calculations have been executed on "ODRA 1305" computer (executing add or subtract operations on integers in about 6 microseconds). The experiment has been divided into four parts:

(a) creation of supply, (b) creation of demand, (c) execution of the algorithm, (d) statistical data processing.

Three random variables (all rectangular distributed) determined the supply of stock rectangles (sheets). They denoted respectively the length, and the width of the sheet and the number of sheets. Maximal number of types of the rectangles and maximal ratio of dimensions have been given.

Sets of orders have been created according to generated supply. Roughly speaking, each sheet on hand stated the basis for the creation of ordered elements in the following manner. Both dimensions (length and width) have been randomly divided into a few parts. Trimming (theoretically) a sheet, respectively to the random numbers, rectangle elements have been obtained. Next, a random number greater than $\frac{1}{2}$ and less than 1 has been generated. It has been treated as a probability of including any rectangle into the set of ordered elements.

Executing OCARE algorithm we have obtained a sample. Each sample has been processed statistically. Tables contain the result of the experiment. We have assumed execution of 100 simulating cutting process samples (for the case of small scale* problems). 8092 slitting patterns have been generated and 1067 accepted by the method.

Nearly 70 per cent of the accepted patterns are characterized by percentage waste less than 10 per cent and more than 50 per cent of the accepted patterns do not exceed 6 per cent waste. It is necessary to note that, additionally, a requirement of exactly fulfilled demand has been included. Thus a comparatively large fraction of unsatisfied patterns has appeared.

Table 1. Cumulative distribution of slitting patterns percentage waste

Percentage waste	Generated patterns	Accepted patterns
0– 2	.037	.141
2– 4	.131	.345
4– 6	.266	.526
6– 8	.390	.621
8– 10	.481	.693
10– 12	.547	.730
12– 14	.606	.758
14– 16	.643	.790
16– 18	.696	.808
18– 20	.709	.828
20– 22	.749	.849
22– 24	.761	.860
24– 26	.784	.874
26– 30	.813	.886
30– 35	.847	.905
35– 40	.866	.913
40– 50	.906	.945
50–100	1.000	1.000

* The supply has been simulated as a stock of 2–5 different types sheets and the demand as a number of different rectangle elements not greater than 40.

Table 2. Cumulative distribution of minimal
percentage waste on one sample

Per cent waste	Distribution
1	0.26
2	0.51
3	0.65
4	0.74
5	0.86
10	0.96
20	0.97

REFERENCES

[1] Gilmore, P. C., Gomory, R. E.: Multistage cutting problems of two and more dimensions, *Opns. Res.* 13 (1965), 94–120.

[2] Gurel, O.: Marker layout problem via graph theory, *Computing Methods in Optimization Problems,* Acad. Press, New York–London 1969, pp. 133–141.

SOME REMARKS ABOUT HOMOMORPHISMS AND GROUP MINIMIZATION PROBLEMS

A. VOLPENTESTA

(Pisa, Italy)

1. INTRODUCTION

Consider an integer linear program of the following kind

$$\begin{cases} \min\ (z=c^T x) \\ Ax=b \\ x\geqq 0 \quad \text{and integer.} \end{cases}$$

Where A is an $n\times m$ matrix, b and c, vectors, all integers; and x an unknown n-vector.

It is well known [3, 7, 2] that such a problem can be reduced to the following group minimization problem

$$\text{(1a)} \qquad \min\left(\sum_{g\in E} c(g)t(g)\right)$$

$$\text{(1b)} \qquad \sum_{g\in E} t(g)\cdot g=g_0$$

$$\text{(1c)} \qquad t(g)\geqq 0 \quad \text{and integer,} \quad \forall g\in E$$

where G is an abelian group generated by the set E.

Let γ be a homomorphism from G onto a group H.

Obviously H is isomorphic to a subgroup of G and it is generated by $Q=\gamma(E)$. Put $h_0=\gamma(g_0)$, $J(h)=\{g\in E: \gamma(g)=h\}$ and $t=(t(g))_{g\in E}$. The function π is defined by $\pi(t)=\hat{t}=(\hat{t}(h))_{h\in Q}$, where $\hat{t}(h)=\sum_{g\in J(h)} t(g)$.

Consider the following group relation:

$$\text{(2a)} \qquad \sum_{h\in Q} t(h)h=h_0$$

$$\text{(2b)} \qquad t(h)\geqq 0, \quad \forall h\in Q.$$

Obviously, if t is a solution of (1ab), then $\hat{t}=\pi(t)$ is a solution of (2ab). Call $P(G,E,g_0)$ and $P(H,Q,h_0)$ the convex hull (corner polyhedron) of the solutions of

(1bc) and (2ab), respectively; thus $\{P(H, Q, h_0) = \emptyset\} = > \{P(G, E, g_0) = \emptyset\}$. Assuming $P(H, Q, h_0) \neq \emptyset$, we will study the relations between $P(G, E, g_0)$ and $P(H, Q, h_0)$.

We say that an integer point $t \in P(G, E, g_0)$ is *weakly irreducible*, iff there exists no other integer point $t' \in P(G, E, g_0)$, such that $t' \leq t$. We will say that an integer point $t \in P(G, E, g_0)$ is *irreducible*, iff for any set of integers $r = (r(g))_{g \in E}$, $r' = (r'(g))_{g \in E}$, the conditions:

(3a)
$$0 \leq r(g) \leq t(g); \quad 0 \leq r'(g) \leq t(g), \quad \forall g \in E$$

(3b)
$$\sum_{g \in E} r(g) t(g) = \sum_{g \in E} r'(g) t(g)$$

imply that $r(g) = r'(g)$, $\forall g \in E$.

It is immediate the following:

Proposition 1. *If it is an irreducible point of $P(G, E, g_0)$, then t is a weakly irreducible point of $P(G, E, g_0)$.*

The following example shows that a weakly irreducible point is not necessarily an irreducible point.

Example 1
$$\begin{cases} 3t_1 + 2t_2 \equiv 1 \pmod{24} \\ t_1, t_2 \geq 0 \text{ and integers.} \end{cases}$$

We have $G = Z/\text{mod } 24$, $E = \{3, 2\}$, $g_0 = 1$. One can easily see that $t = (3, 8)$ is a weakly irreducible point of $P(G, E, g_0)$. Put $r = (2, 0)$ and $r' = (0, 3)$; since r and r' satisfy (3a) and (3b) the point t is not irreducible. One can see in this example that the weakly irreducible points are $(7, 2)$, $(1, 11)$, $(3, 8)$, $(5, 5)$ but the irreducible points are only $(7, 2)$ and $(1, 11)$.

The following is known [5].

Theorem 1. *Every vertex of $P(G, E, g_0)$ is irreducible.*

In general, it is not true that the vertices of $P(G, E, g_0)$ are the only irreducible points. Anyway, if G is the direct sum either of cyclic groups of order 2 or of groups of order 3, the only irreducible points of $P(G, E, g_0)$ are the vertices.

2. MAIN RESULTS

Let K be the kernel of γ and let E' be the group generated by the elements of $k \cap E$. Let $\hat{t} \in P(H, Q, h_0)$ and consider the following relations:

(4a)
$$\sum_{g \in J(h)} t(g) = \hat{t}(h), \quad \forall h \in Q$$

(4b)
$$g_0 - \sum_{g \in E - K} t(g) g \in E'$$

and denote by $R(\hat{t})$ the set of nonnegative integer solutions of (4a) and (4b). The following example shows that $R(\hat{t})$ can be empty for some $\hat{t} \in P(H, Q, h_0)$.

554

Example 2.
$$\begin{cases} 4t_1 + 3t_2 \equiv 2 \pmod{12} \\ t_1, t_2 \geqq 0 \text{ and integers.} \end{cases}$$

In this example we have $G = Z/\text{mod } 12$, $E = \{4, 3\}$, $g_0 = 2$. If we consider the homomorphism γ, from G onto $H = Z/\text{mod } 6$ defined by $\gamma(1) = 1$, we have the following group relation:
$$\begin{cases} 4t_1 + 3t_2 \equiv 2 \pmod{6} \\ t_1, t_2 \geqq 0, \text{ and integers.} \end{cases}$$

Q is the set $\{4, 3\}$ and h_0 is 2. Obviously, $\hat{t} = (2, 0) \in P(H, Q, h_0)$, but it is easily seen that $R(\hat{t}) = \emptyset$.

Theorem 2. *Let $\hat{t}$ be a weakly irreducible point of $P(H, Q, h_0)$ and let $R(\hat{t}) \neq \emptyset$, then for every $t \in R(\hat{t})$ there exists a weakly irreducible point $v(v(g))_{g \notin E}$ of $P(G, E, g_0)$, such that $v(g) = t(g)$, $\forall g \in J(h)$, $h \in Q$.*

Proof. Let $t \in R(\hat{t})$ and put $P(t) = P(G, E, g_0) \cap \{x = (x(g))_{g \in E} : x(g) = t(g) \forall g \in J(h), h \in Q\}$.

Since $t \in R(\hat{t})$, $P(t)$ is not empty. It is sufficient to show that a weakly irreducible point $v = (v(g))_{g \in E}$ of $P(t)$ is a weakly irreducible point v of $P(G, E, g_0)$ too. Let $r = (r(g))_{g \in E}$ in $P(G, E, g_0)$ and such that $r(g) \leqq v(g)$, $\forall g \in E$.

If there exist $g \in E$, $h \in Q$, such that $g \in J(h)$ and $r(g) < v(g)$, consider the point $\Pi(r) = \hat{r} = (\hat{r}(h))_{h \in Q}$, where $\hat{r}(h) = \sum_{g \in J(h)} r(g)$, $\forall h \in Q$.

Since $\hat{r} \in P(H, Q, h_0)$, $r \leqq t$ and $r \neq t$ it is contradicted the assumption of weak irreducibility in $P(H, Q, h_a)$. Besides, if $r(g) = v(g) = t(g) \forall g \in J(h)$, $h \in Q$, and if there exists $g \in K \cap E$, such that $r(g) < v(g)$, it is contradicted the assumption of weak irreducibility of v in $P(t)$. Thus, we have $r = v$. This completes the proof.

Let $\hat{t} \in P(H, Q, h_0)$ and put $P^*(\hat{t}) = P(G, E, g_0) \cap \left\{ x = (x(g))_{g \in E} : \sum_{g \in J(h)} x(g) = \hat{t}(h) \, h \in Q \right\}$.

Theorem 3. *Let $\hat{t}$ be a vertex of $P(H, Q, h_0)$. If v is a vertex of $P^*(\hat{t})$, then v is a vertex of $P(G, E, g_0)$.*

Proof. Let $b = (b_1, \ldots, b_n)$ and $c = (c_1, \ldots, c_n)$ be two points of $P(G, E, g_0)$ such that:
$$\lambda b + (1 - \lambda)c = v \quad \text{for some} \quad \lambda, \quad 0 < \lambda < 1$$

Put $\hat{b} = \pi(b)$ and $\hat{c} = \pi(c)$.

We have $\pi(\lambda b + (1 - \lambda)c) = \pi(v)$ and thus $\lambda \hat{b} + (1 - \lambda)\hat{c} = \hat{t}$; since $\hat{b}, \hat{c} \in P(H, Q, h_0)$ and since $\hat{t}$ is a vertex of $P(H, Q, h_0)$ we have $\hat{b} = \hat{c} = \hat{t}$ and thus $b, c \in P^*(\hat{t})$.

Being v a vertex of $P^*(\hat{t})$ we have $b = c = v$.

This completes the proof.

Corollary 1. *Let $\hat{t}$ be a vertex of $P(H, Q, h_0)$ and assume $R(\hat{t})\neq\emptyset$, then there is a vertex $v=\big(v(g)\big)_{g\in E}$ of $P(G, E, g_0)$ such that $\sum\limits_{g\in J(h)} v(g)=\hat{t}(h)$ $h\in Q$.*

Proof. From $R(\hat{t})\neq\emptyset$ it follows $P^*(\hat{t})\neq\emptyset$, thus $P^*(\hat{t})$ has at least a vertex, which is a vertex of $P(G, E, g_0)$ too.

Corollary 2. *Let $\hat{t}$ be a vertex of $P(H, Q, h_0)$ and let g_h, $h\in Q$ be some elements of E such that $g_h\in J(h)$. Put $E_1=\{g_h\}_{h\in Q}$. $E_2=\{g\in E-k: g\notin E_1\}$. If the point t defined by $t(g_h)=\hat{t}(h)$ $g_h\in E_1$ and $t(g)=0$ $g\in E_2$ is in $R(\hat{t})$, then there exists a vertex $v=\big(v(g)\big)_{g\in E}$ of $P(G, E, g_0)$, such that $v(g)=t(g)$, $\forall g\in E-K$.*

Proof. We must show that a vertex $v=\big(v(g)\big)_{g\in E}$ of $P(t)$ is a vertex of $P^*(\hat{t})$. By assumption $t\in R(\hat{t})$ we have $P(t)\neq\emptyset$. Obviously, $v\in P^*(\hat{t})$.

Let $b=\big(b(g)\big)_{g\in E}$ and $c=\big(c(g)\big)_{g\in E}$ be two points of $P^*(\hat{t})$, such that $\lambda b+(1-\lambda)c=$ $=v$ for some λ, $0<\lambda<1$. Since $v\in P(t)$ we have $b(g)=c(g)=0$ $\forall g\in E_2$; and thus $b(g)=$ $=c(g)=t(g)$ $\forall g\in E_1$, that is to say b, $c\in P(t)$, being v a vertex of $P(t)$ we obtain at last $b=c=v$.

Q.E.D.

Corollary 3. *If $g_0\in E'$ then there exists a vertex $v=\big(v(g)\big)_{g\in E}$ of $P(G, E, g_0)$, such that $v(g)=0$, $\forall g\in J(h)$, $h\in Q$.*

Proof. It is sufficient to remark that the only vertex of $P(H, Q, h_0)$ is the zero and that the assumption $g_0\in E'$ implies $R(0)\neq\emptyset$.

The case of the master polyhedron $P(G, g_0)$ when $E=G-\{0\}$ we call $P(G, E, g_0)$ master polyhedra and we denote it as $P(G, g_0)$.

The connections between faces and vertices of $P(G, g_0)$ are summarized by the following theorem [5]:

Theorem 3. (i) *If $\sum\limits_{g\in E} s(g)t(g)\geqq s_0$ is a facet of $P(G, E, g_0)$, then there exists a facet $\sum\limits_{g\in G} s'(g)t(g)\geqq s_0'$ of $P(G, g_0)$, such that $s'(g)=s(g)$ $\forall g\in E$.* (ii) *The point $t=\big(t(g)\big)_{g\in E}$ is a vertex of $P(G, E, g_0)$, iff the point $t'=\big(t'(g)\big)_{g\in G}$, defined by $t'(g)=0$ $g\notin E$, $t'(g)=t(g)$ $g\in E$, is a vertex of $P(G, g_0)$.*

Remark that it is not true, if $\sum\limits_{g\in G} s'(g)t(g)\geqq s_0'$ is a facet of $P(G, g_0)$, then $\sum\limits_{g\in E} s(g)t(g)\geqq$ $\geqq s_0$ is a facet of $P(G, E, g_0)$. This is shown by the following:

Example 3. $t_1+2t_2+3t_3\equiv 3 \pmod 4$.

We have $G=Z/\text{mod } 4$, $g_0=3$, $t_1+2t_2+3t_3\geqq 3$ is a facet of $P(G, g_0)$, but if put $E=$ $=\{2, 3\}$ then $2t_2+3t_3\geqq 3$ is not a facet of $P(G, E, g_0)$.

In section 2 one can remark that it is very useful to know when (4b) is satisfied by all the nonnegative points which satisfy (4a) however $\hat{t}\in P(H, Q, h_0)$ may be. The following theorem answers to this question.

556

Theorem 4. *Let* $\hat{t} \in P(H, Q, h_0)$. *If* $k \subset E$, *then* (4b) *is satisfied by every point, which satisfies* (4a).

Proof, Let the integers $t(g)$, $g \in J(h)$, $h \in Q$ satisfy (4a); then we have:

$$\gamma \left(g_0 - \sum_{g \in E-K} t(g)g \right) = h_0 - \sum_{h \in Q} \hat{t}(h)h = 0.$$

Because of the equality $k \cap E = k$ we have:

$$\left(g_0 - \sum_{g \in E-K} t(g)g \right) \in E'. \qquad \text{Q.E.D.}$$

Theorems 2, 3 and 4 imply the following:

Corollary 4. *Let* $\hat{t}$ *be a weakly irreducible point of* $P(H, h_0)$; *then for every set of nonnegative integers* $t = (t(g))_{g \notin k}$ *satisfying* (4a) *there exists a weakly irreducible point* $v = (v(g))_{g \in G}$ *of* $P(G, g_0)$, *such that* $v(g) = t(g)$ $g \notin K$.

Corollary 5. *Let* $\hat{t}$ *be a vertex of* $P(H, h_0)$, *then there is a vertex* $v = (v(g))_{g \in G}$ *of* $P(G, g_0)$ *such that* $\sum_{g \in J(h)} v(g) = \hat{t}(h)$, $h \in Q$.

Corollary 6. *Let* $\hat{t}$ *be a vertex of* $P(H, h_0)$ *and let* g_h, $h \in Q$ *be some elements of* G, *such that* $g_h \in J(h)$ $\forall h \in Q$. *Put* $G_1 = \{g_h\}_{h \in Q}$ *and* $G_2 = \{g \in G - K : g \notin G_1\}$, *then there exists a vertex* $v = (v(g))_{g \in G}$, *such that* $v(g_h) = \hat{t}(h)$, $h \in Q$ *and* $v(g) = 0$, $g \in G_2$.

Corollary 7. *If* $g_0 \in K$, *then there exists a vertex* $v = (v(g))_{g \in G}$ *of* $P(G, g_0)$, *such that* $v(g) = 0$, $\forall g \notin K$.

REFERENCES

[1] Burdet, C.-A.: On the Algebra and Geometry of Integer Programming Cuts, Carnegie-Mellon Management Sciences Research, Report, N. 291, October 1972.

[2] Garfinkel, R. S. and Nemhauser, G. L.: Integer Programming, J. Wiley, 1972.

[3] Gomory, R. E.: On the Relation Between Integer and Noninteger Solutions to Linear Programs, *Proc. Nat. Acad. U.S.A.*, 53 (1965), 260–265.

[4] Gomory, R. E.: Integer Faces of a Polyhedron, *Proc. Nath. Acad. Sci. U.S.A.*, 57 (1967), 16–18.

[5] Gomory, R. E.: Some Polyhedra Related to Combinatorial Problems, J. Linear Algebra and its Applications, 2, Am. Elsevier Pub. Co. Oct. 1969, 451–558.

[6] Gomory, R. E. and Johnson, E. L.: Some Continuous Functions Related to Corner Polyhedra I, *Mathematical Programming*, 3 (1972), 23–85.

[7] Hu, T. C.: Integer Programming and Network Flows, Addison–Wesley, 1969.

[8] Volpentesta, A.: Finite Groups and Integer Linear Programs, Pubbl. Series A, N. 20 Dip. Ric. Op. Sc. St. Università di Pisa, Ed. Tecno-Scient., Pisa, 1975.

A NECESSARY AND SUFFICIENT CONDITION FOR THE AGGREGATION OF LINEAR DIOPHANTINE EQUATIONS

F. WEINBERG

(Zürich, Switzerland)

1. INTRODUCTION

Several authors have given sufficient conditions for the weights with which two Diophantine equations are to be multiplied before being aggregated to a new single equation with exactly the same set of integer solutions. In the following article some both sufficient and necessary conditions will be established for an even more general form of linear constraints. In spite of their general validity they are conceived for the case of coefficients and variables taking values of small magnitude only as this may occur in matching, covering and related problems [1].

2. THE SPECTRUM OF A LINEAR FORM

Let $s= \sum_{j=1}^{n} a_j x_j$ be a linear form in the independent variables $x_j \in X_j$, where the a_j are given integers, $j=1, \ldots, n$. $X_j = \{x_j^{(1)}, x_j^{(2)}, \ldots, x_j^{(vi)}\}$ is a given finite non-empty set of pairwise different integers which may be called the *spectrum* of x_j, $j=1, \ldots, n$. The set S of integers, $S=\{s\}$, is then the spectrum of the linear form. This set is in many practical cases of small cardinality as compared to the large number $v_1 \cdot v_2 \cdot \ldots \cdot v_n$ of different generating combinations $\vec{x}=[x_1, \ldots, x_n] \in X_1 \times \times \ldots \times X_n$.

For the calculation of the spectrum the basic operation used is "spectral superposition". It is symbolized by $\oplus$ and generates the set formed by the sum of all element pairs one from each of two spectral sets

$$\{a\} \oplus \{b\} = \{a+b\}.$$

Let X_j be the spectrum of x_j, $j=1, \ldots, n$. Consider the linear form $s= \sum_{j=1}^{n} a_j x_j$, with a_j integer, $j=1, \ldots, n$.

The spectrum S of s is then

$$S=a_1 X_1 \oplus a_2 X_2 \oplus \ldots \oplus a_n X_n.$$

Example

$$X_1=\{3, 0, -1\}, \quad X_2=\{1, 0, -2\}; \quad s=x_1+2x_2;$$
$$S=\{3, 0, -1\} \oplus 2 \cdot \{1, 0, -2\}=\{5, 3, 2, 1, 0, -1, -4, -5\}.$$

Let φ be a number composed of zeros and ones only. φ gives a binary representation of an integer spectrum S by defining

$$k \in S \Leftrightarrow \text{in the } k\text{-th place of } \varphi \text{ there is a 1.}$$

Example

$$S=\{3, 0, -1\} \Leftrightarrow \varphi=1001,1.$$

Let $\beta=\sum\limits_{r=-\infty}^{\infty} B_r \cdot 10^r$, $\gamma=\sum\limits_{r=-\infty}^{\infty} C_r \cdot 10^r$, with $B_r, C_r \in \{0, 1\}$ for all r.
Define

$$\beta \otimes \gamma=\sum_{k=-\infty}^{\infty} \pi_k \cdot 10^k \text{ with } \pi_k=\overset{\infty}{\underset{r=-\infty}{\sigma}} B_{k-r} \cdot C_r,$$

σ being the symbol for a Boolean sum.

Call Φ_j the binary representation of $\{a_j x_j\}=a_j X_j$ with $\Phi_j=1$, if $a_j=0$. Then

$$\varphi=\Phi_1 \otimes \Phi_2 \otimes \ldots \otimes \Phi_n$$

is the binary representation of the spectrum of $s=\sum\limits_{j=1}^{n} a_j x_j$.

Example

$$X_1=\{3, 0, -1\}, \quad X_2=\{1, 0, -2\}; \quad s=x_1+2x_2.$$

Then $\Phi_1=1001,1$, $\Phi_2=101,0001$ and

$$\varphi=\Phi_1 \otimes \Phi_2= \frac{1001.1 \otimes 101.0001}{\begin{array}{l} 10011 \\ \quad 1001\ 1 \\ \qquad\quad 10011 \end{array}}$$
$$101111.10011$$

from which

$$S=\{5, 3, 2, 1, 0, -1, -4, -5\}.$$

There exists a more efficient version of calculation by coding the binary representation by an array of positive numbers, each of which signifies the length of an uninterrupted partial sequence of zeros or ones respectively, zero-sequences being indicated by numbers in brackets. It is clear that long uninterrupted partial sequences of spectral values can be handled very easily this way. For instance the spectrum $S=\{1,000,000,$ $999,999, \ldots, 2, 1, 0\}$ is coded by the sole number $1,000,001$.

3. THE SPECTRAL CORRELATION SET

Let $s_i = \sum_{j=1}^{n} a_{ij}x_j$, $i=1, \ldots, m$, where the a_{ij} are integers and $x_j \in X_j$ as before, and let $S_i = \{s_i\}$, $i=1, \ldots, m$. The spectral correlation set $S_{1,\ldots,m} \subset S_1 \times \ldots \times S_m$ is defined as the set of all m-tuples $[s_1, \ldots, s_m]$ of integers generated by the same vector $\vec{x} = (x_j)$, $x_j \in X_j$, $j=1, \ldots, n$.

The case of $m=2$.

Lemma 1

Let

$$s_i = \sum_{j=1}^{n} a_{ij}x_j, \quad i=1, 2; \ x_j \in X_j, j=1, \ldots, n,$$

and

$$s = \sum_{j=1}^{n} a_{1j}x_j + \lambda \sum_{j=1}^{n} a_{2j}x_j, \quad \text{with} \quad \lambda = \text{integer}.$$

Call S_1, S_2 and S the corresponding spectra.

Define
$$M^- = \left[\sum_{j=1}^{n} a_{1j}x_j \right]_{\substack{\min \\ X_1 \times \ldots \times X_n}}, \quad M^+ = \left[\sum_{j=1}^{n} a_{1j}x_j \right]_{\substack{\max \\ X_1 \times \ldots \times X_n}}.$$

A necessary condition for $\bar{s}_1$ and $\bar{s}_2$ to be correlated: $[\bar{s}_1, \bar{s}_2] \in S_{12}$, is

$$\bar{s}_1 \in S_1, \quad \bar{s}_2 \in S_2, \quad \text{and} \quad \bar{s}_1 + \lambda \bar{s}_2 \in S.$$

If $\lambda \geq M^+ - M^- + 1$, integer, the condition is also sufficient.

For the proof, see [1]. In fact λ does not even need to be an integer.

From this Lemma follows a direct method for the determination of the spectral correlation set of two linear forms:

Method

1. Let $S_i = \{s_i\}$, the spectrum of $s_i = \sum_{j=1}^{n} a_{ij}x_j$, $i=1, 2$; $x_j \in X_j$, $j=1, \ldots, n$, be the bordering entries of a table;

2. Choose $\lambda \geq M^+ - M^- + 1$ and set the interior entries in position (μ, v) of the table equal to $s_1^{(v)} + \lambda s_2^{(u)}$;

3. Calculate S, the spectrum of $s = \sum_{j=1}^{n} a_{1j}x_j + \lambda \sum_{j=1}^{n} a_{2j}x_j$;

4. Wherever $s = \bar{s}$ takes the same value as $s_1^{(v)} + \lambda s_2^{(u)} = \bar{s}_1 + \lambda \bar{s}_2$, the corresponding components $s_1 = \bar{s}_1$ and $s_2 = \bar{s}_2$ are correlated: $[\bar{s}_1, \bar{s}_2] \in S_{12}$.

Numerical example. Let $s_1 = -x_1 + 2x_2$, $s_2 = x_1 - x_2$, and $X_1 = X_2 = \{0, 1\}$. We are interested in the spectral correlation set S_{12}.

Application of the method

$$S_1 = \{-1, 0, 1, 2\}; \quad \Rightarrow M^- = -1, \quad M^+ = +2; \quad \lambda \geq 2+1+1 = 4; \quad \text{choose} \quad \lambda = 4;$$
$$S_2 = \{-1, 0, 1\}.$$
$$s = s_1 + 4s_2 = 3x_1 - 2x_2, \quad S = \{-2, 0, 1, 3\}.$$

s_2 \\ s_1	-1	0	1	2
-1	-5	-4	-3	$\boxed{-2}$
0	-1	$\boxed{0}$	$\boxed{1}$	2
1	$\boxed{3}$	4	5	6

Correlation table: the interior entries are $s_1 + \lambda s_2$ with $\lambda = 4$.

The interior entries of the table which appear also in the spectrum of s, are framed by a rectangle; the corresponding spectral values s_1 ans s_2 are correlated. The whole corresponding triplet $[s, s_1, s_2]$ is a correlation triplet.

4. EQUIVALENCE OF A SYSTEM OF m GENERALIZED LINEAR DIOPHANTINE CONSTRAINTS AND A SINGLE GENERALIZED LINEAR DIOPHANTINE CONSTRAINT

Let $R_i, i = 1, \ldots, m$ be non-empty sets of pairwise different integers. Then any relation

$$\sum_{j=1}^{n} a_{ij} x_j \in R_i, \quad x_j \in X_j, j = 1, \ldots, n,$$

will be called a generalized Diophantine constraint. In fact, this may be an equation, if R_i consists of one element only; it may be an inequality, if R_i contains all integers below or above some fixed value; it may be a system of many inequalities and/or equations representing domains of solutions with many disconnected parts, if R_i contains non-consecutive integers; ot it may be no restriction at all, if

$$\left\{ \sum_{j=1}^{n} a_{ij} x_j \right\} \subset R_i.$$

4.1. The case $m = 2$

Introduce the variables $x_{n+i} \in X_{n+i}$, $i = 1, 2$ with

$$X_{n+i} = -\left[R_i \cap \left\{ \sum_{j=1}^{n} a_{ij} x_j \right\} \right], \quad i = 1, 2.$$

Then

(1) $$\begin{cases} s_i = \sum_{j=1}^{N} a_{ij}x_j = 0, & \text{with} \quad a_{ij} = \begin{cases} 1, & \text{if} \quad j = n+i \\ 0, & \text{otherwise if} \quad j > n \end{cases} \end{cases}, \quad i = 1, 2; \; N = n+2$$

$$x_j \in X_j, \quad j = 1, \ldots, N$$

is the standardized form of the two given constraints.

We are looking for conditions on λ_i, $i = 1, 2$, such that

$$[\lambda_1, \lambda_2] \neq [0, 0], \quad \lambda_1 a_{1j} + \lambda_2 a_{2j} = \text{integer}, \quad j = 1, \ldots, N$$

and

(2) $$\lambda_1 \sum_{j=1}^{N} a_{1j}x_j + \lambda_2 \sum_{j=1}^{N} a_{2j}x_j = 0$$

be equivalent to (1), i.e., every spectral solution of (2) is solution of (1) and vice versa.

Lemma 2

Choose λ_1, λ_2 such that $[\lambda_1, \lambda_2] \neq [0, 0]$, $\lambda_1 a_{1j} + \lambda_2 a_{2j} = integer, j = 1, \ldots, N$. A sufficient and necessary condition for (1) and (2) to be equivalent is

$$\lambda_1 s_1 + \lambda_2 s_2 \neq 0 \quad \text{for} \quad [s_1, s_2] \in S_{12}, \quad [s_1, s_2] \neq [0, 0].$$

Remark. *Lemma 2 may be extended to the case $i = 1, 2, \ldots, m$.*

Proof.

(1) *Sufficiency*

(a) Every spectral solution of (1), is a solution of (2): trivial.

(b) Every spectral solution of (2) is a solution of (1): Let $\vec{\bar{x}}$ be a spectral solution of (2). By this assumption $\bar{x}_j \in X_j$, $j = 1, \ldots, N$. Therefore $\bar{s}_i = s_i(\vec{\bar{x}}) = \sum_{j=1}^{N} a_{ij}\bar{x}_j \in S_i$, $i = 1, 2$, and $[\bar{s}_1, \bar{s}_2] \in S_{12}$. As the condition requests $\lambda_1 s_1 + \lambda_2 s_2 \neq 0$ for all $[s_1, s_2] \neq [0, 0]$, (2) cannot hold for $[s_1, s_2] \neq [0, 0]$. But (2) holds for $\vec{\bar{x}}$. Hence $[s_1, s_2] = [0, 0]$ and (1) must hold.

(2) *Necessity*

Let (1) and (2) be equivalent. We show that the case $[\bar{s}_1, \bar{s}_2] \in S_{12}$, $[\bar{s}_1, \bar{s}_2] \neq [0, 0]$ and $\lambda_1 \bar{s}_1 + \lambda_2 \bar{s}_2 = 0$ with $[\lambda_1, \lambda_2] \neq [0, 0]$ cannot occur.

Indeed, the generating $\vec{\bar{x}}$ cannot be a solution of (1), if $[\bar{s}_1, \bar{s}_2] \neq [0, 0]$. (2) being equivalent to (1), $\vec{\bar{x}}$ cannot be a solution of (2) either. Hence $\lambda_1 \bar{s}_1 + \lambda_2 \bar{s}_2 \neq 0$ for $[\lambda_1, \lambda_2] \neq [0, 0]$. ∎

The graphical consequence is: in the $[s_1, s_2]$-plane a correlation pair $[s_1, s_2] \in S_{12}$ gives a correlation point. Any straight line through the origin which avoids every correlation point other than the origin may serve as new equation (2) as long as $[\lambda_1, \lambda_2] \neq [0, 0]$ and $\lambda_1 a_{1j} + \lambda_2 a_{2j} = \text{integer}, j = 1, \ldots, N$.

Numerical example. In his article "New results on equivalent integer programming formulations", Math. Progr. 8, 1975, F. Glover gives different aggregation coefficients for

$$7x_1+9x_2+5x_3=84 \quad (s_1\text{-equation}),$$
$$6x_1+7x_2+5x_3=72 \quad (s_2\text{-equation}),$$

with $x_j \geqq 0$, integer, $j=1, 2, 3$. Putting $\lambda_1=1$, the coefficient λ_2 is $\geqq 93$ (Mathews) and $\geqq 12$ (F. Glover).

According to Lemma 2 an equivalent aggregation is possible for $\lambda_2=2$ and of course for other pairs of values (Fig. 1).

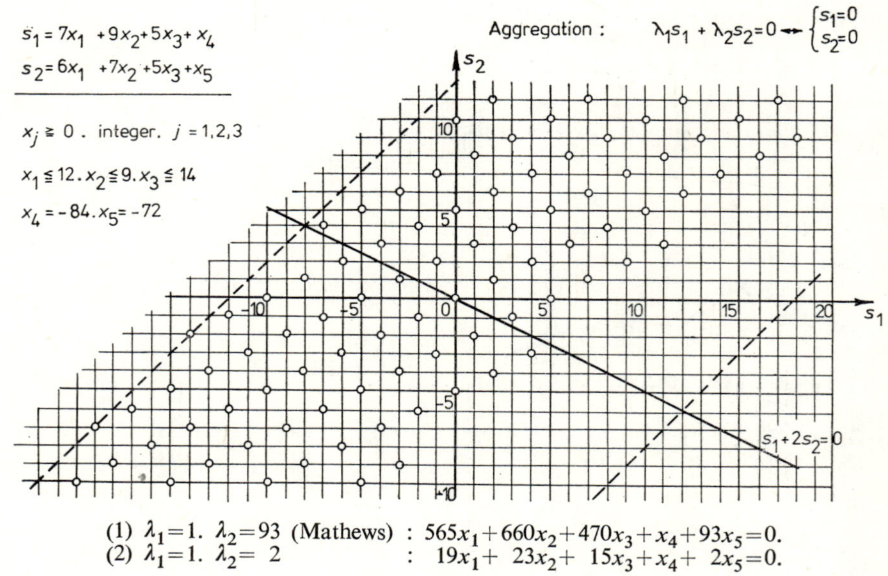

(1) $\lambda_1=1$. $\lambda_2=93$ (Mathews) : $565x_1+660x_2+470x_3+x_4+93x_5=0$.
(2) $\lambda_1=1$. $\lambda_2= 2$: $19x_1+ 23x_2+ 15x_3+x_4+ 2x_5=0$.

Fig 1. Graphical representation of the correlation set S_{12}

4.2. The case $m>2$

If more than two linear Diophantine equations are to be aggregated, any sequential procedure of the above method would give rise to weights λ_i, $i=1, \ldots, m$, larger in their absolute value than necessary. To see this, consider the case of $m=3$ equations $i=1, 2, 3$.

For simultaneous aggregation, according to the extension of Lemma 2 we would have to choose λ_1, λ_2, λ_3 such that

$$[\lambda_1, \lambda_2, \lambda_3] \neq [0, 0, 0], \quad \sum_{i=1}^{3} \lambda_i a_{ij} = \text{integer} \ (j = 1, \ldots, N),$$

and

$$\sum_{i=1}^{3} \lambda_i s_i \neq 0 \quad \text{for} \quad [s_1, s_2, s_3] \in S_{123}, \quad \text{if} \quad [s_1, s_2, s_3] \neq [0, 0, 0].$$

By aggregating sequentially instead, in the first step we would aggregate only two of the equations $\sum_{j=1}^{N} a_{ij} x_j = 0$, $i = 1, 2, 3$, — say those with $i = 1$ and $i = 2$. Doing this, we would forbid the line $\lambda_1 s_1 + \lambda_2 s_2 = 0$ to pass through any point $[s_1, s_2] \neq [0, 0]$ with $[s_1, s_2] \in S_{12}$. This restriction, interpreted as a plane in the R^3, may however be superfluous, if there exists no vector $\vec{\bar{x}} = (\bar{x}_j)$, $\bar{x}_j \in X_j$, $j = 1, \ldots, N$, such that $\bar{s}_3 = 0$ for $\lambda_1 \bar{s}_1 + \lambda_2 \bar{s}_2 + \lambda_3 \bar{s}_3 = 0$, $[\bar{s}_1, \bar{s}_2, \bar{s}_3] \neq [0, 0, 0]$, $[\bar{s}_1, \bar{s}_2, \bar{s}_3] \in S_{123}$, with $\bar{s}_i = s_i(\vec{\bar{x}})$, $i = 1, 2, 3$.

Thus aggregating the equations sequentially is certainly a sufficient but not a necessary procedure. On the other hand, even if a simultaneous method according to Lemma 2 were chosen, the number of correlation tuples $[s_1, s_2, \ldots, s_m]$ would increase with m in an explosive manner, the probability for two or more vectors (x_j), $x_j \in X_j$, $j = 1, \ldots, N$, yielding the same correlation tuple, tending towards zero rapidly as m increases, and the cardinality of $S_{1,2\ldots m}$ tending to that of the generating set $X_1 \times X_2 \times \ldots \times X_N$.

Finally, the determination of some $[\lambda_1, \ldots, \lambda_m]$ satisfying the extended Lemma 2 is a problem for itself.

5. AGGREGATING CONCEPTS

The aggregating concept described in the preceeding paragraph is a row concept because the constraints are aggregated as entities. There exists also the possibility to substitute the constraints each by a new particular equivalent constraint, differing from its predecessor only by one coefficient. The concept underlying such a sequential method is a column concept, because for all constraints just one corresponding coefficient is changed simultaneously. It is always possible to find a new coefficient which is common to all constraints, the final substitute constraint being identical in all coefficients $a'_{ij} = c_j$, $j = 1, \ldots, n$, while the allowed spectrum is the intersection of the calculated allowed spectra R'_i. Details shall be published shortly [2]. It has to be pointed out however that—if the domain of solutions is of complicated nature— its description by one single linear Diophantine constraint will always yield rapidly growing coefficients.

REFERENCES

[1] IFOR-Studienberichte 4 (1976).
[2] IFOR-Studienberichte 5 (1976).

THE ECONOMIC INTERPRETATION
OF DUALITY FOR PRACTICAL MIXED INTEGER
PROGRAMMING PROBLEMS

H. P. WILLIAMS

(Edinburgh, U.K.)

1. INTRODUCTION

It is well known that any Linear Programming (LP) model has a clearly defined dual model. This dual model provides a valuation for each constraint of the original model. These valuations are usually known as *shadow prices*. They are often of considerable economic significance. The alternative names by which they are known reflect the uses to which they can be put. The term *"marginal value"* arises because these valuations give the marginal rate of increase (or decrease) of the objective function with respect to changes in the right-hand-side values taken one at a time. To accountants they are known as *opportunity costs* since they indicate the increased (or decreased) opportunity to make a profit through extra (or fewer) resources. Another term sometimes used by economists is *quasi-rent*. The increasing importance which economists and accountants attach to LP and shadow prices in particular is reflected in a number of books and papers, e.g. Carsberg [8], Salkin and Kornbluth [21], Weingartner [24], Demski [12], Samuels [22] and Baumol and Quandt [6].

Once the optimal set of shadow prices for an LP model is known it generally provides a method of obtaining the optimal solution directly. Such a procedure has its attractions since it provides a method of decentralised planning through a system of prices alone. The procedure is demonstrated by means of the small example below.

$$
\begin{array}{lll}
& \text{Maximise} & 3x_1 + 2x_2 + 4x_3 + 5x_4 \\
\text{(A)} & \text{Subject to} & x_1 + x_2 + x_3 + x_4 \leqq 10 \\
& & x_1 + 2x_2 + x_4 \leqq 11 \\
& & x_2 + 2x_3 + 2x_4 \leqq 13 \\
& & x_1, x_2, x_3, x_4 \geqq 0.
\end{array}
$$

The model could be thought of as one of making four types of product subject to three resource limitations. The unit profits on the four products are given by the objective coefficients and the resource levels by the right-hand-side coefficients. The resource requirements of each product (per unit) give the matrix coefficients. We will suppose that the optimal shadow prices for the constraints are known and are 3, 0 and 1 respectively. An accountant could pursue the following train of reasoning:

(a) Each unit of product 1 uses 1 unit of resource 1 (cost 1×3) and 1 unit of resource 2 (cost 1×0). Hence the imputed internal cost of manufacturing one unit of product 1 (by way of its use of resources) totals 3. This exactly balances its unit profit. We can therefore allow the possibility of manufacturing product 1 (at zero "profit").

Each unit of product 2 uses 1 unit of resource 1 (cost 1×3), 2 units of resource 2 (cost 2×0) and 1 unit of resource 3 (cost 1×1). The total imputed cost of one unit of product 2 is therefore 4 which exceeds its profit. Therefore product 2 should not be made.

Similarly, the imputed cost of product 3 exceeds its profit and it should not be made.

Product 4 breaks even with its imputed cost equalling its profit. Product 4 can therefore be manufactured if desired.

We are therefore left with the possibility of making products 1 and 4 using the resources available.

(b) Resources 1 and 3 have positive shadow prices. They may therefore be regarded as "scarce" and we may assume we use them to capacity. Therefore the first and third constraints may be regarded as binding and treated as equalities.

Resource 2 has a zero shadow price and may therefore be treated as a free good. The second constraint can therefore be ignored.

This leaves us with the following two equations in two variables:

$$\text{(B)} \qquad\qquad\qquad\qquad \begin{aligned} x_1 + x_4 &= 10 \\ 2x_4 &= 13. \end{aligned}$$

These equations are uniquely solvable giving $x_1 = \dfrac{7}{2}$ and $x_4 = \dfrac{13}{2}$.

This is the optimal solution to model (A).

The above procedure for deducing the optimal solution to an LP model by means of shadow prices (valuations on the constraints) does not always work completely. If the model has alternate solutions then the second step of the second paragraph in part (b) is not justified. Not all constraints with zero shadow prices necessarily represent free goods.

In the case of degeneracy in an LP model (the dual concept to alternate solutions) there may be more than one set of shadow prices which lead to the optimal solution. This is the case of so-called "two-valued" shadow prices where the marginal values of increases or decreases in a right-hand-side coefficient are not the same.

The simplicity of the above procedure suggests how valuable shadow prices can be to an accountant. It should be noted, however, that the operational practicality of such a method of planning has been challenged, e.g. Barron [4].

2. THE FAILURE OF A SYSTEM OF PRICES IN INTEGER PROGRAMMING

It would seem useful to seek an analogy to shadow prices for the constraints of an Integer Programming (IP) model. The impossibility of obtaining such a set of prices which would allow the construction of a similar accounting argument to that above is demonstrated by the simple example below.

(C) Maximise $\qquad\qquad\qquad\qquad 15x_1 + 10x_2 + 8x_3$
   Subject to $\qquad\qquad\qquad\quad\ 18x_1 + 14x_2 + 12x_3 \leqq 61$
   $$x_1, x_2, x_3 \geqq 0 \quad \text{and integer.}$$

The optimal solution to this model is $x_1 = 2$, $x_2 = 0$ and $x_3 = 2$ giving an objective value of 46.

If we were to attach a price to the single constraint this would have to be greater than $\frac{5}{7}$ to rule out the possibility of making product 2. But this would also rule out the possibility of making product 3 since its imputed cost would then exceed its unit profit of 8. Clearly the benefit to cost rankings of the products is 1, 2, 3, since

$$\frac{15}{18} > \frac{10}{14} > \frac{8}{12}.$$

But we prefer to make product 3 to product 2 although the benefit/cost of product 3 is less than that of product 2.

Were it not for the integrality (indivisibility) of the products the solution would be simple. We would concentrate on product 1 since it has the greatest benefit to cost ratio $\left(\frac{5}{6}\right)$. This ratio gives the shadow price for the constraint ruling out the manufacture of products 2 and 3 by the argument described in the last section. We would therefore concentrate solely on product 1 making $3\frac{4}{8}$ units of it. In the above integer solution we do not even make as much of product 1 as we can allowing for the integrality requirement. Instead of making 3 of product 1 we only make 2 and sacrifice 1 in order to fit in 2 of product 3.

It is difficult to see how any system of prices alone on the single constraint can suffice to deduce the optimal solution to problem (C). We are concerned not only with the benefit to cost ratios of the products but also how well they "fit in" to the overall production pattern. A product (such as product 3) with a small benefit to cost ratio has this counterbalanced if it uses "small chunks" of the resource.

The differences between shadow prices in LP and any system of prices in IP are summarised below.

(a) No prices can generally be given to the constraints of an IP model which will lead to the direct deduction of the optimal solution. This is what we discussed above.

(b) A constraint may have "slack" but not represent a "free good". In other words, a constraint may have slack yet not be redundant. This is the case in problem (C) above. The optimal solution only uses up 60 units of the resource leaving a slack of 1 unit. Clearly, the constraint is not redundant and provides a limiting factor on production. One would therefore expect any valuation or price which was attached to the constraint to be positive. Clearly step 6 of the argument in Section 1 would fall down here. We could not treat a positively valued constraint as an equality even though it might be "binding" in an intuitive sense. Obviously, there is no analogy to the complementarity principle of LP where a constraint always has either zero slack or a zero shadow price.

The use of a knapsack problem (an IP problem with one constraint) to demonstrate the difficulties of pricing constraints in IP was deliberate. Firstly, it demonstrates the difficulty in a very concrete and obvious fashion. More complex and realistic problems will still contain the difficulty, but there is a greater possibility for "fudging" the issue, (as is sometimes done in accounting journals). Secondly some of the suggested duality procedures described below involve reducing an IP problem to a knapsack problem first. The problem of valuing the single constraint then has to be faced. In this sense a knapsack problem represents a distillation of the difficulty.

3. MOTIVES FOR SEEKING THE DUAL OF AN INTEGER PROGRAMMING MODEL

The reasons why we seek a dual for IP models can be divided into three.

(i) *Mathematical*. A "neat" result such as duality in LP suggests the existence of a counterpart in IP. This motive simply rests on the metaphysical assumption that mathematical theories tend to exhibit an aesthetically satisfying structure.

(ii) *Computational*. The strength of the duality theorem in LP suggests that a duality theory for IP would help with the computational difficulty which IP models present. Indeed, this has already been shown to be the case by the computational success of Lagrangean Relaxation as a method of helping to solve IP models. This is discussed later in this paper. To date most of the work on duality in IP has had a computational motive.

(iii) *Economic*. In view of the usefulness of shadow prices in providing economic and accountancy information in LP it seems worth seeking something analogous in IP, although it must clearly lack many of the useful properties of the LP case. It is obviously meaningless to talk of a "marginal valuation" in many IP models. Both the inputs and outputs of an IP model may come in "discrete lumps". It is impossible to consider continuous changes in the right-hand-side coefficients. The effect of changes on right-hand-side coefficients can only sensibly be discussed in terms of both *allocations* and *prices*, i.e. discrete changes together with their effect on the objective function. In spite of the impossibility of talking about marginal valuations there are other uses to which prices might be put.

570

The economic motive for seeking an IP dual is that which is pursued in this paper.

In order to focus attention on practical problems three examples of IP models will be mentioned where it is necessary to "value" constraints.

Example 1. The Cost of Capital. Considerable attention has been paid in the literature to the problem of project selection when capital is rationed. This problem was considered by Lorie and Savage [18]. It is also discussed by Quirin [19] and Weingartner [24]. One way of formulating the problem is as a pure $0-1$ IP model. The limited capital available provides a constraint. If more than one time period is considered each period provides a separate constraint. The usual methods of project selection such as *Discounted Cash Flow* or *Internal Rate of Return* present difficulties. It is necessary to obtain an *opportunity cost* for the supply of capital to use as the discount rate. For an LP model this would come out naturally as the shadow price on the capital availability constraint. With the IP model we have to value the constraint in some other way.

Example 2. The Allocation of Fixed Costs. It is sometimes necessary to share the fixed costs of a capital investment among the activities which will use it. Often it is by no means obvious how this should be done. If a particular activity has to bear more than a certain amount of the cost it may not be worth carrying out that activity in which case it will contribute nothing. The extra burden will then have to be borne by the other activities. Problems such as this can often be treated by mixed integer programming models. The sharing then implies valuations on some of the constraints. Particular examples of this problem have been considered in the literature where the fixed cost element may be neglected, for example Hirsch and Dantzig [17]. A reformulation of some problems of this type leads to a unimodular IP model which may therefore be solved as on LP model providing well-defined shadow prices. This is described by Rhys [20], and considered later in this paper.

Example 3. The Determination of Electricity Tariff Prices. It is often stipulated that public utilities such as electricity should be sold to consumers at marginal cost. (This is the case in the U.K.) The optimal pattern of generators to meet electricity demand at various times of day can be determined from an IP model. The integer nature of the model arises from the fact that generators cannot be turned on continuously. Either they do not operate or they operate above some threshold level between prescribed limits. Were it not for this an LP model would suffice and the tariff rates would come naturally from the shadow prices on the demand constraints. For the true IP model the charging of any tariff rate implicitly values these constraints in the model. A very simple model of this type is suggested by Garver [14]. Besides the above examples which clearly involve resources coming, or used, in discrete lumps IP is capable of modelling a wide variety of other situations where, for example there are logical conditions, mutually exclusive activities or dependent activities. Valuations on such constraints may themselves have interpretations.

Example 4. A Sample Problem. In order to illustrate some of the later discussion in this paper we will use the following pure IP model.

$$
\begin{array}{lll}
\text{Maximise} & & 15x_1 + 10x_2 + 8x_3 \\
\text{(D)} \quad \text{Subject to} & & 10x_1 + 6x_2 + 10x_3 \leqq 41 \\
& & 4x_1 + 4x_2 + x_3 \leqq 10 \\
& & x_1, x_2, x_3 \geqq 0 \quad \text{and integer.}
\end{array}
$$

It is useful to record the optimal solution to this problem as $x_1 = 2$, $x_2 = 0$ and $x_3 = 2$, giving an objective value of 46.

It is also useful to record the solution to the corresponding continuous problem as $x_1 = 1\frac{29}{30}$, $x_2 = 0$ and $x_3 = 2\frac{2}{15}$ giving an objective value of $46\frac{17}{30}$. The LP shadow prices are $\frac{17}{30}$ and $2\frac{1}{3}$ respectively.

Example 5. Possible Dual Prices for Integer Programming Constraints. One of the earliest attempts to seek an economically meaningful dual for an IP model was made by Gomory and Baumol [16]. They proceed by solving the IP model by a cutting planes algorithm. At stages in the calculation stronger constraints in the form of cutting planes are added to the model. These rule out certain fractional solutions but not integer ones. The model is reoptimised with these stronger constraints. The process is repeated until an integer solution is obtained. As a result shadow prices will be obtained for both the original constraints and the added cutting plane constraints. It is generally very difficult to give these extra constraints any physical meaning and hence make economic sense of their shadow prices. Gomory and Baumol therefore suggest relating these new constraints back to the original ones. The new constraints arise from linear combinations of the original constraints in the course of the calculation. These shadow prices can therefore be imputed back to the original constraints in these same-linear multiples. In this way valuations can be obtained for the original constraints. A severe drawback to this method is that the valuations depend on the route by which optimality was achieved. This need not be unique. Another unsatisfactory result is that it is often necessary to attach a valuation to the non-negativity constraints $x \geqq 0$. It may well be that such a variable comes out at a positive level yet the non-negativity constraint has a positive valuation. This can be regarded as a special case of a constraint with slack having positive valuation described in (b) of Section 2.

We will use a slight variation of this approach and illustrate it by the sample problem of Section 4. Instead of using the Gomory cuts in Gomory and Baumol's paper we could consider the strongest cuts possible and add them all into the model at the beginning. The strongest possible cuts are the facets of the convex hull of integer points defined by the original constraints. In practice it is computationally very expensive to obtain such facets but this approach is worth considering as a theoretical

572

solution to the problem. If the original constraints are replaced by the facet constraints solving as an LP model guarantees an integer solution. In fact Baumol [5] suggests that the dual of this LP model defined by the facet constraints is the only really satisfactory IP dual. It still of course suffers from the fault that there may be little economic meaning to be attached to these "artificial constraints". For the sample problem of Section 4 the facet constraints are:

$$\text{(E)} \qquad \begin{aligned} x_1 + x_2 &\leqq 2 \\ x_1 + x_2 + x_3 &\leqq 4 \\ x_1,\, x_2,\, x_3 &\geqq 0. \end{aligned}$$

(It is a coincidence that there are only two dual constraints apart from the non-negativity constraints. There will usually be many more that the original number.)

If the original objective in (D) is optimised as an LP subject to the facet constraints (E) the integer optimum to (D) is obtained together with shadow prices of 7 and 8 on the constraints in (E). It is now necessary to relate these facet constraints and these shadow prices back to the original constraints. One approach is to get the "minimal representation" of each facet constraint in terms of the original ones. By "minimal representation" we mean expressing $x_1 + x_2$ as:

$$\begin{aligned} &p_1(10x_1 + 6x_2 + 10x_3) \\ &+ p_2(4x_1 + 4x_2 + x_3) \\ &- p_3 x_1 \\ &- p_4 x_2 \\ &- p_5 x_3, \end{aligned}$$

where $p_1, p_2, \ldots, p_5$ are non-negative numbers chosen so as to minimise $41p_1 + 10p_2$. There will be many ways of expressing $x_1 + x_2$ in terms of the expressions on the left of the constraints (including the non-negativity constraints), but minimising gives the representation a uniqueness (up to alternative solutions). As a result we obtain:

$$\text{(F)} \qquad x_1 + x_2 \leqq 2 \text{ arises from } \tfrac{1}{4}(4x_1 + 4x_2 + x_3 \leqq 10) +$$

$$+ \tfrac{1}{4}(-x_3 \leqq 0)$$

$$x_1 + x_2 + x_3 \leqq 4 \text{ arises from } \tfrac{3}{34}(10x_1 + 6x_2 + 10x_3 \leqq 41)$$

$$\text{(G)} \qquad + \tfrac{2}{17}(4x_1 + 4x_2 + x_3 \leqq 10)$$

$$+ \tfrac{6}{17}(-x_1 \leqq 0).$$

The shadow prices of 7 and 8 on the facet constraints therefore give imputed prices of:

$$8 \times \frac{3}{34} = \frac{12}{17} \quad \text{to} \quad 10x_1 + 6x_2 + 10x_3 \leqq 41$$

$$7 \times \frac{1}{4} + 8 \times \frac{2}{17} = 2\frac{47}{68} \quad \text{to} \quad 4x_1 + 4x_2 + x_3 \leqq 10$$

$$8 \times \frac{6}{17} = 2\frac{14}{17} \quad \text{to} \quad -x_1 \leqq 0$$

$$7 \times \frac{1}{4} = 1\frac{3}{4} \quad \text{to} \quad -x_3 \leqq 0.$$

These imputed prices can be used to decide which products should not be made using the argument exhibited in (a) of Section 1.

The prices on the non-negativity constraints can be regarded as subsidies which products 1 and 3 should be given to compensate for their integral nature.

Unfortunately, it is not possible to proceed to part [b] of the argument in Section 1 since positive prices do not imply that a constraint is satisfied as an equation. If, however, a constraint receives a zero price it does generally represent a free good and is a redundant constraint. This last remark does, however, have to be qualified in the same way as in the LP case for alternate solutions. The reverse property does not necessarily hold. A free good constraint may receive a non-zero price. Alcaly and Klevorick [1] attempt to remedy this. They also point out that the new shadow prices should only really apply to the portions of the resources used, e.g. in the expressions (F) we should subtract 2 from the right-hand-side value 10 in deriving the facet constraint. The price should not apply to this portion of the resource when calculating the total value of the inputs (right-hand-sides times their prices). If this is done the first part of the duality theorem of LP can be preserved. The total value of the outputs (optimal objective value) equals the total value of the inputs.

One property of the Gomory/Baumol prices does make them plausible in certain situations. For a model with n variables if the IP optimum arises from the same set of n constraints as the LP optimum, then the Gomory/Baumol shadow prices will be the same as the shadow prices for the corresponding constraints of the LP model. It has been the author's experience with a model of this kind described in example 3 of Section 3 that this property is useful. Where the IP optimum is fairly close to the LP optimum e.g. it can be obtained by rounding variables up and down this result may well apply.

A very full description of the Gomory/Baumol shadow prices is given by Weingartner [24]. This method works for both pure and mixed IP models.

Another early approach to duality in IP is that of Balas [3]. He expresses a mixed IP model in a form which contains both the original variables and the dual variables. This form of the model has a fairly natural dual, again containing both types of variables. If the dual variables are interpreted as prices on the constraints then the

574

continuous variables of the original model can enter the solution only if they have "zero profit" in the sense already described. In this way the prices behave in exactly the same fashion as exhibited in step (a) of Section 1. For integer variables, however, the "inputed cost" may under- or overshoot the unit profit. The amount by which it does, this is weighted by the value of the integer variable. Part of the effect of the optimisation is to minimise the sum total of these weighted deviations.

Balas' prices have an economic appeal. Each constraint has a price but integer variables may require subsidies or penalties as well, of the amount by which the imputed cost differs from the unit profit. Free goods receive a zero price but unfortunately some resources which do not represent free goods also receive a zero price. The system of prices does not allow a deduction of the optimal IP solution in the manner exhibited for LP in Section 1.

This system of prices does not make much sense for pure IP models.

A recent and highly attractive set of prices for pure IP problems has been described by Forgo [13]. In IP it is often possible to add constraints together in suitable multiples and create a constraint having exactly the same effect as the original constraints. Clearly this is not possible in LP. If all constraints are equations it has been shown by Bradley [2] that it is *always* possible to add them all together in suitable multiples to create a single constraint. Forgo has suggested finding a set of multiples so that the resulting knapsack problem has the same optimal solution as the original problem. In order to create a sensible problem these multiples will be taken as integers and the set of multiples found which minimise the corresponding linear combination of the right-hand-side coefficients. In order to demonstrate the approach we will consider the sample problem of Section 4. If we make the constraints equations we obtain:

$$\begin{aligned}
\text{Maximise} \quad & 15x_1 + 10x_2 + 8x_3 \\
\text{(H)} \quad \text{Subject to} \quad & 10x_1 + 6x_2 + 10x_3 + u_1 = 41 \\
& 4x_1 + 4x_2 + x_3 + u_2 = 10 \\
& x_1, x_2, x_3, u_1, u_2 \geqq 0 \quad \text{and integer.}
\end{aligned}$$

We seek non-negative integers π_1 and π_2 such that $41\pi_1 + 10\pi_2$ is minimised but that the following knapsack problem has the same optimal solution as (H).

$$\begin{aligned}
\text{Maximise} \quad & 15x_1 + 10x_2 + 8x_3 \\
\text{(I)} \quad \text{Subject to} \quad & \\
& (10\pi_1 + 4\pi_2)x_1 + (6\pi_1 + 4\pi_2)x_2 + (10\pi_1 + \pi_2)x_3 + \pi_1 u_1 + \pi_2 u_2 = 41\pi_1 + 10\pi_2.
\end{aligned}$$

It can be shown that in this case the values of π_1 and π_2 are 1 and 2 respectively. These are the (relative) prices that Forgo would suggest for problem (H).

This method of allocating prices to the constraints of a pure IP model has great economic appeal. It seems reasonable to say that the second constraint of (H) is "twice as important" as the first in obtaining the optimal solution. Clearly this method cannot provide an absolute price for each constraint. It can only give the ratios of the prices. After the prices have been determined we are left with a knapsack problem having the same optimal solution as the original model. It is still necessary to "value" the single constraint of this problem. In the above example we obtain the knapsack problem:

$$\text{Maximise} \qquad 15x_1 + 10x_2 + 8x_3$$
$$\text{(J)} \quad \text{Subject to} \qquad 18x_1 + 14x_2 + 12x_3 + u_1 + 2u_2 = 61$$
$$x_1, x_2, x_3, u_1, u_2 \geqq 0 \quad \text{and integer.}$$

For this particular example it was unnecessary first to convert the inequalities of the sample problem to equations. Clearly, problem (J) is the same as the knapsack problem which was discussed in Section 2.

Part of the motive for discussing the problem of valuing the constraint of a knapsack problem in Section 2 was that Forgo's method results in such a problem. If absolute rather than relative valuations are required then an ultimate valuation of the knapsack constraint is necessary.

One attractive feature of Forgo's method is that a constraint receives a zero price if and only if it is a redundant constraint (e.g. represents a free good). The method can be regarded as providing a limited form of decentralized planning. If an accountant were equipped with these relative prices for the resources he could reduce the problem of finding the optimal production pattern to that of simply solving a knapsack problem.

Clearly, Forgo's method could be applied to the multiperiod project selection problem discussed in example 1 in Section 3. Relative valuations for capital in different periods would be obtained giving possible discount rates relative to each other in successive years.

There is another approach to IP which involves reducing a model to a knapsack problem. Unlike Forgo's method it could be applied to a general mixed IP model. This method is due to Dantzig and Eaves [11]. By applying the dual of Fourier–Motzkin elimination they successively eliminate constraints from an IP model. They will ultimately be left with a knapsack problem. This method is complex and requires a change of variables at each stage. It will not be discussed further here but the author feels it once again serves to emphasise that the difficulty of pricing the constraint of a knapsack problem encompasses the whole problem of obtaining dual valuations for constraints of general IP models. In this case the final knapsack problem will contain both continuous and integer variables.

A highly successful computational approach to solving IP problems is Lagrangean Relaxation. This method is described very fully by Geoffrion [15]. This approach involves attaching non-negative prices to some of the constraints of the IP model and subtracting these constraints, multiplied by their prices, from the objective function. The remaining IP model will be a "relaxation of the original model and may provide a useful bound on the optimal value of the objective function. However "correct" the choice of prices is they will generally not lead to the optimal IP solution in the way that was exhibited for an LP problem in Section 1. There will be a so-called "duality gap". It does, however, seem worth seeking these prices on constraints which minimise this gap. We will try to elucidate the discussion by means of the sample problem introduced in Section 4.

Suppose we subtract a non-negative multiple λ, of the first constraint from the objective function and optimise subject to the second constraint. We obtain the problem:

$$
\begin{array}{lll}
& \text{Maximise} & (15-10\lambda_1)x_1+(10-6\lambda_1)x_2+(8-10\lambda_1)x_3 \\
\text{(K)} & \text{Subject to} & 4x_1+4x_2+x_3\leqq10 \\
& & x_1,\,x_2,\,x_3\geqq0 \quad \text{and integer.}
\end{array}
$$

The "relaxed" problem will generally give larger objective values than the IP optimum of 46. Such a solution to a relaxed problem will of course be infeasible relative to the constraints of the original problem. In order to minimise the amount by which the objective value of the relaxed problem exceeds that of the original IP problem we wish to solve the following problem:

$$
\text{(L)} \quad \underset{\lambda_1\geqq0}{\text{Minimise}} \left[
\begin{array}{l}
\text{Max } (15-10\lambda_1)x_1+(10-6\lambda_1)x_2+(8-10\lambda_1)x_3+41\lambda_1 \\
\text{Subject to } 4x_1+4x_2+x_3\leqq0 \\
\quad x_1,\,x_2,\,x_3\geqq0 \text{ and integer}
\end{array}
\right].
$$

The optimal objective value turns out to be $46\frac{17}{30}$. The value of λ_1 which gives rise to this solution is $\frac{17}{30}$.

In practice the way in which one partitions the constraints of the IP problem into those which are incorporated into the objective function and those which are not is dictated by the economic meaning of the problem. In the artificial problem presented here this is not possible. Geoffrion discusses this in relation to a number of practical examples. Cornuejols, Fisher and Nemhauser [10] apply this approach to a particular class of problems. This is discussed further in the next Section.

Geoffrion shows that what one is really doing in a Lagrangean relaxation is insisting that some of the constraints are effectively replaced by the (stronger) facet constraints of their convex hull of integer points. The remaining constraints play their conventional LP role. When this strengthened problem is solved as on LP it is the shadow prices on these second set of constraints which we are interested in. These shadow prices, in a loose sense, indicate the marginal valuations on the constraints when the remaining constraints retain the strength implied by their integrality requirement. In certain situations this makes economic sense. We may be interested in marginal valuations on certain resources while retaining, as far as possible, the strength of major policy decisions implied by other integer constraints.

One possible, general, way of using Lagrangean relaxation to obtain a price on a constraint is to take the single constraint, incorporate it into the objective and solve the analogous problem to (L) using all the other constraints. If this is done for each constraint in turn a price is obtained for each representing, in a sense, its marginal valuation with respect to the rest of the problem. This is what we have done for the sample problem. For completeness we will also consider the Lagrangean relaxation in which the second constraint is incorporated into the objective function and optimised with respect to the first constraint. This results in an objective value of 46. The value of λ_2 (the multiple of the second constraint subtracted from the objective function) giving rise to this solution is $\frac{7}{3}$. Clearly, in this case, the relaxation is sufficiently strong to lead to the IP optimum.

A redundant constraint representing a free good will receive a zero valuation using this method.

The idea of Lagrangean relaxation has also been considered using the group theoretic reformulation of IP problems by Shapiro [23] and Bell and Shapiro [7].

Finally, the possibility of reformulating an IP model so as to lead to a more sensible set of prices on the constraints should not be overlooked. It has already been pointed out that this is always theoretically possible by taking the set of facet constraints defining the convex hull of feasible integer points.

The reformulated model can then be solved as an LP and shadow prices obtained for the facet constraints. Unfortunately, these facet constraints usually have little physical meaning and their shadow prices are hard to interpret. There are, however, certain types of problem where a reformulation of this kind is fairly obvious and physically meaningful. One such problem which arises in open cast mining is described by Williams [26]. This gives rise to a pure IP model. One formulation produces a so called *unimodular* matrix of coefficients which guarantees that the LP optimum will be an integer solution. The constraints of this formulation define the convex hull of feasible integer points and have sensible shadow prices. Another problem with rather similar constraints is described by Rhys [20]. This is the problem of sharing the fixed costs of various capital facilities among the activities using those facilities in an optimal fashion. Again there is a unimodular formulation of this problem providing an economically sensible set of shadow prices. Both these problems are considered in more detail in the next Section since their constraints are of a type which arises in other contexts such as the depot location problem.

6. TYPICAL INTEGER PROGRAMMING CONSTRAINTS AND THE ECONOMIC MEANING OF THEIR VALUATIONS

Most of the constraints that involve integer variables in IP models are of very special types. The general sort of constraint as used in the sample problem of part 4 is not at all common (apart from capital budgeting and project selection problems). We will therefore consider some of the common special types of integer constraint which arise in practical models and see what economic interpretation might be attached to their prices however these prices be obtained.

The following type of constraint often arises in *fixed charge* problems.

(1)
$$x - M\delta \leqq 0.$$

The variable x represents the level at which some activity is to be carried out. It may be a continuous or integer variable. δ is a $0-1$ integer variable indicating whether the activity is, or is not, to be carried out at all. M is an upper bound on the level of x. x usually has a coefficient in the objective function representing its unit marginal cost (or profit). δ has a coefficient representing some fixed cost (such as setting up a machine) which is incurred only if the activity is carried out at a positive level. If we think of the problem as a maximisation the coefficient of x will generally be positive (a profit) and that of δ negative (a cost). If a price π is attached to the constraint (1) and this multiple of the constraint subtracted from the objective function (in the manner of Lagrangean relaxation), then the effect is to simultaneously subtract πM from the fixed cost making it less and add π to the unit marginal cost of the activity thereby decreasing its unit profit. The variable δ will be "compensated" for some or all of its fixed cost and activity x will be charged a proportion $\dfrac{x}{M}$ of this compensation.

The price π therefore has a very obvious interpretation. The more "realistic" it is in fixing the correct rate of compensation the closer will be the optimal solution of the Lagrangean relaxation to the optimal operating policy. If a more conventional method of accounting is to be used to solve this problem rather than IP there is then a sense in which we can speak of the "correct price", although a system of accountancy prices alone cannot be relied on to produce the best solution.

Depot Location problems frequently contain the following type of constraint.

(2)
$$\delta_1 + \delta_2 + \delta_3 + \ldots + \delta_n - n\delta \leqq 0$$

δ is a $0-1$ integer variable indicating whether a depot should or should not be built. δ_i are $0-1$ integer variables indicating whether or not particular customers should be supplied from the depot. For a maximisation form of the problem δ will normally have a negative (cost) coefficient in the objective function and the δ_i variables positive (benefits) coefficients. A price π attached to the constraints will indicate a compensation πn for the cost of the depot which should be split equally and charged to all the customers.

Constraint (2) can be rewritten in another way as the series of constraints:

(3)
$$\delta_1 - \delta \leqq 0$$
$$d_2 - \delta \leqq 0$$
$$\cdot$$
$$\cdot$$
$$\cdot$$
$$\delta_n - \delta \leqq 0.$$

These constraints have exactly the same effect as (1) when the δ are $0-1$ integer variables. It is well known, however, that (3) provides a "tighter" associated LP problem. The LP optimal solution to this form of the problem will therefore be closer to the IP optimum. Any prices which might be attached to the constraints (3) would tend to be more useful than that attached to (2). Certainly, prices attached to (3) can be more discriminating. If we attach prices $\pi_1, \pi_2, \ldots, \pi_n$ to these constraints they can be interpreted as successive compensations for the cost of the depot which have to be borne by each customer in turn. This provides a more flexible accounting system. Not every customer need be charged the same.

It is worth pointing out that in LP the more constrained we make the primal problem the less constrained the dual problem becomes. A similar situation occurs here. The more constrained version of the constraint (2) as a series of constraints (3) has a more flexible system of prices associated with it.

Cornuejols, Fisher and Nemhauser (10) have applied Lagrangean relaxation to an application of a mathematically similar problem to the depot location problem. This is the problem of a company strategically locating bank accounts to maximise clearing times. By attaching suitable prices to some of the constraints they produce a heuristic algorithm for producing a feasible solution to the problem.

A number of other problems produce similar type constraints to (2). Williams [26] describes an open-cast mining problem where a volume of ore and overburden is to be extracted so as to maximise total profit. Because the sides of the pit cannot exceed a certain slope it is only possible to extract a portion of ore if a suitable volume above it is also extracted. The problem can be tackled by thinking of the volume broken up into blocks. In Fig. 1 block A can only be extracted if blocks B, C, D and E above it are also extracted. $0-1$ integer variables used to indicate whether a particular block is or is not extracted. Some blocks (rich ore) produce positive profit while other blocks (overburden) may produce negative profit. It is, however, sometimes necessary to extract overburden in order to get at the richer ore underneath.

The necessity of taking out blocks B, C, D and E in order to get at A can be represented by the constraint:

$$(4) \qquad 4\delta_A - \delta_B - \delta_C - \delta_D - \delta_E \leqq 0.$$

A price π on this constraint can be interpreted as a direction to take 4π off the profit for block A and add π to the profit of each of the blocks B, C, D and E. B, C, D and E are being compensated for the fact that they must be extracted to make way for A.

An alternative formulation is to split constraint (4) into the constraints:

$$(5) \qquad \begin{aligned} \delta_A - \delta_B &\leqq 0 \\ \delta_A - \delta_C &\leqq 0 \\ \delta_A - \delta_D &\leqq 0 \\ \delta_A - \delta_E &\leqq 0. \end{aligned}$$

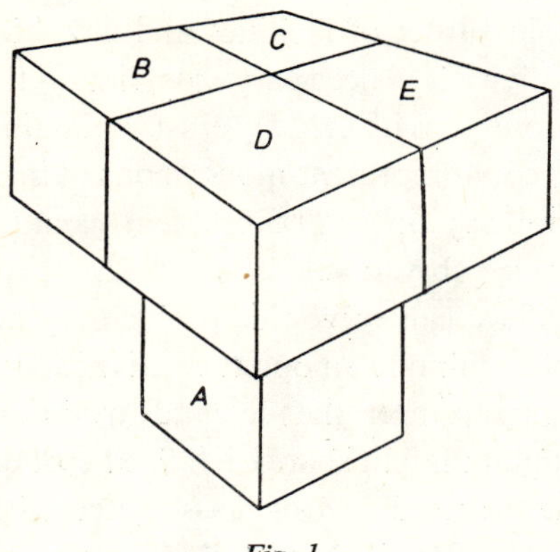

Fig. 1

If prices π_B, π_C, π_D and π_E are attached to these constraints then block A is penalised by having a total of $\pi_B + \pi_C + \pi_D + \pi_E$ taken off its profit. This is used to compensate B, C, D and E by π_B, π_C, π_D and π_E respectively.

All the constraints of this problem will be of the form (4) or of the form (5) in the second formulation. The second formulation is a unimodular one. The LP optimal solution will be integer. We therefore can use the LP shadow prices in the way described above to reallocate costs between blocks. After this reallocation has been completed only the blocks with a zero resultant profit should be extracted. All other blocks will have a negative profit and should not be extracted. In this case, it would be possible for an accountant to use a system of cost reallocation to deduce the optimal extraction pattern.

Another problem which also leads to a reformulation as a unimodular IP model is described by Rhys [20]. This is the problem of sharing the fixed costs of certain facilities optimally among the activities which make use of them. If each activity needs at most two of the facilities the problem can be considered graphically, although the class of problems is more general than this representation allows. Fig. 2 illustrates an expository example.

Nodes A, B, C, D and E are facilities with fixed costs associated with them. The arcs AB, AC, AD, BD, CD, DE and CE are activities with revenues attached to them.

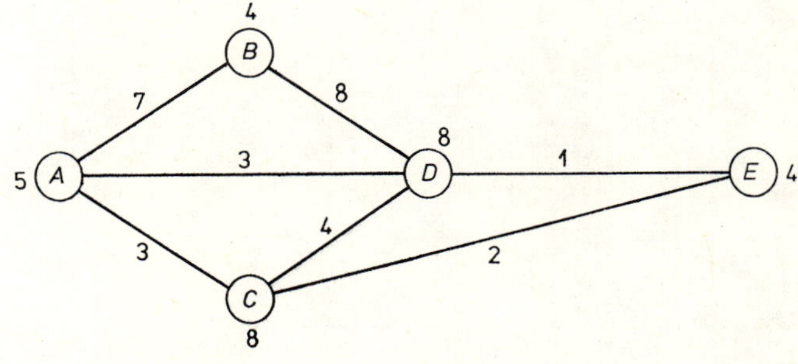

Fig. 2

The problem is to find that subset of facilities and activities which maximises total profit. To carry out an activity it is necessary to maintain the facilities supporting it. For example, if we carry out activity *AB* we must maintain both facilities *A* and *B*. As an example, the nodes could represent money consuming railway stations and the arcs revenue producing railway lines. The problem would be to prune the railway system to its most profitable subsystem.

It is interesting to see if we can solve this problem by accountancy. We want to share out the fixed cost of each node among the activities using it. If an activity does not produce enough revenue to meet the demands made on it, then it should be cut out. A conservative accountant might share each fixed cost equally between all activities using it. For example the fixed cost of 8 associated with node *D* would be split four ways between arcs *AD*, *BD*, *DE* and *CD*. Pursuing this analysis all arcs and nodes are cut out except the following shown in Fig. 3.

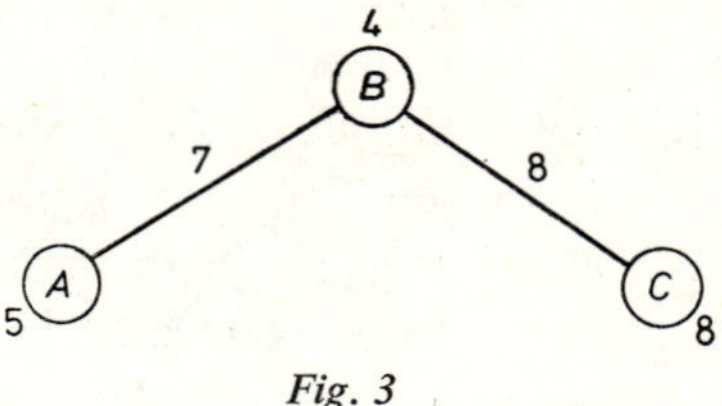

Fig. 3

Pursuing the analysis again *BD* now ceases to be viable since it now has to bear the full cost of *D*. Cutting out *BD* makes *AB* no longer viable reducing the network to nothing.

In fact this is not the best solution. By splitting the prices differently as shown in Fig. 4 we can obtain the best solution shown in Fig. 5.

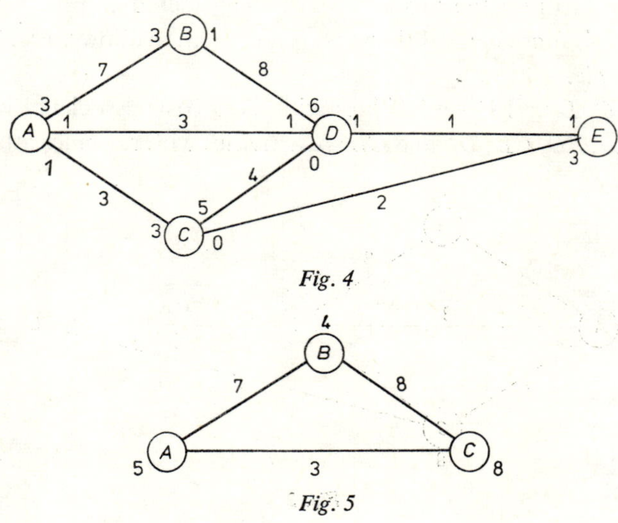

Fig. 4

Fig. 5

This problem can be formulated as an IP model in more than one way. We can write constraints such as:

$$(6) \qquad -\delta_i - \delta_j + 2\delta_{ij} \leqq 0$$

δ_i are $0-1$ integer variables indicating the discarding or retaining of nodes. δ_{ij} are $0-1$ integer variables indicating the discarding or retaining of arcs. Constraint (6) stipulates that if an ij is retained then nodes i and j must also be retained. In the same way that constraint (2) can be reexpressed as constraints (3) we may rewrite (6) as the two constraints (7).

$$(7) \qquad \begin{aligned} -\delta_i + \delta_{ij} &\leqq 0 \\ -\delta_j + \delta_{ij} &\leqq 0. \end{aligned}$$

This is again a unimodular model which can be solved as an LP to provide a set of shadow prices. The shadow price on the first constraint in (7) provides the portion of the cost of node i which should be borne by arc ij.

Both the last two problems admit unimodular formulations which are dual LP models to network flow models. The dual prices therefore come out naturally as the values of flows in networks. Charnes and Cooper (9) have examined how networks can be used to characterise certain accounting problems.

A slight variant of the depot location constraint (2) arises when we have n possible customers for a depot but can only supply r of them. A natural way of writing the constraint is:

$$(8) \qquad \delta_1 + \delta_2 + \ldots + \delta_n - r\delta \leqq 0.$$

A price π on such a constraint has the effect of reallocating πr of the cost of the depot in equal proportions $\dfrac{\pi r}{n}$ to the customers.

The "tightest" form of constraint (8) could be expected to yield the most valuable method of cost reallocation. The facet constraints associated with the convex hull of feasible integer points for constraint (8) are (rather surprisingly):

$$(9) \qquad \begin{aligned} \delta_1 && -\delta &\leqq 0 \\ \delta_2 && -\delta &\leqq 0 \\ &\vdots \\ \delta_n - \delta &\leqq 0 \\ \delta_1 + \delta_2 + \ldots + \delta_n - \delta &\leqq r-1 \end{aligned}$$

A system of prices associated with the first n of these constraints reallocates different costs from the depot to the customers. Finally, a price associated with the last constraint shares an extra element of the cost equally among the n potential customers.

Another problem which admits an improved formulation as an IP is described by Williams [27]. This problem concerns decisions of whether to operate particular mines in successive years. Mines may be either "open" or "closed". If a mine is open royalties are payable on it and it may or may not be worked. If a mine is closed no royalties are payable, but it may not be worked and it must remain closed in future years. The following constraints arise in the first formulation of this problem as an IP:

$$(10) \qquad\qquad t\delta_t - \gamma \leqq 0$$

δ_t is a $0-1$ variable indicating whether a particular mine is, or is not, worked in year t.

γ is a general integer variable giving the number of years the mine should be open. Clearly, constraints (10) make it impossible to work the mine after γ years.

The δ_t variables have positive (profit) coefficients and the variables negative (royalty cost) coefficients.

If the prices on constraints (10) for successive values of t ($t = 1, 2, 3$, etc.) are π_1, π_2, π_3 etc. their effect will be to remove these costs from the yearly royalty cost associated with variable γ. To compensate extra costs π_1, $2\pi_2$, $3\pi_3$, etc. will be subtracted from the profit coefficients associated with the variables δ_t for successive years. Clearly the nature of the problem suggests that a decision to work a mine in a later year should be penalised by a greater cost than the decision to work it in an earlier year since for later work it is necessary to keep the mine open (and pay royalty costs) in all earlier years as well. On the other hand, if a mine is worked in an early year it can be closed the next year and royalty costs saved from then on.

Williams [27] suggests a reformulation of this problem in order to make it easier to solve. Instead of the single (general) integer variable γ a series of $0-1$ integer variables γ_t are used to indicate whether the mine is, or is not, closed in particular years. In order to ensure that once a mine is closed it remains closed the following constraints are introduced:

$$(11) \qquad\qquad -\gamma_t + \gamma_{t+1} \leqq 0.$$

Constraints (10) are replaced by the following:

$$(12) \qquad\qquad \delta_t - \gamma_t \leqq 0.$$

It can be shown that the LP problem corresponding to this reformulation is more constrained. One might therefore expect any prices associated with (11) and (12) to be more realistic. Prices associated with (11) have the effect of reallocating yearly royalty costs on to later years. Prices associated with (12) then reallocate these new royalty costs by subtracting them off the profits in the corresponding years. This clearly provides a more flexible system for reallocating costs although the effect is still to move costs away from earlier years and charge them against the operating profits for later years.

584

REFERENCES

[1] Alcaly, R. F. and Klevorick, A. K.: A Note on the Dual Prices of Integer Programs, *Econometrica* 34 (1966), 206–214.

[2] Bradley, G. H.: Transformation of Integer Programs to Knapsack Problems, *Discrete Mathematics* 1 (1971), 29–45.

[3] Balas, E.: Duality in Discrete Programming, Reprint No. 519 (1970) Graduate School of Industrial Administration, Carnegie-Mellon University.

[4] Barron, M. J.: The Application of Linear Programming Dual Prices in Management Accounting — Some Cautionary Observations, *Journal of Business Finance,* 4 (1972), 51–69.

[5] Baumol, W. J.: *Economic Theory and Operations Analysis,* (Prentice-Hall, New Jersey, (1965).

[6] Baumol, W. J. and Quandt, R. E.: Investment and Discount Rates under Capital Rationing; A Programming Approach, in S. H. Archer and C. A. D'Ambrosio, eds., *Theory of Business Finance: A Book of Readings* (Macmillan, New York, 1967), 540–552.

[7] Bell, D. E. and Shapiro, J. F.: A Finitely Convergent Duality Theory for Zero-One Integer Programming, RM-7S-33 (1975) International Institute for Applied Systems Analysis, Laxenburg, Austria.

[8] Carsberg, B. V.: *An Introduction to Mathematical Programming for Accountants* (Allen and Unwin, London, 1969).

[9] Charnes, A. and Cooper, W. W.: Some Network Characterisations for Mathematical Programming and Accounting Approaches to Planning and Control, *Acc. Review* 44 (1969), 467–481.

[10] Cornuejols, G., Fisher, M. L. and Nemhauser, G. L.: An Analysis of Heuristics and Relaxations for the Uncapacitated Location Problem, Technical Report No. 271 (1975), Department of Operations Research, Cornell University.

[11] Dantzig, G. B. and Eaves, B. C.: Fourier–Motzkin Elimination and its Dual, *J. Combinational Theory* 14 (1973), 288–297.

[12] Demski, J. S.: An Accounting System Structured in a Linear Programming Model, *Acc. Review* 42 (1967), 701–712.

[13] Forgó, F.: Shadow Prices and Decomposition for Integer Programs, DM 74–6 (1974), Department of Mathematics, Karl Marx University of Economics, Budapest.

[14] Garver, L. L.: Power Scheduling by Integer Programming, *AIEE Transactions, Power Apparatus and Systems* 81 (1963), 730–735.

[15] Geoffrion, A. M.: Lagrangean Relaxation for Integer Programming, *Math. Prog. Studies* 2 (1974), 82–136.

[16] Gomory, E. R. and Baumol, W. J.: Integer Programming and Pricing, *Econometrica* 28 (1960), 521–550.

[17] Hirsch, W. M. and Dantzig, G. B.: The Fixed Charge Problem, RM-1383 (1954), The RAND Corporation.

[18] Lorie, J. H. and Savage, L. J.: Three Problems in Rationing Capital, in E. Solomon, ed., *The Management of Corporate Capital* (Free Press of Glencoe, New York, 1959).

[19] Quirin, G. D.: *The Capital Expenditure Decision,* (Irwin, Homewood, Illinois, 1967).

[20] Rhys, J. M.W.: A Selection Problem of Shared Fixed Costs and Network Flows, *Mgmt. Sci.* 17 (1970), 200–207.

[21] Salkin, G. and Kornbluth, J.: *Linear Programming in Financial Planning and Accounting,* Haymarket Publishing, London, 1973).

[22] Samuels, J. M.: Opportunity Costing: An Application of Mathematical Programming, *Journal of Acc. Res.* 1 (1963), 182–191.

[23] Shapiro, J. F.: Generalised Lagrange Multipliers in Integer Programming, *Op. Res.* 19 (1971), 68–76.

[24] Weingartner, H. M.: *Mathematical Programming and the Analysis of Capital Budgeting Problems*, (Academic Press, London, 1974).

[25] Whinston, A.: Price Guides in Decentralised Organisations, in W. W. Cooper, H. J. Leavitt and M. W. Shelley, eds., *New Perspectives in Organisational Research* (Wiley, New York, 1964) 405–448.

[26] Williams, H. P.: Experiments in the Formulation of Integer Programming Problems; *Math. Prog. Studies* 2 (1974), 180–197.

[27] Williams, H. P.: The Reformulation of Two Mixed Integer Programming Problems, *Math. Prog.* 14 (1978), 325–331.

CONTENTS

VOLUME 1

PART 1

LINEAR PROGRAMMING AND THEORY OF GAMES

PART 2

NONLINEAR PROGRAMMING. THEORY

PART 2

NONLINEAR PROGRAMMING. METHODS

588

VOLUME 3

PART 9

MATHEMATICAL PROGRAMMING, SOFTWARE

PART 10

TEACHING OF MATHEMATICAL PROGRAMMING

PART 11

APPLICATIONS IN INDUSTRY AND ENGINEERING DESIGN

589

PART 12

APPLICATIONS IN HUMAN AND NATURAL SCIENCES